**Zeolites and Catalysis**

*Edited by*
*Jiří Čejka, Avelino Corma, and Stacey Zones*

## *Related Titles*

Blaser, H.-U., Federsel, H.-J. (eds.)

**Asymmetric Catalysis on Industrial Scale**

**Challenges, Approaches and Solutions**

2010
ISBN: 978-3-527-32489-7

Mizuno, N. (ed.)

**Modern Heterogeneous Oxidation Catalysis**

**Design, Reactions and Characterization**

2009
ISBN: 978-3-527-31859-9

Barbaro, P., Bianchini, C. (eds.)

**Catalysis for Sustainable Energy Production**

2009
ISBN: 978-3-527-32095-0

Crabtree, R. H. (ed.)

**Handbook of Green Chemistry - Green Catalysis**

**Three Volumes**

2009
ISBN: 978-3-527-31577-2

Ozkan, U. S. (ed.)

**Design of Heterogeneous Catalysts**

**New Approaches based on Synthesis, Characterization and Modeling**

2009
ISBN: 978-3-527-32079-0

Swiegers, G.

**Mechanical Catalysis**

**Methods of Enzymatic, Homogeneous, and Heterogeneous Catalysis**

Hardcover
ISBN: 978-0-470-26202-3

Jackson, S. D., Hargreaves, J. S. J. (eds.)

**Metal Oxide Catalysis**

2009
ISBN: 978-3-527-31815-5

Ding, K., Uozumi, Y. (eds.)

**Handbook of Asymmetric Heterogeneous Catalysis**

2008
ISBN: 978-3-527-31913-8

Ertl, G., Knözinger, H., Schüth, F., Weitkamp, J. (eds.)

**Handbook of Heterogeneous Catalysis**

**Eight Volumes**

2008
ISBN: 978-3-527-31241-2

# Zeolites and Catalysis

Synthesis, Reactions and Applications

*Volume 1*

*Edited by*
*Jiří Čejka, Avelino Corma, and Stacey Zones*

WILEY-VCH Verlag GmbH & Co. KGaA

**The Editor**

*Prof. Dr. Jiří Čejka*
Academy of Sciences of the Czech Republic
Heyrovský Institute of Physical Chemistry
Dokjškova
Dolejskova 3
182 23 Prague 8
Czech Republic

*Prof. Dr. Avelino Corma*
University Politecnica de Valencia
Institute de Tecnologia Quimica
Avenida de los Naranjos s/n
46022 Valencia
Spain

*Prof. Dr. Stacey I. Zones*
Chevron Texaco Energy Research
and Technology Company
100 Chevron Road
Richmond, CA 94802
USA



**Library of Congress Card No.:** applied for

**British Library Cataloguing-in-Publication Data**
A catalogue record for this book is available from the British Library.

**Bibliographic information published by the Deutsche Nationalbibliothek**
The Deutsche Nationalbibliothek lists this publication in the Deutsche Nationalbibliografie; detailed bibliographic data are available on the Internet at http://dnb.d-nb.de.



**Composition** Laserwords Private Limited, Chennai
**Printing and Bookbinding** strauss GmbH, Mörlenbach
**Cover Design** Formgeber, Eppelheim

Printed in the Federal Republic of Germany
Printed on acid-free paper

**ISBN:** 978-3-527-32514-6

# Contents to Volume 1



*Zeolites and Catalysis, Synthesis, Reactions and Applications. Vol. 1.*
Edited by Jiří Čejka, Avelino Corma, and Stacey Zones

ISBN: 978-3-527-32514-6

## Contents to Volume 2

# Preface

One can safely say that the impact of zeolites in science and technology in the last 50 years has no precedents in the field of materials and catalysis. Although the first description of zeolites dates back up to 250 years ago, the last five decades experienced an incredible boom in zeolite research activities resulting in the successful synthesis of almost 200 different structural types of zeolites, numerous excellent scientific papers on the synthesis of zeolites, characterization of their properties, and applications of zeolites in adsorption and catalysis that have revolutionized the petrochemical industry. In addition, based on the knowledge of zeolites several other areas of porous materials have recently emerged including mesoporous materials, hierarchic systems, metal-organic frameworks (cationic-periodic polymers) and mesoporous organosilicas. All these materials have substantially increased the portfolio of novel porous materials possessing new interesting properties, but this topic is not covered in this book.

This book consists of two volumes. The first one is mostly concentrated on recent advances in the synthesis of zeolites and understanding of their properties while the second volume describes recent achievements in the application of zeolites mostly in catalysis.

More specifically, the first volume starts with a chapter by P. Cubillas and M.W. Anderson (Chapter 1) discussing mechanisms of the synthesis of zeolites and zeotypes, including nucleation and crystal growth, employing various microscopic techniques. This is followed by a chapter of K. Strohmaier (Chapter 2) providing a detailed survey on the synthesis of novel zeolites and different layered precursors incorporating different metal ions into the framework, and applying ever increasing number of structure-directing agents. A new approach to the synthesis of zeolites and other porous materials by ionothermal synthesis combining ionic liquids as the solvent together with the structure-directing agent is presented by R. Morris (Chapter 3). Zeolite synthesis can also be controlled by a simultaneous use of two different templates providing new tool for creative chemistry Nas discussed by the group of J. Pérez-Pariente (Chapter 4). Morphological control of zeolite crystals is one of the key issues to understand the mechanism of zeolite crystallization as well as to control the performance of zeolites in various applications as it is outlined by S.-E. Park and N. Jiang in Chapter 5. Introduction of other elements than silicon into the zeolite framework can be done not only via synthesis but also in the

*Zeolites and Catalysis, Synthesis, Reactions and Applications. Vol. 1.*
Edited by Jiří Čejka, Avelino Corma, and Stacey Zones

ISBN: 978-3-527-32514-6

postsynthesis steps as highlighted for deboronation followed by realumination as described by C.Y. Chen and S.I. Zones (Chapter 6). P.A. Wright and G.M. Pearce show how the individual zeolite structures are built from basic secondary building units. The authors focus not only on general aspects of zeolite structures but also on the description of structures of zeolites determined very recently (Chapter 7).

Structural and textural characterization of zeolites starts in Chapter 8, written by E. Stavitski and B.M. Weckhuysen, providing good examples of application of vibrational spectroscopy under static conditions that can drive into *in situ* catalytic investigations. The group of K. de Jong (Chapter 9) makes an effort to evaluate different physicochemical methods used for textural characterization of zeolites. Gas physisorption, mercury porosimetry, electron microscopy (including 3D experiments), various NMR techniques up to *in situ* optical and fluorescence microscopy are discussed in detail. The location, coordination, and accessibility of framework aluminum are of key importance for acid-catalyzed reactions in zeolites and these issues are addressed by J.A. van Bokhoven and N. Danilina in Chapter 10. Theoretical background of zeolite reactivity employing different computational approaches and models is covered in Chapter 11 by E.A. Pidko and R.A. van Santen. S. Calero Diaz presents an overview of current developments in modeling of transport and accessibility in zeolites showing some recent models and simulation methods that are applied for systems of environmental and industrial interests (Chapter 12). The final chapter of the first volume is written by the group of F. Kapteijn (Chapter 13), in which diffusion in zeolites starting from basic models of diffusion up to the role of diffusion in adsorption and catalytic processes is discussed.

The second volume starts with a chapter of the group of J. Coronas concentrating on special applications of zeolites including green chemistry, hybrid materials, medicine, veterinary, optical- and electrical-based applications, multifunctional fabrics, and nanotechnology (Chapter 14). After that K.B. Yoon presents the opportunities to organize zeolite microcrystals into two- and three-dimensionally organized structures and the application of these organized entities in membranes, antibacterial functional fabrics, supramolecularly organized light-harvesting systems, and nonlinear optical films (Chapter 15).

The remaining chapters are exclusively devoted to the application of zeolites in catalysis. G. Bellussi opens this part with a broad overview of current industrial processes using zeolites as key components of the catalysts and further challenges in this area (Chapter 16). Generation, location, and characterization of catalytically active sites are discussed in depth by M. Hunger showing different aspects of shape selectivity and structural effect on the properties of active sites (Chapter 17). M. Rigutto (Chapter 18) stresses the importance of zeolites and the main reasons for their application in cracking and hydrocracking, the largest industrial processes employing zeolites as catalysts. Further, C. Perego and his coworkers focus on reforming and upgrading of diesel fractions, which with gasoline are by far the most important and valuable key fractions produced by petroleum refineries (Chapter 19). Transformation of aromatic compounds forms the heart of petrochemical processes with zeolites as key components of all catalysts. S. Al-Khattaf, M.A. Ali, and J. Čejka highlight the most important recent achievements in application of zeolites in various alkylation, isomerization, disproportionation, and transalkylation reactions of aromatic hydrocarbons (Chapter 20). With decreasing supply of oil, natural gas obtains more and more importance. A. Martinez and his coauthors discuss in some detail different ways of methane upgrading into valuable fuels and chemicals (Chapter 21). Methanol, which can be obtained from natural gas, could be one of the strategic raw materials in future. Novel processes transforming methanol into olefins or gasoline are covered in Chapter 22 by M. Stöcker. Incorporation of catalytically active species into zeolite frameworks or channel systems for oxidation reactions is covered in Chapter 23 by T. Tatsumi. The main attention is devoted to Ti-silicates. G. Centi and S. Perathoner focus on increasing applicability of zeolites in environmental catalysis with a particular attention to conversion of nitrogen oxides (Chapter 24). K.L. Yeung and W. Han describe the emerging field of application of zeolites in fuel cells for clean energy generation. The authors show that zeolites can play an important role in hydrogen production, purification, conditioning, and storage (Chapter 25). The final chapter by the authors from the group of A. Corma presents possibilities of application of zeolite as catalysts in the synthesis of fine chemicals. The examples discussed include, for example, acylation, hydroxyalkylation, acetalization, isomerization, Diels–Alder, and Fischer glucosidation reactions.

Bringing together these excellent chapters describing the cutting edge of zeolite research and practice provides an optimistic view for the bright future of zeolites. The number of new synthesized zeolites is ever increasing and particularly novel extra-large pore zeolites or even chiral zeolitic materials will surely be applied in green catalytic processes enabling to transform bulkier substrates into desired products. In a similar way, application of zeolites in adsorption or separation is one of the most important applications of this type of materials saving particularly energy needed for more complex separation processes if zeolites were not available to do the job. Fast development of experimental techniques enables deeper insight into the structural and textural properties of zeolites, while particularly spectroscopical methods provide new exciting information about the accessibility of inner zeolite volumes and location and coordination of active sites. Catalysis is still the most promising area for application of zeolites, in which novel zeolitic catalysts with interesting shape-selective properties can enhance activities and selectivities not only in traditional areas such as petrochemistry but also in environmental protection, pollution control, green chemistry, and biomass conversion. Last but not least, novel approaches in the manipulation and modification of zeolites directed to fuel cells, light harvesting, membranes, and sensors clearly evidence a large potential of zeolites in these new areas of application. The only limitation in zeolite research is the lack of our imagination, which slows down our effort and attainment of new exciting achievements.

It was our great pleasure working with many friends and excellent researchers in the preparation of this book. We would like to thank sincerely all of them for their timely reviews on selected topics and the great effort to put the book together. We believe that this book on zeolites will be very helpful not only for experienced researchers in this field but also students and newcomers will find it as a useful reference book.

*Jiří Čejka* Prague October 2009 | *Avelino Corma Canos* Valencia | *Stacey I. Zones* Richmond

## List of Contributors

***Michael W. Anderson***
University of Manchester
School of Chemistry
Centre for Nanoporous Materials
Oxford Road
Manchester M13 9PL
UK

***Sofía Calero Diaz***
University Pablo de Olavide
Department of Physical,
Chemical, and Natural Systems
Ctra. Utrera km. 1
41013 Seville
Spain

***Cong-Yan Chen***
Chevron Energy Technology
Company
100 Chevron Way
Richmond, CA 94802
USA

***Pablo Cubillas***
University of Manchester
School of Chemistry
Centre for Nanoporous Materials
Oxford Road
Manchester M13 9PL
UK

***Nadiya Danilina***
ETH Zurich HCI
Department of Chemistry and
Applied Biosciences
Wolfgang-Pauli-Str. 10
8093 Zurich
Switzerland

***Krijn P. de Jong***
Utrecht University
Department of Chemistry
Inorganic Chemistry and
Catalysis Group
Debye Institute for
Nanomaterials Science
Sorbonnelaan 16
3584 CA Utrecht
The Netherlands

***Petra E. de Jongh***
Utrecht University
Department of Chemistry
Inorganic Chemistry and
Catalysis Group
Debye Institute for
Nanomaterials Science
Sorbonnelaan 16
3584 CA Utrecht
The Netherlands

*Zeolites and Catalysis, Synthesis, Reactions and Applications. Vol. 1.*
Edited by Jiří Čejka, Avelino Corma, and Stacey Zones

ISBN: 978-3-527-32514-6

***Raquel García***
Instituto de Catálisis y
Petroleoquímica
CSIC
Marie Curie 2
Cantoblanco
28049 Madrid
Spain

***Jorge Gascon***
TU Delft
DCT
Catalysis Engineering
Julianalaan 136
2628 BL Delft
The Netherlands

***Luis Gómez-Hortigüela***
Instituto de Catálisis y
Petroleoquímica
CSIC
Marie Curie 2
Cantoblanco
28049 Madrid
Spain

***Nanzhe Jiang***
Inha University
Department of Chemistry
Laboratory of Nano-Green
Catalysis and Nano Center for
Fine Chemicals Fusion
Technology
Incheon 402-751
Korea

***Freek Kapteijn***
TU Delft
DCT
Catalysis Engineering
Julianalaan 136
2628 BL Delft
The Netherlands

***Russell E. Morris***
University of St Andrews
EaStCHEM School of Chemistry
Purdie Building
St Andrews KY16 9ST
Scotland

***Sang-Eon Park***
Inha University
Department of Chemistry
Laboratory of Nano-Green
Catalysis and Nano Center for
Fine Chemicals Fusion
Technology
Incheon 402-751
Korea

***Gordon M. Pearce***
University of St Andrews
School of Chemistry
Purdie Building
St Andrews
Fife KY16 9ST
UK

***Joaquin Pérez-Pariente***
Instituto de Catálisis y
Petroleoquímica
CSIC
Marie Curie 2
Cantoblanco
28049 Madrid
Spain

***Evgeny A. Pidko***
Eindhoven University of
Technology
Department of Chemical
Engineering and Chemistry
Molecular Heterogeneous
Catalysis
P.O. Box 513
5600 MB Eindhoven
The Netherlands

***Ana Belén Pinar***
Instituto de Catálisis y
Petroleoquímica
CSIC
Marie Curie 2
Cantoblanco
28049 Madrid
Spain

***Eli Stavitski***
Utrecht University
Inorganic Chemistry and
Catalysis group
Debye Institute for
Nanomaterials Science
Sorbonnelaan 16
3584 CA Utrecht
The Netherlands

***Karl G. Strohmaier***
ExxonMobil Research and
Engineering Company
1545 Route 22 East
Annandale
NJ 08801-3096
USA

***Jeroen A. van Bokhoven***
ETH Zurich HCI
Department of Chemistry and
Applied Biosciences
Wolfgang-Pauli-Str. 10
8093 Zurich
Switzerland

***Johan van den Bergh***
TU Delft
DCT
Catalysis Engineering
Julianalaan 136
2628 BL Delft
The Netherlands

***Adri N.C. van Laak***
Utrecht University
Department of Chemistry
Inorganic Chemistry and
Catalysis Group
Debye Institute for
Nanomaterials Science
Sorbonnelaan 16
3584 CA Utrecht
The Netherlands

***Rutger A. van Santen***
Eindhoven University of
Technology
Department of Chemical
Engineering and Chemistry
Molecular Heterogeneous
Catalysis
P.O. Box 513
5600 MB Eindhoven
The Netherlands

***Bert M. Weckhuysen***
Utrecht University
Inorganic Chemistry and
Catalysis group
Debye Institute for
Nanomaterials Science
Sorbonnelaan 16
3584 CA Utrecht
The Netherlands

***Paul A. Wright***
University of St Andrews
School of Chemistry
Purdie Building
St Andrews
Fife KY16 9ST
UK

***Lei Zhang***
Utrecht University
Department of Chemistry
Inorganic Chemistry and
Catalysis Group
Debye Institute for
Nanomaterials Science
Sorbonnelaan 16
3584 CA Utrecht
The Netherlands

***Stacey I. Zones***
Chevron Texaco Energy Research
and Technology Company
100 Chevron Road
Richmond, CA 94802
USA

# 1
# Synthesis Mechanism: Crystal Growth and Nucleation

*Pablo Cubillas and Michael W. Anderson*

## 1.1
## Introduction

Crystal growth pervades all aspects of solid-state materials chemistry and the industries that rely upon the functionality of these materials. In the drive toward greener, more efficient processes crystal engineering is an increasingly important requirement in materials such as catalysts; semiconductors; pharmaceuticals; gas-storage materials; opto-electronic crystals; and radio-active waste storage materials. In order to impart this desired functionality it is crucial to control properties such as crystal perfection, crystal size, habit, intergrowths, chirality, and synthesis cost [1].

The issues that concern crystal growth for nanoporous materials are similar to those that concern all crystal growths. Crystal habit and crystal size are of vital importance to the efficient functioning of these, and any other crystals, for real application. In the extreme case, single-crystal nanoporous films will require substantial skewing of both habit and size from normal bounds – this is currently impossible for zeolites but is being realized to some extent for metal organic framework (MOF) materials. Less extreme is the modification of crystal aspect ratio, for example, in hexagonal crystal systems where the pore architecture is often one-dimensional, growth of tablet-shaped crystals is usually preferred over more common needle-shaped crystals, particularly when molecular diffusion is important. All crystals incorporate both intrinsic and extrinsic defects, but whereas the presence of the latter may be easily controlled through purity of synthesis conditions, control of the former requires a deep knowledge of the underlying crystal growth mechanism. By defect we mean a well-defined aperiodic interruption in the periodic crystal structure. First, it is important to understand the nature of the defect, which normally requires a form of microscopy. Transmission electron microscopy (TEM) is the principal method used for this, but scanning probe microscopy is also useful. Owing to the structural complexity of framework crystals, each crystal system tends to display a unique defect structure that must be individually characterized. An extension of the same phenomenon is the incorporation of intergrowth and twin structures. Such defects are introduced during the crystal growth stage usually as a result of competing crystallization pathways that are near

*Zeolites and Catalysis, Synthesis, Reactions and Applications. Vol. 1.*
Edited by Jiří Čejka, Avelino Corma, and Stacey Zones

ISBN: 978-3-527-32514-6

energy equivalent. By understanding the growth mechanism it should be possible to identify the crucial step that controls this fork in the crystal growth, determine the energetic considerations, and predict modifications to growth conditions so as to enhance the probability of forming one particular crystal over another. This is crucial, for instance, for the preparation of chiral crystals that are assembled from a spiral stacking of achiral units [2, 3].

The advent of atomic force microscopy (AFM) (Figure 1.1) has opened up new possibilities to investigate the molecular events that occur during crystal growth and dissolution/recrystallization. The technique can be used both *in situ* and *ex situ* with each method suited to particular problems. *Ex situ* operation allows a vast array of synthetic parameters to be varied without concern for the delicacies of the AFM operation. In this respect, careful quenching experiments whereby the state of the nanoscopic features at the crystalline surface may be frozen rapidly before transfer to the AFM can be performed. This is crucial to prevent secondary processes caused by changing growth conditions upon crystal cooling and extraction from the mother-liquor. Rates and energies of crystal growth processes can be determined via such *ex situ* experiments through modeling both crystal topology and habit. *In situ* AFM gives a more direct approach to determining growth and dissolution rates. Further, surface structures that are inherently less stable may not be seen in *ex situ* analysis. Consequently, where possible, *in situ* analysis is preferred. The

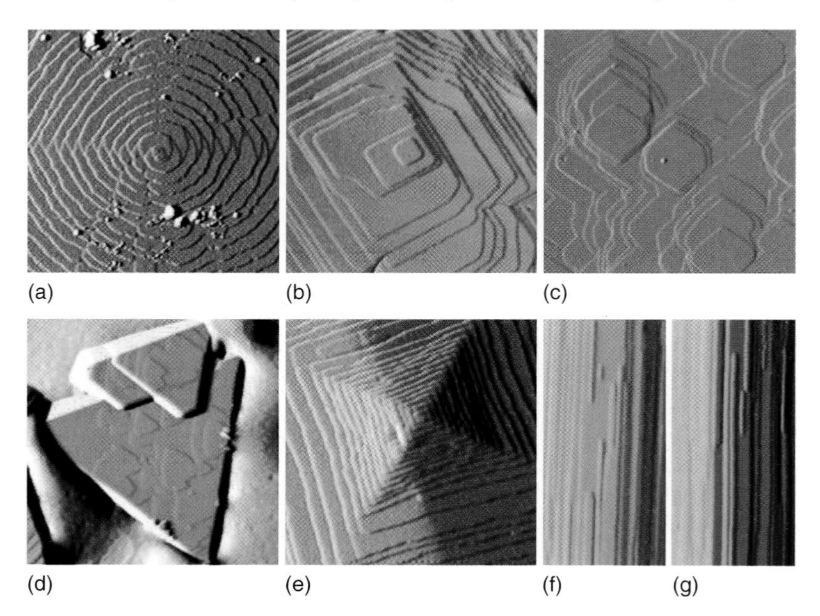


**Figure 1.1** (a) Interlaced spiral on aluminophosphate STA-7; (b) zeolite A reducing supersaturation; (c) metal organic framework ZIF-8; (d) *in situ* $ZnPO_4$-FAU growth structure; (e) interlaced spiral $ZnPO_4$-SOD growth structure; and (f, g) *in situ* dissolution of zeolite L.

structural details leading to the observed crystal growth, defect, and intergrowth structure can also be probed using electron microscopy, and by slicing crystals open we can look at the consequences of structural growth decisions in the heart of the crystals. To probe the solution chemistry from which the crystals have evolved, a combination of the speciation delineation of nuclear magnetic resonance (NMR) with the speed and sensitivity of mass spectrometry is increasing our knowledge substantially. Both these techniques also probe the extent of oligomerization in the buildup to nucleation that can be further probed using cryo-TEM methods.

## 1.2 Theory of Nucleation and Growth

### 1.2.1 Nucleation

The formation of a new crystalline entity from a solution starts through the nucleation process. *Nucleation* is defined as the series of atomic or molecular processes by which the atoms or molecules of a reactant phase rearrange into a cluster of the product phase large enough as to have the ability to grow irreversibly to a macroscopically larger size. The *cluster* is defined as nucleus [4] or critical nuclei.

Nucleation can be homogeneous, in the absence of foreign particles or crystals in the solution, or heterogeneous, in the presence of foreign particles in the solution. Both types of nucleation are collectively known as *primary nucleation*. Secondary nucleation takes place when nucleation is induced by the presence of crystals of the same substance.

### 1.2.2 Supersaturation

The driving force needed for the nucleation and growth of a crystal is referred to as *supersaturation* and is defined as the difference in chemical potential between a molecule in solution and that in the bulk of the crystal phase:

$$\Delta\mu = \mu_{\mathrm{s}} - \mu_{\mathrm{c}} \tag{1.1}$$

where $\mu_{\mathrm{s}}$ is the chemical potential of a molecule in solution and $\mu_{\mathrm{c}}$ is the chemical potential of the molecule in the bulk crystal. Following thermodynamics Eq. (1.1) can be expressed as

$$\Delta\mu = kT \ln S \tag{1.2}$$

where $k$ is the Boltzmann constant, $T$ is the absolute temperature, and $S$ is the supersaturation ratio. When $\Delta\mu > 0$ the solution is said to be supersaturated, meaning that nucleation and/or growth is possible, whereas when $\Delta\mu < 0$ the solution will be undersaturated and dissolution will take place. The form of the supersaturation ratio will change depending on the system considered (i.e., gas/solid, solution/solid, melt/solid). For nucleation and growth from solutions it takes the following form:

$$S = \frac{\Pi a_i^{n_i}}{\Pi a_{i,e}^{n_i}} \tag{1.3}$$

where $n_i$ is the number of $i$th ions in the molecule of the crystal, and $a_i$ and $a_{i,e}$ the actual and equilibrium activities of the $i$ molecule in the crystal.

### 1.2.3 Energetics

According to nucleation theory, the work necessary to form a cluster of $n$ number of molecules is the difference between the free energy of the system in its final and initial states [4, 5] plus a term related to the formation of an interface between nucleus and solution. This can be expressed by (assuming a spherical nucleus):

$$\Delta G_T = -n\Delta\mu + 4\pi \cdot r^2\sigma \tag{1.4}$$

where $r$ is the radius of the nucleus and $\sigma$ is the surface free energy. If each molecule in the crystal occupies a volume $V$, then each nucleus will contain $(4/3)\pi \cdot r^3/V$ molecules. Eq. (1.4) will then take the following form:

$$\Delta G_T = -\frac{4}{3}\pi \cdot \frac{r^3}{V}\Delta\mu + 4\pi \cdot r^2\sigma \tag{1.5}$$

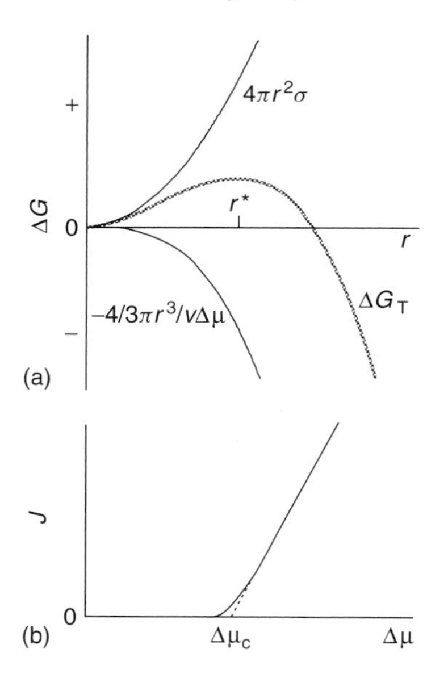


**Figure 1.2** (a) Total free energy versus cluster size. (b) Nucleation rate as a function of supersaturation (showing the critical supersaturation).

Figure 1.2a shows a plot of $\Delta G_T$ as a function of $r$; it can be seen how the function reaches a maximum, which represents the energetic barrier that needs to be surpassed to achieve nucleation ($\Delta G^*$). The *value of r at this maximum ($r^*$)* is defined as the critical radius or nucleus size [4, 5]. Its value is defined by

$$r^* = \frac{2\sigma \cdot V}{kT \ln S} \tag{1.6}$$

It has been proved that the value of $r^*$ decreases (as well as that of $\Delta G^*$) as the supersaturation increases [6], meaning that the probability of having nucleation in a given system will be higher, the higher the supersaturation.

### 1.2.4
### Nucleation Rate

The rate of nucleation (i.e., the number of nuclei formed per unit time per unit volume) can be expressed by an Arrhenius-type equation [5]:

$$J = A \exp\left(\frac{-\Delta G^*}{kT}\right) \tag{1.7}$$

where $A$ also depends on supersaturation. A typical plot of $J$ as a function of supersaturation ($S$) is depicted in Figure 1.2b. It can be seen in this plot that the nucleation rate is virtually zero until a critical value of supersaturation is achieved, after which the rate increases exponentially. This critical supersaturation ($\Delta\mu_c$) defines the so-called metastable zone where crystal growth can proceed without concomitant nucleation taking place.

### 1.2.5
### Heterogeneous and Secondary Nucleation

Equations (1.5) and (1.6) shows that both $\Delta G^*$ and $r^*$ depend heavily on the surface free energy ($\sigma$), so any process that modifies this value will have an effect on the possible viability of the nucleation process. It has been proved that in the presence of a foreign substrate the decrease in the value of $\sigma$ therefore reduces the values of $\Delta G^*$ and $r^*$ at constant supersaturation [6], that is, making nucleation more favorable. A decrease in $\sigma$ will also decrease the value of the critical supersaturation ($\Delta\mu_c$), since the nucleation rate is also dependent on the surface energy (Eq. (1.7)). This will make heterogeneous nucleation more viable than homogeneous nucleation at low supersaturation conditions. The reduction of the surface energy will be the highest when the best match between the substrate and the crystallizing substance is achieved. This situation is created, of course, when both the substrate and the crystallizing substance are the same, referred to as *secondary nucleation*. This mechanism will be more favorable than both heterogeneous and homogeneous nucleation and thus produced at lower supersaturation.

### 1.2.6 Induction Time

*Induction time* is defined as the amount of time elapsed between the achievement of a supersaturated solution and the observation of crystals. Its value will thus depend on the setting of $t = 0$ and the technique used to detect the formation of crystals. The induction period can be influenced by factors such as supersaturation, agitation, presence of impurities, viscosity, and so on. Mullin [5] defined the induction time as

$$t_i = t_r + t_n + t_g \tag{1.8}$$

The induction time is separated into three periods: $t_r$ is the relaxation time, required for the systems to achieve a quasi-steady-state distribution of molecular clusters; $t_n$ is the time required for the formation of a nucleus; and $t_g$ is the time required for the nucleus to grow to a detectable size.

### 1.2.7 Crystal Growth

Crystal growth is the series of processes by which an atom or a molecule is incorporated into the surface of a crystal, causing an increase in size. These different processes can be summarized into four steps [7, 8] illustrated in Figure 1.3:

1) transport of atoms through solution;
2) attachment of atoms to the surface;
3) movement of atoms on the surface;
4) attachment of atoms to edges and kinks.

The first process is the so-called transport process, whereas 2–4 are referred to as *surface processes* (and may involve several substeps). Since these different steps normally occur in series, the slowest process will control the overall crystal growth. Therefore, growth can be transport (when step 1 is the slowest) or surface controlled (when steps 2–4 are the slowest).

### 1.2.8 Crystal Surface Structure

Crystal growth theories are based on considerations of the crystal surface structure. One of the most commonly used models was that provided by Kossel [9]. This model envisions the crystal surface as made of cubic units (Figure 1.4) which form layers of monoatomic height, limited by steps (or edges). These steps contain a number of kinks along their length. The area between steps is referred to as a *terrace*, and it may contain single adsorbed growth units, clusters, or vacancies. According to this model, growth units attached to the surface will form one bond, whereas those attached to the steps and kinks will form two and three bonds, respectively. Hence, kink sites will offer the most stable configuration. Growth will then proceed by

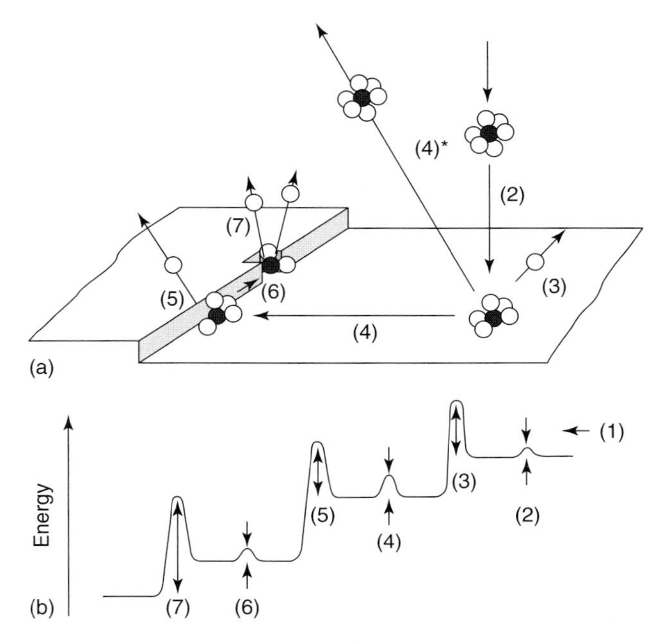


**Figure 1.3** (a) Schematic representation of processes involved in the crystal growth: (1) Transport of solute to a position near the crystal surface; (2) diffusion through boundary layer; (3) adsorption onto crystal surface; (4) diffusion over the surface; (4*) desorption from the surface; (5) attachment to a step or edge; (6) diffusion along the step or edge; (7) Incorporation into kink site or step vacancy. (b) Associated energy changes for the processes depicted in (a). Figure modified from Elwell *et al.* [7].

the attachment of growth units to kink sites in steps. The kink will move along the step producing a net advancement of the step until this step reaches the face edge. Then, a new step will be formed by the nucleation of an island of monolayer height (or two-dimensional (2D) nucleus) on the crystal surface. This mechanism of growth is normally referred to as *layer growth* or *single nucleation growth* and is represented in Figure 1.5. A variation of this growth mechanism occurs when the nucleation rate is faster than the time required for the step to cover the whole crystal surface. In this case, 2D nuclei will form all over the surface and on top

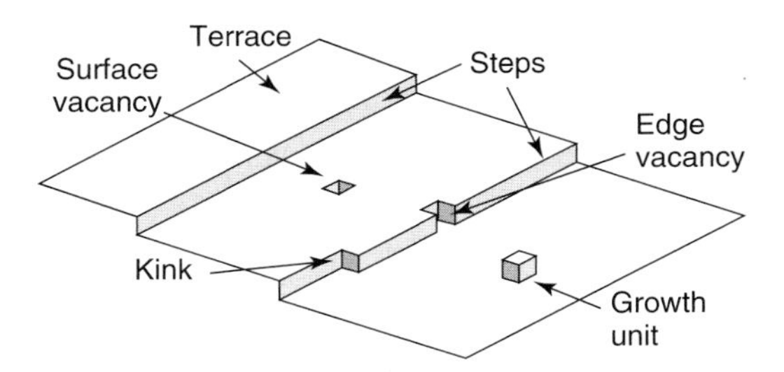


**Figure 1.4** Kossel model of a crystal surface.

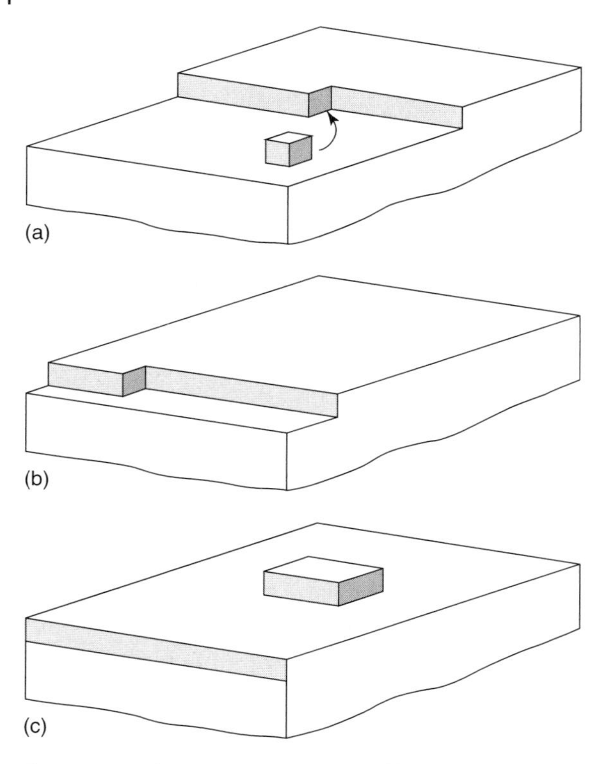


**Figure 1.5** Schematic representation of layer growth. (a) Incorporation of growth units into step. (b) The step has almost advanced to the edge of the crystal. (c) Formation of 2D nucleus.

of other nuclei. These nuclei will spread and coalesce forming layers. This growth mechanism is normally referred to as *multinucleation multilayer growth* or *birth and spread* [10].

### 1.2.9
### 2D Nucleation Energetics

The total free energy change due to the formation of a 2D nucleus of height $h$ and radius $r$ can be calculated by using Eq. (1.5):

$$\Delta G_{T-2D} = -\pi \cdot \frac{hr^2}{V}\Delta\mu + 2\pi \cdot rh\sigma \tag{1.9}$$

The maximum of this function defines the value of the critical radius which is given by

$$r_{2D}{}^{*} = \frac{\sigma \cdot V}{kT \ln S} \tag{1.10}$$

It can be seen that the value of $r_{2D}^*$ is half of the nucleus size for homogeneous nucleation (Eq. (1.6)).

### 1.2.10
### Spiral Growth

The energetics of layer growth predict that growth takes place at relatively high supersaturation (needed to overcome the energy barrier associated with 2D nucleation). Nevertheless, it has been observed that crystals can still grow at lower supersaturation than predicted [11]. This dilemma was solved by Frank [12] who postulated that crystal surfaces are intercepted by dislocations. These dislocations will create steps in the surface, obviating the necessity for 2D nucleation. Figure 1.5 shows a schematic diagram on the formation and development of spiral growth. In the initial stage the dislocation creates a step (Figure 1.6a). Growth units attach to the step making it advance and thus generating a second step (Figure 1.6b). This second step will not advance until its length equals $2r_{2D}^*$; this is because any growth of a step with a smaller size is not thermodynamically favored. Once the second step starts advancing, it will generate a third step which in turn will not start moving until its length equals $2r_{2D}^*$ (Figure 1.6c), then a fourth step will appear, and so on (Figure 1.6d). This will generate a spiral pattern around the dislocation core, and a self-perpetuating source of steps where growth requires less energy than a layer mechanism (therefore, it can proceed at smaller supersaturation). In the case of a curved step, the spiral will be rounded and its curvature will be determined by the $r_{2D}^*$ value at the specific supersaturation conditions in which

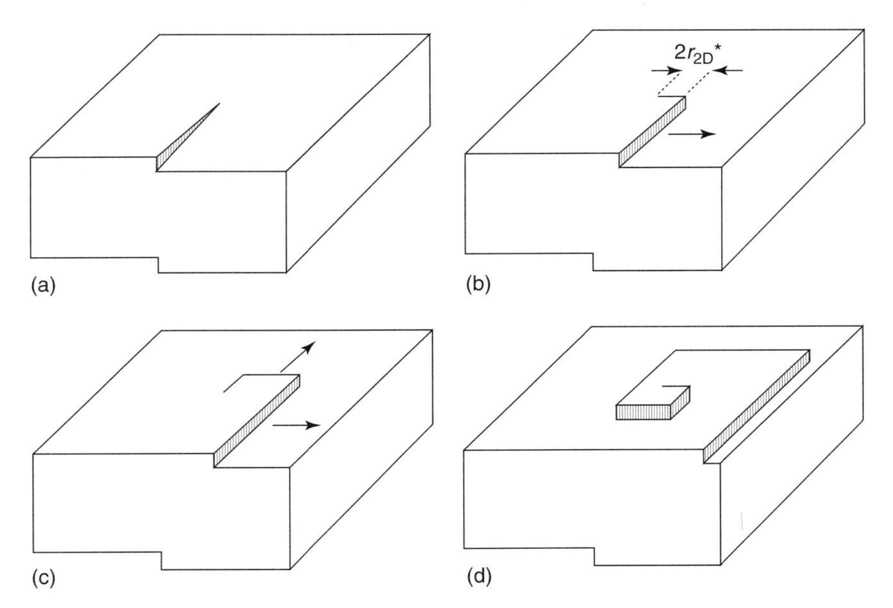


**Figure 1.6** Development of a polygonal spiral.

the crystal grows. The theory of crystal growth by spiral dislocation was further refined by Burton, Cabrera, and Frank [13], giving rise to what is known as the *BCF theory*.

### 1.2.11 Interlaced Spirals

Interlaced spirals are the result of a periodic stacking of differently oriented growth layers, each having a different lateral anisotropy of step velocity [14]. In other words, in this type of spiral a step with unit cell may dissociate in substeps symmetrically related but crystallographically different. The dissociation is the result of a different growth anisotropy by each step due to its different crystallography. Figure 1.7 shows a schematic diagram showing the formation of a spiral of this type according to van Enckevort [14]. The surface in the figure is produced by two distinct types of steps (I and II), of height 1/2 $d_{hkl}$, emanating from a central point O. Layers of type I are bound by steps a and b, whereas layers from type II are bound by steps c and d. Steps a and d move fast and steps b and c move slowly. This results in steps a of layer I catching up with steps c from layer II, producing a double step of unit-cell height. The same process is observed in steps d joining the slow steps b. The result is a pattern of unit-cell height steps with interlaced crossovers formed by lower steps of height 1/2 $d_{hkl}$. Interlaced spirals have been observed in numerous systems, including barite [15], molecular crystals [16, 17], silicon carbide [18], GaN [19], and sheet silicates [20].

### 1.2.12 Growth Mechanisms: Rough and Smooth Surfaces

The growth mechanisms can be classified into three types depending on the interface structure. If the surface is rough the growth mechanism will be of adhesive type, whereas if the surface is smooth growth will take place by either

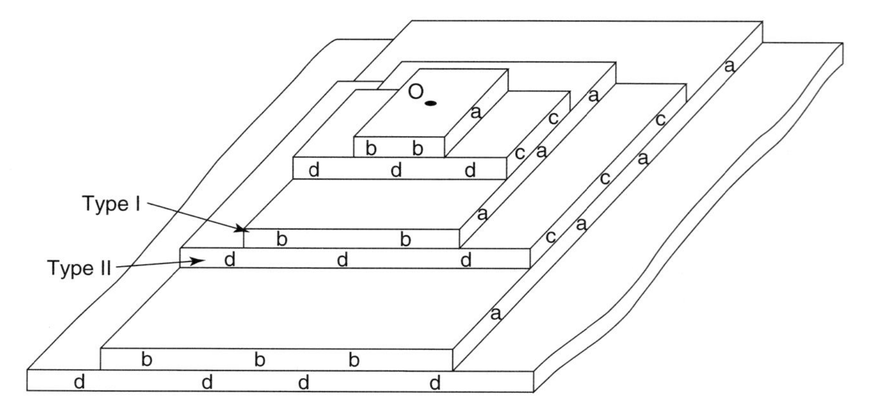


**Figure 1.7** Interlaced spiral formation. Figure modified from van Enckevort *et al.* [14].

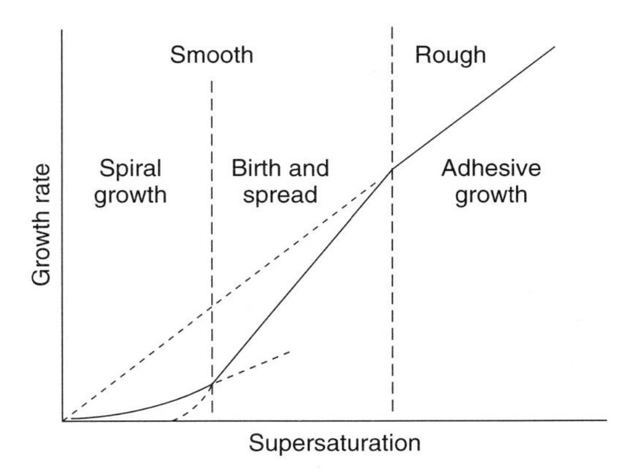


**Figure 1.8** Mechanisms of growth as a function of supersaturation.

birth and spread, or spiral growth. A surface will transform from smooth to rough at high driving force conditions (high supersaturation). Figure 1.8 shows the different growth mechanisms as a function of supersaturation. At a low supersaturation, the interface is smooth and spiral growth is the mechanism of growth. After reaching a critical supersaturation for 2D nucleation, birth and spread dominates the growth. In these two domains crystals are bound by crystallographically flat faces with polyhedral morphologies. At high supersaturation the surface transforms to a rough interface, and adhesive-type growth dominates. In the adhesive-type regime the energetics of growth unit attachment are the same regardless of the crystallographic direction, giving rise to crystals bounded by rounded noncrystallographic surfaces, producing spherulitic, fractal, and dendritic patterns.

## 1.3 Nucleation and Growth in Zeolites

### 1.3.1 Overview

Zeolite and zeotype synthesis is well known to be a complex process. The rate of crystallization, types of products formed, and their particulate properties (habit, morphology, and crystal size distribution) depend on a large number of parameters [21]. These parameters encompass the crystallization conditions (temperature, stirring, seeding, and gel aging) as well as composition-dependent parameters (pH, water content, and the ratio between framework-forming elements, template concentration, and ionic strength). Nevertheless, a typical zeolite/zeotype synthesis will involve the following steps [22]:

1) A mixture of amorphous reactants which contains the structure-forming ions (such as Si, Al, P, Ga, Zn, etc.) in a basic medium (although a few zeolite synthesis can also take place in acidic medium [23]). This leads to the formation

of a heterogeneous, partly reacted phase, which has been referred to as the *primary amorphous phase* [22, 24]. The nature of this amorphous phase ranges from gel-like to colloidal in the so-called clear solution synthesis [25, 26].

2) Heating of the reaction mixture (above 100 °C) at autogenic pressures in metal autoclaves. Prior to this the reaction mixture may be left for aging for a period of time (hours to a few days).
3) Formation of a "secondary amorphous phase" at pseudo-equilibrium with a solution phase [22]. Evidence exists that this phase possesses some short range order due to the structuring effect of cations in the solution [24, 27–30].
4) After an induction period the formation of nuclei takes place. This induction time can be related to the definition for simple systems given in Eq. (1.8) [31]. The relaxation time ($t_r$) would be the time required for steps 1–3 to take place, that is, for the formation of the quasi-steady state amorphous solid, whereas $t_n$ and $t_g$ have the same meaning.
5) Growth of the zeolite material at the expense of the amorphous solid.

These steps are well defined for a multitude of zeolite and zeotype syntheses, but in many cases it may be difficult to differentiate them. This could be because some of the steps overlap or because the experimental difficulties in studying the synthesis are excessive [22]. Steps 1–3 have been studied in the last several years by many authors. It is not the purpose of this chapter to deal in detail on this subject but the reader is directed to the review by Cundy and Cox [22] for additional information and references.

Figure 1.9 shows the typical shape of a crystallization curve for a zeolite synthesis where both the nucleation rate and the evolution of the crystal-length or crystallinity in the system are plotted as a function of the synthesis time. It can be seen that nucleation only takes place after an induction time, that is, after steps 1–3 have taken place. The rate of nucleation increases rapidly but then decreases to zero. After a certain number of nuclei have formed crystal growth takes place. Initially the

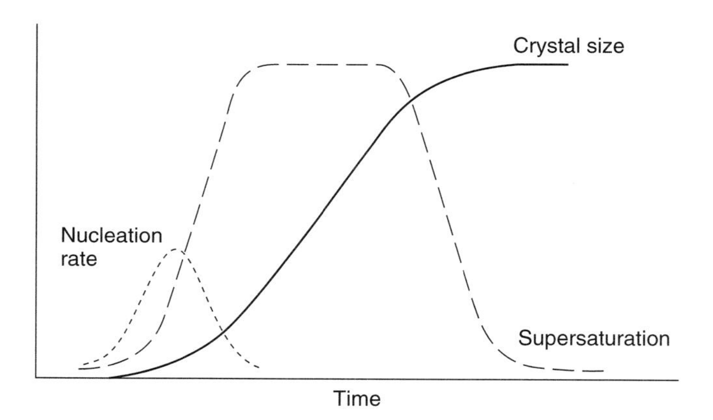


**Figure 1.9** Schematic representation of the zeolite synthesis process showing the evolution of nucleation and growth rates, as well as supersaturation, as a function of time.

growth rate increases exponentially but rapidly achieves a steady state before finally decreasing to zero when the nutrients are exhausted. The synthesis process can also be followed according to the theoretical supersaturation curve (superimposed in Figure 1.9). The supersaturation increases initially giving rise to the nucleation and growth phase, then it levels off, as the growth rate achieves a steady state, and finally decreases to zero as all the nutrients in the solution are incorporated into the growing phase.

### 1.3.2
### Zeolite Nucleation

Zeolite nucleation is a complex problem, since it implies the transformation of an initially amorphous or random structure into a crystalline framework. As observed before, during the formation of the secondary amorphous phase there is an increase in the structural order although of very short range. From this step a random number of structured areas may achieve the size of a nucleus and start to grow into a macroscopic crystal.

The use of traditional nucleation theory for studying zeolite nucleation has been employed in the past, for example, in calculating the nucleation rate as the inverse of the induction time [32]. Nevertheless it has been observed that there may be important differences between zeolite crystallization and that of more condensed phases. One of these differences may stem from the high internal surface area of zeolites [33].

The process of zeolite nucleation has proved to be difficult to study and analyze owing to the experimental difficulties in making *in situ* measurements. Various authors have obtained information by using the size distribution and using mathematical models [34] to infer the growth and nucleation rate [21, 35, 36]. Other studies have looked at the effect of aging and seeding on zeolite nucleation. Aging of the initial solution has proved to have influence on the final crystal distribution [22], and hence it can provide valuable information on the nucleation mechanism as has been demonstrated for zeolite A [37] and silicalite [38]. The use of seeds can be the factor used to differentiate between primary and secondary nucleation [22].

Some of the most heated debate on the study of nucleation on zeolite and zeotype synthesis has been centered on the nucleation mechanism, whether this is homogeneous [36, 39] or heterogeneous [40] (primary nucleation) or even secondary [41] (crystal induced). Difficulties in discerning one mode or other stem in part due to the very nature of the gel phase, especially in the so-called clear solution system where it is physically difficult to separate colloid-sized gel particles from the aqueous phase [22]. Nevertheless, there appears to be a growing body of studies supporting the idea that nucleation occurs mainly in a gel phase, specifically at the solution–gel interface [29, 42] where the nutrient concentration gradients are probably the highest. Recent studies on clear solution synthesis have also demonstrated that nucleation actually takes place inside the colloid-sized gel particles [26, 43–45].

The mechanistic aspects of the nucleation process have also been extensively discussed and it is accepted that the process of progressive ordering inside the gel is conducted through a reversible mechanism of breaking and remaking the chemical bonds in the framework catalyzed by hydroxyl ions [46, 47]. Cations and organic-structure-directing agents also have a crucial role in the nucleation process by surrounding themselves with metal-oxide species in preferred geometries owing to electrostatic and van der Waals interactions [48, 49].

### 1.3.3
### Crystal Growth on Zeolites and Zeotypes

A multitude of studies have been carried out to study how zeolite and zeotype crystals grow. Surprisingly, in spite of the vast number of structure types, framework composition, and synthesis procedures, it has been generally found that zeolite growth increases linearly during most of the crystallization process. This is true for both gel synthesis [1, 50–55] and clear solution synthesis [56–59], although studies on the latter system have shown a dependence of the growth rate on the crystal size for values below 15–20 nm [60]. Zeolite growth has been found to be affected by a multitude of parameters such as temperature, gel composition, agitation, and aging, and many studies have been devoted to these topics [1, 22]. In general, measured growth rates of zeolites are consistently lower than those of more dense phases (such as ionic crystals), which has been postulated as due to the more complex assembling mechanism of the open-polymeric structure of zeolite and zeotypes [22].

The fact that the growth rate proceeds linearly almost to the end of the synthesis has been used to support the idea of a surface-controlled mechanism [21, 61]. This idea has been supported by the measured values of activation energies for the growth process, which vary between 45 and 90 kJ $mol^{-1}$ [21, 62–64]. These values are much higher than those corresponding to a diffusion-controlled mechanism [21].

The quest for understanding the true growth mechanism and its fundamentals has been recently aided by the development of new high-resolution surface-sensitive techniques such as AFM [65–67], high-resolution scanning electron microscopy (HRSEM), [68], and high-resolution transmission electron microscopy (HRTEM) [69, 70]. Additionally, new developments in liquid- and solid-state NMR as well as mass-spectrometry techniques shed new light on the physical and chemical nature of the growing units [70]. Furthermore, advanced modeling techniques and theoretical studies have been used to further validate the experimental observation and to provide more insight into the molecular aspects of crystal growth [67].

AFM studies on zeotypes and zeolites were initially limited to the study of natural zeolites [71–75]; nevertheless, in the last several years, numerous studies on synthetic materials have been published [65–67, 76–82], including some on zeotypes [83–85]. Most of these studies have been *ex situ* where crystals were removed from the solution used for growth before analysis but a few *in situ* dissolution studies have also been performed [86, 87].

Initial AFM studies were focused on natural zeolites obtaining high-resolution images of the zeolite surface, so its porous structure could be observed at the surface termination [71, 72, 75]. Nevertheless, in 1998, Yamamoto *et al.* [73] published images of natural heulandite crystals that showed the presence of steps suggesting a possible birth-and-spread mechanism. On the synthetic front, Anderson *et al.* [88] published the first AFM study of zeolite Y, which showed also the formation of steps and terraces on the crystal surface. These studies have been followed by others dealing mainly with zeolite A [67, 77, 87, 89], and X/Y [78, 90] and silicalite [81]. From these investigations, detailed information on the surface termination, the mode of growth, dissolution, and the possible identity of the growth units has been inferred. Until recently, most AFM studies have only revealed the presence of steps and terraces, making some authors to conclude that birth and spread may be the preferential mode of growth for these materials [22]. Nevertheless, a recent study on zeolite A [91] and stilbite [74] shows the formation of spirals. Also, results highlighted in this chapter show that spiral growth in zeotype structures may be more prevalent than originally thought.

## 1.4 Techniques

### 1.4.1 The Solid Crystal

#### 1.4.1.1 AFM

The AFM is a surface-scanning technique invented by Binning *et al.* [92] in 1986 as a development of the scanning tunneling microscope (STM) [93]. The AFM provides three-dimensional images of surfaces by monitoring the force between the sample and a very sharp tip (a few nanometers wide). This is in contrast to the STM, which relies on the formation of a tunneling current between the tip and the sample. Therefore, the AFM can be used to scan the surface of virtually any kind of material. Typically, the sample is mounted on a piezoelectric scanner, which moves the sample in $x$–$y$ and $z$. The lateral resolution is limited by the tip radius, which normally varies between 10 and 30 nm (although it may be as low as 3 nm), whereas the vertical resolution is around 1 Å, making it ideal to observe small surface details such as steps or 2D nuclei. The end of the tip is attached to a cantilever, which bends when the force between the sample and the tip changes. The deflection of the cantilever is monitored by shining a laser on its top surface, which is reflected back to a photodiode detector. The output signal of the photodiode is then transmitted to a computer. A feedback control system informs the piezoelectric scanner of any changes in force between the tip and the sample, allowing it to alter the tip–sample separation to maintain the force at a constant value.

There are different imaging modes available when using AFM depending on the motion of the tip over the sample. In contact mode, the tip is raster scanned over the sample while the cantilever deflection is kept constant using feedback control. Intermittent contact mode utilizes an oscillating tip and monitors the phase and amplitude of the cantilever [94]. In this mode the contact between the tip and sample is minimized, and hence may be advantageous for softer samples (e.g., biological samples). Some AFMs can monitor both the vertical and lateral forces (and friction) [95], known as *friction force microscopy* (FFM). In this case, information on the adhesion, friction, or other mechanical properties of the sample can be obtained [96].

The use of AFM has revolutionized the study of crystal growth in the last several years, not only due to its high vertical resolution, but because of its ability to scan in fluids, thus making it possible to monitor *in situ* dissolution and growth processes. AFM has been extensively used in crystal growth studies of macromolecular crystals [97], minerals [98–100], ionic crystals [101, 102], organic semiconductors [103], thin films [104], hierarchical porous materials [105], and many other crystal systems. The use of AFM in studying synthetic zeolites and zeotypes has been slow to come, mainly due to two reasons: (i) the limitations of studying micrometer-sized crystals using a conventional top-head low magnification ($< 20\times$) optics makes locating good crystals very time consuming; (ii) the fact that most zeolites crystallize at temperatures above 100 °C. Nevertheless, the development of tip-scanning AFMs coupled with high-magnification (up to $100\times$) inverted optical microscopes plus new developments in temperature-controlled fluid cells have immensely increased the range of observations that can be made using these systems. This is further illustrated in the six case studies presented in this chapter.

#### 1.4.1.2 HRSEM

Scanning electron microscopy is a well known and used technique for the characterization of microporous materials. It has long been used in conjunction with X-ray diffraction (XRD) as one of the main tools in the studies of crystal growth in zeolites [1, 22]. Its use though has been limited to study the habit, morphology, and size of the synthesized materials but not to characterize the surface detail. This is due to resolution limitations owing to excessive charging on steps at the crystal surface. In recent years, the development of low voltage field emission electron sources for SEMs (FE-SEM) has significantly reduced this issue. For example, in 2005 Wakihara *et al.* [89] reported SEM images of zeolite A where steps could be observed. This new kind of SEM has been dubbed HRSEM and has been shown to be able to observe steps as small as 1.2 nm with ease [68]. The use of this technique promises to open a new chapter in the study of crystal growth of nanoporous materials, complementing the AFM and offering nanometer-scale resolution in areas where an AFM tip cannot have easy access (such as twins, intergrowths, and rough surfaces). Some examples of the full potential of this technique are highlighted in some of the following case studies.

#### 1.4.1.3 Confocal Microscopy

During crystal growth macroscopic defect structures often form, leading to intergrowths, twins, and other more exotic extended structures. Knowledge of these internal structures of the crystals gives a lot of information about the growth mechanism. Very often these macroscopic growth features can be observed with the resolution of an optical microscope. By operating with confocal optics, often with additional fluorescence from probe molecules selectively adsorbed within the nanoporous crystal, these macroscopic features can be illuminated [106, 107].

## 1.4.2 Solution Chemistry – Oligomers and Nanoparticles

### 1.4.2.1 Nuclear Magnetic Resonance

In order to understand how crystals grow, it is necessary not only to understand how the solid phase grows but also to understand the chemistry occurring in the solution phase. Unlike molecular crystals, such as those used in the pharmaceutical industry, the building units for nanoporous materials are in constant interchange in solution. The ephemeral nature of the species makes the task of unraveling, not only what is present in solution but also which are the rate-determining steps in the kinetics of this constant interchange, a very daunting task. The most powerful tool for speciation in the solution state is NMR, and this can be used to good effect to monitor $^{29}Si$ in silicates [108–119], $^{31}P$ in phosphates, and $^{19}F$ in fluorides [120]. These are spin one-half nuclei and, as a consequence, tend to yield highly resolved spectra that are amenable to both one- and two-dimensional spectroscopy. By determining connectivities by INADEQUATE and COSY NMR, a large number of species have now been identified (Figure 1.10). The experiment must be carefully set up in order to ensure quantitation [121] and by inserting chemical probes the course of crystallization may be followed by monitoring pH by NMR [122]. $^{29}Si$ is only about 4% abundant and therefore two-dimensional spectra can take some time to acquire, precluding the ability to follow rapid temporal changes. This problem is further compounded, as spin half nuclei in solution often have long relaxation times that also substantially slow data acquisition. Nonetheless, $^{29}Si$ spectra of silicate solutions in particular can reveal an enormous amount of information on a plethora of species. Operated in a multinuclear fashion the complete chemistry of a crystallization such as silicoaluminophosphate (SAPO)-34 [123] may be followed. Quadrupolar nuclei, such as $^{27}Al$ [124] and $^{17}O$ [125], can also play a role in understanding solution chemistry. The spectra, although often less well resolved in terms of revealing multiple speciation, can be collected very rapidly and therefore dynamic information may be extracted. Care must also be taken regarding the degree of condensation of species, because as soon as nanoparticulates/colloids are formed the restricted motion of species starts preventing spectral averaging, which is vital for high-resolution spectra. Not only does the lack of motion broaden spectra but also a continuum of complex species with slightly different T–O–T angles will result in a continuum of chemical shifts causing a broadening that cannot be removed even by magic-angle spinning. However, this apparent downside of NMR can also be used to good advantage. For spin one-half nuclei such as $^{29}Si$ the spectral broadening is often not so substantial to render the resonances invisible. It is therefore very

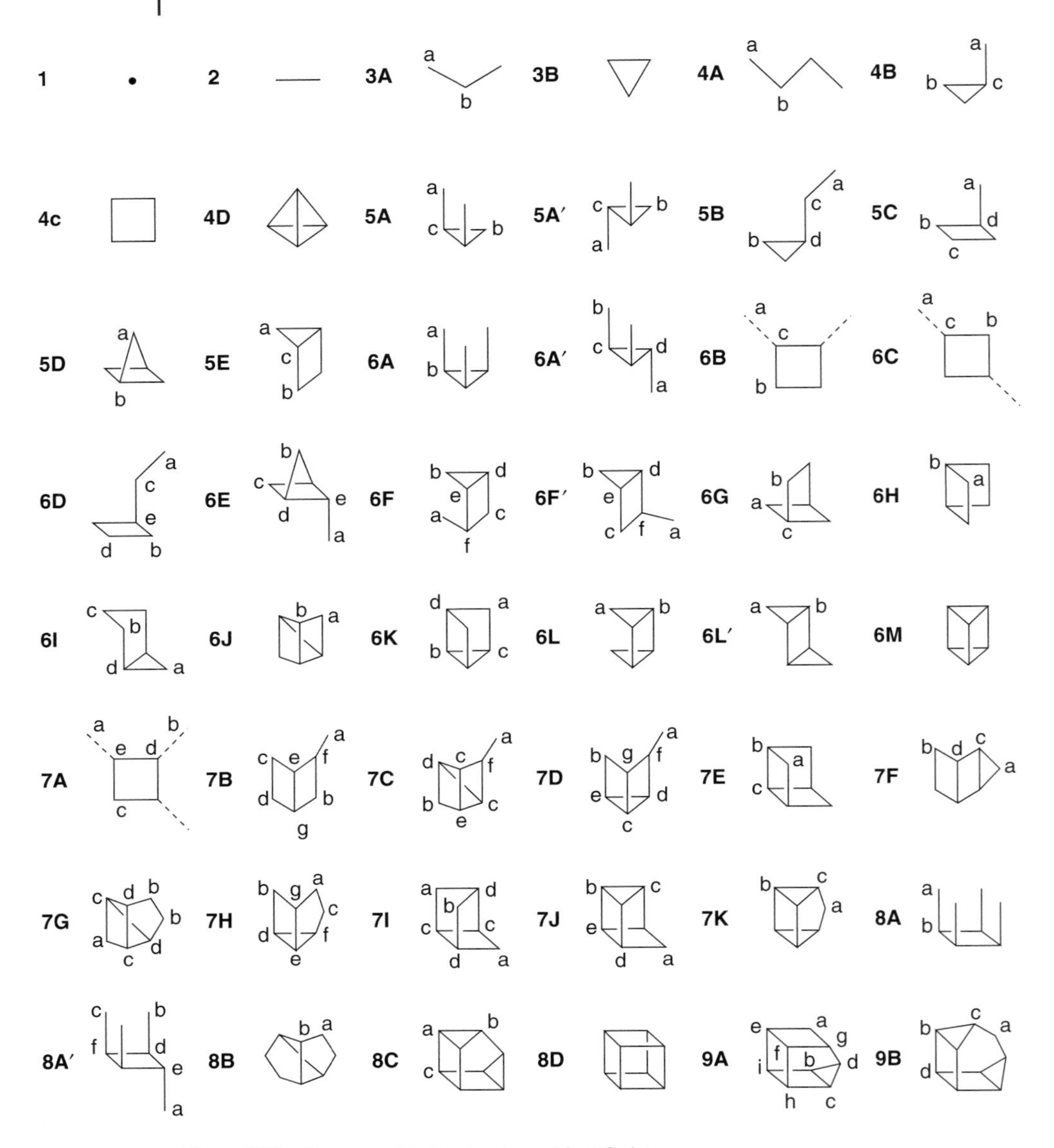


**Figure 1.10** Aqueous silicate structures identified in concentrated alkaline solution by $^{29}Si-^{29}Si$ COSY NMR [116].

easy to distinguish between small oligomers, giving rise to sharp resonances, and nanoparticles giving broad resonances. Further, because for spin one-half nuclei quantitation of spectra is straightforward, it is a simple exercise to determine the relative concentration of oligomeric and nanoparticulate species [118, 119] (Figure 1.11).

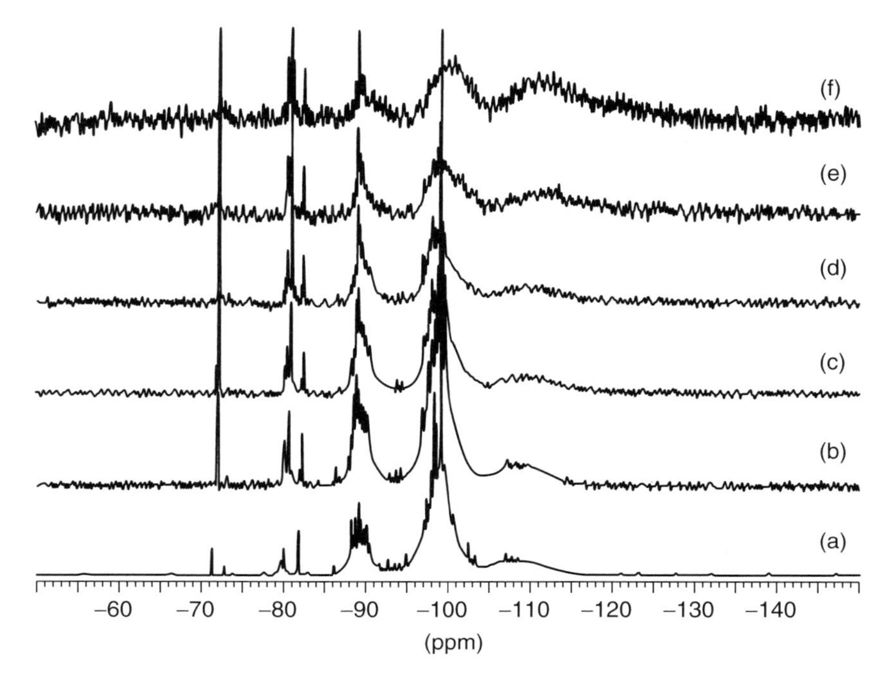


**Figure 1.11** Normalized $^{29}$Si NMR spectra of clear solutions with TEOS : TPAOH : $H_2O$ molar ratio of 25 : 9 : $x$ where (a) $x$ 1/4 152; (b) $x$ 1/4 400; (c) $x$ 1/4 900; (d) $x$ 1/4 1900; (e) $x$ 1/4 4000; and (f) $x$ 1/4 9500 [119].

#### 1.4.2.2 **Mass Spectrometry**

In order to increase sensitivity and temporal resolution, mass spectrometry is becoming increasingly used in the study of solution speciation. Modern mass spectrometers utilizing soft ionization procedures such as electrospray ionization can readily yield parent ions in relatively high mass/charge ratio. Although the speciation is not as unique as that determined by NMR, through isotopic distribution analysis the possibilities may be narrowed considerably. The enormous advantage over NMR is the sensitivity and rapid data collection that permits *in situ* analysis of dynamic events [126–132]. The power of this technique is most aptly demonstrated in a recent work on the interconversion between silicate oligomers [132]. In a clever experiment, a solution enriched with $^{29}$Si and a solution of naturally abundant $^{28}$Si containing cubic octamers were mixed. The mass spectrometry clearly revealed that the first exchange species contained equal amounts of the two isotopes indicating a concerted exchange mechanism involving four silicon nuclei. A similar experiment with triangular prismatic hexamers showed a concerted exchange of three silicon nuclei (Figure 1.12). Such concerted exchange mechanisms are probably omnipresent in the chemistry of silicates and are likely to play an important role in zeolite crystal nucleation and growth.

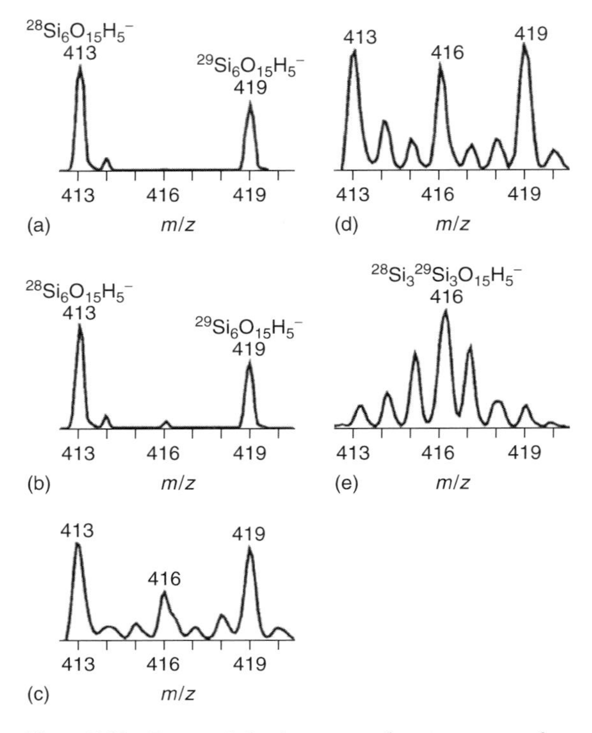


**Figure 1.12** Temporal developments of mass spectra after mixing equally concentrated solutions containing the prismatic hexamer with naturally abundant silicon and $^{29}Si$-enriched silicon [132].

#### 1.4.2.3 Cryo-TEM

The transition from solution to nuclei and finally to crystals is a complicated process to monitor experimentally. Scattering techniques such as dynamic light scattering and X-ray or neutron scattering are able to determine dynamically the presence of important nanoparticulates during these crucial early stages of the birth of crystals. However, recently a new and powerful technique has been added to the arsenal, namely, cryo-TEM. By rapid freezing of the growth medium the nucleation and growth processes may be stopped and the sample transferred, while it remains cold to be analyzed by the electron microscope. This permits high-resolution electron microscopy to be performed on more-or-less unperturbed crystallization media. Recent results on the silicalite system [133] prepared from TEOS and TPA show clearly that the initially formed 5-nm nanoparticulates are amorphous in nature before they agglomerate and crystallize by an intraparticulate re-organization. By performing these experiments in the electron microscope as opposed to utilizing x-ray scattering techniques it is possible to discern the presence or absence of structural order even at these very early stages.

## 1.4.3
## Modeling

### 1.4.3.1 Monte Carlo Modeling of Crystal Growth

Since the advent of AFM and high-resolution scanning electron microscopy we have a new window to follow the nanoscopic details of crystal growth. In order to interpret these data new tools are required to model not only the crystal morphology but also the details of the surface topology. Monte Carlo techniques provide a possible route to simulation of crystal morphology and topology whereby the structure is developed according to a set of thermodynamic rules. The first problem is to decide which unit to choose as the growth unit. For a molecular crystal the answer is straightforward as the indivisible element in the growth is a single molecule, and Monte Carlo techniques have been successfully used, for instance, for urea [134–136]. For nanoporous materials such as zeolites it is necessary to coarse grain the problem in a way that makes the calculation manageable. This is readily done by realizing that the rate-determining steps in the crystal growth process are related to closed-cage structures capped with $Q_3$ groups at the surface. Open-cage structures exposing $Q_2$ and $Q_1$ groups are highly susceptible to dissolution and as a consequence do not persist at the surface. Having selected the coarse grain as a closed-cage structure, a network of closed cages is constructed and the probabilities for growth and dissolution determined according to the energetics at each site. The methodology chosen is essentially that of Boerrigter *et al.* [137] whereby each site at the surface of a crystal is assigned an energy relative to the bulk phase (see Figure 1.13 for description of energy levels). Growth from solution then occurs through an activated complex, essentially desolvation followed by adsorption, but the energy which defines the relative rate, or probability $P$, of growth and dissolution, $P_{i2j}^{\text{growth}}/P_{i2j}^{\text{etch}}$, is a combination of the energy of the site (termed *site i2j* in Meekes terminology), relative to the bulk phase, $(\Delta U_{i2j} - \Delta U)$and the supersaturation, $\Delta\mu$ (Eq. (1.11)) [138–140]. The ordering of the energy levels to the first approximation follows the order of the nearest neighbor connectivity, with second-order connectivity resulting in smaller energy differences.

$$\frac{P_{i2j}^{\text{growth}}}{P_{i2j}^{\text{etch}}} = \exp(\beta(\Delta U_{12j} - \Delta U) + \beta\Delta\mu) \tag{1.11}$$

By treating this problem numerically by computer the real meaning of supersaturation and equilibrium becomes immediately apparent. The Monte Carlo treatment is conducted so that growth and dissolution events occur according to Eq. (1.11). The supersaturation may be treated as a constant or variable. By setting up a virtual solution phase which increases in concentration for an etch event and decreases in concentration for a growth event, the supersaturation is allowed to drop freely as the crystal grows. Equilibrium is established when the crystal stops growing and the number of growth and etch events over time are equal. At this point the value of $\Delta\mu$ is determined. Conventionally, $\Delta\mu$ is zero

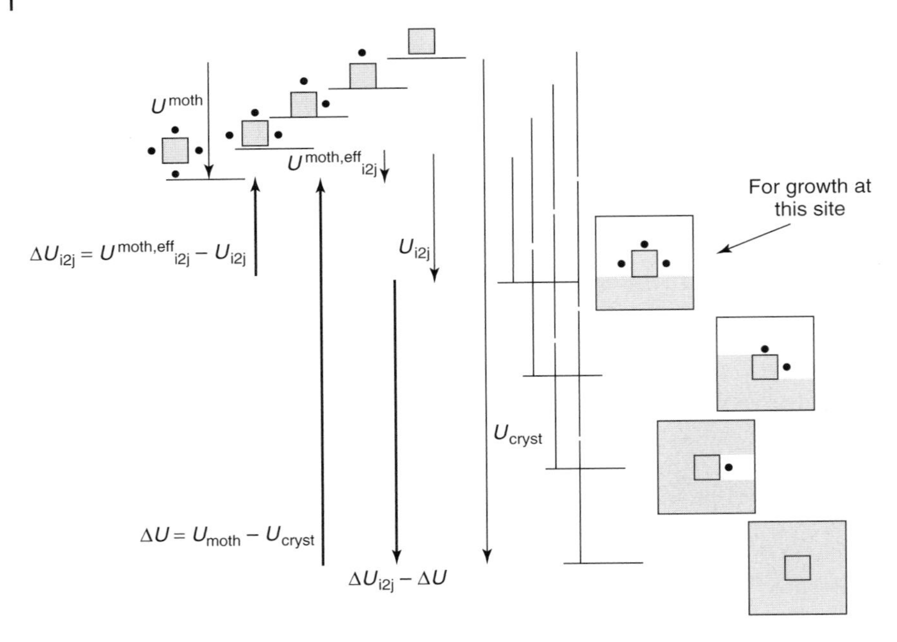


**Figure 1.13** Energy level descriptions for growth of crystal from solution. At left is the solvated growth unit in solution. At top is a fully desolvated growth unit in vacuum and at the bottom, the growth unit fully condensed in the bulk crystal. Energy nomenclature corresponding to [137] is given, which relates to the probabilities for growth and dissolution given by Eq. (1.11).

at equilibrium as the conventional definition of supersaturation is the difference between solution concentration and that recorded at equilibrium. However, a much more powerful definition of equilibrium is achieved by considering the value of $\Delta\mu$ relative to the energies of each growth site, $(\Delta U_{i2j} - \Delta U)$. At equilibrium some sites will be in undersaturation (the low-coordinate sites) and some sites will be in supersaturation (the high-coordinate sites). Equilibrium is just the balance point, the center of gravity, of energies of all the sites taking into consideration the number of each of those sites. The Monte Carlo approach finds this center of gravity equilibrium state that always lies within the range of the kink sites. It is not a problem that has a ready analytical solution because the number of sites of each type depends upon the specific connectivity of the crystal – in essence it depends on the crystallography of the crystal. For studies of nanoporous materials, this gives a route to study the problem experimentally without worrying about the extremely difficult problem of determining conventional supersaturation of the solution. Crystals may be taken to the equilibrium condition and then compared against the Monte Carlo calculation for the same condition. Figure 1.14 shows examples of crystals simulated under different conditions.

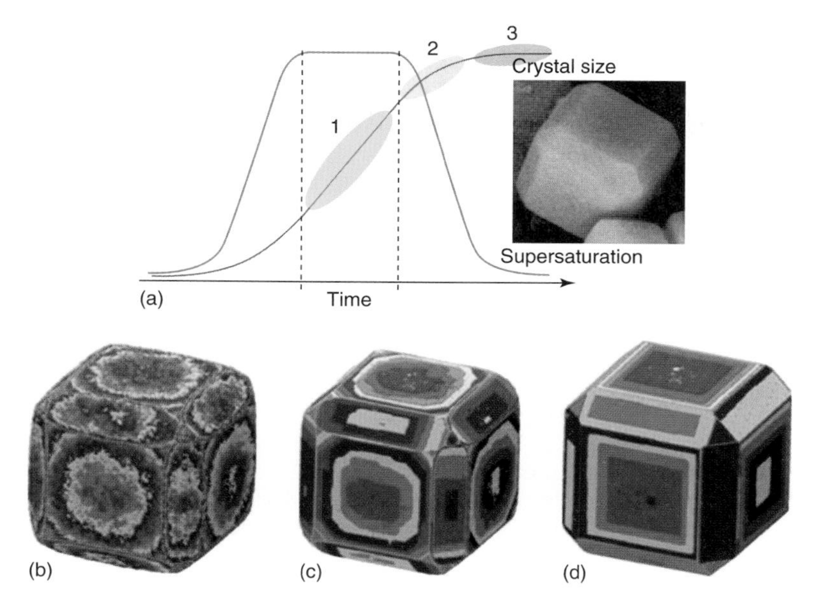


**Figure 1.14** Typical crystal size and supersaturation evaluation as a function of time for zeolite A synthesis. Inset SEM picture (image size 2.5 μm) shows typical zeolite A crystal produced in the synthesis. b–d show about $0.3 \times 0.3 \times 0.3 \mu m^3$ sized zeolite A crystals simulated at time interval 1, 2, and 3, respectively [138].

## 1.5 Case Studies

### 1.5.1 Zeolite A

Zeolite A is one of the most widely used zeolites due to its ion-exchange capabilities [141]. Zeolite A has a framework structure known as *Linde type A* (LTA) [142] consisting of sodalite (SOD) cages linked via four rings, and an Si to Al ratio of 1 : 1. The SOD units join to produce an $\alpha$-cage with a diameter of 11.4 Å (the large cavity in the center of the structure), and two channel systems which are connected to allow motion of the $Na^+$ ions and water molecules. Its empirical formula is $Na_{12}[Al_{12}Si_{12}O_{48}]\cdot 216H_2O$. Investigations on zeolite A have been numerous and varied with a multitude of studies focusing on the effect of different parameters in its growth, such as the addition of organic molecules to the synthesis mixture [143, 144], seeding [145], aging [37], and clear solution [25].

Zeolite A has also been the subject of various AFM studies [67, 146, 147], including the first *in situ* dissolution study in a zeolite [87]. These observations were limited to the study of the {100} face in crystals extracted at the end of the synthesis (i.e., a low supersaturation condition), with no information available on the crystal growth mechanisms under other conditions. A more thorough study has been carried out to study the evolution of the surface features of zeolite A at different synthesis times by means of AFM and HRSEM. Zeolite A crystals were prepared following the preparation of Thompson and Huber [148]. The synthesis was carried out at 60 °C and at the following synthesis times: 2.5, 4, 8, 20, 30, and 50 hours. A second preparation following the method by Petranovskii *et al.* [144] includes the addition of diethanolamine (DEA), an organic molecule whose main effect is to increase the size of the crystals and which also has an effect on the morphology of the crystals. In this experiment, the reaction was carried out at 90 °C and the synthesis times were 12, 16, 20, 24, 28, 32, 36, 40, 44, 48, 72, 96, 120, 168, 336, and 504 hours. The reason behind the much longer experimental times was to observe any possible change in surface topography at long "equilibrium times" in order to allow for surface and habit rearrangement.

#### 1.5.1.1 Thompson Synthesis

Figure 1.15 shows the evolution of zeolite A synthesized as a function of time. It can be seen that after 8 hours the crystallization is almost complete. At 2.5 and 4 hours the growth rate of the crystals is at its maximum, which also corresponds to the highest supersaturation achieved in the system (Figure 1.9). At 8 hours, however, the growth rate has probably started to decrease, reaching almost zero at 20 hours.

Figure 1.16 shows the corresponding HRSEM of crystals extracted after 2.5, 4, 8, and 20 hours, whereas Figure 1.17 shows the AFM images after 4, 8, and 20 hours. When comparing both images a more detailed picture of the crystallization process emerges. At 2.5 hours (Figure 1.16a) it can be seen that crystals are small (a few nanometers to 400 nm) and rounded, and there is also strong evidence of intergrowth formation and aggregation of crystals. Although no AFM images could be obtained on these round crystals, the HRSEM resolution allows us to see that the crystal surfaces are very rough. All these features are evidence of an adhesive mechanism of growth typical of high supersaturation conditions. At 4

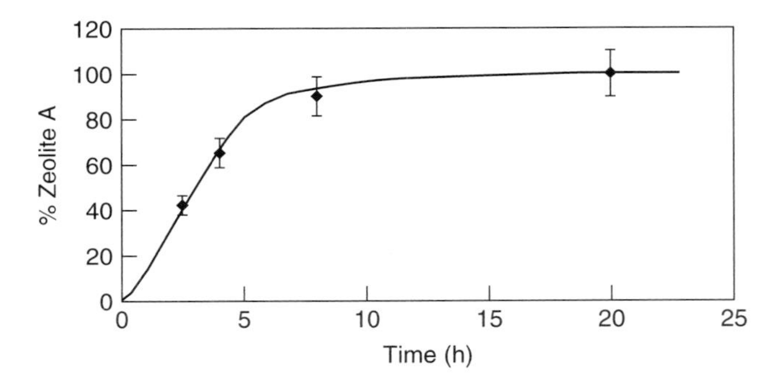


**Figure 1.15** Percentage of zeolite A crystallized as a function of time. Superimposed is a theoretical growth curve for this kind of system.

**Figure 1.16** HRSEM photomicrographs of zeolite A crystals after (a) 2.5 hours; (b) 4 hours; (c) 8 hours; and (d) 20 hours of synthesis.

hour HRSEM (Figure 1.16b) and AFM (Figure 1.17a) images show a similar picture. At this point of time, crystals are much bigger (up to 1 μm) and still somewhat rounded but they start to show the formation of facets. This can also be seen in the AFM image (Figure 1.17a) where steps start to be visible, although evidence of active 2D nucleation is also present. Eight hours into the synthesis the crystals have grown even more, reaching sizes up to 1.5 μm. HRSEM images (Figure 1.16c) show that the crystals are more faceted than before, although still with rounded edges. The corresponding AFM image (Figure 1.17b) shows well-defined steps and terraces with nucleation limited to the central terrace. This indicates that supersaturation has now dropped and growth takes place by a layer growth or single nucleation mechanism (Section 1.2.12). At 20 hours the crystals are fully developed (Figure 1.16d) and show the typical faceted morphology for this synthesis [149] bounded by {100} and {110} and {111} faces. AFM taken on these samples on the {100} face (Figure 1.16d) agrees very well with what has been published before [67, 89] and displays single square terraces at the center of the face that grow by step advancement toward the edge of the crystal. The steps are almost perfectly rectilinear as well. Figure 1.16c shows the AFM image of a {110} face; it can be seen that in this face the shape of the terrace is rectangular, displaying a faster growth along the <100> direction and slower on the <110>. This situation is indicative of a very low supersaturation and has been successfully replicated in Monte Carlo simulations.

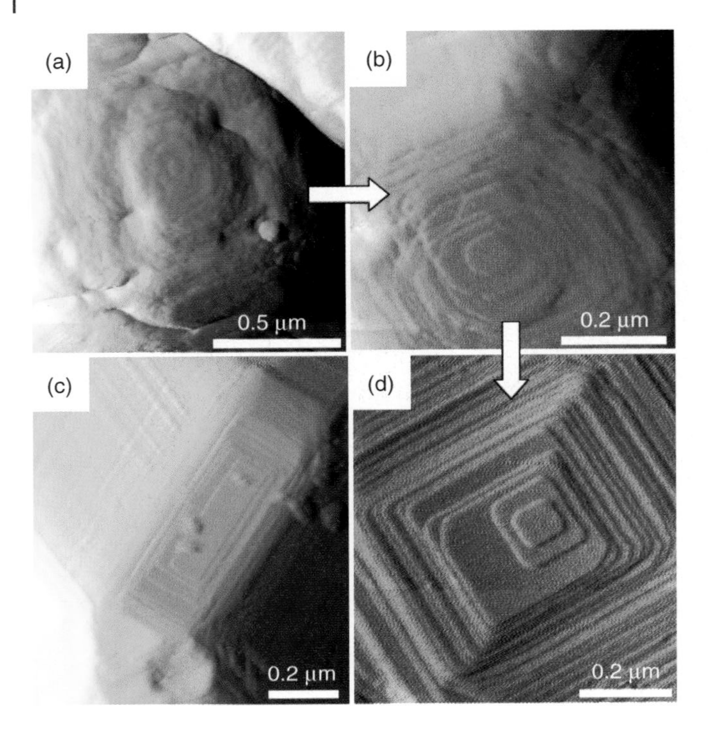


**Figure 1.17** AFM deflection images of zeolite A crystals after (a) 4 hours; (b) 8 hours; and (c and d) 20 hours of synthesis. a, b, and d show {100} faces, whereas (c) shows the {110} face.

#### 1.5.1.2 **Petranovskii Synthesis**

Figure 1.18 shows two SEM micrographs of the end product of the synthesis of zeolite A in the presence of DEA at 90 °C. The end product in this case displays a much larger size (up to 15 μm) and the crystals are bound only by the {100} and {110} faces. Also the relative size of the {110} face in comparison to the {100} face is higher for this synthesis.

A detailed AFM study was undertaken to analyze the surface topography of all the samples synthesized. Figure 1.19 shows eight AFM images taken on crystals extracted at 44, 48, 72, 96, 120, 168, 336, and 504 hours after the start of the synthesis. At 44 hours (Figure 1.19a), it can be seen that the crystal surface is still quite rough, indicating that 2D nucleation is still happening at a faster rate. Still SEM analysis on the evolution of the length of the crystals with synthesis time shows that at this time the reaction is almost complete, so supersaturation has already started to decrease. This is reinforced by the fact that at 48 hours, 2D nucleation rate has decreased to the point that the steps and growing nuclei can easily be resolved with the AFM (Figure 1.19b), although nucleation still occurs on narrow terraces. At 72 hours, AFM analysis (Figure 1.19c) shows fewer nuclei at

**Figure 1.18** SEM photomicrographs of zeolite A crystals extracted after 120 hours of synthesis.

the surface in accordance with the gradual decrease in supersaturation expected at this point of the synthesis. Note that all the terraces are bounded by straight steps with sharp corners. Figure 1.19d–f show the central area of different {100} faces. In the three images a similar situation is observed, where nucleation has almost stopped but the terraces still possess sharp corners. Further into the synthesis, at 336 hours, there is a change in the shape of the terraces, as can be seen in Figure 1.19g. Now the corners of the square terraces are not sharp but curved. At 504 hours, terraces were found to be also curved at the corners (Figure 1.19h). This effect can be explained by means of the Monte Carlo simulation and corresponds to the equilibrium situation explained in Section 1.4.3.1. Since the "center of gravity" equilibrium state is located around energies of the kink sites, steps would "evolve" toward a shape composed mainly of these sites, resulting in some dissolution and rounding of the terrace corners.

Zeolite A has been dissolved *in situ* using a mild sodium hydroxide solution [87, 150]. Under *in situ* conditions, it is often possible to capture the less-stable surfaces, which tend to be absent if the crystal is removed from the mother-liquor. All *ex situ* measurements on zeolite A have revealed a terrace height of 1.2 nm. However, these terraces dissolve in two steps according to different mechanisms and on different timescales. Figure 1.20 shows an AFM image selected from an *in situ* series which captures both structures. A 0.9-nm terrace is observed dissolving by terrace retreat and a 0.3-nm terrace remains undissolved on the same timescale. The explanation of these results is that SOD cages, 0.9-nm high, which are interconnected, dissolve in a correlated fashion by terrace retreat. One cage must dissolve before the next one can. Single four rings, 0.3 nm, dissolve on a different timescale in an uncorrelated fashion as they are not connected to each other. These results further reveal the importance of closed-cage structures, two of which are seen in this experiment, the SOD cage, and the double four ring.

**Figure 1.19** AFM deflection (a, b) and height images (c–h) of zeolite A crystals after (a) 44 hours, (b) 48 hours, (c) 72 hours, (d) 96 hours, (e) 120 hours, (f) 168 hours, (g) 336 hours, and (h) 504 hours.

### 1.5.2 Silicalite

Silicalite is often used in fundamental studies of zeolites owing both to the importance of the MFI structure for catalysis through ZSM-5 and also because of the simplification afforded by having a pure silica framework with no charge-compensating

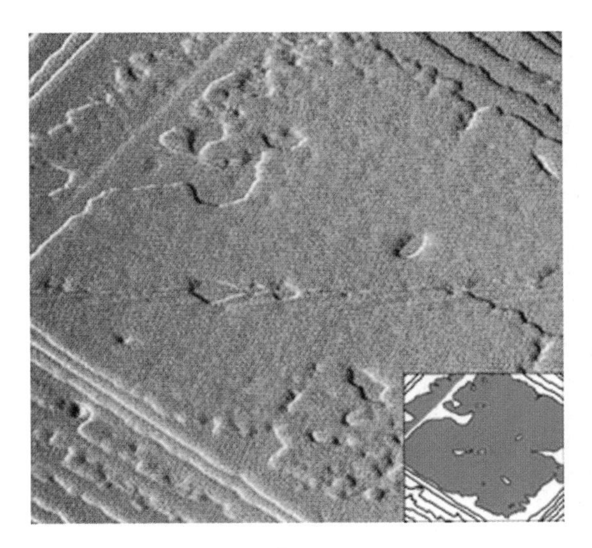

**Figure 1.20** 4.3 × 4.3 $mm^2$ deflection AFM micrograph of a zeolite A crystal under static solution of 0.5 M NaOH from *in situ* measurement after 33 minutes. Inset blue area shows 0.9 nm terrace and white area 0.3 nm terrace.

cations. From a crystal growth point of view it also presents an opportunity to investigate (i) the role of templates as the tetrapropyl ammonium cation is such a strong structure director and (ii) the role of intergrowth formation to the MEL, silicalite-2, structure.

Both MFI and MEL structures are composed of connected pentasil chains that can be connected either via a mirror plane or an inversion center. Different choices result in the two structures and this is most readily controlled via structure directing agents that register with the resulting channel system. All AFM measurements to date, on either system, show that the fundamental growth step height is 1 nm, consistent with the pentasil chain unit, thereby conferring an important status upon this closed structure on the growth mechanism.

In a study on the effect of supersaturation, a series of silicalite samples were prepared in a semicontinuous synthesis [151]. Untwinned seed crystals in their spent mother-liquor were placed in a continuous feed reactor. After the slurry reached equilibrium, the nutrient feed was switched on and varied to maintain a constant crystal growth of 0.4 $\mu m\ h^{-1}$. The reaction was continued for a total of 64 hours. During the reaction time, the nutrient feed was stopped for periods of about 16 hours each at 4.3, 9.6, and 16.4 hours, and the supersaturation allowed to drop such that the growth of the crystal more-or-less ceased. The AFM was recorded on each of these crystals, and this is shown in Figure 1.21. During the periods of constant growth the supersaturation level is relatively high and the surface of the crystal is characterized by having a high density of growth nuclei. The crystal grows exclusively via a birth-and-spread mechanism. As the supersaturation drops, when the nutrient feed is turned off the terraces continue to spread but surface

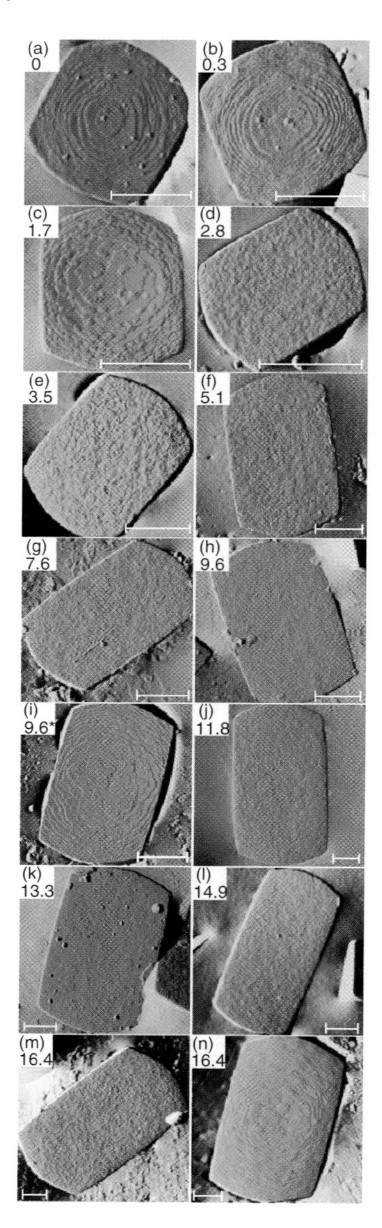


**Figure 1.21** AFM deflection images of {010} faces of the silicalite samples. Crystal shown in (a) corresponds to the seed crystal; (b–h) and (j–m) are crystals that were recovered from the reactor under continuous feed. Crystals shown in (i) and (n) were recovered from the reactor after periods of 16 hours when the nutrient feed was switched off and are denoted by 9.6* and 16.4* hours. Time is expressed in hours. The scale bars represent 1 mm [151].

nucleation more-or-less ceases. The reason for this is that surface nucleation requires the highest energy, see Eq. (1.11), or highest value of supersaturation. Near equilibrium the relative rate of growth to dissolution, $P_{i2j}^{growth}/P_{i2j}^{etch}$, will be less than unity, and consequently as soon as an entity grows on the surface, it will immediately dissolve back into solution. Conversely, terrace spread is dependent upon lower coordinate, edge and kink sites for which $P_{i2j}^{growth}/P_{i2j}^{etch}$ will be closer to or above unity permitting the terraces to continue spreading. The process is reversible as surface nucleation and crystal growth continue when the supersaturation is subsequently raised.

This supersaturation control experiment demonstrates the ability to switch on and off specific crystal growth processes. Such a phenomenon could be used to control defects and intergrowths in zeolites. Zeolite intergrowths in zeolites often result from layer growth, whereby a new layer has more than one choice with similar energy (e.g., an ABA stacking rather than an ABC stacking). If there is a high density of surface nuclei present on any given surface, then there occurs a high probability that some of these are of the slightly less favored stacking sequence. Now, one given layer will have a number of C layers mixed in with the A layers, and as the terraces spread a defect will result when they merge as they will be incompatible. A possible route to overcome such defects would be to lower the rate of nucleation, by lowering the supersaturation such that terrace spreading is still very rapid, but the low nucleation density means that the probability of having nuclei of a different stacking sequence is minimized. Hence, the defect density should decrease. Conversely, by working at very high supersaturation the nucleation density will be high and the probability for stacking sequence incompatibilities will also be high leading to a high defect density. Nanoporous materials such as faujasite (FAU), BEA, and ETS-10 grow according to such rules.

In silicalite, stacking sequence problems arise when pentasil chains attach by an inversion symmetry rather than a mirror symmetry. This will result in a switch from the MFI to the MEL structure when the mistake is on the (100) face, but on the (010) face the pentasil chain cannot connect to the ensuing crystal. This slows crystal growth, and large terrace fronts stack up along the foreign pentasil chain until eventually the defect is overgrown leaving a high density of undercoordinated $Q_3$ silicons within the structure [81].

Lessons on the role of templates can be learned through competitive templating. TPA is known to be a strong structure director for the MFI structure. By increasing the length of the hydrocarbon chain by one unit tetra butylammonium (TBA) cations will be directed toward the MEL structure. However, the TBA cation is too large to fit at every channel intersection, and consequently there is a topological dilution of the template at the growing crystal surface. The effect of this is immediately apparent in mixed TPA/TBA preparations. Greater than 90% TBA is required in the synthesis before there is a substantial incorporation of TBA over TPA with a corresponding substantial switch from MFI to MEL. Indeed even with 98% TBA and 2% TPA there is still a substantial frustration in the growth of the MEL structure (Figure 1.22). In order to have a smoother transition from MFI to MEL, a smaller template is required and *N*, *N*-diethyl 3,5-dimethyl piperidinium iodide

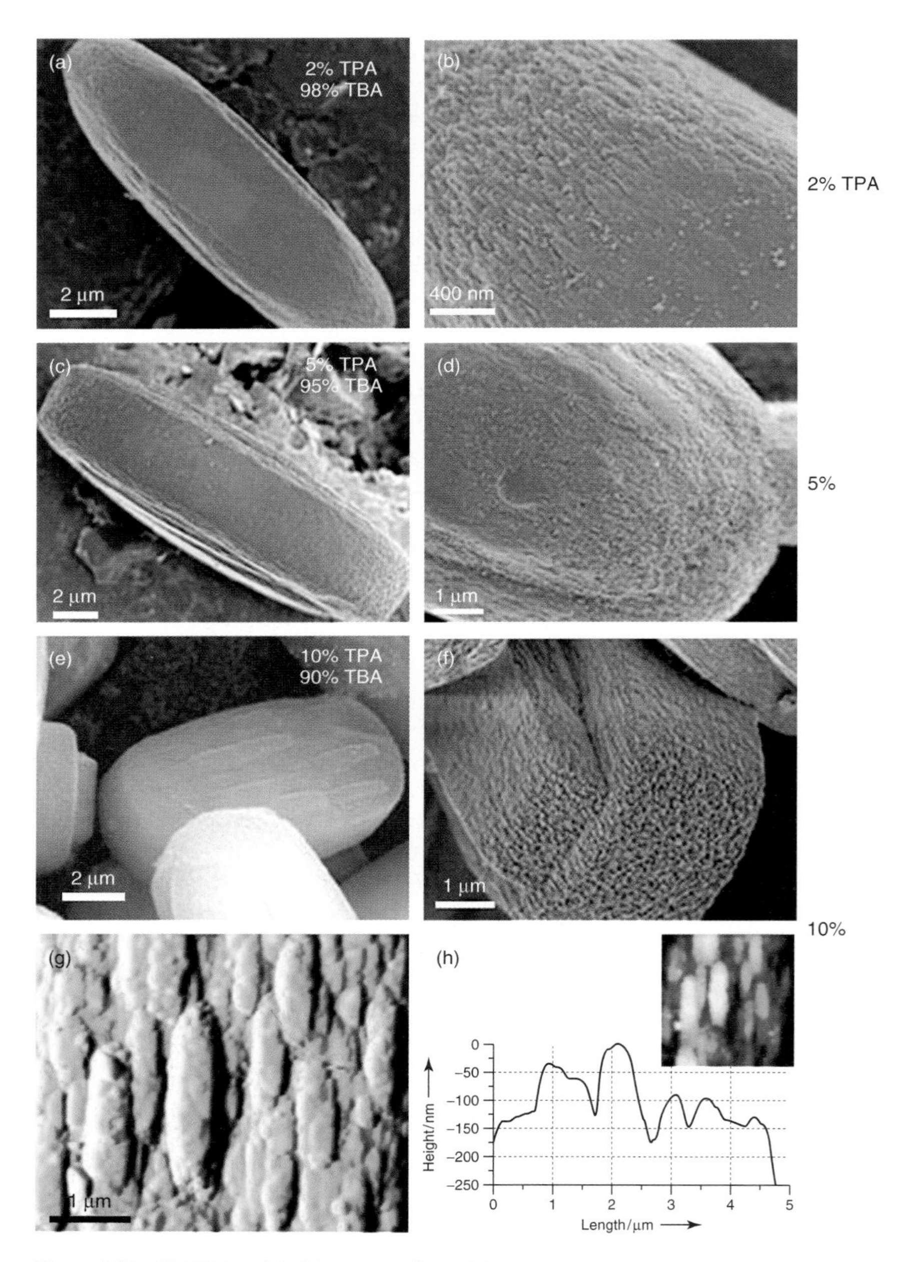


**Figure 1.22** HRSEM and AFM images of MFI/MEL intergrowths showing frustration of MEL growth by the incorporation of as little as 2% TPA in the growth swamped with TBA.

(DEDMPI) serves this role. DEDMPI can be accommodated at every intersection and produces MEL crystals that are much less susceptible to stacking sequence problems and the inherent defects.

### 1.5.3 LTL

Zeolite L has a unidimensional channel structure that typically grows as long hexagonal prismatic crystals with the channels running along the long axis. From a catalytic point of view this is undesirable as the intracrystalline path length for reactants and products is maximized resulting in restricted diffusion. As a consequence, there have been a number of attempts to modify the normal habit to yield tablet-like crystals with a short *c*-dimension. AFM studies of zeolite L [152] reveal the reasons for this crystal habit. Figure 1.23 shows the AFM image recorded on both the hexagonal (001) face and the sidewall (100) face of the crystal. On the top (001) facet the crystal grows via a layer growth, and the smallest height of the terrace is equivalent to the height of one cancrinite cage. The sidewalls show more interesting behavior. Long, thin, straight terraces are observed which are elongated in the *c*-direction of the crystal. Two significantly different heights are measured for these terraces. The narrowest terraces are always 1.2 nm in height

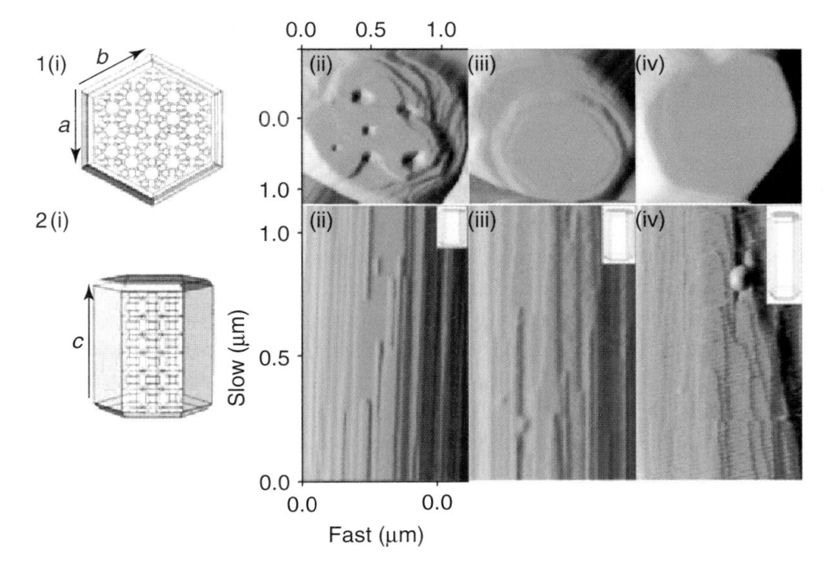


**Figure 1.23** Error signal AFM images of zeolite L with different aspect ratios. (1) The hexagonal face down the [001] direction of the crystal and (2) the sidewalls down the [100] direction of the crystal. For each face, (i) shows a schematic framework of the crystal, and (ii), (iii), and (iv) crystals with aspect ratios 1.5, 2.3, and 5.1, respectively.

and the wider terraces are always 1.6 nm in height. The reason for this is that all the narrowest terraces correspond to a single cancrinite column, which grows very rapidly along the *c*-direction of the crystal but is highly frustrated to growth across the side-wall in the *a*- or *b*-direction. Wider terraces are 1.6-nm high because neighboring cancrinite columns are connected by a further cancrinite column that acts as a bridge across the large 12-ring channel. This result illustrates a generic problem to grow crystals with large pores. Circumventing the large pores can result in very unfavorable processes which are more-or-less akin to fresh nucleation. The bridging cancrinite column is more likely to dissolve back into the solution rather than persist until it is secured by the cancrinite column at the other side of the bridge – again according to Eq. (1.11). The problem lies in the fact that two, not one, cage structure is required to circumvent the large 12-ring pore. Growing both cancrinite column structures before one dissolves is an unlikely event, and therefore the kinetics are slow. The traditional way to build nanoporous structures over large void spaces is to add an organic templating agent which facilitates the process; in other words the kinetics for the process are improved. Zeolite L, however, is an example of a wide-pore zeolite that grows readily in the absence of organic-structure-directing agents. Consequently, zeolite L is a good demonstration of how large pore structures can be incorporated into a structure without expensive organic additives, and at the same time this system illustrates where the kinetics are severely frustrated in the process. This frustrated growth also results in the typical long prismatic crystals and the kinetics would need to be adjusted in order to successfully grow low-defect tablet-shaped crystals. Most methods reported to create shorter *c*-dimension zeolite L crystals operate under high supersaturation conditions, where the crystals have a high density of defects and the aspect ratio is only altered as a result of the interrupted growth at macro-defects. Careful adjustment of crystal aspect ratio, while maintaining a low defect density, is yet to be achieved.

Zeolite L is also a very interesting system to investigate, *in situ* by AFM, the mechanism of dissolution. Figure 1.24 shows a series of images recorded as a function of time as the crystal dissolves under mild basic conditions. The micrographs have been recorded in lateral deflection mode, which monitors lateral twist of the cantilever during scanning. This mode is normally used for nanotribology studies as local friction will cause a twist of the cantilever. Three things are immediately apparent from the images. First, the terraces dissolve very rapidly along the *c*-direction of the crystal and very slowly in the lateral direction. Second, the place where the crystal dissolves is bright white, indicating a high degree of lateral twist. In essence, the AFM illuminates where the chemistry on the crystal occurs. This twist is also observed during the growth of nanoporous crystals. Third, it is observed that the bright white region on the dissolving terrace is substantially larger at the top of the terrace when the AFM tip is scanning down the crystal and vice versa. This demonstrates that the tip is aiding the dissolution. The tip in effect warms the crystal. By recording the images at different cantilever loads and temperatures it is possible to make a series of Arrhenius plots, which when extrapolated to zero load (i.e., no effect of the tip) yield the activation energy

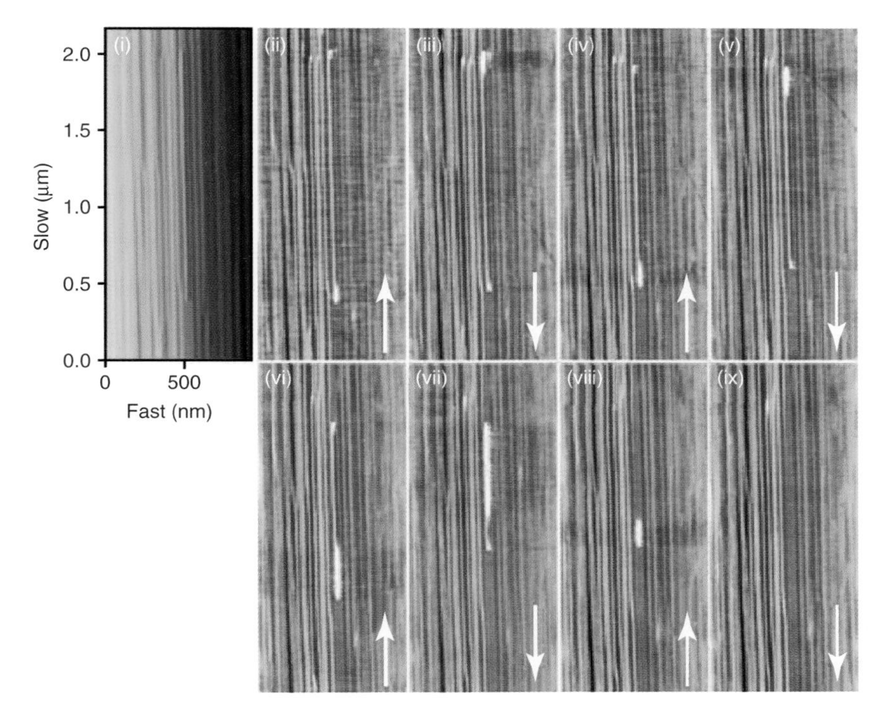


**Figure 1.24** Atomic force micrographs of the (100) face of zeolite L during dissolution in 0.2 M NaOH. The image in (i) shows a vertical deflection micrograph at the beginning of the experiment and ii–ix show lateral deflection micrographs on subsequent scans over the crystal surface. The time between each image was approximately 4 minutes. The white "lights" indicate a change in friction experienced by the AFM tip, shown as a high contrast change on the image. The white arrows indicate the direction of scanning.

for this fundamental dissolution process, which is $23 \pm 6$ kJ mol$^{-1}$. The cause of the high lateral force is not entirely clear. It could be due to a change in friction when the crystal is undergoing rapid dissolution. However, it also might be due to high local energy being imparted to or from the tip as a result of a large energy change during dissolution. This latter explanation would be consistent with the fact that the phenomenon is observed during both growth and dissolution.

### 1.5.4 **STA-7**

STA-7 is a SAPO [153] with the structure type SAV [142]. SAV belongs to a group of four framework structures, composed of double 6-rings (D6Rs), which also includes the frameworks CHA, AEI, and KFI. The only difference between these structures is the arrangement of the D6R along the *x*, *y*, and *z* axes [154]. The SAPO STA-7 structure belongs to the space group $P\,4/n$ and contains two types of cages,

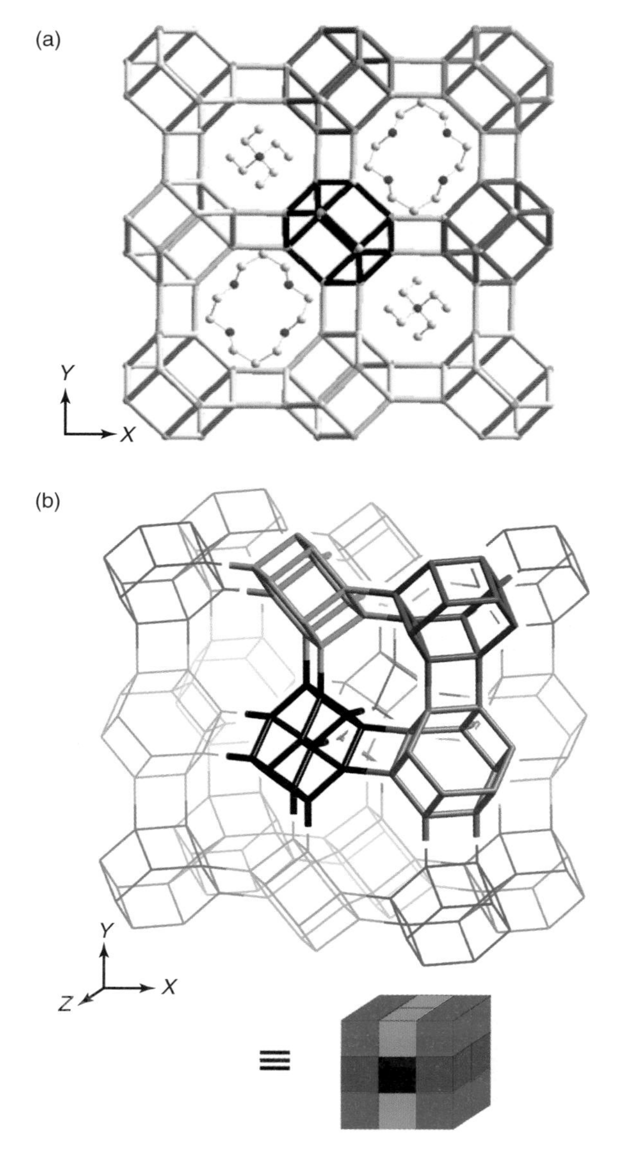


**Figure 1.25** (a) STA-7 structure on the {001} face showing the position of the cyclam and TEA molecules inside the two different cages. The four different orientations of the D6R are also highlighted in different colors. (b) 3D view of the STA-7 structure. The four possible orientation of the D6R are highlighted in red, green, black, and blue. The unit cell is highlighted in bold. It can be seen how the units alternate orientation along *x* and *y* directions but not along the *z* direction. A simplified color model of the structure is also shown for comparison.

A and B, connected three dimensionally by eight-ring windows. The larger cage, B, is templated by cyclam and the smaller one, A, by the co-template tetraethylammonium (TEA), as shown in Figure 1.25. In the structure, the D6R units have four different orientations (highlighted in different colors in Figure 1.25) and are related to each other by fourfold symmetry axes and an n-glide plane perpendicular to the <001> direction. Hence, alternating D6R units can be found along the <001> and <010> directions. In contrast, D6Rs form chains along the <001> direction. Consequently, a unit cell consists of two D6Rs along $x$ and $y$ axes, and one along the $z$ axis, as can be seen in Figure 1.25b (highlighted in bold). For simplicity the STA-7 structure can be represented (hereinafter) as a series of cubes of different colors, each one representing a D6R of different orientation (Figure 1.25b).

STA-7 crystals were prepared from SAPO-based gels treated hydrothermally (190 °C for 3 and 10 days). Further details on the characterization can be found in Castro *et al.* [153], but it is important to point out that the formation of STA-7 has proven to be dependent on the presence of the two templates (cyclam and TEA). SEM observations at the end of the synthesis reveal the formation of crystals with a well-defined tetragonal prismatic morphology and with a typical size of 30–35 mm (Figure 1.26a). Therefore, the crystals are bound by two crystallographically distinct

**Figure 1.26** (a) Scanning electron micrograph of STA-7 crystals after the end of the synthesis. (b) Optical micrograph of an STA-7 crystal showing the {100} face. (c) Optical micrograph of the {001} face.

faces, {100} and {001} (Figure 1.26b and c, respectively). Both types of faces were characterized by *ex situ* AFM.

#### 1.5.4.1 {001} Faces

Figure 1.27a shows a representative AFM image of a {001} face. It can be seen that the surface is covered by multiple nearly isotropic spirals, with all of the scanned surfaces showing a similar dislocation density of approximately 1–2 dislocations per 10 $\mu m^2$. No evidence of 2D nucleation was observed, indicating that the system was near to equilibrium when the crystals were extracted. The isotropic morphology of the spirals in this face indicates that there are no preferential growth directions at low supersaturation. Height analysis shows that the step height at the dislocation core (Burgers vector) is always $0.9 \pm 0.1$ nm (Figures 1.27b,c), which corresponds to the $d_{001}$ spacing, that is, to the height of a D6R along the <001> direction.

Figure 1.8d shows a simplified block diagram of the STA-7 structure, with the different orientations of the D6R highlighted in different colors as described in Figure 1.25b. Rows of similarly oriented D6Rs run parallel to the $z$ direction. The original step structure produced by the dislocation is shown by the lightly colored blocks, assuming that the dislocation also runs parallel to the $z$ axis with a Burgers vector of 0.9 nm. It can be seen that the sequence of D6Rs on both sides of the dislocation does not change with respect to the normal sequence (green–red–green in the diagram), that is, the bonding through the dislocation between the D6R units is the same as in the undisturbed (nondefective) crystal. Therefore, growth units (assumed to be a D6R) could attach to the step created without modifying the alternation sequence of D6R units along any direction, as represented by the darker colored blocks, perpetuating the STA-7 structure through spiral growth.

#### 1.5.4.2 {001} Faces

AFM observations on the {100} faces reveal the formation of two very distinct types of spirals. The first, which is the more numerous one, has an elongated shape, with the long axis, or faster growth direction, parallel to the <001> direction. Height analysis shows that this type of spiral is produced by a dislocation with a Burgers' vector of $\approx 0.9$ nm (Figure 1.28a). The second type corresponds to an interlaced spiral (Figure 1.28b,c). The step splitting in this interlaced spiral produces the characteristic "saw tooth" pattern (Section 1.2.11) extending from the dislocation, which in this case is parallel to the <100> direction (highlighted by the white box in Figure 1.28c). Height analysis along this pattern reveals that the height of the steps is half a unit cell, that is, $0.9 \pm 0.1$nm (red line). On the contrary, a cross section from the dislocation center but parallel to the <001> direction (blue line) reveals that the most common step height is one unit cell, that is, $1.8 \pm 0.1$ nm (i.e., two monolayers). Figure 1.28b shows a higher resolution image of the spiral center, where it can be seen that two single steps ($0.9 \pm 0.1$ nm) emanate from the dislocation, hence the Burgers vector of the dislocation is equal to a unit cell, that is, $1.8 \pm 0.1$ nm. As explained in Section 1.2.11, an interlaced spiral is produced when each of the different monolayers emanating from the dislocation possesses a different speed anisotropy. In the case of STA-7 this anisotropic growth

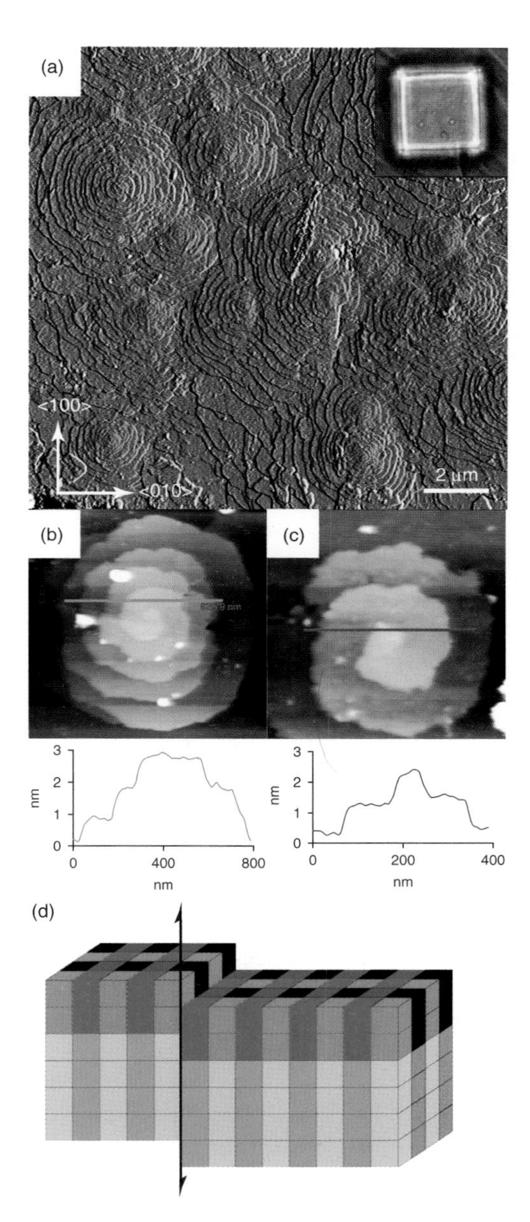


**Figure 1.27** (a) AFM amplitude image of a {001} face of an STA-7 crystal. Isotropic spirals in the surface are clearly visible. Inset shows the corresponding optical microscopy image. (b) AFM height image of the lower spiral and cross section. (c) Height image of the upper spiral and cross section. The cross sections were taken along the red and blue lines in the images. (d) Simplified spiral structure after the two new layers have grown. The structure of the newly grown layers matches that of the underlying substrate.

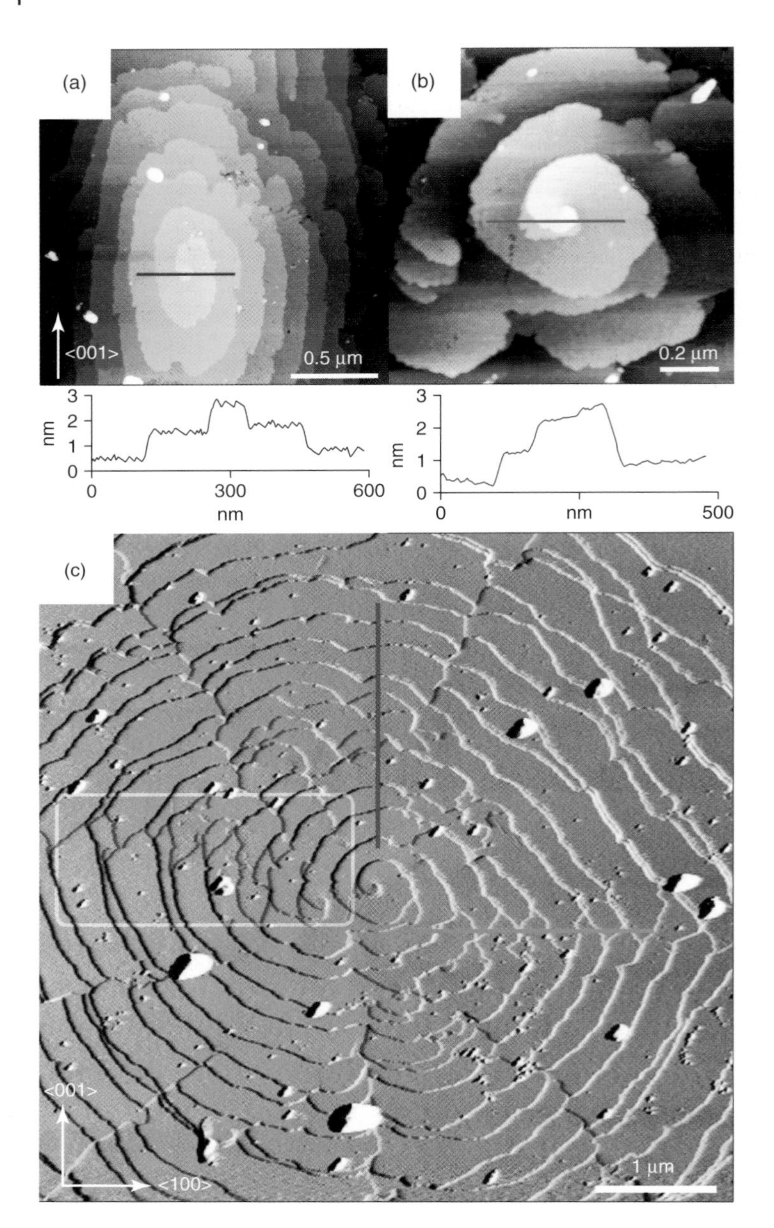


**Figure 1.28** (a) AFM height images of an elliptical spiral with cross section. (b) Height image showing details of the central area of the interlaced spiral shown in (c). Two substeps emanating from the dislocation are clearly seen. The cross section confirms the single layer (i.e., 0.9-nm high) nature of these steps. (c) Amplitude image of an interlaced type spiral. Highlighted box contains the characteristic "sawtooth" pattern observed in these types of spirals.

is symmetry-induced due to the presence of an n-glide plane perpendicular to the {100} face. Recently, van Enckevort and Bennema [14] demonstrated that interlacing will be expected when a screw axis and/or a glide plane are perpendicular to the growing surface. The presence of the n-glide plane also determines the shape of the spiral to be symmetrical at both sides of the <100> direction, which marks the intersection of the n-glide plane with the {100} face. Also, the spiral is symmetrical along <001> direction owing to the symmetry axis. By taking into account these symmetry constraints, it is possible to deconstruct the growth anisotropies of the individual substeps. In the case studied, the two substeps show a difference in the growth rates along [001] and [00$\bar{1}$] directions. One substep will grow faster along the [001] direction than along [00$\bar{1}$], whereas the other will be opposite. This situation is summarized in Figure 1.29. Figure 1.29a,b shows a simplified diagram of the two substeps as they would grow if no interference occurred. The anisotropic growth along <001> direction for each substep is clearly evident. In Figure 1.29c, the trajectories of the two spirals are superimposed, demonstrating clearly how they can form the observed interlacing pattern (Figure 1.28c).

The cause of the anisotropic growth can be explained by the tilting of the D6R units with regard to the <001> direction. This tilt can be seen more clearly in Figure 1.30a, which shows a cross section of the STA-7 structure perpendicular to <100> direction. In one layer, D6R units are tilted toward [001] direction, whereas on the following they will all be tilted in the opposite direction, [00$\bar{1}$]. This tilt creates two different step geometries, one acute (red box) and one obtuse (blue box), one of which may favor the attachment/docking of the template preferentially over the

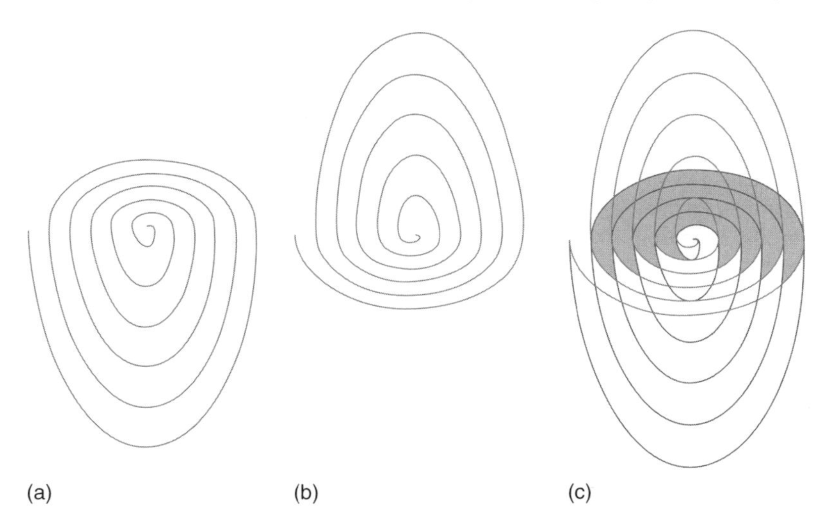


**Figure 1.29** Simplified diagrams showing the formation of an interlaced spiral. a and b The assumed shape of each substep if they could grow freely. (c) The overlapping of the two substeps. Here the interlaced pattern is readily observable.

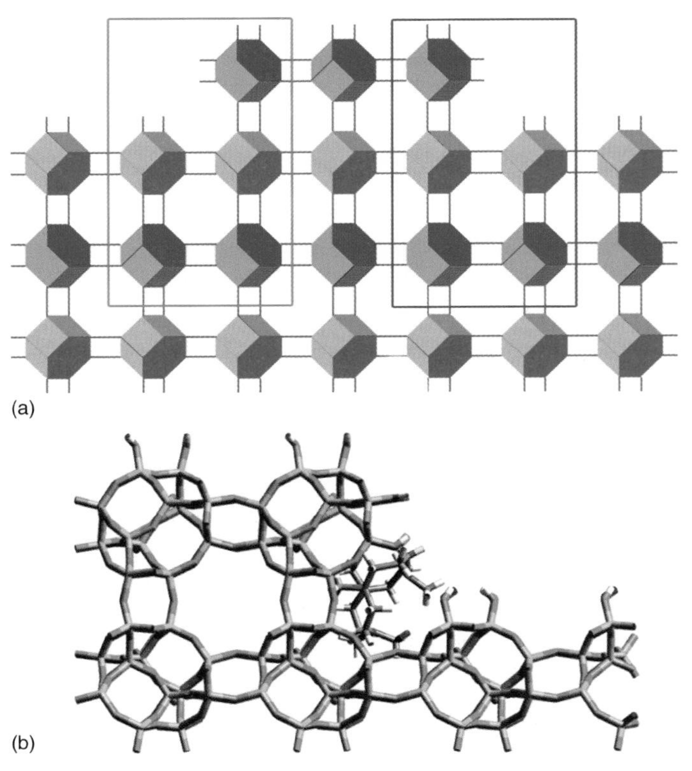


**Figure 1.30** (a) Simplified cross section perpendicular to <100> direction, showing the step structure in a {100} face. The two distinct step geometries, acute and obtuse, are highlighted in red and blue. (b) Cross section of the acute step showing the most stable docking configuration for the cyclam molecule after simulation.

other and hence favor the growth along one direction. A similar situation has been observed in calcite crystals where steps are also nonequivalent and ion sorption depends on their geometry [155]. To test this hypothesis, template adsorption at the surface was simulated using an adapted version of the ZEBEDDE program [156] to perform a Monte Carlo simulated annealing (MCSA) [157]. The template adsorption energies (nonbonding) from the simulations suggest that while TEA can adsorb at all adsorption sites with approximately equal energy, the larger cyclam can only adsorb favorably into the large cage site on the "acute" side of the step. Therefore, it is presumed that growth will be favored on the acute step side, where the cyclam accelerates the rate of growth unit attachment relative to the TEA-only mechanism on the obtuse step. As the position of the acute steps alternates between layers as the {100} face grows, the fastest growth direction follows, creating the interlaced pattern. Figure 1.30b shows a detailed cross section of the step structure with the cyclam molecule attached in its more favorable position.

## 1.5.5 Zincophosphates

Zincophosphate open-framework materials are a class of zeotype materials. In some cases, they show framework types same as those found on zeolites, such as SOD [158] and FAU [159]. However, some possess unique framework types, such as the chiral zincophosphate (CZP) framework, which have no aluminosilicate analog. The zinc phosphates with SOD and FAU structures were first synthesized by Nenoff *et al.* [158] under very mild pH and temperature conditions. However, the synthetic conditions are much milder in the case of the zinc phosphates [160]. These milder conditions are particularly well suited for performing *in situ* experiments on the AFM, as compared to aluminosilicate zeolites. Following results from *in situ* growth experiments on ZnPO-SOD and ZnPO-FAU are discussed.

### 1.5.5.1 $ZnPO_4$-Sodalite

SOD zinc phosphate, which is synthesized at temperatures ranging from room temperature to 50 °C [158], has a primitive cubic framework, with a unit cell constant $a = 0.882$ nm and belongs to the *P*-43*n* space group. The system contains a 1 : 1 mixture of tetrahedral zinc and phosphorous units that alternates within the framework, giving a stoichiometry of $Na_6(ZnPO_4)_6{\cdot}8H_2O$ [158]. ZnPO-SOD was synthesized following the original room temperature recipe by Nenoff [158]. This produced highly intergrown crystals which were not suitable for AFM experiments. To solve this issue, the synthesis was performed at 6 °C with the goal of decreasing growth and nucleation rates, which may produce better quality single crystals. The synthesis did produce single crystals, as well as intergrowth and the ZnPO-CZP. SOD crystals had dimensions ranging from a few microns up to 15 μm. Crystals were bound by {100}, {110}, and {111} faces. These crystals were attached to a resin and brought in contact with low supersaturated solutions.

*In situ* experiments performed on the {100} face in contact with low supersaturated solutions revealed the formation of spirals. Figure 1.31a shows one of these spirals. The angles between steps are slightly distorted since the growth rate is too fast for the scan speed to catch up. The overall morphology observed should be a square. Still it can be clearly observed that the spiral formed is of the interlaced type, such as those observed on STA-7, with two monolayers spreading out from the dislocation. In this case, however, the splitting has a fourfold symmetry, which, of course, agrees with that of the ZnPO-SOD crystal studied. Also, it has a polygonal shape, contrary to the rounded contours on the STA-7. Height analysis reveals that each monolayer height is about 0.45 nm, which corresponds to half a unit cell of ZnPO-SOD. Figure 1.31b shows the simplified 3D structure for ZnPO-SOD where a monolayer (highlighted) has started to grow.

The real insight on the crystal growth process that can be achieved by monitoring the process *in situ* is highlighted in Figure 1.32. In this figure, a sequence of lateral force AFM images shows two interlaced spirals growing, and the interlacing-inducing growth anisotropy for each substep can be clearly observed. In Figure 1.32a the two spiral centers are highlighted by white circles. Because of the

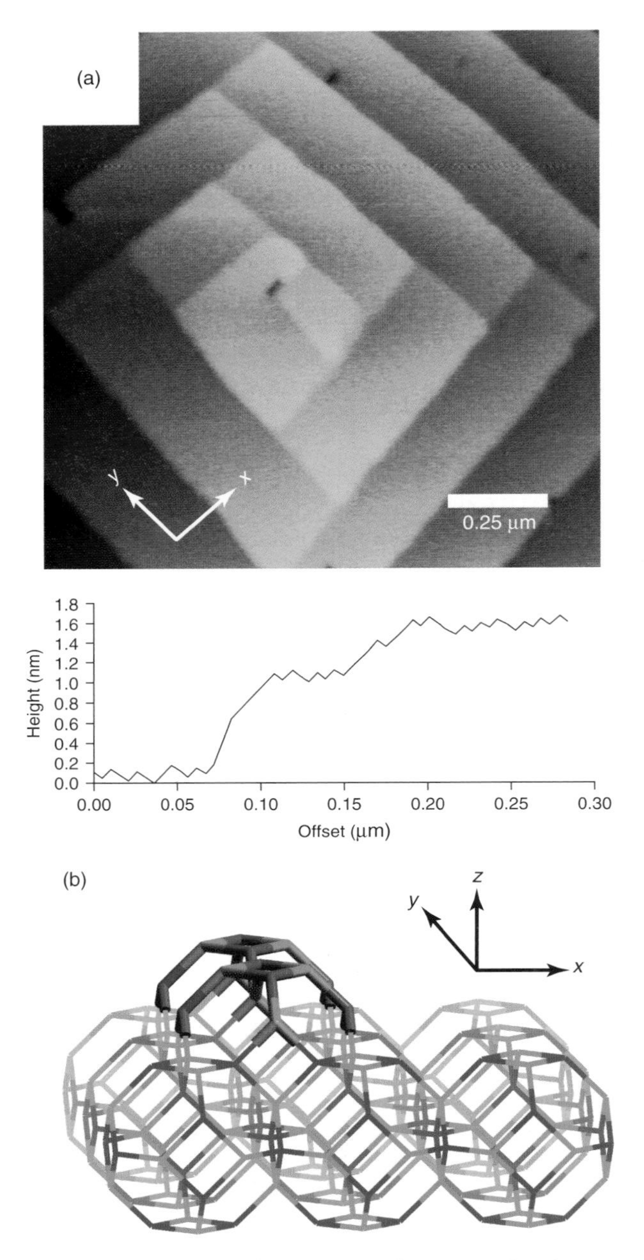


**Figure 1.31** (a) AFM deflection image showing interlaced spiral growth on a ZnPO-SOD crystal and associated cross section. (b) Simplified 3D structure of ZnPO-SOD showing the 1/2 unit cell step (highlighted in bold).

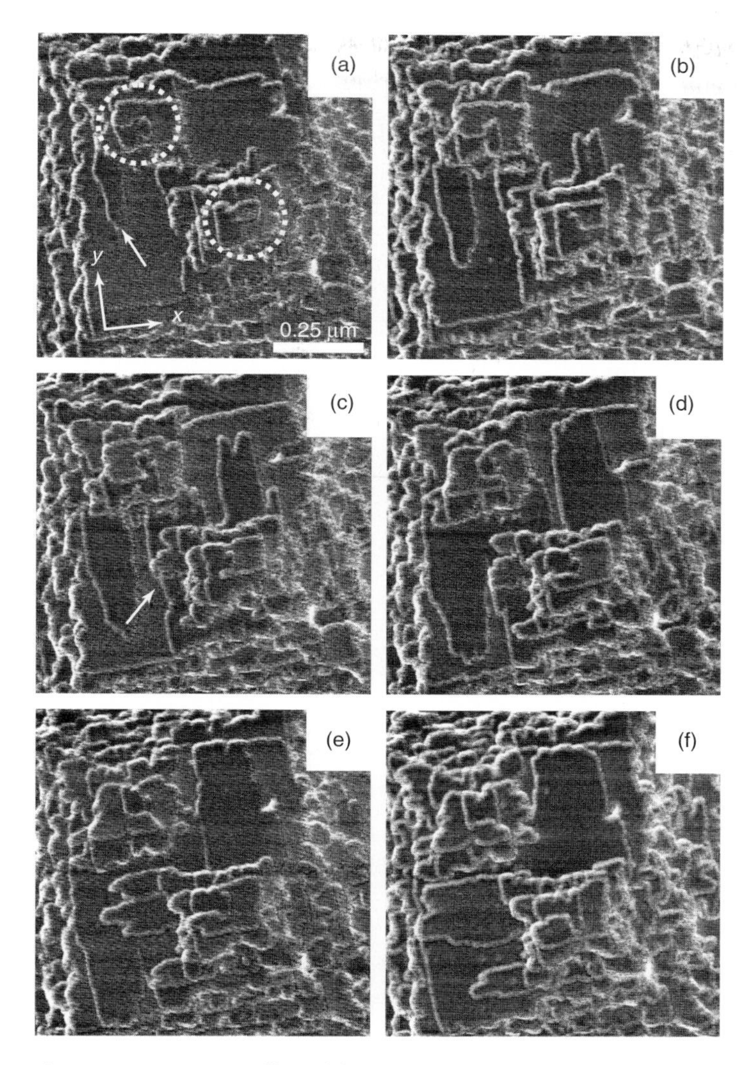


**Figure 1.32** Sequence of lateral force AFM images showing the growth of two interlaced spirals.

way in which both spirals interact; two clear square areas are created in between. A white arrow in Figure 1.32a signals the position of a step which can be seen advancing in the following images. The advancing speed of this step is much higher along <010> direction than along <100>. On the contrary, the next step to growth on top (highlighted with white arrow in Figure 1.32c) advances much faster along <100> direction than along <010>, as can be seen on Figure 1.32d–f. Figure 1.33 shows the two interlaced patterns of each step as if they could grow freely (in a similar fashion as in Figure 1.29 for STA-7). It can be seen that the shape of each step would be rectangular because of the anisotropic growth, and how successive

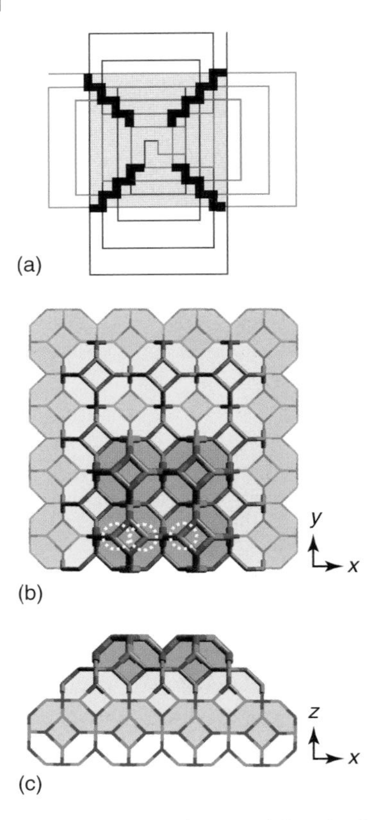


**Figure 1.33** (a) Theoretical free development of steps in the absence of interference. (b and c) Simplified structure of ZnPO–SOD showing three different steps (highlighted in different colors) at various stages of growth.

steps will alternate between fast and slow directions. The interference pattern created by two such substeps is the actual pattern observed in the experiments.

The reason behind this anisotropic growth is not fully understood, but it may have to do with the alternation of Zn and P positions in the ZnPO-SOD and the rates of condensation of the two different elements into the structure. Figure 1.33b shows a schematic ZnPO-SOD structure where three monolayers have grown (highlighted in three different colors). The red and blue colors represent Zn and P tetrahedra, respectively. Figure 1.33c shows a cross section of the structure, highlighting the fact that each monolayer is of half unit-cell height. Looking at the top monolayer (shaded in blue) in Figure 1.33b, it can be seen that the top Zn tetrahedra in the monolayer (inside white circles) are oriented in a line along <100> direction, whereas in the underlying monolayer (shaded in pink) they lie in a direction parallel to <010>. Correspondingly, the top P tetrahedra in each monolayer also alternate position. If the rate-determining step in the formation of a half SOD cage (necessary for step advancement) depends on the identity of the atom in the structure to which it bonds a difference in growth rates as a function of direction can be envisioned.

#### 1.5.5.2 $ZnPO_4$-Faujasite

FAU zincophosphate was synthesized for the first time by Gier and Stucky [161]. ZnPO-FAU belongs to the space group Fd-3, and has a unit cell constant $a = 25.1991$ Å. Its formula is $Na_{67}TMA_{12}Zn(ZnPO_4)\cdot 192H_2O$, where TMA stands for tetramethylammonium. For this investigation, crystals were prepared using the original synthesis at 4 °C [161]. The crystals produced have the typical octahedral morphology [83] and a size of a few microns. The solution used for crystal growth in the experiments was the clear mother-liquor produced as the synthesis takes place and the crystals form. It was taken after just 4 hours of synthesis.

*In situ* observations of FAU-type ZnPO crystals at low supersaturation conditions showed a "birth and spread" growth mechanism (Figure 1.34). The shape of the 2D nuclei formed is triangular, in accordance with previous *ex situ* observations [147]. At high supersaturation conditions, growth takes place by the advancement of macrosteps with a height of tenths of nanometers.

### 1.5.6 Metal Organic Frameworks

Nanoporous MOF crystals have bonding halfway between that of weak hydrogen bonding in molecular crystals and strong covalent bonding in zeolites. The first *in situ* AFM images of growing nanoporous crystals were reported on an MOF as the conditions for growth are less aggressive for the microscope than for zeolites. Figure 1.35 shows a series of AFM images of the important copper trimesate, $Cu_3(C_9H_3O_6)_2(H_2O)_3$ (HKUST-1) [162], which is a significant crystalline nanoporous MOF [163] built from $Cu_2(H_2O)_2$ units and benzene-1,3,5-tricarboxylate (BTC) groups and used to form a cubic framework with a three-dimensional nanoporous channel system. The crystals exhibit only (111) facets and the sample has been prepared by growing under ambient conditions in an oriented manner on gold substrates functionalized with self-assembled monolayers

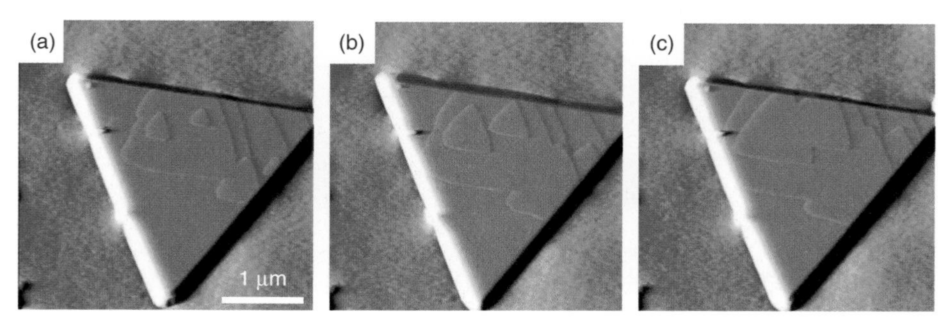


**Figure 1.34** Sequence of AFM deflection images showing the growth of nuclei in a ZnPO–FAU crystal.

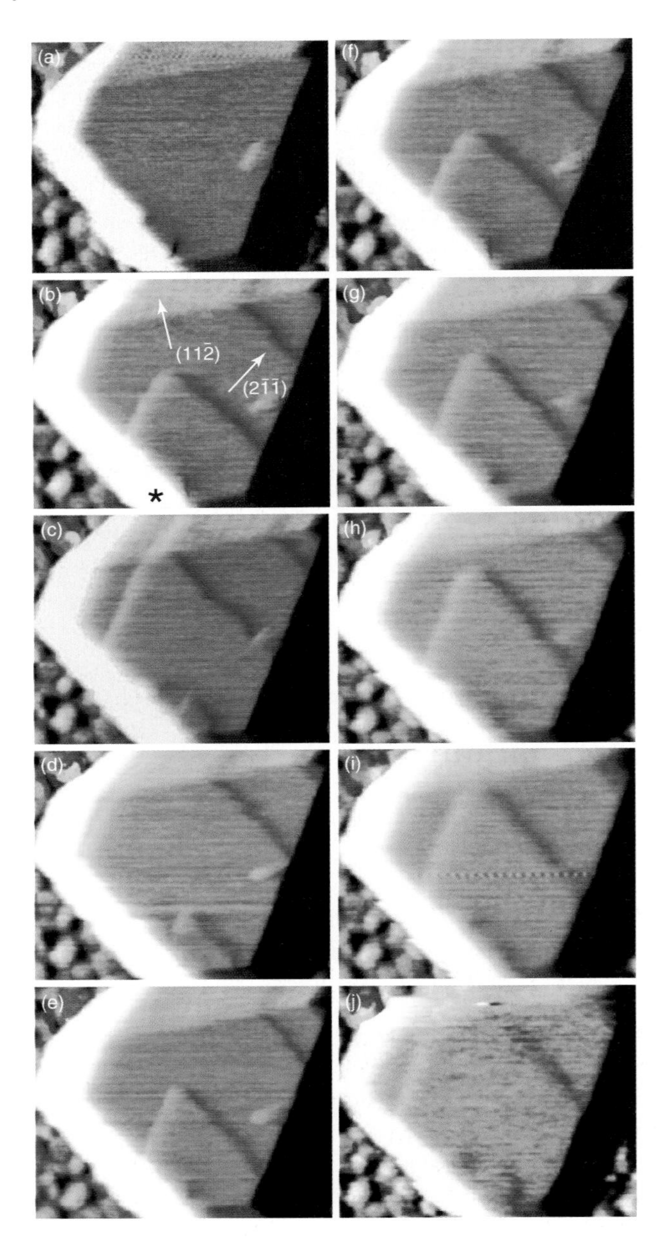


**Figure 1.35** Real-time deflection AFM images of the growing {111} facet of a HKUST-1 crystal at (a) 56, (b) 77, (c) 79, (d) 82, (e) 85, (f) 88, (g) 91, (h) 94, (i) 97, (j) 108 minutes, after injection of the growth solution. (Times refer to the end of each scan). Image sizes are $0.763 \times 0.613\ \mu m^2$.

(SAMs) of 16-mercapto-1-hexadecanol [162]. This provides a unique platform for *in situ* AFM studies, since the crystals are firmly anchored by direct attachment to a gold-coated glass substrate, but more importantly, the orientation of the crystals can be tuned by using different functional groups for surface functionalization, such that the growth of the {111} face can be monitored directly.

The crystal growth could be clearly monitored 56 minutes after the injection of the solution used for growth. The image at 56 minutes (Figure 1.35a) reveals an extremely flat and relatively defective-free crystal surface exemplifying the utility of this synthetic protocol to produce high-quality crystal surfaces. In the subsequent images, growth of the surface is seen to proceed by a 2D crystal growth mechanism in which each new crystal layer nucleates at the same point on the crystal surface, indicated by an asterisk in Figure 1.35b. It is likely that a defect at this point on the crystal facet is acting as a nucleation center. Cross-sectional analyses of height images at each time during the growth reveals that the vast majority of growth steps have heights of $1.5 \pm 0.1$ nm corresponding to the 1.5 nm $d_{111}$ crystal spacing of the HKUST-1 structure, but also half-step $d_{222}$ crystal spacings are observed. Interestingly, the triangular terraces exhibit a linear growth until the apex of the terrace reaches the edge of the crystal. At that point the growth slows considerably, illustrating that the abundance of kink sites near the apex of the triangle is dominant for the propagation of the terrace.

The results suggest that a layer grows by initial attachment of BTC and copper species onto a stably terminated crystal surface to form a small volume of a $d'_{222}$ step with metastable termination. The attachment of additional reagent to the crystal occurs more rapidly at this newly created metastable termination, creating a new $d_{111}$ step with a stable terminated surface.

## 1.6 Conclusions and Outlook

By applying a number of novel techniques to the problem of crystal growth in nanoporous materials, it is now possible to understand the mechanism at the molecular scale. In particular, the advent of AFM has opened a new window on the subject. Crystals are observed to grow by birth-and-spread mechanism as well as by spiral growth. The effects of supersaturation, temperature, chemical speciation, and structure are becoming apparent at this fundamental scale. It can be hoped that in the near future it should be possible to control crystal shape and habit, defects, and intergrowths through careful control of these growth parameters.

In particular, it can be expected that scanning probe microscopies will develop apace. Most AFMs operate at 1 atm pressure with modest variations of temperature under solution conditions. Hydrothermal AFM has been realized to operate at $P =$ 10 bar and $T = 150^{\circ}$C [164], however, current designs do not permit the location of micron-sized crystals via optical microscopy techniques, and consequently some technique is required to be developed in order to realize this goal. Nevertheless, we can expect such breakthroughs in the near future that will extend the range of applicability of AFM to more zeolite systems.

In AFM, the cantilever deflection is determined optically using a laser light source that is reflected off the back of the cantilever surface. For crystal growth measurements this poses many difficult choices. The most obvious is that in a solution it is imperative that there is no turbidity present. The solution may be colored but should not contain any particulates that scatter light. Many crystallizations operate in a chemistry regime where substantial light scattering may be expected. Also, in order to monitor crystallization on micron-sized crystals, it is essential to combine the AFM with a high-resolution optical microscope in order to locate the cantilever on the desired crystal facet. Modern AFM design goes to some lengths in order to organize the geometry of the AFM/optical microscope tandem arrangement in order to accommodate the laser path of the AFM and the optical path of the optical microscope. These complications can be overcome if it was possible to determine the tip displacement or force by measuring the resistance change of a piezoelectric material on the cantilever. There are a number of groups who are developing such techniques since the first images were recorded by Tortonese *et al.* [165] If it becomes possible to realize nonoptical methods for cantilever detection under crystal growth conditions, this would considerably expand the application of AFM in this field.

Very recently, the first ever video-rate AFM recorded under solution conditions has been reported [166]. This work is based around a resonant scanning system. Topographic information is then obtained by deflection of the cantilever determined optically. Video-rate AFM until this work has been confined to samples in air, but this new development is particularly interesting for the study of crystal growth or dissolution where it is difficult to bracket the kinetics within a typical AFM frame rate.

Finally, there has been considerabe progress recently to improve the lateral resolution of AFM to atomic, or even subatomic resolution [167]. Applied to the problem of crystal growth, this might permit the direct observation of template molecules at surfaces.

## References

1. Subotic, B. and Bronic, J. (2003) in *Handbook of Zeolite Science and Technology* (eds S.M. Auerbach, K.A. Carrado, and P.K. Dutta), Marcel Dekker, New York, p. 1184.
2. Anderson, M.W., Terasaki, O., Ohsuna, T., Philippou, A., Mackay, S.P., Ferreira, A., Rocha, J., and Lidin, S. (1994) *Nature*, **367**, 347–351.
3. Newsam, J.M., Treacy, M.M.J., Koetsier, W.T., and De Gruyter, C.B. (1988) *Proc. R. Soc. London Ser. A: Math. Phys. Eng. Sci.*, **420**, 375–405.
4. Kashchiev, D. (2000) *Nucleation: Basic Theory with Applications*, Butterworth-Heinemann, Oxford.
5. Mullin, J.W. (2001) *Crystallization*, 4th edn, Butterworth-Heinemann, Oxford.
6. Kashchiev, D. and van Rosmalen, G.M. (2003) *Cryst. Res. Technol.*, **38**, 555–574.
7. Elwell, D. and Scheel, H.J. (1975) *Crystal Growth from High-temperature Solutions*, Academic Press, New York.
8. Lasaga, A.C. (1998) *Kinetic Theory in the Earth Sciences*, Princeton University Press, Princeton, p. 811.

9. Kossel, W. (1934) *Annal. Phys.*, **21**, 457–480.
10. Ohara, M. and Reid, R.C. (1973) *Modeling Crystal Growth Rates from Solution*, Prentice-Hall International Series in the Physical and Chemical Engineering Sciences, Prentice-Hall, p. 272.
11. Volmer, M. and Schultz, W. (1931) *Z. Phys. Chem.*, **156**, 1–22.
12. Frank, F.C. (1949) *Discuss. Faraday Soc.*, **5**, 48–54.
13. Burton, W.K., Cabrera, N., and Frank, F.C. (1951) *Philos. Trans. R. Soc. London*, **A243**, 299–358.
14. van Enckevort, W.J.P. and Bennema, P. (2004) *Acta Crystallogr., Sect. A*, **60**, 532–541.
15. Pina, C.M., Becker, U., Risthaus, P., Bosbach, D., and Putnis, A. (1998) *Nature*, **395**, 483–486.
16. Astier, J.P., Bokern, D., Lapena, L., and Veesler, S. (2001) *J. Cryst. Growth*, **226**, 294–302.
17. Aquilano, D., Veesler, S., Astier, J.P., and Pastero, L. (2003) *J. Cryst. Growth*, **247**, 541–550.
18. van der Hoek, B., van der Eerden, J.P., and Tsukamoto, K. (1982) *J. Cryst. Growth*, **58**, 545–553.
19. Zauner, A.R.A., Aret, E., van Enckevort, W.J.P., Weyher, J.L., Porowski, S., and Schermer, J.J. (2002) *J. Cryst. Growth*, **240**, 14–21.
20. Baronnet, A., Amouric, M., and Chabot, B. (1976) *J. Cryst. Growth*, **32**, 37–59.
21. Barrer, R.M. (ed.) (1982) *Hydrothermal Chemistry of Zeolites*, Academic Press, London, p. 360.
22. Cundy, C.S. and Cox, P.A. (2005) *Microporous Mesoporous Mater.*, **82**, 1–78.
23. Flanigen, E.M. and Patton, R.L. (1978) Silica polymorph 76-726744 4073865, 19760927.
24. Nicolle, M.A., Di Renzo, F., Fajula, F., Espiau, P., and Des Courieres, T. (1993) **1**, 313–320.
25. Mintova, S., Olson, N.H., Valtchev, V., and Bein, T. (1999) *Science*, **283**, 958–960.
26. Mintova, S. and Valtchev, V. (1999) *Stud. Surf. Sci. Catal.*, **125**, 141–148.
27. Walton, R.I. and O'Hare, D. (2001) *J. Phys. Chem. Solids*, **62**, 1469–1479.
28. Yang, H., Walton, R.I., Antonijevic, S., Wimperis, S., and Hannon, A.C. (2004) *J. Phys. Chem. B*, **108**, 8208–8217.
29. Kosanovic, C., Bosnar, S., Subotic, B., Svetlicic, V., Misic, T., Drazic, G., and Havancsak, K. (2008) *Microporous Mesoporous Mater.*, **110**, 177–185.
30. Wakihara, T., Kohara, S., Sankar, G., Saito, S., Sanchez-Sanchez, M., Overweg, A.R., Fan, W., Ogura, M., and Okubo, T. (2006) *Phys. Chem. Chem. Phys.*, **8**, 224–227.
31. Cundy, C.S. and Forrest, J.O. (2004) *Microporous Mesoporous Mater.*, **72**, 67–80.
32. Cundy, C.S. and Cox, P.A. (2003) *Chem. Rev.*, **103**, 663–701.
33. Pope, C.G. (1998) *Microporous Mesoporous Mater.*, **21**, 333–336.
34. Bransom, S.H., Dunning, W.J., and Millard, B. (1949) *Discuss. Faraday Soc* 83–95.
35. Giaya, A. and Thompson, R.W. (2004) *AIChE J.*, **50**, 879–882.
36. Thompson, R.W. and Dyer, A. (1985) *Zeolites*, **5**, 202–210.
37. Gora, L. and Thompson, R.W. (1997) *Zeolites*, **18**, 132–141.
38. Čižmek, A., Subotić, B., Kralj, D., Babić-Ivančić, V., and Tonejc, A. (1997) *Microporous Mesoporous Mater.*, **12**, 267.
39. Brar, T., France, P., and Smirniotis, P.G. (2001) *Ind. Eng. Chem. Res.*, **40**, 1133–1139.
40. Bronic, J. and Subotic, B. (1995) *Microporous Mesoporous Mater.*, **4**, 239–242.
41. Warzywoda, J., Edelman, R.D., and Thompson, R.W. (1991) *Zeolites*, **11**, 318–324.
42. Cundy, C.S., Lowe, B.M., and Sinclair, D.M. (1990) **100**, 189–202.
43. Kajcsos, Z., Kosanovic, C., Bosnar, S., Subotic, B., Major, P., Liszkay, L., Bosnar, D., Lazar, K., Havancsak, H., Luu, A.T., and Thanh, N.D. (2009) *Mater. Sci. Forum*, **607**, 173–176.
44. Erdem-Senatalar, A. and Thompson, R.W. (2005) *J. Colloid Interface Sci.*, **291**, 396–404.

45. Mintova, S., Fieres, B., and Bein, T. (2002) *Stud. Surf. Sci. Catal.*, **142A**, 223–229.
46. Chang, C.D. and Bell, A.T. (1991) *Catal. Lett.*, **8**, 305.
47. Flanigen, E.M., Bennett, J.M., Grose, R.W., Cohen, J.P., Patton, R.L., Kirchner, R.M., and Smith, J.V. (1978) *Nature*, **271**, 512–516.
48. Burkett, S.L. and Davis, M.E. (1995) *Chem. Mater.*, **7**, 920–928.
49. Wakihara, T. and Okubo, T. (2005) *Chem. Lett.*, **34**, 276–281.
50. Zhdanov, S.P. and Samulevich, N.N. (1980) Proceedings of the 5th International Conference on Zeolites, pp. 75–84.
51. Bosnar, S. and Subotic, B. (1999) *Microporous Mesoporous Mater.*, **28**, 483–493.
52. Iwasaki, A., Hirata, M., Kudo, I., Sano, T., Sugawara, S., Ito, M., and Watanabe, M. (1995) *Zeolites*, **15**, 308–314.
53. Cundy, C.S., Henty, M.S., and Plaisted, R.J. (1995) *Zeolites*, **15**, 353–372.
54. Cundy, C.S., Henty, M.S., and Plaisted, R.J. (1995) *Zeolites*, **15**, 400–407.
55. Bosnar, S., Antonic, T., Bronic, J., and Subotic, B. (2004) *Microporous Mesoporous Mater.*, **76**, 157–165.
56. Gora, L., Streletzky, K., Thompson, R.W., and Phillies, G.D.J. (1997) *Zeolites*, **18**, 119–131.
57. Kalipcilar, H. and Culfaz, A. (2000) *Cryst. Res. Technol.*, **35**, 933–942.
58. Schoeman, B.J. (1997) *Progress in Zeolite and Microporous Materials, Parts A-C*, pp. 647–654.
59. Caputo, D., Gennaro, B.D., Liguori, B., Testa, F., Carotenuto, L., and Piccolo, C. (2000) *Mater. Chem. Phys.*, **66**, 120–125.
60. Schoeman, B.J. (1997) *Zeolites*, **18**, 97–105.
61. Cundy, C.S., Lowe, B.M., and Sinclair, D.M. (1993) *Faraday Discuss.*, **95**, 235–252.
62. Schoeman, B.J., Sterte, J., and Otterstedt, J.E. (1994) *Zeolites*, **14**, 568–575.
63. Kacirek, H. and Lechert, H. (1975) *J. Phys. Chem.*, **79**, 1589–1593.
64. Kacirek, H. and Lechert, H. (1976) *J. Phys. Chem.*, **80**, 1291–1296.
65. Anderson, M.W. (2001) *Curr. Opin. Solid State Mater. Sci.*, **5**, 407–415.
66. Anderson, M.W., Ohsuna, T., Sakamoto, Y., Liu, Z., Carlsson, A., and Terasaki, O. (2004) *Chem. Commun.*, 907–916.
67. Anderson, M.W., Agger, J.R., Meza, L.I., Chong, C.B., and Cundy, C.S. (2007) *Faraday Discuss.*, **136**, 143–156.
68. Stevens, S.M., Cubillas, P., Jansson, K., Terasaki, O., Anderson, M.W., Wright, P.A., and Castro, M. (2008) *Chem. Commun (Camb.)*, 3894–3896.
69. Terasaki, O. and Ohsuna, T. (2003) *Top. Catal.*, **24**, 13–18.
70. Slater, B., Ohsuna, T., Liu, Z., and Terasaki, O. (2007) *Faraday Discuss.*, **136**, 125–141.
71. Komiyama, M. and Yashima, T. (1994) *Jpn. J. Appl. Phys.*, **33**, 3761–3763.
72. Komiyama, M., Tsujimichi, K., Oumi, Y., Kubo, M., and Miyamoto, A. (1997) *Appl. Surf. Sci.*, **121-122**, 543–547.
73. Yamamoto, S., Sugiyama, S., Matsuoka, O., Honda, T., Banno, Y., and Nozoye, H. (1998) *Microporous Mesoporous Mater.*, **21**, 1–6.
74. Voltolini, M., Artioli, G., and Moret, M. (2003) *Microporous Mesoporous Mater.*, **61**, 79–84.
75. Yamamoto, S., Matsuoka, O., Sugiyama, S., Honda, T., Banno, Y., and Nozoye, H. (1996) *Chem. Phys. Lett.*, **260**, 208–214.
76. Sugiyama, S., Yamamoto, S., Matsuoka, O., Honda, T., Nozoye, H., Qiu, S., Yu, J., and Terasaki, O. (1997) *Surf. Sci.*, **377**, 140–144.
77. Sugiyama, S., Yamamoto, S., Matsuoka, O., Nozoye, H., Yu, J., Zhu, G., Qiu, S., and Terasaki, I. (1999) *Microporous Mesoporous Mater.*, **28**, 1–7.
78. Wakihara, T., Sugiyama, A., and Okubo, T. (2004) *Microporous Mesoporous Mater.*, **70**, 7–13.
79. Anderson, M.W., Agger, J.R., Hanif, N., and Terasaki, O. (2001) *Microporous Mesoporous Mater.*, **48**, 1–9.
80. Agger, J.R., Hanif, N., and Anderson, M.W. (2001) *Angew. Chem. Int. Ed.*, **40**, 4065–4067.

81. Agger, J.R., Hanif, N., Cundy, C.S., Wade, A.P., Dennison, S., Rawlinson, P.A., and Anderson, M.W. (2003) *J. Am. Chem. Soc.*, **125**, 830–839.
82. Dumrul, S., Bazzana, S., Warzywoda, J., Biederman, R.R., and Sacco, A. Jr. (2002) *Microporous Mesoporous Mater.*, **54**, 79–88.
83. Singh, R., Doolittle, J., George, M.A., and Dutta, P.K. (2002) *Langmuir*, **18**, 8193–8197.
84. Meza, L.I., Agger, J.R., Logar, N.Z., Kaucic, V., and Anderson, M.W. (2003) *Chem. Commun.*, 2300–2301.
85. Warzywoda, J., Yilmaz, B., Miraglia, P.Q., and Sacco, A. Jr. (2004) *Microporous Mesoporous Mater.*, **71**, 177–183.
86. Yamamoto, S., Sugiyama, S., Matsuoka, O., Kohmura, K., Honda, T., Banno, Y., and Nozoye, H. (1996) *J. Phys. Chem.*, **100**, 18474–18482.
87. Meza, L.I., Anderson, M.W., Slater, B., and Agger, J.R. (2008) *Phys. Chem. Chem. Phys.*, **10**, 5066–5076.
88. Anderson, M.W., Agger, J.R., Thornton, J.T., and Forsyth, N. (1996) *Angew. Chem. Int. Ed.*, **35**, 1210–1213.
89. Wakihara, T., Sasaki, Y., Kato, H., Ikuhara, Y., and Okubo, T. (2005) *Phys. Chem. Chem. Phys.*, **7**, 3416–3418.
90. Agger, J.R. and Anderson, M.W. (2002) *Impact of Zeolites and Other Porous Materials on the New Technologies at the Beginning of the New Millennium, Parts A and B*, pp. 93–100.
91. Walker, A.M., Slater, B., Gale, J.D., and Wright, K. (2004) *Nat. Mater.*, **3**, 715–720.
92. Binnig, G., Quate, C.F., and Gerber, C. (1986) *Phys. Rev. Lett.*, **56**, 930–933.
93. Binnig, G., Rohrer, H., Gerber, C., and Weibel, E. (1982) *Phys. Rev. Lett.*, **49**, 57–61.
94. Zhong, Q., Inniss, D., Kjoller, K., and Elings, V.B. (1993) *Surf. Sci.*, **290**, L688–L692.
95. Mate, C.M., McClelland, G.M., Erlandsson, R., and Chiang, S. (1987) *Phys. Rev. Lett.*, **59**, 1942–1945.
96. Szlufarska, I., Chandross, M., and Carpick, R.W. (2008) *J. Phys. D Appl. Phys.*, **41**, 123001.
97. McPherson, A., Malkin, A.J., and Kuznetsov, Y.G. (2000) *Annu. Rev. Biophys. Biomol. Struct.*, **29**, 361–410.
98. Bose, S., Hu, X., and Higgins, S.R. (2008) *Geochim. Cosmochim. Acta*, **72**, 759–770.
99. Higgins, S.R., Boram, L.H., Eggleston, C.M., Coles, B.A., Compton, R.G., and Knauss, K.G. (2002) *J. Phys. Chem. B*, **106**, 6696–6705.
100. Teng, H.H., Dove, P.M., and De Yoreo, J.J. (2000) *Geochim. Cosmochim. Acta*, **64**, 2255–2266.
101. Maiwa, K., Nakamura, H., Kimura, H., and Miyazaki, A. (2006) *J. Cryst. Growth*, **289**, 303–307.
102. Radenovic, N., van Enckevort, W., Kaminski, D., Heijna, M., and Vlieg, E. (2005) *Surf. Sci.*, **599**, 196–206.
103. Moret, M., Campione, M., Caprioli, S., Raimondo, L., Sassella, A., Tavazzi, S., and Aquilano, D. (2007) *J. Phys.*, **61**, 831–835.
104. Richter, A. and Smith, R. (2003) *Cryst. Res. Technol.*, **38**, 250–266.
105. Loiola, A.R., da Silva, L.R.D., Cubillas, P., and Anderson, M.W. (2008) *J. Mater. Chem.*, **18**, 4985–4993.
106. Karwacki, L., Stavitski, E., Kox, M.H.F., Kornatowski, J., and Weckhuysen, B.M. (2008) *Stud. Surf. Sci. Catal.*, **174B**, 757–762.
107. Kox, M.H.F., Stavitski, E., Groen, J.C., Perez-Ramirez, J., Kapteijn, F., and Weckhuysen, B.M. (2008) *Chem. Eur. J.*, **14**, 1718–1725.
108. Harris, R.K., Knight, C.T.G., and Hull, W.E. (1981) *J. Am. Chem. Soc.*, **103**, 1577–1578.
109. Harris, R.K. and Knight, C.T.G. (1983) *J. Chem. Soc. Faraday Trans. 2: Mol. Chem. Phys.*, **79**, 1525–1538.
110. Harris, R.K. and Knight, C.T.G. (1983) *J Chem. Soc. Faraday Trans. 2: Mol. Chem. Phys.*, **79**, 1539–1561.
111. Knight, C.T.G. (1988) *J. Chem. Soc. Dalton Trans. Inorg. Chem.*, 1457–1460.
112. Kinrade, S.D., Knight, C.T.G., Pole, D.L., and Syvitski, R.T. (1998) *Inorg. Chem.*, **37**, 4278–4283.
113. Kinrade, S.D., Donovan, J.C.H., Schach, A.S., and Knight, C.T.G.

(2002) *J. Chem. Soc., Dalton Trans.*, 1250–1252.
114. Knight, C.T.G. and Kinrade, S.D. (2002) *J. Phys. Chem. B*, **106**, 3329–3332.
115. Knight, C.T.G., Wang, J., and Kinrade, S.D. (2006) *Phys. Chem. Chem. Phys.*, **8**, 3099–3103.
116. Knight, C.T.G., Balec, R.J., and Kinrade, S.D. (2007) *Angew. Chem. Int. Ed.*, **46**, 8148–8152.
117. Haouas, M. and Taulelle, F. (2006) *J. Phys. Chem. B*, **110**, 22951.
118. Aerts, A., Follens, L.R.A., Haouas, M., Caremans, T.P., Delsuc, M.-A., Loppinet, B., Vermant, J., Goderis, B., Taulelle, F., Martens, J.A., and Kirschhock, C.E.A. (2007) *Chem. Mater.*, **19**, 3448–3454.
119. Follens, L.R.A., Aerts, A., Haouas, M., Caremans, T.P., Loppinet, B., Goderis, B., Vermant, J., Taulelle, F., Martens, J.A., and Kirschhock, C.E.A. (2008) *Phys. Chem. Chem. Phys.*, **10**, 5574–5583.
120. Serre, C., Corbiere, T., Lorentz, C., Taulelle, F., and Ferey, G. (2002) *Chem. Mater.*, **14**, 4939–4947.
121. Gerardin, C., Haouas, M., Lorentz, C., and Taulelle, F. (2000) *Magn. Reson. Chem.*, **38**, 429–435.
122. Gerardin, C., In, M., Allouche, L., Haouas, M., and Taulelle, F. (1999) *Chem. Mater.*, **11**, 1285–1292.
123. Vistad, O.B., Akporiaye, D.E., Taulelle, F., and Lillerud, K.P. (2003) *Chem. Mater.*, **15**, 1639–1649.
124. Shi, J., Anderson, M.W., and Carr, S.W. (1996) *Chem. Mater.*, **8**, 369–375.
125. Egger, C.C., Anderson, M.W., Tiddy, G.J.T., and Casci, J.L. (2005) *Phys. Chem. Chem. Phys.*, **7**, 1845–1855.
126. Bussian, P., Sobott, F., Brutschy, B., Schrader, W., and Schuth, F. (2000) *Angew. Chem. Int. Ed.*, **39**, 3901–3905.
127. Schuth, F., Bussian, P., Agren, P., Schunk, S., and Linden, M. (2001) *Solid State Sci.*, **3**, 801–808.
128. Schuth, F. (2001) *Curr. Opin. Solid State Mater. Sci.*, **5**, 389–395.
129. Pelster, S.A., Schrader, W., and Schuth, F. (2006) *J. Am. Chem. Soc.*, **128**, 4310–4317.
130. Pelster, S.A., Kalamajka, R., Schrader, W., and Schuth, F. (2007) *Angew. Chem. Int. Ed.*, **46**, 2299–2302.
131. Pelster, S.A., Schueth, F., and Schrader, W. (2007) *Anal. Chem.*, **79**, 6005–6012.
132. Pelster, S.A., Weimann, B., Schaack, B.B., Schrader, W., and Schueth, F. (2007) *Angew. Chem. Int. Ed.*, **46**, 6674–6677.
133. Kumar, S., Wang, Z., Penn, R.L., and Tsapatsis, M. (2008) *J. Am. Chem. Soc.*, **130**, 17284–17286.
134. Piana, S., Reyhani, M., and Gale, J.D. (2005) *Nature*, **438**, 70–73.
135. Piana, S., and Gale, J.D. (2005) *J. Am. Chem. Soc.*, **127**, 1975–1982.
136. Piana, S., and Gale, J.D. (2006) *J. Cryst. Growth*, **294**, 46–52.
137. Boerrigter, S.X.M., Josten, G.P.H., van de Streek, J., Hollander, F.F.A., Los, J., Cuppen, H.M., Bennema, P., and Meekes, H. (2004) *J. Phys. Chem. A*, **108**, 5894–5902.
138. Umemura, A. (2009) PhD thesis, The University of Manchester, Manchester.
139. Anderson, M.W., Meza, L.I., Agger, J.R., Attfield, M.P., Shoaee, M., Chong, C.B., Umemura, A., and Cundy, C.S. (2008) *Turning Points in Solid-State, Materials and Surface Science*, 95–122.
140. Umemura, A., Cubillas, P., Anderson, M.W., and Agger, J.R. (2008) *Stud. Surf. Sci. Catal.*, **174A**, 705–708.
141. Breck, D.W. (1974) *Zeolite Molecular Sieves*, John Wiley & Sons, Ltd, New York.
142. Baerlocher, C., Meier, W.M., and Olson, D.H. (2007) *Atlas of Zeolite Structure Types*, 7th edn, Elsevier, Amsterdam.
143. Charnell, J.F. (1971) *J. Cryst. Growth*, **8**, 291–294.
144. Petranovskii, V., Kiyozumi, Y., Kikuchi, N., Hayamisu, H., Sugi, Y., and Mizukami, F. (1997) *Stud. Surf. Sci. Catal.*, **105A**, 149–156.
145. Gora, L. and Thompson, R.W. (1995) *Zeolites*, **15**, 526–534.
146. Agger, J.R., Pervaiz, N., Cheetham, A.K., and Anderson, M.W. (1998) *J. Am. Chem. Soc.*, **120**, 10754–10759.
147. Wakihara, T. and Okubo, T. (2004) *J. Chem. Eng. Jpn.*, **37**, 669–674.

148. Thompson, R.W. and Huber, M.J. (1982) *J. Cryst. Growth*, **56**, 711–722.
149. Yang, X.B., Albrecht, D., and Caro, E. (2006) *Microporous Mesoporous Mater.*, **90**, 53–61.
150. Meza, L.I., Anderson, M.W., and Agger, J.R. (2007) *Chem. Commun (Camb.)*, 2473–2475.
151. Meza, L.I., Anderson, M.W., Agger, J.R., Cundy, C.S., Chong, C.B., and Plaisted, R.J. (2007) *J. Am. Chem. Soc.*, **129**, 15192–15201.
152. Brent, R. and Anderson, M.W. (2008) *Angew. Chem. Int. Ed.*, **47**, 5327–5330.
153. Castro, M., Garcia, R., Warrender, S.J., Slawin, A.M.Z., Wright, P.A., Cox, P.A., Fecant, A., Mellot-Draznieks, C., and Bats, N. (2007) *Chem. Commun.*, 3470–3472.
154. Cubillas, P., Castro, M., Jelfs, K.E., Lobo, A.J.W., Slater, B., Lewis, D.W., Wright, P.A., Stevens, S.M., and Anderson, M.W. (2009) *Crys. Growth Des.*
155. Paquette, J. and Reeder, R.J. (1995) *Geochim. Cosmochim. Acta*, **59**, 735–749.
156. Lewis, D.W., Willock, D.J., Catlow, C.R.A., Thomas, J.M., and Hutchings, G.J. (1996) *Nature*, **382**, 604–606.
157. Jelfs, K.E., Slater, B., Lewis, D.W., and Willock, D.J. (2007) *Zeolites to Porous MOF Materials*, Vol. 170B, Elsevier B.V., pp. 1685–1692.
158. Nenoff, T.M., Harrison, W.T.A., Gier, T.E., and Stucky, G.D. (1991) *J. Am. Chem. Soc.*, **113**, 378–379.
159. Harrison, W.T.A., Gier, T.E., and Stucky, G.D. (1991) *J. Mater. Chem.*, **1**, 153–154.
160. Nenoff, T.M., Harrison, W.T.A., Gier, T.E., Calabrese, J.C., and Stucky, G.D. (1993) *J. Solid State Chem.*, **107**, 285–295.
161. Gier, T.E., and Stucky, G.D. (1991) *Nature*, **349**, 508–510.
162. Shoaee, M., Agger, J.R., Anderson, M.W., and Attfield, M.P. (2008) *Cryst EngComm*, **10**, 646–648.
163. Chui, S.S.Y., Lo, S.M.F., Charmant, J.P.H., Orpen, A.G., and Williams, I.D. (1999) *Science*, **283**, 1148–1150.
164. Higgins, S.R., Eggleston, C.M., Knauss, K.G., and Boro, C.O. (1998) *Rev. Sci. Instrum.*, **69**, 2994–2998.
165. Tortonese, M., Barrett, R.C., and Quate, C.F. (1993) *Appl. Phys. Lett.*, **62**, 834–836.
166. Picco, L.M., Bozec, L., Ulcinas, A., Engledew, D.J., Antognozzi, M., Horton, M.A., and Miles, M.J. (2007) *Nanotechnology*, **18**, 044030-1–044030-4.
167. Giessibl, F.J. and Quate, C.F. (2006) *Phys. Today*, **59**, 44–50.

# 2
# Synthesis Approaches

*Karl G. Strohmaier*

## 2.1
## Introduction

Zeolites were first identified as natural minerals having unique physical absorption properties. Early researchers tried to reproduce the geological conditions in an attempt to make them synthetically. From the years 1845 to 1937 many researchers investigated the hydrothermal conversion and synthesis of silicates. Although there were a number of claims that zeolites had been made synthetically, they were unsubstantiated as identification was based upon chemical analysis and optical observations. Later attempts at reproducing these early experiments did not give identifiable zeolites. It was not until the 1940s that Professor Richard M. Barrer found the necessary conditions for making zeolites synthetically and he identified them by powder X-ray diffraction. By reacting the powder minerals leucite and analcime with aqueous solutions of barium chloride at temperatures of 180–270 °C for two to six days, a synthetic chabazite was obtained. In 1948, Professor Barrer was able to synthesize zeolites entirely from synthetic solutions of sodium aluminate, silicic acid, and sodium carbonate [1]. A dry powdered gel was first obtained by drying the mixture at 110 °C, and then crystallizing it to a synthetic mordenite by reacting the gel with water at higher temperatures. In the 1950s, Donald Breck and researchers at the Union Carbide Company began to synthesize zeolites from reactive aluminosilicate gels prepared from sodium aluminate and sodium silicate solutions. This led directly to the discovery of the first two new synthetic zeolites, Linde A [2] and Linde X ($Si/Al = 1.0-1.5$) [3], both having new framework structures, **LTA** and **FAU**, respectively. Linde A was found to have excellent ion exchange and gas separation properties. Later, a new, higher silica containing composition of **FAU**, designated Linde Y [4] ($Si/Al = 1.5-3.0$), was synthesized and was found to be an excellent cracking catalyst for the conversion of distilled crude oil to gasoline products. The discovery of these two industrially important zeolites caught the attention of many scientists and kick started the field of zeolite synthesis research.

*Zeolites and Catalysis, Synthesis, Reactions and Applications. Vol. 1.*
Edited by Jiří Čejka, Avelino Corma, and Stacey Zones

ISBN: 978-3-527-32514-6

Until this time, synthetic zeolites were made entirely from inorganic reagents and many experiments were performed to determine the effects of various alkali and alkali earth metals. The next major advancement in zeolite synthesis came in 1961 when Professor Barrer began experimenting with substituting part of the alkali and alkali earth cations with organic cations such as tetramethylammonium ($TMA^+$) [5]. Using $TMA^+$, he was able to make the zeolite N-A (Si/Al = 1.2), the high silica composition of Linde A (Si/Al = 1.0). Chemists at the Mobil Oil Company also began to utilize organic cations in zeolite syntheses and, in 1967, they became the first research group to make new zeolite structures with organic cations, which are also referred to as *templates* or *structure-directing agents* (SDAs). Two new commercially important zeolites, beta and ZSM-5, were made with tetraethylammonium ($TEA^+$) and tetrapropylammonium ($TPA^+$) cations, respectively. Both beta and ZSM-5 are the first synthetic zeolites prepared with Si/Al ratios greater than 5. Because most natural zeolites have Si/Al ratios of one to about five, earlier researchers did not thoroughly investigate high-silica-containing gels. This new approach of utilizing organic SDAs and high-silica-containing gels led to the discovery of a large number of new zeolite structures by Mobil scientists and other groups over the next two decades. Some of the more important ones are ZSM-11 (**MEL**), ZSM-12 (**MTW**), ZSM-22 (**TON**), ZSM-23 (**MTT**), ZSM-48 (***MRE**), ZSM-57 (**MFS**), EU-1 (**EUO**), and NU-87 (**NES**). Even to this day the use of new SDAs remains the primary strategy for discovering new zeolite structures.

## 2.2 Aluminophosphates

Another major breakthrough in zeolite synthesis occurred with the discovery of aluminophosphates (AlPOs) and, shortly thereafter, silicoaluminophosphates (SAPOs) molecular sieves. A number of scientists recognized that aluminum phosphate framework structures were known in nature in which phosphorus and aluminum atoms had tetrahedral coordination similar to that of aluminum and silicon in zeolites. While several groups worked on being the first to make synthetic AlPOs, it was the researchers at Union Carbide in the early 1980s who found the right reagents (pseudoboehmite alumina and phosphoric acid with organic SDAs) and conditions (typically no inorganic cations) necessary for easily making them [6]. Technically, the microporous AlPOs are not zeolites because they do not have aluminosilicate compositions, but nonetheless they have similar or identical frameworks to zeolites. The synthesis of SAPOs, metalloaluminophosphates (MeAPOs, where Me = a metal or combination of metals such as Mg, Ti, Cr, Mn, Fe, Co, Ni, Zn, B, and Ge capable of tetrahedra coordination) and metallosilicoalumino-phosphates (MeAPSOs) quickly followed.

The structures of AlPOs and SAPOs are such that there are alternating aluminum-phosphorus tetrahedra throughout the framework. As with zeolites, Lowenstein's rule prevails in AlPO and SAPO materials also, such that aluminum does not have aluminum as a nearest neighbor. It was also found that phosphorus does not have phosphorus or silicon as a nearest neighbor. For this reason, it was believed that all phosphate structures have only even number of rings in their frameworks, that is, 4-, 6-, 8-, and 12-member rings. This results in the absence of AlPO materials having structures with odd-numbered rings such as ZSM-5, mordenite, beta, and many others. Other structures with only even-member rings such as ERI, FAU, and CHA frameworks exist in both the aluminosilicate and phosphate compositions. It was not until the discovery of ECR-40 that it was realized that it was possible to synthesize odd-numbered ring containing structures having phosphate compositions. By adding silicon to an AlPO synthesis with $(C_2H_5OH)_x(CH_4)_{4-x}NOH$ ($x = 2, 3$) as SDAs, it was found that small amounts of a SAPO material having the **MEI** structure could be obtained at relatively long crystallization times (three to four weeks) [7]. The **MEI** framework contains three-, five-, and seven-member rings, which, as stated above, are not attainable due to alternating aluminum and phosphorus atoms in the structure because a five-membered ring always has either Al–Al or P–P nearest neighbors. In the ECR-40 structure, it was found that the Al, P, and Si were ordered such that there where no Si–P neighbors and an *anti*-Lowenstein Al–O–Al bond. For this to be possible, the SAPO had to have the specific composition $Al_{16}P_{12}Si_6O_{72}$, unlike all other SAPO materials, which can usually be prepared with a range of Si levels down to Si/Al = 0, that is, an AlPO. With this information, it was now straightforward to optimize the syntheses by using the structural composition, $Al_{16}P_{12}Si_6O_{72}$, in the gel synthesis to prepare the material in high yield [8]. This example represents the use of structural information to develop a strategy for optimizing the synthesis of a zeolitic material.

## 2.3 Mineralizers

An important feature of conventional zeolite synthesis is the use of high hydroxide concentrations to assist the mineralization of silicate and aluminate species in the reactant gel. It is important to have the right concentration of hydroxide ion as there is equilibrium between solution species and solid species in the gel. A very high concentration shifts the equilibrium very far toward the solution and prevents crystallization because the solubility of the silica and alumina is too high. On the other hand, a very low concentration does not solubilize the species enough to cause their transport to the growing crystals, resulting in an amorphous product.

The concentration of hydroxide can also effect the transformation of metastable phases to dense phases. In most of the cases, the desired zeolite phase is metastable and the first phase that crystallizes in a synthesis. At longer times, these metastable phases can transform in to undesirable dense phases such as zeolites P, sodalite, analcime, and quartz. Therefore, it is important to determine the correct hydroxide ion level so that the desired phase can be recovered before denser impurities begin to form.

In addition to the aforementioned importance of hydroxide concentration, the level of hydroxide can also have a large effect on the crystallization rate, crystal size, the Si/Al ratio of the product, and the ultimate product formed. An excellent example of the latter is a recent study that observes the effects of hydroxide concentration on syntheses performed using 1,4-bis(*N*-methylpyrrolidinium) butane as an SDA [9]. At a constant Si/Al ratio of 30, ZSM-12 (**MTW**) was produced from gels having a $OH^-/Si$ ratio $\leq 0.6$, TNU-9 (**TUN**, a new 10-ring zeolite having a three-dimensional channel framework similar to ZSM-5) was produced from gels having a $OH^-/Si$ ratio $= 0.73$ and TNU-10 (a new high silica form of stilbite, **STI**) was produced from gels having a $OH^-/Si$ ratio $= 1.0$. The synthesis of IM-5 (**IMF**) is another example of a zeolite prepared with a relatively high $OH^-$ ratio ($OH^-/Si = 0.6$). A recent study by Jackowski *et al.* [10] looked further at the effect of $OH^-$ levels in zeolite syntheses with a series of heterocyclic-substituted diquaternary ammonium compounds as SDAs. These studies illustrate the importance of evaluating a large $OH^-/Si$ range when trying to synthesize new zeolites.

Besides hydroxide, fluoride can also be used as an effective mineralizer for the synthesis of zeolites and phosphate materials. Flanigen and Patton were the first to use fluoride in a zeolite synthesis to prepare a highly siliceous MFI material called *silicalite* [11]. The incorporation of fluoride into zeolites was found to reduce the number of defect sites in highly siliceous zeolites and improve their overall stability. Before the general use of fluoride to prepare zeolites was undertaken, it was used first as a mineralizer in phosphate synthesis by Guth and Kessler [12]. A number of new materials were discovered by the use of fluoride in phosphate systems. Cloverite (**-CLO**) was synthesized with fluoride in the gallium phosphate system using quinuclidine as an SDA [13]. Another gallium phosphate material having the **LTA** framework was also prepared with HF [14]. It was found that $GaPO_4$-LTA could not be prepared in the absence of fluoride and that the fluoride was located at the center of the double 4-ring (D4R) in the structure [15]. Although other gallium phosphate materials, such as ULM-[16] and $Ga_5(PO_4)_5F_42[N_2C_4H_{12}]$ [17], have been prepared in the presence of fluoride, they usually are found to contain $GaO_4F$ trigonal bipyramids and $GaO_4F_2$ octahedra in their structures. In addition to synthesizing new $GaPO_4$ structures, HF was also found to be effective in synthesizing new AlPO compositions and structures such as $AlPO_4$-**CHA** [18, 19], UiO-6 (**OSI**) [20], and UiO-7 (**ZON**) [21].

The use of fluoride was also found to improve the distribution of silicon substitution into SAPO materials. The way silicon substitutes for phosphorus or aluminum-phosphorus pairs in AlPO synthesis is very important to the acidity of the final product, as silicon is the source of its Brønsted acidity [22]. At low silicon levels, isolated silicon ($Si\text{-}Al_4Si_0$), having four aluminum nearest neighbors, is present. As Al has relatively low electronegativity, the isolated Si atoms generate weak acid sites. As more silicon substitutes in the framework, silicon islands begin to form where Si can have one silicon neighbor ($Si\text{-}Al_3Si_1$) at the corners of the islands, two Si neighbors ($Si\text{-}Al_2Si_2$) along the edges of the islands, and three ($Si\text{-}Al_1Si_3$) on the inside corners of the islands. Because of the higher electronegativity of silicon, Si sites with more Si nearest neighbors have stronger acid site strength. Si atoms located in the interior of the islands have four Si neighbors (Si-$Al_0Si_4$) and generate no acid sites, similar to the absence of acid sites in a pure-silica zeolite. Therefore, to maximize the acidity (both concentration and strength) of SAPO materials, it is desirable to introduce as much silicon into the structure as possible to form a large number of small Si islands having Si sites along the edges and corners, but not too much silicon where most of the Si ends up in the center of the islands to give nonacidic Si-$Al_0Si_4$ sites. The difficulty in doing this arises from the fact that it is not easy to regulate the silicon incorporation at near neutral pH conditions. Much work has been done to find the best way to introduce silicon into the framework of AlPO materials. One strategy for introducing Si into SAPO materials is the use of fluoride as the mineralizing or complexing agent.

G.H. Kuehl at Mobil was the first to use fluoride in SAPO syntheses as a way to control the silicate concentration to gradually release silicate to the growing crystals as it is used up during crystallization [23]. Using ammonium fluoride, he was able to prepare a SAPO-20 (**SOD**) material with high silicon substitution into the framework, which was confirmed by $^{29}$Si NMR.

## 2.4
## Dry Gel Conversion Syntheses

Conventional zeolite syntheses are performed from a reaction medium containing an aqueous phase of dissolved reagents and solid species, which may exist as gel particles and be suspended in solution. Recently, new methods of zeolite syntheses have been studied in which the solid species are kept separated from the aqueous phase such that the solids never come in direct contact with water. These methods are generally referred to as dry gel conversion (DGC) syntheses [24] which include the vapor phase transport [25] (VPT) and steam-assisted conversion (SAC) [26] methods. Xu *et al.* [27] were the first to make ZSM-5 from a dry gel by contact with vapors of water and volatile amines. In the VPT method, amorphous powders were obtained by drying sodium aluminosilicate gels of various compositions. They were transformed into ZSM-5 by contact with vapors of amines and water, or water alone. Amines were found to participate in the crystallization process by absorbing into the reacting, hydrous phase and elevating the pH. Whereas in the VPT method the SDA is physically separate from the aluminosilicate gel, in the SAC method the gel is prepared containing the SDA before the drying step. The dry gel is crushed and then suspended in a specially designed autoclave above a small reservoir of water. Using this method, Rao *et al.* [28] were able to make self-bonded pellets of high silica beta (Si/Al up to 365) after only 12 hours of crystallization. The SAC method was recently used to make pure-silica MCM-68 (**MSE**) [29]. After heating for five days at 150 °C, a dry gel containing the MCM-68 SDA was transformed into a siliceous product, designated YNU-2P. The XRD pattern of this material showed it to be MCM-68, but it was found to lose crystallinity upon calcination. Rietveld refinement of the powder diffraction data showed that some of the tetrahedral framework sites were not fully occupied resulting in defects, which was confirmed by $^{29}$Si MAS NMR. After a simple postsynthesis silylation step using tetramethyl orthosilicate and HCl, the new material, now designated YNU-2, was found to have significantly fewer defect sites. The repaired material was now stable to calcination to remove the organic SDA.

## 2.5
## Low Water Syntheses

Although the use of fluoride to prepare silica-**MFI** has been known for some time [11], it was not until years later that its use was exploited to its full potential after an important discovery was made at the laboratories of Prof. Corma at the Polytechnic University of Valencia. By combining the use of HF with highly concentrated gels, Prof. Corma and his colleagues were able to prepare a number of low-framework density zeolites in their purely siliceous form. In this technique, the gel was not separated from the aqueous phase as in the DGC techniques. Conventional aluminosilicate zeolite syntheses use water levels having the ratios of $H_2O : SiO_2 > 25$ to allow easy homogenization and handling of the synthesis gels. By using very low levels of water, that is, $H_2O : SiO_2 < 10$, in the presence of fluoride it was found that siliceous zeolite beta (**BEA**) could be prepared without the use of seeds [30]. Using the synthesis method described in the original patent, zeolite beta could only be prepared with $Si/Al = 5-100$ [31]. It was found by $^{29}Si$ MAS NMR that siliceous beta had no connectivity defects. Until this time, low-framework density silica zeolites (<15 T-atoms per cubic angstrom) were known only in materials having high aluminum framework content, such as zeolites chabazite, faujasite, and zeolite A. A few years after the low-water, fluoride synthesis of beta was disclosed, the synthesis of pure-silica chabazite was published [32]. Unlike silica beta, silica chabazite was found to contain a small number of $Si(OSi)_3OH$ defects. A review by Camblor *et al.* describes the syntheses of a number of all-silica and high-silica zeolites from fluoride-containing, low-water syntheses [33]. The effect of water level on the synthesis was found to have a profound effect on the product that was formed. With the same gel, merely varying the water level can produce a different zeolite. The general trend that lower $H_2O/SiO_2$ ratios in the synthesis gel form zeolites with lower framework densities was observed. Using the trimethyladamantammonium SDA in the presence of HF, low-framework-density CHA was made at $H_2O/SiO_2 = 3$, medium-density SSZ-23 (**SST**) was made at $H_2O/SiO_2 = 7.5$, and high-density SSZ-31 (***STO**) [34] made at $H_2O/SiO_2 = 15$ (Figure 2.1).

Zones *et al.* at Chevron Research Laboratories have also used this approach to prepare many other frameworks with all-silica compositions in the presence of fluoride [35, 36]. Besides the usefulness of this new synthesis technique to prepare the all-silica form of known aluminosilicate and borosilicate zeolites, it has also led to a number of *new* zeolite frameworks which are tabulated in Table 2.1.

Zeolite A (**LTA**) was first synthesized over 50 years ago, and remained until recent times the only aluminosilicate that contained D4Rs in its structure. While there are a number of phosphate-containing frameworks, such as **ACO**, **AFY**, **AST**, **-CLO**, and **DFO**, that have D4Rs in their structure [45], for a considerable period of time, no new

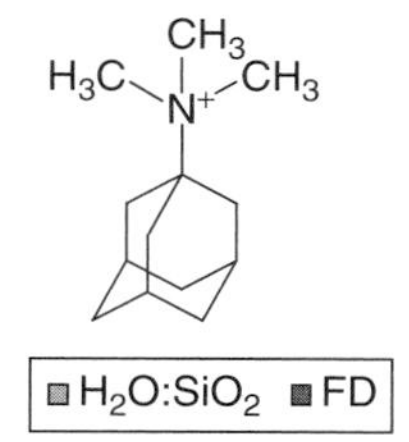


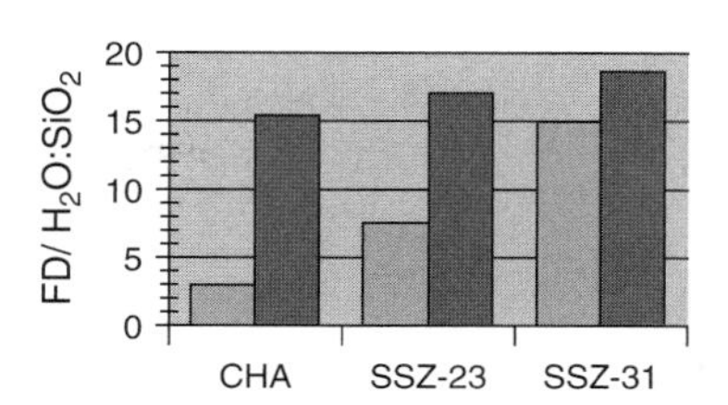


**Figure 2.1** The effect of water level on the framework density (FD) of the zeolite formed [33]. FD units are in T-atoms per cubic angstrom and the water level is the $H_2O/SiO_2$ ratio. The structure of the trimethyladamantammonium template is shown above the chart legend.

**Table 2.1** New silica frameworks discovered from the fluoride syntheses route.

| Zeolite | Framework code | Framework density | Synthesis $H_2O/SiO_2$ | References |
|---|---|---|---|---|
| Octadecasil | **AST** | 16.7 | 8.3 | [37] |
| ITQ-3 | **ITE** | 16.3 | 7.7 | [38] |
| ITQ-4 | **IFR** | 17 | 15 | [39] |
| ITQ-7 | **ISV** | 15.4 | 5.4 | [40] |
| ITQ-12 | **ITW** | 16.3 | 7 | [41] |
| ITQ-13 | **ITH** | 17.8 | 7 | [42] |
| ITQ-27 | **ITV** | 15.7 | 3 | [43] |
| SSZ-74 | **-SVR** | 17.1 | 3–7 | [44] |

silica or aluminosilicate frameworks containing D4Rs were made. The discoveries of octadecasil (**AST**), ITQ-7, ITQ-12, and ITQ-13 zeolites represent the first examples of high-silica-containing frameworks with D4Rs. The synthesis of all these materials was made possible by the use of fluoride as discussed above. Subsequent characterization studies by X-ray diffraction and MAS NMR suggest that the fluoride anion exerts a structure-directing effect toward frameworks containing D4Rs and other small-cage secondary building units (SBUs) containing a high density of 4-member rings (4MRs) [33]. In some materials, it was found that fluoride strongly interacts with framework silica to give pentacoordinate $[SiO_4F]^-$ species [46]. In D4R frameworks, the fluoride anion was often found at the center of the D4R cube.

## 2.6 Germanium Zeolites

A new synthesis strategy, first utilized by Yaghi to synthesize ASU-7 [47] and then exploited by researchers at Stockholm University, Polytechnic University of Valencia, and University de Haute Alsace, utilizes germanium to promote D4R-containing structures. As discussed above, the addition of fluoride to a synthesis gives it the ability to crystallize high silica frameworks and frameworks containing D4Rs. Analysis of pure-silica zeolite frameworks have found the most common Si–O–Si bond angle value to be 148°, slightly less than the mean value of 154 ± 9° [48]. In silica frameworks, the D4R interatomic bond angles are inclined to be smaller because of the 90° Si–Si–Si angle of the D4R cube. The presence of germanium in zeolite syntheses increases the likelihood of forming D4R-containing frameworks due to the smaller Ge–O–Ge angle (130°) compared to the larger Si–O–Si angle. Theoretical energy calculations have shown that the substitution of Ge for Si in the D4R stabilizes this SBU [49, 50]. Using this new strategy, Prof. Corma's laboratory and other groups have synthesized a large number of new frameworks, many having desirable multidimensional, extra-large pores. These new frameworks are summarized in Table 2.2. Although some have argued that the D4R in all-silica frameworks are strained, force-field calculations have shown that the strain is minor [51], suggesting the possibility that these materials may be synthesized without germanium or fluoride.

Germanium is recovered as by-product from zinc ores and is very expensive. Its worldwide production was about 100 t in 2007 [66]. Its cost has varied between US$400 and US$1400 $kg^{-1}$ in the past decade, making its use in a petrochemical catalyst highly unlikely. Not only is germanium prohibitively expensive, its presence in a framework reduces the zeolite's stability. Corma's laboratory has been successful, in some cases, in removing Ge from the synthesis by seeding. Germanium-free ITQ-24 has been prepared by seeding with Ge-containing ITQ-24 crystals in a boron-containing gel [67], but the final products always contain a

**Table 2.2** New D4R zeolite frameworks discovered with the use of germanium.

| Zeolite | Framework code | Framework density | Channels | Reference |
|---|---|---|---|---|
| ASU-7 | **ASV** | 17.9 | 1D 12 ring | [47] |
| ITQ-17/FOS-5 | **BEC** | 13.9 | 3D 12 ring | [52, 53] |
| ITQ-15/ITQ-25/IM-12 | **UTL** | 15.5 | 14 × 12 ring | [54–56] |
| ITQ-21 | – | 13.6 | 3D 12 ring | [57] |
| ITQ-22 | **IWW** | 16.6 | 12 × 10 × 10 ring | [58] |
| ITQ-24 | **IWR** | 15.6 | 12 × 10 × 10 ring | [59] |
| ITQ-26 | **IWS** | 14.4 | 3-D 12 ring | [60] |
| ITQ-33 | – | 12.9 | 18 × 10 × 10 ring | [61] |
| ITQ-34 | **ITR** | 17.4 | 10 × 10 × 9 ring | [62] |
| SU-15 | **SOF** | 16.4 | 10 × 9 × 9 ring | [63] |
| SU-32 | **STW** | 15.2 | 2D 8 ring | [63] |
| IM-10 | **UOZ** | 16.7 | 6 ring | [64] |
| IM-16 | **UOS** | 17.9 | 1D 10 ring | [65] |

significant amount of germanium from the seeds. Because many of these new D4R frameworks have desirable multidimensional, large, and extra-large pore structures that do not exist in other frameworks, the challenge is to synthesize them without germanium.

One approach to synthesizing Ge-free D4R structures may be to choose a suitable template. Zeolite A is a D4R-containing structure synthesized at low Si/Al ratios. By using supramolecular self-assembled molecules as SDAs, a pure-silica **LTA** material, designated ITQ-29, was prepared [68]. It was found that $\pi-\pi$-type interactions cause the assembly of two julolidine-derived cations with proper rigidity and polarity properties to template the **LTA** framework. The key to the success of this synthesis was also the realization that the LTA framework has *sod* cages [45], which could be templated by the addition of a second SDA, $TMA^+$.

## 2.7 Isomorphous Substitution

Researchers have also investigated the substitution of other metals, monovalent, divalent, and trivalent ones, for trivalent aluminum. A number of elements in the periodic table have been claimed to substitute, at least in small amounts, for aluminum in many *known* zeolite frameworks, particularly **MFI**. While a detailed discussion of these materials is beyond the scope of this chapter, the use of isomorphous substitution of aluminum to discover *new* frameworks is briefly discussed. Elements other than aluminum (Al–O bond length is about 1.74 Å) have different tetrahedral atom–oxygen bond lengths, T–O–T angles, and/or charges, which could stabilize other SBUs and induce the formation of new structures.

Gallium is right below aluminum in the periodic table and easily substitutes for it in most of the zeolite structures. The notable exception is zeolite A, likely because it has D4Rs, which may be too strained with the larger Ga–O bond length (about 1.82 Å). A study by Newsam and Vaughan [69], on the other hand, showed a general decrease in T–O–T bond angles with gallium substitution in a number of zeolite frameworks, which would likely stabilize the D4R as germanium does. Two new frameworks have been discovered with the use of gallium, **CGS** and **ETR**. The gallosilicate TsG-1 (**CGS**), having no aluminosilicate analog, was synthesized in 1985 by Krutskaya *et al.* from potassium gallosilicate gels [70]. In 1999, it was determined that TsG-1 had the same structure as CoGaPhosphate-6, whose structure was determined the year before [71]. A second gallosilicate zeolite, ECR-34 (**ETR**) having no aluminosilicate analog was discovered by Strohmaier and Vaughan [72]. Its synthesis requires three cations, $Na^+$, $K^+$, and $TEA^+$. ECR-34 and ITQ-33 [61] are the only silicate zeolites to have 18-ring pore openings. It is interesting to note that both the **CGS** and **ETR** frameworks can be built from the same SBU and that this SBU is not seen in any other aluminosilicate frameworks, indicating a structure-directing effect of gallium and potassium for this particular SBU (Figure 2.2).

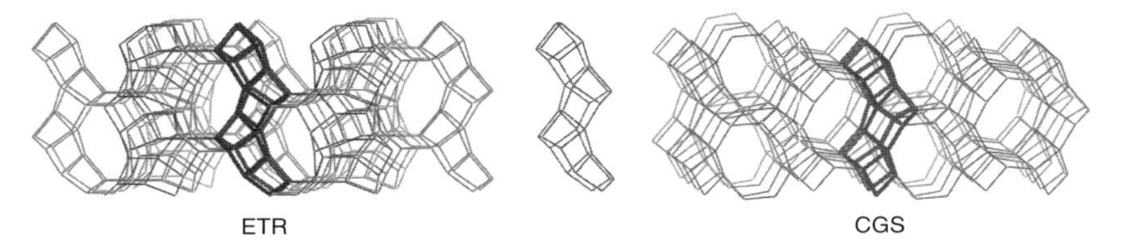


**Figure 2.2** Chain of connected open hexagonal prisms – the secondary building unit found only in the gallosilicate ETR and CGS frameworks.

Boron is known to substitute for aluminum in zeolites to form borosilicates (sometimes referred to as *boralites*). The boron–oxygen tetrahedral bond length (about 1.46 Å) is significantly shorter than the aluminum–oxygen bond length (about 1.74 Å). Also, the B–O–Si bond angle is found to be, on average, smaller than that of either Si–O–Si or Al–O–Si bond lengths in tetrahedral framework materials [73]. It would therefore be expected that boron substitution for aluminum would induce the formation of new zeolite structures. While early work did not show any significant substitution of boron into zeolite frameworks [74], later studies showed that it could be inserted into high-silica frameworks such as **MFI**, ***BEA**, **MWW** (as ERB-1) **MTW**, and others [75]. Zeolite Nu-1, synthesized by Imperial Chemical Industries (ICI) in 1977 [76], was first made as an aluminosilicate, but it was found that it could be made more reproducibly by using boron in the synthesis [77]. Its structure (**RUT**) was solved later by Gies and Rius [78]. Zones *et al.* prepared an aluminosilicate, designated SSZ-26, using the hexamethyl[4.3.3.0] propellane-8,11-diammonium SDA [79]. Later they used the *N,N,N*-trimethyl-8-ammonium tricylclo[5.2.1.0]decane SDA to prepare a borosilicate called *SSZ-33*, having a similar powder XRD pattern to that of SSZ-26 [80]. Sometime later, Lobo and Davis used a third template, *N,N,N*-trimethyl *cis*-myrtanyl ammonium, to prepare the borosilicate CIT-1 [81]. CIT-1 was found to have a 12 × 12 × 10 ring pore structure of the pure end-member polymorph B (**CON**), while the SSZ-26 and SSZ-33 structures, were found to be intergrowths of two closely related polymorphs, A and B [82]. Using boron instead of aluminum, scientists from Chevron, Ruhr University, ExxonMobil, and Stockholm University have discovered a number of new frameworks in the past 10 years. These new borosilicate zeolites are listed in Table 2.3. The synthesis of a new borosilicate, SSZ-82, was recently published, but its framework has not yet been determined [83]. Millini *et al.* reviewed the synthesis and characterization of borosilicate zeolites in 1999 [75].

Besides the trivalent boron and gallium analogs of aluminosilicates, divalent zinc, beryllium, and monovalent lithium are known to substitute into siliceous frameworks. Zinc was found to direct the synthesis of three new silicates, VPI-8 (**VET**) [93], VPI-9 (**VNI**) [94], and RUB-17 (**RSN**) [95], all three zeolites being prepared with the $TEA^+$ SDA. The Zn–O bond length is very long (about 1.94 Å) and, while zinc is claimed to promote the formation of low density, 3R (three-ring)-containing framework structures, only two of the three zincosilicate frameworks (**VNI, RSN**) have 3Rs. Like zinc, beryllium is a divalent cation capable of tetrahedral coordination

**Table 2.3** Borosilicate zeolites having new frameworks.

| Code | Type material | Framework composition | Reference |
|---|---|---|---|
| CON | CIT-1 | $B_2Si_{54}O_{112}$ | [81] |
| RTH | RUB-13 | $B_2Si_{30}O_{64}$ | [84] |
| RUT[a] | RUB-10 | $B_4Si_{32}O_{72}$ | [78] |
| [b] | MCM-70 | $B_{0.6}Si_{11.4}O_{24}$ | [85] |
| SFE | SSZ-48 | $Si_{13.8}B_{0.2}O_{28}$ | [86] |
| SFG | SSZ-58 | $Si_{72}B_2O_{148}$ | [87] |
| SFH | SSZ-53 | $Si_{62.4}B_{1.6}O_{128}$ | [88] |
| SFN | SSZ-59 | $Si_{15.6}B_{0.4}O_{32}$ | [88] |
| SFS | SSZ-56 | $Si_{54.7}B_{1.3}O_{112}$ | [89] |
| SOS | SU-16 | $B_8Ge_{16}O_{148}$ | [90] |
| SSF | SSZ-65 | $B_{1.5}Si_{52.5}O_{108}$ | [91] |
| SSY | SSZ-60 | $Si_{27}BO_{56}$ | [92] |

[a]The RUT framework was first discovered as the aluminosilicate, NU-1.
[b]Note – The structure of MCM-70 has been determined, but the framework has not yet been approved by the Inter. Zeolite Assoc. Structure Commission.

and there are a number of beryllium-containing natural zeolites such as lovdarite and nabesite that have 3R in their structures. Two 3R-containing zeolites, OSB-1 (**OSO**) and OSB-2 (**OBW**) [96], have been prepared with beryllium. The ionic radius of lithium (0.73 Å) is about the same as $Zn^{2+}$, but its substitution in a silicate would generate an even higher charge than divalent tetrahedral atoms. Nonetheless, two lithosilicates, RUB-23 [97] and RUB-29 [98], have been synthesized to form tetrahedral framework structures with lithium-containing strain-free 3Rs. No lithosilicate structure has yet been approved by the International Zeolite Association Structure Commission.

## 2.8 Structure-Directing Agents

As mentioned before, the use of organic SDAs is one of the primary approaches to discovering new zeolites. It is therefore important to understand how to design new SDAs that will give new porous frameworks. The size, geometry, rigidity, and hydrophobicity are very important features of SDAs that determine their ability to form zeolite structures. SDAs that are very small and spherical in shape tend to form small cages and give nonporous clathrasil products, while larger ones tend to form porous, large-pore zeolites. Linear molecules such as long diquaternary ammonium cations typically give one-dimensional medium pore zeolites such as EU-1 (**EUO**) and ZSM-48 (***MRE**). Even though they are large, these linear molecules are quite flexible and tend to give one-dimensional zeolites. More rigid templates have been found to be an important factor in forming large-pore three-dimensional frameworks such as CIT-1 (**CON**) [99] and MCM-68 (**MSE**) [100]. As large- and extra-large-pore zeolites are desirable, they can be made with even larger SDAs. As the size of the organic molecule increases, its hydrophobicity also increases, which limits its solubility in aqueous medium and also its ability to form solvated cations. The interaction of solvated cations and condensing silicate species in solution has a profound effect on the crystalline product that can form and is the basis for the structure-directing effect of organic cations. Zones *et al.* [101] have discussed the effect of the carbon to nitrogen ratio ($C/N^+$) of organic molecules on their effectiveness to crystallize porous zeolites. They concluded that $C/N^+$ ratios between 11 and 15 were optimal for the formation of high-silica molecular sieves. Molecules that are moderately hydrophobic give the best match between the SDA and the silica precursors to induce the formation of zeolitic-building units. When the SDA has excessive hydrophilicity, the silicate species are not strong enough to interrupt the extensive hydration sphere around the hydrophilic SDA [102]. To study the effect of the hydrophobic/hydrophilic character of a number of SDAs, Zones *et al.* have determined the ability of charged organic molecules to partition between an aqueous solution and an organic chloroform phase. The molecules that partition well between both aqueous and organic phases had the right balance between hydrophobicity and hydrophilicity to function as good SDAs [101]. By surveying the relationship of SDA characteristics and the type of silicates that they can form, Lobo *et al.* [73] have summarized the important trends of high silica syntheses (Table 2.4).

The simple organic molecules such as $TMA^+$, $TEA^+$, $TPA^+$, and $TBA^+$ (tetrabutylammonium) were the first to be thoroughly investigated as promoters to synthesize new zeolite structures. These molecules were readily available commercial reagents. Later on, simple linear diquaternary cations such as hexamethonium (HM) (*N,N,N,N′,N′,N′*-hexamethylhexane-1,6-diammonium) were investigated and

**Table 2.4** Trends between SDAs and zeolite structure types [73].

| SDA type | Zeolite type |
|---|---|
| None | Dense phases |
| Small globular molecules | Clathrasils |
| Excess alkali metal cations | Layered structures, mordenite |
| Linear molecules | One-dimensional medium-pore zeolites |
| Branched molecules | Three-dimensional medium-pore zeolites |
| Large polycyclic molecules | Three-dimensional large-pore zeolites |
| Large globular molecules | One-dimensional large-pore zeolites |
| Large globular molecules + Al | Three-dimensional large-pore zeolites |
| Large globular molecules + B | Three-dimensional large-pore zeolites |
| Large globular molecules + Zn | VPI-8 a one-dimensional large-pore zeolite |

were also found to be excellent SDAs. To design and make more complex SDAs, early researchers began to make their own organic molecules by taking readily available primary, secondary, and tertiary amines and performing a simple alkylation to quaternize them. For example, by alkylating substituted piperidines and adamantanes, Nakagawa *et al.* [103] have discovered a large number of new zeolites such as SSZ-23, SSZ-35, SSZ-39, and SSZ-44. To make more complex linear diquaternary molecules, workers began to substitute pyrrolidine and piperidine on the end of the linear diamines instead of methyl, ethyl, and so on, to discover some new structures and compositions such as IM-5, TNU-9, and TNU-10, as discussed above. In addition to pyrrolidine and piperidine, Jackowski *et al.* [10] also investigated the use of *n*-methyl homopiperidine, *m*-methyl tropane, and quinuclidine as end-group substituents on linear diamines. With the addition of HF as mentioned above (Table 2.1), they were able to synthesize zeolite SSZ-74, a new two-dimensional medium pore zeolite, using the hexamethylene-1,6-bis(*N*-methyl-*N*-pyrrolidinium) cation as the SDA. Alkylation of substituted imidazoles has also been investigated and has been found to give mainly one-dimensional zeolites such as **MTW**, **TON**, and **MTT** [104]. Later, substituted imidazoles were found to make three new zeolites, ITQ-12 (**ITW**) in the presence of HF [41], SSZ-70 (structure unknown) in the presence of boron [105], and IM-16 (**UOS**) in the presence of germanium [65]. The CoAPO, SIZ-7 [106] (**SIV**), was also prepared with an imidazolium template.

In an effort to discover new zeolites, more complex SDAs were designed to prepare organic molecules more diverse than those obtainable by the simple alkylation of readily available amines. By using Diels–Alder chemistry, Zones *et al.* [107] prepared a family of tricyclodecane derivatives to make **MOR**, **MTW**, **CON**, SSZ-31, and SSZ-37 zeolites. Calabro *et al.* [100] used the Diels–Alder reaction to prepare the *N,N,N′,N′*-tetraalkyl-*exo,exo*-bicyclo[2.2.2]oct-7-ene-2,3:5,6-dipyrrolidinium cation, which was used to prepare MCM-68 (**MSE**), the first zeolite to have an intersecting $12 \times 10 \times 10$ ring channel system [108]. Using the Beckmann rearrangement reaction, Lee *et al.* [109] made a family of polycyclic SDAs to make a number of zeolites including SSZ-50, which has the **RTH** framework. Other multistep organic syntheses have been used, which include reduction of alkyl nitriles to make the SSZ-53 (**SFH**) and SSZ-55 templates [110], catalytic hydrogenation of substituted quinoline to give the bicyclic SDA for making SSZ-56 [111] (**SFS**), reductive amination of ketones to give the SSZ-57 [112] and SSZ-58 (**SFG**) templates [87], and the amination of an acyl halide to make the SSZ-59 (**SFN**) template [88] and SSZ-65 (**SSF**) template [91]. Both SSZ-53 and SSZ-59 are notable because their structures contain extra-large 14-ring pores. While these complex organic molecules are very proficient in making new and interesting zeolite frameworks, their cost may be initially too expensive for them to be used as commercial catalysts. The challenge remains to find ways to make these zeolites with cheaper SDAs.

Besides using organic tetraalkylammonium cations as SDAs, other types of molecules have been used to make porous zeolites. Small amines have been found to be useful templates for many zeolite syntheses, although they tend to be nonspecific for a given framework [113]. Cyclic ethers have also been employed by a number of researchers to make zeolites. Dioxane was used by De Witte *et al.* [114] to prepare **MAZ** zeolite and trioxane was used to prepare ECR-1 (**EON**) [115]. Larger cyclic crown ethers, 15-crown-5 and 18-crown-6, which are able to efficiently coordinate to alkali metal cations, have been used to prepare both high silica **FAU** and EMC-2 (**EMT**) zeolites, respectively [116]. Like the complex tetraalkylammonium SDAs described above, these crown ethers are expensive to use as templates and other routes to high-silica **FAU** and **EMT** are known, which use the more economical TPA [117] and triethylmethylammonium [118] SDAs, respectively. Zeolite UTD-1 was the first 14-ring zeolite to be synthesized and that used the organometallic bis(pentamethylcyclo-pentadienyl) cobalt(III) cation as a template [119]. Tetraalkylphosphonium templates have also been shown to be effective SDAs. Both ITQ-26 (**IWS**) [120] and ITQ-27 (**IWV**) [43] are new multidimensional large-pore zeolites prepared with phosphonium templates. Because both structures contain D4R, they require either fluoride or germanium in their syntheses as discussed above.

## 2.9 SDA Modeling

In addition to the efforts put into the syntheses of organic molecules, many studies have been undertaken to understand the energy of interaction between the SDA molecule and the structure of the zeolite. The reason for these studies is to rationalize which organic molecules promote the formation of a given zeolite framework or, better yet, to predict which SDA makes a hypothetical structure. Using Monte Carlo and energy minimization calculations, Lewis *et al.* have calculated the stability and location of the SDAs in a number of zeolites [121]. The effects of aluminum content on the effect of SDA–zeolite interactions and the SDA–SDA interactions in the zeolite pores have also been studied by Sastre *et al.* [122] using the General Utility Lattice Program (GULP) energy minimization code. They were able to explain the role the SDA has in determining the Si/Al range in the synthesis of some zeolite structures. They also used GULP to determine the effect on SDA–framework interactions by germanium incorporation in the **EUO**, **ITH**, **IWW**, and **IWR** frameworks. All four of these zeolites are made with the HM SDA. They concluded that the low rigidity of the HM and relatively low SDA–framework interaction resulted in the lower selectivity of this template to promote the formation of one structure over another. The zeolite that formed was therefore determined by the effects of germanium and aluminum on the stability of a given framework. The **EUO** framework was the subject of another study to understand the catalytic differences between EU-1 materials made with different templates. By using molecular modeling to dock the HM and dibenzyldimethylammonium (DBDMA) SDAs in the **EUO** framework, they were able to locate the position of these two templates in the pores. By rationalizing that the electrostatic interactions would be most favorable when the positive charge of the SDA was located near the aluminum atoms in the structure, they were then able to determine the location of the active Al sites. With this information, they were able to explain the selectivity of products in the *n*-decane cracking test. The active sites for HM-prepared EU-1 were determined to be in both the 10-member ring channels and the larger side pockets, while the active sites were only located in the side pockets of the DBDMA-prepared sample. ZSM-18 (**MEI**) is a one-dimensional 12-ring zeolite originally prepared with an expensive triquaternary SDA. By using molecular dynamics, Schmitt *et al.* [123] were able to predict potential ZSM-18 SDAs that were easier to synthesize. This was the first time that molecular modeling was successfully used to design a template for a specific framework.

In the previous paragraph, the importance of being able to predict which SDA would promote a known or hypothetical structure was introduced. Hypothetical structures were first generated manually by early zeolite researchers who observed the chains, layers, and polyhedra found in known zeolite structures. By connecting these units in different manners, or by applying symmetry operators, several thousand new structures were generated. Joe Smith from the University of Chicago enumerated and began to keep a database of these structures [124]. Recently, several groups have developed strategies for automating the generation of new structures. With the recent advancement of computational speed, millions of hypothetical structures have been identified. Using a symmetry-constrained intersite bonding search method, Treacy *et al.* [125] have generated over 100 000 plausible zeolite frameworks. Simulated annealing was used by Earl and Deem [126] to generate over 4 million hypothetical structures, of which 450 000 were found to be plausible structures by comparing their lattice energies, calculated by the GULP program, to the energies of known zeolite structures [127]. Other researchers have used the tiling theory to generate new zeolite structures [128] and have systematically enumerated 1761 both simple and quasi-simple uninodal and simple binodal and trinodal frameworks. Of these, 176 were found to be chemically feasible by using calculated lattice energy, framework density, and other structural parameters [129]. These databases can now be searched for structures that could be candidates as useful absorbents for separating molecules or as shape selective catalysts for industrially important chemical processes. In particular, multidimensional, extra-large pore (pore openings with greater than 12 tetrahedral atoms) frameworks can be identified. One way to directly generate these type of structures with defined pore geometry is the method of constrained assembly of atoms [130]. By defining a forbidden zone, where the framework atoms cannot reside, zeolite frameworks can be generated with a given pore size and dimensionality. While these new structures have generated considerable attention, they remain a *Gedanken* experiment until someone discovers a way to synthesize them.

The syntheses of inorganic zeolites have been thoroughly investigated for decades, and it is now rare that a new zeolite is discovered without the use of a SDA. A more recent notable example is MCM-71, a 10-ring zeolite of the mordenite group, obtained in the K–Al–Si system [131]. The ultimate goal is to be able to design an SDA to promote the crystallization of a given hypothetical structure. A close match between the size and shape of the organic molecule and a zeolite pore can help stabilize the growing crystal. As discussed above, molecular modeling may help if a database of organic molecules is available for evaluation of the energy of interaction between the SDA and the zeolite framework. This strategy is limited by the ability to be able to visually propose an SDA, which has a good fit between its energy minimized conformation and the shape of the pores and/or cages of the target framework. A *de novo* strategy has been developed by Lewis *et al.* [132] in which organic molecules are compositionally grown right in the pore to be modeled. Starting from a seed molecule placed inside the targeted zeolite host, organic fragments from a library are subsequently attached to systematically build up the size and shape of the developing template to match the size and shape of the pore. A cost function based on the overlap of van der Waals spheres was used to control the growth of the new template. While the authors were able to predict a suitable SDA for the known LEV structure, it still remains to be seen if this new computer algorithm, called ZEBEDDE, can predict a suitable organic SDA to template one of the many hypothetical structures discussed above.

## 2.10
## Co-templating

The idea of cotemplating has been in existence for some time. The concept involves using two SDAs, each one templating a different SBU in the target structure. This technique was first used by Flanigen *et al.* to synthesize SAPO-37 (**FAU**) [133]. It was found that SAPO-37 could only be prepared in the presence of two SDAs, $TPA^+$ and $TMA^+$. Apparently the large $TPA^+$ cation stabilizes the large **FAU** supercage and the smaller $TMA^+$ cation stabilizes and promotes the formation of the smaller *sod* cages. The $TMA^+$ cation was also used to help stabilize the *sod* cage in the synthesis of ITQ-29 as discussed previously. In a more recent study, Wright *et al.* at the University of St. Andrews used molecular modeling to rationalize the synthesis of two new SAPO materials, STA-7 (**SAV**) and STA-14 (**KFI**) [134]. Both frameworks have two types of cages, which are templated by two different SDAs. In the former case, the SAPO composition of **SAV** was prepared for the first time by using cyclam and $TEA^+$ as SDAs. In the latter case, the **KFI** framework was prepared for the first time with a phosphate composition by using a mixture of azaoxacryptand template and $TEA^+$. The $TEA^+$ cation was found by single crystal structure determination to be located in the smaller *mer* cage and the larger azaoxocrypd and molecule was believed to be in the larger *alpha* cage. The authors of this work suggested that this cotemplating approach may be a way of rationalizing the synthesis of new hypothetical structures that contain more than one cage.

Besides promoting the synthesis of structures having more than one cage, the dual templating approach may also be used to economize the synthesis of a zeolite that uses an expensive template. Zones and Nakagawa developed a synthesis method where a small amount of expensive SDA cation is utilized to nucleate the formation of a given structure and a second inexpensive amine is added to fill up and stabilize the cages or pores of the structure during crystallization. Using this approach, they were able to synthesize a number of zeolites requiring complex templates, such as SSZ-25, SSZ-28, SSZ-32, and SSZ-35, by using a smaller amount of the expensive SDA and by adding the second amine such as isopropyl amine or isobutylamine. The key feature is that the amine is smaller than the organic SDA and the organic SDA is used in an amount less than that required to crystallize the zeolite in the absence of the amine [135]. Later, Zones and Huang found that the mixed template system can accelerate the crystallization rate of zeolite syntheses and also nucleate the synthesis of a new zeolite named *SSZ-47* [136]. A more thorough discussion of cotemplating in zeolites is given in Chapter 13 of this book.

One type of cotemplating strategy was developed by UOP is called charge density mismatch (CDM) [137]. In this technique, a precursor gel is first prepared in which there is a charge density mismatch between the organic cations (low density) and the potential aluminosilicate (high density at low Si/Al ratios) material that is expected to form. One part of this strategy is to use readily available and cheap SDAs to see if they could be induced to make new zeolites and another part of this strategy is to use a mixture of the two SDAs. The mixture is heated for some time at a suitable temperature, but crystallization from this initial CDM solution is difficult or impossible. These solutions were found to be stable at 100–150 °C for a few days to a few weeks. After a given amount of time, a second high charge density cation, such as sodium and/or potassium, or even a small organic SDA such as $TMA^+$, was added, which then allowed the mixture to crystallize into a zeolite. The one new zeolite discovered using the CDM approach using both $TMA^+$ and $TEA^+$ templates was UZM-5 (**UFI**) [138], a new two-dimensional framework with D4R and alpha cages like those in the **LTA** framework. Other materials made with the CDM method so far are compositional variants of known zeolites and it is yet to be shown that this method is a productive approach for discovering new framework materials. However, there have been a number of new compositions of known frameworks that have allowed these materials to be studied as adsorbents. For example, new compositions with the **MEI** and **BPH** frameworks have been found that are stable to calcination (Figure 2.3).

## 2.11 Layered Precursors

Whereas zeolites are three-dimensional crystalline materials containing tetrahedral framework atoms, clays have two-dimensional structures containing both octahedral and tetrahedral coordination. The two-dimensional building units can be regarded as layers that are aligned along the third dimension by ionic or hydrogen bonds. Researchers in the late 1970s realized that the interlayer inorganic cations of

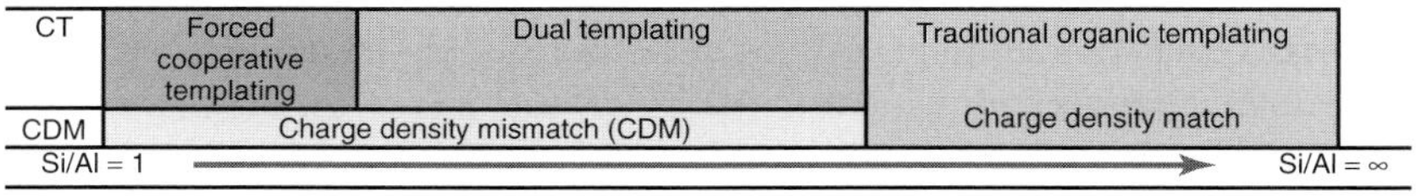

<table>
<tr><td>CT</td><td>Forced cooperative templating</td><td>Dual templating</td><td rowspan="2">Traditional organic templating<br>Charge density match</td></tr>
<tr><td>CDM</td><td colspan="2">Charge density mismatch (CDM)</td></tr>
<tr><td colspan="4">Si/Al = 1 ⟶ Si/Al = ∞</td></tr>
</table>

**Figure 2.3** Defined templating regions for the CDM approach to zeolite synthesis [137].

smectite clays could be exchanged and replaced with larger oligomeric polycations such as the Keggin-like chlorohydrol cation, $[Al_{13}O_4(OH)_{24}(H_2O)_{12}]^{7+}$ [139, 140]. Upon calcination, the cations are converted to alumina pillars that keep the layers separated and provide for an interlayer microporous network. Large polycationic zirconia and titania species were also found to be effective for pillaring clays.

Some zeolite structures can be described as being built from layers of tetrahedral aluminum and silicon oxide. While these layers are fully connected with covalent Si–O and/or Al–O bonds in the third dimension, it has been sometimes found that during its synthesis a precursor is formed where the layers of the zeolite are not connected by covalent bonding, but are connected by hydrogen bonding and/or kept propped open by the organic SDA. This precursor can, in some cases, be isolated as a stable solid, which can, in many cases, irreversibly transform to the fully tetrahedrally connected zeolite upon calcination. The first unequivocally identified zeolite precursor was for the structure **MWW** and was designated MCM-22(P) [141]. This was exploited for the preparation of the first pillared zeolite MCM-36, which is a high-activity micro/mesoporous composite, and which launched the discovery of the extensive family of distinct zeolites based on different packings of MWW layers [142, 143]. By using the fluoride synthesis method and a bulky 4-amino-2,2,6,6-tetramethylpiperidine template, Schreyeck *et al.* [144] were able to prepare PREFER, a layered microporous aluminosilicate with (100) layers of ferrierite. When the layers in PREFER condense to form the **FER** structure, they are aligned in a manner to form a mirror plane parallel between the layers to give *Immm* symmetry. It has been found that other SDAs can force the FER layers to arrange in a different manner so as to form a new framework structure designated **CDO**. MCM-65 was synthesized using both quinuclidinium$^+$ and TMA$^+$ as cotemplates in the presence of sodium [145]. The powder XRD pattern of the as-synthesized material gave a *C*-centered orthorhombic structure. Upon calcination, the XRD pattern showed a reduction of 4.5 Å of the longest axis, indicating that a new structure had formed, possibly by condensation of a layered precursor. The framework of calcined MCM-65 was determined by model building after examination of the electron diffraction patterns. It has *Cmcm* symmetry and is built from the same layers as those found in **FER**, but with no mirror plane parallel to the layers [146] (Figure 2.4). Examination of the powder XRD and electron diffraction patterns of ZSM-52, ZSM-55, both prepared with choline SDA, and MCM-47, prepared with bis(*N*-methylpyrrolidinium)-1,4 butane, indicated that these materials have very similar structures to as-synthesized MCM-65. The structure of as-synthesized MCM-47 was determined by Burton *et al.* [147] from powder XRD data using the FOCUS and ZEFSaII algorithms. The location of the template was found to be located between the FER layers. While MCM-65 can be readily calcined to form a condensed crystalline structure, MCM-47 cannot. Two other materials, CDS-1 [148], prepared with TMA$^+$ and dioxane, and UZM-25 [149], prepared from ethyltrimethylammonium or dimethyldiethylammonium, were also found to have the **CDO** structure.

Other zeolite structures were also found to be formed through layered precursors. Nu-6(2), having the **NSI** framework, was formed from the Nu-6(1) precursor made with the 4,4′-bipyridyne SDA [150]. Another related material, called *EU-20b*, formed

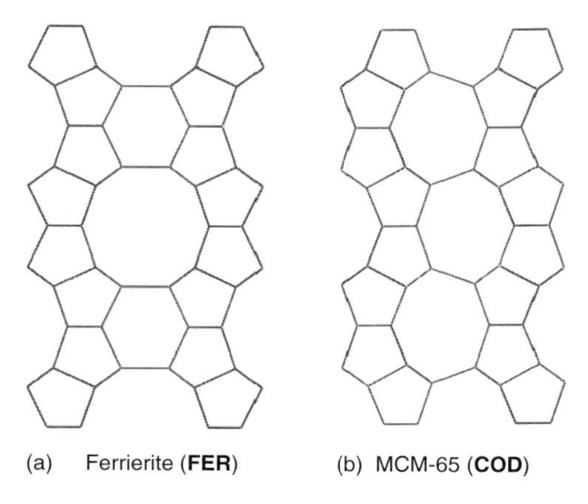


**Figure 2.4** Frameworks of (a) ferrierite and (b) MCM-65, showing the different manner in which the layers are connected.

from calcining its precursor, EU-19 prepared with a piperazinium SDA, was found to form an intergrowth material composed of **NSI** and **CAS** framework layers [151]. Acetic acid treatment was found to decrease the distance between the layers of RUB-15 from 14 to 7.7 Å, allowing the conversion of the material to the pure-silica **SOD** framework upon calcination [152]. The key to the success of the acetic acid treatment was that it allowed the layers to shift one half unit cell along the *c*-axis to align the layers, which does not occur by treatment with hydrochloric acid. The RUB-18 layered material was found to transform into RUB-24 zeolite having the two-dimensional eight-ring **RWR** framework [153]. Although RUB-24 was not fully crystalline and contained some structural defects, its crystallinity was good enough to determine its structure. The two-dimensional zeolite, RUB-41, having a $10 \times 8$ ring **RRO** framework channel system, was formed by calcining the precursor, RUB-39, prepared with the dimethyldipropylammonium template [154]. One of the more important zeolites to be prepared from a layered precursor is MCM-22, which was first prepared in 1984 by BASF as the impure phase, PSH-3 [155]. The conditions for making it without impurities were later found by Mobil Oil Corporation. [156], who designated their material, MCM-22. The structure of the calcined material was determined by Leonowicz *et al.* [157] in 1994, and was found to have the **MWW** framework, which has two-dimensional pores of 10 rings, but with 12-ring pockets on the surface of the crystal. It was found that MCM-22 is a good catalyst for the alkylation reaction of benzene and propylene to make cumene, which was determined to occur on the surface of the crystals in the 12-ring pockets [158].

The realization of the importance of surface chemistry on the MCM-22 catalyst performance and the fact that this and a number of other zeolites are formed from layered precursors led to the development of new synthesis approaches aimed at maximizing the surface chemistry of these materials. By increasing the size of the pore opening to the site, or by exposing the active sites that were originally inside the pores to the surface of the crystal, larger molecules could gain access to the catalytically active sites. One approach was to first swell and then pillar the layers in a manner similar to that used to make pillared clays. Kresge *et al.* [159] were able to swell the MCM-22 layered precursor, MCM-22(P), by heating with a solution of cetyltrimethylammonium and TPAOH. The swollen material, MCM-36, could then be pillared with the D4R TMA silicate or tetraethylorthosilicate (TEOS) to give a porous material after calcination, having greatly increased surface area and activity [160]. Later, in a similar manner, Chica *et al.* [161] were able to swell and pillar the PREFER material to give ITQ-36, a pillared ferrierite. A new method of postalkoxysilyation on zeolitic lamellar precursors was recently developed by Wu *et al.* [162]. By treatment with diethoxydimethylsilane, they were able to directly expand the pore apertures between the layers of **MWW**, **FER**, **CDO**, and MCM-47 structures under low pH conditions without first swelling the layers. The expanded structures containing Si–O–Si linkages between the layers were stable to calcination or reflux in water. These new materials, designated IEZs (interlayer expanded zeolites), showed catalytic activities indicative of structures having larger pores than those obtained by direct calcination of their nonsilylated precursors.

A second approach for maximizing the surface area of zeolites formed from layered precursors is by delamination of the layers to give very thin sheets of the crystals. By first using the method of Kresge *et al.* [159], researchers from Polytechnic University of Valencia, PQ Corporation and Shell International Chemicals, B. V., swelled the **MWW** layers of an MCM-22 material and then delaminated them by a simple one-hour ultrasonic treatment [163]. The delamination was confirmed by TEM imaging, nitrogen absorption, and by the increased presence of surface hydroxyl groups compared to the MCM-22 zeolite as determined by IR. This new material, called *ITQ-2*, was found to have greatly increased the cracking activity for diisopropylbenzene and vacuum gas oil. Using the same technique, Corma *et al.* [164] were able to obtain similar results with PREFER. This material, called *ITQ-6*, showed diisopropylbenzene cracking activity four times that of conventional ferrierite. The titanium form of ITQ-6 was found to be more active than either TiFER or Ti-Beta for the epoxidation of 1-hexene with hydrogen peroxide. Likewise, NU-6(2) was delaminated to give MCM-39-Si [165] and ITQ-18 [166]. Delamination of NU-6(2) was confirmed by nitrogen absorption, $^{29}$Si MAS NMR, and IR spectroscopy. Yet another material, ITQ-20 [167], was prepared by the delamination of its layered precursor, PREITQ-19. Zeolite PREITQ-19 apparently gives the **CDO** structure upon calcination. By modification of the synthesis of MCM-22, a new material called *MCM-56* was made [168]. The structure of this zeolite is believed to be similar to MCM-22, but with layers that are not connected in a regular manner [143]. Evaluation of MCM-56 showed surface area and catalytic activity intermediate between ITQ-2 and MCM-22.

A summary of these layered precursors and their related materials are given in Table 2.5. The continuing discovery of new layered zeolite precursors and especially of the long-known structures like FER and SOD led to a postulate of a new fundamental principle applicable to many, if not all, microporous structures, that is, that they can be generated by two pathways: a direct assembly in 3D or, indirectly, via the 2D layered precursor route [142].

**Table 2.5** Zeolites formed from layered precursors.

| Material | Description | Reference |
|---|---|---|
| Layered precursors | | |
| PREFER | Ferrierite (**FER**) precursor | [144] |
| MCM-22(P) | MCM-22 (**MWW**) precursor | [156] |
| SSZ-25 as syn | SSZ-25 calcined (**MWW**) precursor | [169] |
| ERB-1 as syn | ERB-1 calcined (**MWW**) precursor | [170] |
| NU-6(1) | NU-6(2) (**NSI**) precursor | [150] |
| EU-19 | EU-20b (**CAS-NSI** intergrowth) precursor | [151] |
| RUB-15 | Silica **SOD** precursor | [152] |
| RUB-18 | RUB-24 (**RWR**) precursor | [153] |
| RUB-39 | RUB-41 (**RRO**) precursor | [154] |
| PREITQ-19 | ITQ-19 (**CDO**) precursor | [171] |
| MCM-65 as syn | MCM-65 calcined (**CDO**) precursor | [145] |
| PLS-1 | CDS-1 (**CDO**) precursor | [148] |
| UZM-13 | UZM-25 (**CDO**) precursor | [149] |
| UZM-17 | UZM-25 (**CDO**) precursor | [149] |
| UZM-19 | UZM-25 (**CDO**) precursor | [149] |
| HLS | Ga-**SOD** precursor | [172] |
| [F, Tet-A]-AlPO-1 | AlPO-41 (**AFO**) precursor | [173] |
| SAPO-34 prephase | SAPO-34 precursor | [174] |
| Delaminated zeolites | | |
| ITQ-2 | Delaminated **MWW** | [163] |
| MCM-56 | Disordered/delaminated **MWW** | [175] |
| ITQ-6 | Delaminated **FER** | [164] |
| ITQ-18 | Delaminated **NSI** | [166] |
| ITQ-20 | Delaminated **CDO** | [167] |
| Pillared/expanded zeolites | | |
| MCM-36 | Pillared **MWW** | [159] |
| ITQ-36 | Pillared **FER** | [161] |
| IEZ | Expanded **MWW**, **FER**, **CDO**, and **MCM-47** | [162] |
| Others | | |
| PSH-3 | Impure **MWW** | [155] |
| ZSM-52 | **CDO** precursor that does not fully condense | [176] |
| ZSM-55 | **CDO** precursor that does not fully condense | [177] |
| MCM-47 | **CDO** precursor that does not fully condense | [147] |
| MCM-49 | **MWW** directly synthesized | [178] |

## 2.12 Nonaqueous Solvents

Water is the solvent of choice for synthesizing zeolites because of its ability to solubilize the components needed to form porous crystalline materials. Other solvents, such as alcohols, amines, ammonia, and ionic liquids have been used to make zeolites and MeAlPOs with limited success. Bibby and Dale [179] were the first to report the synthesis of zeolite in an organic solvent in 1985. Using ethylene glycol or propanol, they were able to synthesize silica-sodalite for the first time by using TMA as SDA. Later, the pentasils silicalite (**MFI**), ZSM-39 (**MTN**), and ZSM-48 (*__MRE__) were made using ethylene glycol, butanol, and glycerol as solvents [180]. Kuperman *et al.* [181] made giant crystals of **MTN**, **FER**, and **MFI** zeolites using pyridine/HF and triethylamine solvent systems. Using ethylene glycol as a solvent, Huo and Xu were able to prepare the **AFI**, **AEL**, and **AWO** AlPO materials [182]. A few years later, Huo *et al.* prepared a new material, designated JDF-20, with glycol solvents [183]. The structure of this material was found to have 20 T-atom pore openings like those of cloverite (**-CLO**), and an interrupted framework with an Al : P ratio of 5 : 6. These and other nonaqueous zeolite syntheses were reviewed by Morris and Weigel in 1997 [184]. Ethylene glycol was found to be a particularly useful solvent for making new gallium phosphate materials such as the **CGF** and **CGS** frameworks [71].

Researchers at Mobil used a two-liquid (aqueous and immiscible organic) reaction system to slowly introduce Si to the growing crystals of SAPO materials [185]. Phase-transfer of the organic silicon reagent, tetraethylorthosilicate, from the organic hexanol phase and its hydrolysis at the interphase with the aqueous phase are suggested as critical steps in the substitution of silicon into the framework of the SAPO MCM-9 (**VFI**) [186]. More recent work has shown the advantage of using a soluble organic cosolvent over the insoluble hexanol reagent. Highly active SAPO-11 (**AEL**) material, named *ECR-42*, could be prepared using ethanol as a cosolvent with an organic source of silicon, such as tetramethylorthosilicate or TEOS. It was proposed that a monomeric source of silica is the optimum way to introduce silicon into the framework of AlPO materials [187]. Increased isomerization rates of *n*-decane and cracking rates of hexane and hexene along with $^{29}$Si NMR characterization confirmed the increased activity and better Si distribution within the crystals of ECR-42 over conventionally prepared SAPO-11 [188].

Because liquid and aqueous ammonia are known to solubilize alkali metal cations, its use in zeolite syntheses were investigated by Vaughan and Strohmaier [189]. While no crystalline materials were formed in pure ammonia, aqueous ammonia yielded high-silica cancrinite (**CAN**) with Si/Al ratios up to 1.96 using preformed silica–alumina gels with sodium hydroxide. The high-silica-containing framework was confirmed by $^{29}$Si MAS NMR. Lithium cations gave **ABW**, **EDI**, and **PHI/MER** products, while potassium gave **EDI**, **GIS**, and **PHI/MER** materials. More open framework materials were not observed. Later, they were able to prepare a nanocrystalline **LTL** ($<$200 Å) with aqueous ammonia using potassium hydroxide and a silica–alumina cracking catalyst as a starting reagent [190]. A short time later, Garces *et al.*, from Dow Chemical Co., used liquid ammonia ($NH_3/H_2O = 9$) to prepare Na-P1(**GIS**), dodecasil 3C (**MTN**), octadecasil (**AST**), and DSM-8, a pure-silica beta (*__BEA__) [191].

More recently, ionic liquids and eutectic mixtures have been investigated by the University of St. Andrews as alternate solvents for the synthesis of AlPO zeolite analogs [192]. Ionic liquids are salts that are fluid at ambient temperatures and contain ionic species, while eutectic mixtures are higher melting organic salts to which other compounds are added to suppress their melting point temperature. Common ionic liquids are dialkylimidazolium salts, and a mixture of choline chloride with urea is an example of a eutectic mixture. Ionic liquids have low vapor pressures and the advantage that, at common crystallization temperatures (100–200 °C), the reaction can be performed in open vessels as compared to the sealed autoclaves normally employed in the hydrothermal syntheses of zeolites and AlPOs. It was found that ionic liquids not only solubilize the starting reagents but can also act as a template. Using these new solvent systems, the group at St. Andrews prepared three known AlPO frameworks, SIZ-3 (**AEL**), SIZ-4 (**CHA**), and SIZ-5 (**AFO**). In the cobalt AlPO system, they were able to prepare two known frameworks, SIZ-8 (**AEI**) and SIZ-9 (**SOD**), and one new framework, SIZ-7 (**SIV**) [106]. The structure of SIZ-7 is similar to that of the related three-dimensional, eight-ring frameworks **PHI**, **MER**, and **GIS**. Ionothermal synthesis, or the use of ionic liquids and both solvents and templates, for the synthesis of zeolite analogs and metal-organic frameworks (MOFs) has been recently reviewed by Parnham and Morris [193]; a more detailed description of the approach is discussed in Chater 6 of this book.

## 2.13
## Summary and Outlook

The primary approaches for discovering new zeolite materials have been discussed. While the earlier synthesis work focused on the effects of alkali metals, later work shifted to utilizing organic SDAs. Later, new framework T-atoms, such as B, Ga, Ge, P, Zn, and Be, were evaluated and were found to be successful in finding many new zeolite frameworks. While many of the these new frameworks have unique pore structures, especially the germanium-containing ITQ materials, their compositions are such that they are expensive and not as thermally stable as conventional aluminosilicates. This may limit their use as catalysts; therefore, a method has to be found to synthesize these new and interesting structures as aluminosilicate compositions. The realization that some zeolites form layered precursors has been exploited and new, active analogs have been prepared by pillaring, expansion, and delamination techniques. Furthermore, it may be a general phenomenon that many, if not all, frameworks can be assembled by two pathways: direct 3D growth or via a layered precursor. Several such pairs are known and are being discovered. In recent years, the effects of using fluoride as a mineralizer, low water ratios, and new solvent systems, such as ionic liquids, have been investigated. These techniques have generated a number of new materials, especially new high-silica compositions of known zeolites and new low-density frameworks with multidimensional pores. Much research has also been carried out in supramolecular templating to prepare zeolites with meso and macropores; however, these researches have not been useful for discovering new frameworks and hence have not been discussed. Stirring and seeding effects are also important variables in crystallization, having a large effect on purity, crystal size, and morphology, but they are generally used to optimize the synthesis of a zeolite after it has been discovered.

There are a large number of variables to explore in a zeolite synthesis program and many experiments (typically thousands) have to be performed before a new zeolite is discovered. In the past 10 years, new, automated, high-throughput techniques, which allow many more experiments to be performed in a given amount of time, have been disclosed. Researchers at Sintef [194], Purdue University [195], Max Planck Institute [196], Jilin University [197], Polytechnic University of Valencia [198], and others have all described high-throughput equipment (HTE) for synthesizing microporous materials. Although many groups now have HTE, the rate of zeolite discovery has not increased significantly in the past 10 years.

The discovery of new zeolites and porous materials has always been initiated by new compositional ideas and synthesis approaches. There will definitely be new ideas to exploit in the future, which will lead to new frameworks and new compositional and structural variants of known materials. High-throughput techniques will allow these new ideas to be exploited at a much faster rate. As we continue to more thoroughly understand how zeolites crystallize, characterize their structures, and find new methods to modify their properties, we will continue to increase their utility and find new applications for these fascinating materials.

## Acknowledgments

The help of Wieslaw J. Roth in preparing this manuscript is greatly appreciated. The support of ExxonMobil Research and Engineering Company, Corporate Strategic Research Laboratories is also acknowledged.

## References

1. Barrer, R.M. (1948) *J. Chem. Soc.*, 2158.
2. Reed, T.B. and Breck, D.W. (1956) *J. Am. Chem. Soc.*, **78**, 5972–5977.
3. Milton, R.M. (1959) US Patent 2,882,244.
4. Breck, D.W. (1974) US Patent 3,130,007.
5. Barrer, R.M. and Denny, P.J. (1961) *J. Chem. Soc.*, 971–982.
6. Wilson, S.T., Lok, B.M., Messina, C.A., Cannan, T.R., and Flanigen, E.M. (1982) *J. Am. Chem. Soc.*, **104**, 1146–1147.
7. Vaughan, D.E.W. (1999) US Patent 5,976,491.
8. Afeworki, M., Dorset, D.L., Kennedy, G.J., and Strohmaier, K.G. (2004) *Stud. Surf. Sci. Catal.*, **154**, 1274–1281.
9. Hong, S.B. (2008) *Catal. Surv. Asia*, **12**, 131–144.
10. Jackowski, A., Zones, S.I., Hwang, S.-J., and Burton, A.W. (2009) *J. Am. Chem. Soc.*, **131**, 1092–1100.
11. Flanigen, E.M. and Patton, R.L. (1978) US Patent 4,073,865.
12. Guth, J.L., Kessler, H., Higel, J.M., Lamblin, J.M., Patarin, J., Seive, A., Chezeau, J.M., and Wey, R. (1989) in *Zeolite Synthesis*, ACS Sympoiusm Series, Vol. 398 (eds M. Occelli and H. Robson), Oxford Univ. Press, New York, p. 176.

13. Estermann, M., McCusker, L.B., Baelocher, Ch., Merrouche, A., and Kessler, H. (1991) *Nature*, **353**, 320–323.
14. Merrouche, A., Patarin, J., Kessler, H., and Anglerot, D. (1992) European Patent Application 0 497 698.
15. Simmen, A., Patarin, J., and Baerlocher, Ch. (1993) in *Proceedings of 9th International Zeolite Conference* (eds R. von Ballmoos, J.B. Higgins, and M.M.J. Treacy), Butterworth-Heinemann, Boston, pp. 443–440.
16. Loiseau, T., Retoux, R., Lacorre, P., and Férey, G. (1994) *J. Solid State Chem.*, **111**, 427–436.
17. Bonhomme, F., Thomas, S.G., Rodriguez, M.A., and Nenoff, T. (2001) *Microporous Mesoporous Mater.*, **47**, 185–194.
18. Guth, F. (1989) PhD thesis, *Synthesis and Characterizations of Crystallized Microporous Solids Containing AL, P and Sl*, University de Haute Alsace, Mulhouse.
19. Halvorsen, E.N. (1996) PhD thesis, *Synthesis and Characterization of Aluminophosphate Molecular Sieves*, University of Oslo, Norway.
20. Akporiaye, D.E., Fjellvåg, H., Halvorsen, E.H., Haug, T., Karlsson, A., and Lillerud, K.P. (1996) *Chem. Commun.*, 1553.
21. Akporiaye, D.E., Fjellvåg, H., Halvorsen, E.H., Hustveit, J., Karlsson, A., and Lillerud, K.P. (1996) *Chem. Commun.*, 601.
22. Barthomeuf, D. (1994) *Zeolites*, **14**, 394–401.
23. Keuhl, G.H. (1988) US Patent 4,786,487.
24. Matsukata, M., Nishiyama, N., and Ueyama, K. (1993) *Microporous Mesoporous Mater.*, **1**, 219.
25. Kim, M.H., Li, H.X., and Davis, M.E. (1993) *Microporous Mesoporous Mater.*, **1**, 191.
26. Matsukata, M., Ogura, M., Osaki, T., Rao, P.R.H.P., Nomura, M., and Kikuchi, E. (1999) *Top. Catal*, **9**, 77.
27. Xu, W., Dong, J., Li, J., Li, W., and Wu, F. (1990) *J. Chem. Soc., Chem. Commun.*, 755.
28. Rao, P.R.H.P., Ueyama, K., and Matsukata, M. (1998) *Appl. Catal. A: Gen.*, **166**, 97–103.
29. Koyama, Y., Ikeda, T., Tatsumi, T., and Kubota, Y. (2008) *Angew. Chem. Int. Ed. Engl.*, **47**, 1042–1046.
30. Camblor, M., Corma, A., and Valencia, S. (1996) *Chem. Commun.*, 2365.
31. Wadlinger, R.L., Kerr, G.T., and Rohrbaugh, W.J. (1967) US Patent 3,308,069.
32. Díaz-Cabañas, M.J., Barrett, P.A., and Camblor, M.A. (1998) *Chem. Commun.*, 1881.
33. Camblor, M.A., Villaescusa, L.A., and Díaz-Cabañas, M.J. (1999) *Top. Catal.*, **9**, 59–76.
34. Lobo, R.F., Tsapatsis, M., Freyhardt, C.C., Chan, I., Chen, C.-Y., Zones, S.I., and Davis, M.E. (1997) *J. Am. Chem. Soc.*, **119**, 3732–3744.
35. Zones, S.I., Hwang, S.-J., Elomari, S., Ogino, I., Davis, M.E., and Burton, A.W. (2005) *C. R. Chim.*, **8**, 267–282.
36. Zones, S.I., Burton, A.W., Lee, G.S., and Olmstead, M.M. (2007) *J. Am. Chem. Soc.*, **129**, 9096–9079.
37. Caullet, P., Guth, J.L., Hazm, J., Lamblin, J.M., and Gies, H. (1991) *Eur. J. Solid State Chem.*, **28**, 345.
38. Camblor, M.A., Corma, A., Lightfoot, P., Villaescusa, L.A., and Wright, P.A. (1997) *Chem. Commun.*, 2659–2661.
39. Barrett, P.A., Camblor, M.A., Corma, A., Jones, R.H., and Villaescusa, L.A. (1997) *Chem. Mater.*, **9**, 1713–1715.
40. Villlaescusa, L.A., Barrett, P.A., and Corma, A. (1999) *Angew. Chem. Int. Ed. Engl.*, **38**, 1997–2000.
41. Barrett, P.A., Boix, T., Puche, M., Olson, D.H., Jordan, E., Koller, H., and Camblor, M.A. (2003) *Chem. Commun.*, 2114–2115.
42. Corma, A., Puche, M., Rey, F., Sankar, G., and Teat, S.J. (2003) *Angew. Chem. Int. Ed. Engl.*, **42**, 1156–1159.
43. Dorset, D.L., Kennedy, G.J., Strohmaier, K.G., DíazCabañas, M.J., Rey, F., and Corma, A. (2006) *J. Am. Chem. Soc.*, **128**, 8862–8867.
44. Zones, S.I., Burton, A.W., and Ong, K. (2007) International Patent WO 2007/079038.

45. Baerlocher, Ch., McCusker, L.B., and Olson, D.H. (2007) Atlas of zeolite framework types, *IZA Structure Commission*, 6th edn, Elsevier, Amsterdam. 16–111
46. Koller, H., Wölker, A., Eckert, H., and Panz, C. (1997) *Angew. Chem. Int. Ed. Engl.*, **36**, 2823.
47. Li, H. and Yaghi, O.M. (1998) *J. Am. Chem. Soc.*, **120**, 10569–10570.
48. Wragg, D.S., Morris, R.E., and Burton, A.W. (2008) *Chem. Mater.*, **20**, 1561.
49. Blasco, T., Corma, A., Díaz-Cabañas, M.J., Rey, F., Vidal-Moya, J.A., and Zicovich-Wilson, C.M. (2002) *J. Phys. Chem. B*, **106**, 2634–2642.
50. Zwijnenburg, M.A., Bromley, S.T., Jansen, J.C., and Maschmeyer, T. (2004) *Microporous Mesoporous Mater.*, **73**, 171.
51. Sastre, G. and Corma, A. (2006) *J. Phys. Chem. B*, **110**, 17949–17959.
52. Corma, A., Navarro, M.T., Rey, F., Rius, J., and Valencia, S. (2001) *Angew. Chem. Int. Ed. Engl.*, **40**, 2277–2280.
53. Conradsson, T., Dadachov, M.S., and Zou, X.D. (2000) *Microporous Mesoporous Mater.*, **41**, 183–191.
54. Corma, A., Díaz-Cabañas, M.J., Rey, F., Nicolopoulus, S., and Boulahya, K. (2004) *Chem. Commun.*, 1356–1357.
55. Corma, A., Díaz-Cabañas, M.J., and Rey, F. International (2005) Patent Application 2005/108526.
56. Paillaud, J.-L., Harbuzaru, B., Patarin, J., and Bats, N. (2004) *Science*, **304**, 990–992.
57. Corma, A., Díaz-Cabañas, M.J., Martinez-Triguero, J., Rey, F., and Ruiz, J. (2002) *Nature*, **418**, 514–517.
58. Corma, A., Rey, F., Valencia, S., Jorda, J.L., and Rius, J. (2003) *Nat. Mater.*, **2**, 493–497.
59. Castaneda, R., Corma, A., Fornes, V., Rey, F., and Ruiz, J. (2003) *J. Am. Chem. Soc.*, **125**, 7820–7821.
60. Dorset, D.L., Strohmaier, K.G., Kliewer, C.E., Corma, A., Díaz-Cabañas, M.J., Rey, F., and Gilmore, C.J. (2008) *Chem. Mater.*, **20**, 5325–5331.
61. Corma, A., Díaz-Cabañas, M.J., Jordá, J.L., Martíez, C., and Moliner, M. (2006) *Nature*, **443**, 842.
62. Corma, A., Díaz-Cabañas, M.J., Jordá, J.L., Rey, F., Sastre, G., and Strohmaier, K.G. (2008) *J. Am. Chem. Soc.*, **130**, 16482–16483.
63. Tang, L., Shi, L., Bonneau, C., Sun, J., Yue, H., Ojuva, A., Lee, B.-L., Kritikos, M., Bell, R.G., Bacsik, Z., Mink, J., and Zou, X. (2008) *Nat. Mater.*, **7**, 381–385.
64. Mathieu, Y., Paillaud, J.-L., Caullet, P., and Bats, N. (2004) *Microporous Mesoporous Mater.*, **75**, 13–22.
65. Lorgouilloux, Y., Dodin, M., Paillaud, J.-L., Caullet, P., Michelin, L., Josien, L., Ersen, O., and Bats, N. (2009) *J. Solid State Chem.*, **182**, 622–629.
66. U.S. Geological Survey (2008) *Germanium-Statistics and Information*, U.S. Geological Survey, Mineral Commodity Summaries.
67. Cantin, A., Corma, A., Diaz-Cabanas, M.J., Jorda, J.L., and Moliner, M. (2006) *J. Am. Chem. Soc.*, **128**, 4216–4217.
68. Corma, A., Rey, F., Rius, J., Sabater, M.J., and Valencia, S. (2004) *Nature*, **431**, 287–290.
69. Newsam, J.M. and Vaughan, D.E.W. (1986) *Stud. Surf. Sci. Catal.*, **28**, 457.
70. Krutskaya, T.M., Kolyshev, A.N., Morozkova, V.E., and Berger, A.S. (1985) *Russ. J. Inorg. Chem.*, **30**, 438–442.
71. Chippendale, A.M. and Cowley, A.R. (1998) *Microporous Mesoporous Mater.*, **21**, 271–279.
72. Strohmaier, K.G. and Vaughan, D.E.W. (2003) *J. Am. Chem. Soc.*, **125**, 16035–16039.
73. Lobo, R.F., Zones, S.I., and Davis, M.E. (1995) *J. Inclusion Phenom. Mol. Recognit. Chem.*, **21**, 47–78.
74. Breck, D.W. (1974) *Zeolite Molecular Sieves*, John Wiley & Sons, Ltd, New York, p. 370.
75. Millini, R., Perego, G., and Bellussi, G. (1999) *Top. Catal*, **9**, 13–34.
76. Whittam, T.V. and Youll, B. (1977) US Patent 4,060,590.
77. Bellussi, G., Millini, R., Catati, A., Maddinelli, G., and Gervasini, A. (1990) *Zeolites*, **10**, 642.
78. Gies, H. and Rius, J. (1995) *Z. Kristallogr.*, **210**, 475–480.

79. Zones, S.I., Santilli, D.S., Ziemer, J.N., Holtermann, D.L., Pecoraro, T.A., and Innes., R.A. (1990) US Patent 4,910,006.
80. Zones, S.I. (1990) US Patent 4,963,337.
81. Lobo, R.F. and Davis, M.E. (1995) *J. Am. Chem. Soc.*, **117**, 3766–3779.
82. Lobo, R.F., Pan, M., Chan, I., Li, H., Medrud, R.C., Zones, S.I., Crozier, P.A., and Davis, M.E. (1993) *Science*, **262**, 1543–1546.
83. Burton, A.W. (2009) US Patent Application 2009/0060813.
84. Vortmann, S., Marler, B., Gies, H., and Daniels, P. (1995) *Microporous Mesoporous Mater.*, **4**, 112–121.
85. Dorset, D.L. and Kennedy, G.J. (2005) *J. Phys. Chem. B.*, **109**, 13891–13898.
86. Wagner, P., Terasaki, O., Ritsch, S., Nery, J.G., Zones, S.I., Davis, M.E., and Hiraga, K. (1999) *J. Phys. Chem. B*, **103**, 8245–8250.
87. Burton, A., Elomari, S., Medrud, R.C., Chan, I.Y., Chen, C.-Y., Bull, L.M., and Vittoratos, E.S. (2003) *J. Am. Chem. Soc.*, **125**, 1633–1642.
88. Burton, A., Elomari, S., Chen, C.-Y., Medrud, R.C., Chan, I.Y., Bull, L.M., Kibby, C., Harris, T.V., Zones, S.I., and Vittoratos, E.S. (2003) *Chem. Eur. J.*, **9**, 5737–3748.
89. Elomari, S., Burton, A., Medrud, R.C., and Grosse-Kunstleve, R. (2009) *Microporous Mesoporous Mater.*, **118**, 325–333.
90. Li, Y. and Zou, X. (2005) *Angew. Chem. Int. Ed. Engl.*, **44**, 2012–2015.
91. Elomari, S., Burton, A.W., Ong, K., Pradhan, A.R., and Chan, I.Y. (2007) *Chem. Mater.*, **19**, 5485–5492.
92. Burton, A. and Elomari, S. (2004) *Chem. Commun.*, 2618–2619.
93. Freyhardt, C.C., Lobo, R.F., Khodabandeh, S., Lewis, J.E. Jr., Tsapatsis, M., Yoshikawa, M., Camblor, M.A., Pan, P., Helmkamp, M.W., Zones, S.I., and Davis, M.E. (1996) *J. Am. Chem. Soc.*, **118**, 7299–7310.
94. McCusker, L.B., Grosse-Kunstleve, R.W., Baerlocher, Ch., Yoshikawa, M., and Davis, M.E. (1996) *Microporous Mesoporous Mater.*, **6**, 295–309.
95. Röhrig, C. and Gies, H. (1995) *Angew. Chem. Int. Ed. Engl.*, **34**, 63–65.
96. Cheetham, A.K., Fjellvåg, H., Gier, T.E., Kongshaug, K.O., Lillerud, K.P., and Stucky, G.D. (2001) *Stud. Surf. Sci. Catal.*, **135**, 158.
97. Park, S.H., Daniels, P., and Gies, H. (2000) *Microporous Mesoporous Mater.*, **37**, 129–143.
98. Park, S.-H., Parise, J.B., Gies, H., Liu, H., Grey, C.P., and Toby, B.H. (2000) *J. Am. Chem. Soc.*, **122**, 11023–11024.
99. Kubota, Y., Helmkamp, M.M., Zones, S.I., and Davis, M.E. (1996) *Microporous Mesoporous Mater.*, **6**, 213–229.
100. Calabro, D.C., Cheng, J.C., Crane, R.A. Jr., Kresge, C.T., Dhingra, S.S., Steckel, M.A., Stern, D.L., and Weston, S.C. (2000) US Patent 6,049,018.
101. Zones, S.I., Nakagawa, Y., and Rosenthal, J.W. (1994) *Zeoraito*, **11**, 81.
102. Goretsky, A.V., Beck, L.W., Zones, S.I., and Davis, M.E. (1999) *Microporous Mesoporous Mater.*, **28**, 387–393.
103. Nakagawa, Y., Lee, G.S., Haris, T.V., Yuen, L.T., and Zones, S.I. (1998) *Microporous Mesoporous Mater.*, **22**, 69–85.
104. Zones, S.I. (1989) *Zeolites*, **9**, 483.
105. Zones, S.I. and Burton, A.W. Jr. (2006) US Patent 7,108,843.
106. Parnham, E.R. and Morris, R.E. (2006) *J. Am. Chem. Soc.*, **128**, 2204–2205.
107. Zones, S.I., Nakagawa, Y., Yuen, L.T., and Haris, T.V. (1996) *J. Am. Chem. Soc.*, **118**, 7558–7567.
108. Dorset, D.L., Weston, S.C., and Dhingra, S.S. (2006) *J. Phys. Chem. B*, 2045–2050.
109. Lee, G.S. and Zones, S.I. (2002) *J. Solid State Chem.*, **167**, 69–85.
110. Elomari, S.A. and Zones, S.I. (2001) *Stud. Surf. Sci. Catal.*, **135**, 479.
111. Elomari, S.A., Burton, A., Medrud, R.C., and Grosse-Kunstleve, R. (2009) *Microporous Mesoporous Mater.*, **118**, 325–333.
112. Elomari, S. (2003) US Patent 6,616,911.
113. Rollman, L.D., Schlenker, J.L., Lawton, J.L., Kennedy, C.L., Kennedy, G.J., and Doren, D.L. (1999) *J. Phys. Chem. B*, **103**, 7175–7183.

114. De Witte, B., Patarin, J., Guth, J.L., and Cholley, T. (1997) *Microporous Mater.*, **10**, 247–257.
115. Keijsper, J.J. and Mackay, M. (1994) US Patent 5,275,799.
116. Delprato, F., Delmottte, L., Guth, J.L., and Huve, L. (1990) *Zeolites*, **10**, 546–552.
117. Vaughan, D.E.W. and Strohmaier, K.G. (1990) US Patent 4,931,267.
118. Vaughan, D.E.W. (1989) US Patent 4,879,103.
119. Freyhardt, C.C., Tsapatsis, M., Lobo, R.F., Balkus, K.J., and Davis, M.E. (1996) *Nature*, **381**, 295–298.
120. Dorset, D.L., Strohmaier, K.G., Kliewer, C.E., Corma, A., Díaz-Cabañas, M.J., Rey, F., and Gilmore, G.J. (2008) *Chem. Mater.*, **20**, 5325–5331.
121. Lewis, D.W., Freeman, C.F., and Catlow, C.R.A. (1995) *J. Phys. Chem. B*, **99**, 11194.
122. Sastre, G., Leiva, S., Sabater, M.J., Gimenez, I., Rey, F., Valencia, S., and Corma, A. (2003) *J. Phys. Chem. B*, **107**, 5432–5440.
123. Schmitt, K.D. and Kennedy, G.J. (1994) *Zeolites*, **14**, 635.
124. Han, S. and Smith, J.V. (1999) *Acta Crystallogr.*, **A55**, 332–382.
125. Treacy, M.M.J., Rivin, I., Balkovski, E., Randall, K.H., and Foster, M.D. (2004) *Microporous Mesoporous Mater.*, **74**, 121–132.
126. Earl, D.J. and Deem, M.W. (2006) *Ind. Eng. Chem. Res.*, **45**, 5449–5454.
127. Deem, M.W. (2008) ICMR Workshop on Design and Synthesis of New Materials, August 1-2, Santa Barbara.
128. Friedrichs, O.D., Dress, A.W.M., Huson, D.H., Klinowski, J., and Mackay, A.L. (1999) *Nature*, **400**, 644.
129. Majda, D., Almeida Paz, F.A., Friedrichs, O.D., Foster, M.D., Simperler, A., Bell, R.G., and Klinowski, J. (2008) *J. Phys. Chem. C*, **112**, 1040–1047.
130. Li, Y., Yu, J., Liu, D., Yan, W., Xu, R., and Xu, Y. (2003) *Chem. Mater.*, **15**, 2780–2785.
131. Dorset, D.L., Roth, W.J., Kennedy, G.J., and Dhingra, S.S. (2008) *Z. Kristallogr.*, **223**, 456–460.
132. Lewis, D.W., Willock, D.J., Catlow, C.R.A., Thomas, J.M., and Hutchings, G.J. (1996) *Nature*, **382**, 604.
133. Lok, B.M., Messina, C.A., Patton, R.L., Gajec, R.T., Cannan, T.R., and Flanigen, E.M. (1984) *J. Am. Chem. Soc.*, **106**, 6092–6093.
134. Castro, M., Garcia, R., Warrender, S.J., Slawin, A.M.Z., Wright, P.A., Cox, P.A., Fecant, A., Mellot-Draznieks, C., and Bats, N. (2007) *Chem. Commun.*, 3470–3472.
135. Zones, S.I. and Nakagawa, Y. (1998) US Patent 5,785,947.
136. Zones, S.I. and Hwang, S.-J. (2002) *Chem. Mater.*, **14**, 313–320.
137. Lewis, G.J., Miller, M.A., Moscoso, J.G., Wilson, B.A., Knight, L.M., and Wilson, S.T. (2004) *Stud. Surf. Sci. Catal.*, **154**, 364–372.
138. Blackwell, C.S., Broach, R.W., Gatter, M.G., Holmgren, J.S., Jan, D.-Y., Lewis, G.J., Mezza, B.J., Mezza, T.M., Miller, M.A., Moscoso, J.G., Patton, R.L., Rohde, L.M., Schoonover, M.W., Sinkler, W., Wilson, B.A., and Wilson, S.T. (2003) *Angew. Chem. Int. Ed. Engl.*, **23**, 1737–1740.
139. Vaughan, D.E.W., Lussier, J., and Magee, J.S. (1979) US Patent 4,176,090.
140. Pinnavaia, T.J. (1983) *Science*, **220**, 365–371.
141. Lawton, S.L., Fung, A.S., Kennedy, G.J., Alemany, L.B., Chang, C.D., Hatzikos, G.H., Lissy, D.N., Rubin, M.K., and Timken, K.C. (1996) *J. Phys. Chem.*, **100**, 3788.
142. Roth, W.J. (2007) *Stud. Surf. Sci. Catal.*, **168**, 221.
143. Roth, W.J. (2005) *Stud. Surf. Sci. Catal.*, **158**, 19–26.
144. Schreyeck, L., Caullet, P., Mougenel, J.C., Guth, J.L., and Marler, B. (1996) *Microporous Mesoporous Mater.*, **6**, 249.
145. Dhingra, S., Kresge, C.T., and Casmer, S.G. (2005) US Patent 6,869,587.
146. Dorset, D.L. and Kennedy, G.J. (2004) *J. Phys. Chem. B*, **108**, 15216–15222.
147. Burton, A., Accardi, R.J., Lobo, R.F., Falconi, M., and Deem, M.W. (2000) *Chem. Mater.*, **12**, 2936.
148. Ikeda, T., Akiyama, Y., Oumi, Y., Kawai, A., and Mizukami, F. (2004)

*Angew. Chem. Int. Ed. Engl.*, **43**, 4892–4896.

149. Lewis, G.J., Knight, L.M., Miller, M.A., and Wilson, S.T. (2005) US Patent Application 2005/0065016.
150. Zanardi, S., Alberti, A., Cruciani, G., Corma, A., Fornés, V., and Brunelli, M. (2004) *Angew. Chem. Int. Ed. Engl.*, **43**, 4933–4937.
151. Marler, B., Camblor, M.A., and Gies, H. (2006) *Microporous Mesoporous Mater.*, **90**, 87–101.
152. Moteki, T., Chaikittisilp, W., Shimojima, A., and Okubo, T. (2008) *J. Am. Chem. Soc.*, **130**, 15780–15781.
153. Marler, B., Ströter, N., and Gies, H. (2005) *Microporous Mesoporous Mater.*, **83**, 201–211.
154. Wang, Y.X., Gies, H., and Lin, J.H. (2007) *Chem. Mater.*, **19**, 4181–4188.
155. Puppe, L. and Weisser, J. (1984) US Patent 4,439,409.
156. Rubin, M.K. and Chu, P. (1990) US Patent 4,954,325.
157. Leonowicz, M.E., Lawton, J.A., Lawton, S.L., and Rubin, M.K. (1994) *Science*, **264**, 1910–1913.
158. Degnan, T.F. Jr. (2003) *J. Catal.*, **216**, 32–46.
159. Kresge, C.T., Roth, W.J., Simmons, K.G., and Vartuli, J.C. (1993) US Patent 5,229,341.
160. Roth, W.J., Kresge, C.T., Vartuli, J.C., Leonowicz, M.E., Fung, S.B., and McCullen, S.B. (1995) *Catal. Microporous Mater. Stud. Surf. Sci. Catal.*, **94**, 301.
161. Chica, A., Corma, A., Fornés, V., and Díaz, U. (2003) US Patent 6,555,090.
162. Wu, P., Ruan, J., Wang, L., Wu, L., Wang, Y., Liu, Y., Fan, W., He, M., Terasaki, O., and Tatsumi, T. (2008) *J. Am. Chem. Soc.*, **130**, 8178–8187.
163. Corma, A., Fornés, V., Pergher, S.B., Maesen, Th.L.M., and Buglass, J.G. (1998) *Nature*, **396**, 353.
164. Corma, A., Diaz, U., Domine, M.E., and Fornés, V. (2000) *Angew. Chem. Int. Ed. Engl.*, **38**, 1499–1501.
165. Kresge, C.T. and Roth, W.J. (1993) US Patent 5,266,541.
166. Corma, A., Fornes, V., and Diaz, U. (2001) *Chem. Commun.*, 2642–2643.
167. Corma, A., Díaz, U., and Fornes, V. (2006) US Patent 7,008,611.
168. Fung, A.S., Lawton, S.L., and Roth, W.J. (1994) US Patent 5,363,697.
169. Zones, S.I., Hwang, S.J., and Davis, M.E. (2001) *Chem. Eur. J.*, **7**, 1990.
170. Millini, R., Perego, G., Parker, W.O. Jr., Bellussi, G., and Carluccio, L. (1995) *Microporous Mesoporous Mater.*, **4**, 221.
171. Corma, A., Díaz, U., and Fornes, V. (2006) US Patent 7,008,651.
172. Kiyozumi, Y., Ikeda, T., Hasegawa, Y., Nagase, T., and Mizukami, F. (2006) *Chem. Lett.*, **35**, 672.
173. Wheatley, P.S. and Morris, R.E. (2006) *J. Mater. Chem.*, **16**, 1035.
174. Vistad, Ø.B., Akporiaye, D.E., and Lillerud, K.P. (2001) *J. Phys. Chem. B*, **105**, 12437–12447.
175. Fung, A.S., Lawton, S.L., and Roth, W.J. (1994) US Patent 5,362,697.
176. Chu, P., Herbst, J.A., Klocke, D.J., and Vartuli, J. (1991) US Patent 4,985,223.
177. Rubin, M.K. (1991) US Patent 5,063,037.
178. Bennett, J.M., Chang, C.D., Lawton, S.L., Leonowicz, M.E., Lissy, D.N., and Rubin, M.K. (1993) US Patent 5,236,575.
179. Bibby, D.M. and Dale, M.P. (1985) *Nature*, **317**, 157–158.
180. Huo, Q., Feng, S., and Xu, R. (1988) *J. Chem. Soc., Chem. Commun.*, 1486–1487.
181. Kuperman, A., Nadimi, S., Oliver, S., Ozin, J.A., Garcés, J.M., and Olken, M.M. (1993) *Nature*, **365**, 239–242.
182. Huo, Q. and Xu, R. (1990) *J. Chem. Soc., Chem. Commun.*, 783.
183. Huo, Q., Xu, R., Li, S., Ma, Z., Thomas, J.M., Jones, R.H., and Chippindale, A.M. (1992) *J. Chem. Soc., Chem. Commun.*, 875.
184. Morris, R.E. and Weigel, S. (1997) *J. Chem. Soc. Rev.*, **26**, 309–317.
185. Derouane, E.G., Valyocsik, E.W., and Von Ballmoos, R. (1990) European Patent 146384.
186. Derouane, E.G., Maistriau, L., Gabelica, Z., Tuel, A., Nagy, J.B., and Von Ballmoos, R. (1989) *Appl. Catal.*, **51**, L13–L20.

187. Strohmaier, K.G. and Vaughan, D.E.W. (2001) US Patent 6,303,534.
188. Strohmaier, K.G., Afeworki, M., Chen, T.J., and Vaughan, D.E.W. (2007) 15th International Zeolite Conference, Bejing, Recent Research Reports. R-09–03.
189. Vaughan, D.E.W. and Strohmaier, K.G. (1992) in *Proceedings of 9th International Zeolite Conference* (eds R. von Ballmoos, J.B. Higgins, M.M.J. Treacy), Butterworth-Heinemann, Boston, pp. 197–206.
190. Vaughan, D.E.W. and Strohmaier, K.G. (1994) US Patent 5,318,766.
191. Garces, J.M., Millar, D.M., and Howard, K.E. (1996) US Patent 5,589,153.
192. Cooper, E.R., Andrews, C.D., Wheatley, P.S., Webb, P.B., Wormald, P., and Morris, R.E. (2004) *Nature*, **430**, 1012–1016.
193. Parnham, E.R. and Morris, R.E. (2007) *Acc. Chem. Res.*, **40**, 1005–1013.
194. Akporiaye, D.E., Dahl, I.M., Karlsson, A., and Wendelbo, R. (1998) *Angew. Chem. Int. Ed. Engl.*, **37**, 609–611.
195. Choi, K., Gardner, D., Hilbrandt, N., and Bein, T. (1999) *Angew. Chem. Int. Ed. Engl.*, **38**, 2891–2894.
196. Klein, J., Lehmann, C.W., Schmidt, H.-W., and Maier, W.F. (1998) *Angew. Chem. Int. Ed. Engl.*, **37**, 3369–3372.
197. Song, Y., Yu, J., Li, G., Li, Y., Wang, Y., and Xu, R. (2002) *Chem. Commun.*, 1720.
198. Moliner, M., Serra, J.M., Corma, A., Argente, E., Valero, S., and Botti, V. (2005) *Microporous Mesoporous Mater.*, **78**, 73–81.

# 3
# Ionothermal Synthesis of Zeolites and Other Porous Materials

*Russell E. Morris*

## 3.1
## Introduction

Innovation in zeolite synthesis remains an important aspect in the search for new framework materials with potential applications. The driver for the search for new zeolites and related solids is not only the need to provide materials for new and emerging applications [1] but is also the desire to understand how these fascinating materials are made, and ultimately how to control their architectures. Given that the applications of zeolites (and other porous solids such as metal organic frameworks) are intimately connected with their architecture, new synthetic methods that aim to understand how their structure can be controlled are very important.

The core strategy that has been exercised over recent years has been the development of new organic compounds that can be used as structure directing agents (SDAs or templates). Simply preparing new SDAs has led to a significant increase in the numbers of zeolite structures over recent years. This is still a method that produces some remarkable new materials, such as IM-12 [2]. However, other more innovative methods have also made their impact, both in terms of finding new ways to recycle templates [3] and in completely new synthesis concepts such as the use of fluoride mineralizers and charge density mismatch solutions [4–7]. The use of fluoride as a mineralizing agent to improve the solubility of the starting reagents and to catalyze the formation of bonds in the target frameworks has been exploited by several groups to produce several new materials over recent years [4–6]. Charge density mismatch solutions, developed by workers at UOP, have also provided routes to new solids [7]. In this process, stable solutions of the inorganic starting materials are prepared by using organic cations that do not make good SDAs because their charge density does not match that of the chemical composition of the inorganic framework which will be formed. Crystallization of the framework is then initiated by addition of another SDA, often in quite small amounts. High-throughput methods have also been applied with some distinct success, particularly by Corma's group in Valencia [8]. In our laboratory, we have pioneered the use of ionic liquids (ILs) as both the solvent and SDA simultaneously [9]. The change from a molecular solvent, such as water or organic molecules, to

*Zeolites and Catalysis, Synthesis, Reactions and Applications. Vol. 1.*
Edited by Jiří Čejka, Avelino Corma, and Stacey Zones

ISBN: 978-3-527-32514-6

an ionic one changes the chemistry of the system markedly. We have given the name ionothermal synthesis to this method to delineate it from hydrothermal or solvothermal synthesis.

Over the last few years, ILs have received great attention in many fields [10]. Most of the studies have focused on the "green" chemistry [11] potential of these compounds, with particular emphasis on the drive to replace organic solvents in homogeneous catalysis [12]. The particular property of ILs that makes them environmentally suitable for these purposes is their low vapor pressure [13], which has significant advantages when replacing highly volatile organic solvents. However, there are many other uses of ILs in diverse areas of technology ranging from electrolytes in batteries and fuel cells [14], as electrodeposition solvents [15] to the use of supported IL as catalysts [16]. In some reactions, the ILs act only as inert solvents and in others the liquid plays a more active role in the reactions that take place.

The broadest definition of an IL is any material in the liquid state that consists predominantly of ionic species [10]. Any ionic salt that can be made molten can therefore be classified as an "IL," always assuming that the ionic components of the solid remain intact on melting. There are many examples in the literature of molten salts being used as the medium in which inorganic materials have been prepared [17]. Usually, these synthetic procedures take place at highly elevated temperatures, producing dense phase solids. For example, alkali metal hydroxide molten salts can be used as the molten phase, often contained in sealed inert (such as silver) vessels in the synthesis of many inorganic solids. In general, such molten salt synthesis methods have been used as direct replacements for traditional solid-state synthesis techniques [17]. However, the modern definition of ILs tends to concentrate on those compounds that are liquid at relatively low temperatures and that contain organic components [18]. Room temperature ionic liquids (RTILs) are, as the name suggests, liquid at room temperature, while near room temperature ionic liquids (nRTILs) are often defined as being liquid below a certain temperature, often 100 °C, although this varies depending on the application envisaged for the liquids. For ionothermal synthesis, *nRTILs* are often defined as being liquid below about 200 °C, the temperatures traditionally used in hydrothermal synthesis [9]. In modern usage, the term *ionic liquid* is almost exclusively reserved for liquids that contain at least one organic ion. The organic components of ILs tend to be large and often quite asymmetric, which contribute to their low melting points by making efficient packing in the solid state more difficult [10].

ILs show a range of properties that make them suitable for use as media for the preparation of inorganic and inorganic–organic hybrid materials. They can be relatively polar solvents, ensuring reasonably good solubility of inorganic precursors [19, 20]. Many (but not all) ILs have good thermal stability, enabling them to be used at elevated temperatures.

Deep eutectic solvents (DESs) are a related class of IL, produced as a mixture of two or more compounds that has a lower melting point than either of its constituents [21]. Eutectic mixtures display unusual solvent properties that are very similar to those shown by the ILs. High solubility can be observed (depending on the eutectic mixture used) for inorganic salts, salts that are sparingly soluble in water, aromatic acids, amino acids, and several metal oxides [22]. Advantages of eutectic mixtures over other ILs are their ease of preparation in pure state and their relative nonreactivity with water. Many are biodegradable and the toxicology of the components maybe well characterized. Eutectic mixtures based on relatively available components such as urea and choline chloride are also far cheaper than some other ILs.

Fundamentally, there is of course no real difference between an IL and a molten salt, except perhaps that the organic nature of the components of an IL introduces much more scope for introducing functionality into the solvents. In the following feature chapter, the focus is on the use of the nRTILs containing organic components as the media for materials synthesis, consistent with modern usage of the terminology. In particular, the focus is on the synthesis of templated crystalline materials such as zeolites and metal organic frameworks where the IL cation acts to direct the structure of the resultant inorganic or inorganic–organic hybrid material.

## 3.2
## Hydrothermal, Solvothermal, and Ionothermal Synthesis

Broadly speaking, the synthesis of crystalline solid-state materials can be split into two main groups; those where the synthesis reaction takes place in the solid state and those where it takes place in solution. The solid-state method usually requires rather high temperatures to overcome difficulties in transporting the reactants to the sites of the reaction. The high temperatures of solid-state reactions also tend to provide routes to the thermodynamically more favored phases in the systems of interest. Typically, this method is used to prepare solid-state oxides.

Transport in the liquid phase is obviously much easier than in solids, and syntheses require much lower temperatures (often less than 200 °C). The most commonly used of this type of preparative technique is hydrothermal synthesis, where the reaction solvent is water [23]. The most common method of accomplishing hydrothermal synthesis is to seal the reactants inside Teflon-lined autoclaves so that there is also significant autogenous hydrothermal pressure produced, often up to 15 bar. The lower temperatures required for hydrothermal synthesis often lead to kinetic control of the products formed, and it is much easier to prepare metastable phases using this approach than it is using traditional solid-state approaches. The important reaction and crystallization processes in hydrothermal synthesis do not necessarily take place in solution (although, of course, they can) but can occur at the surfaces of gels present in the mixtures.

Solvothermal synthetic methods refer to the general class of using a solvent in the synthesis of materials [24]. Of course water is by far the most important solvent, hence the special usage of the term *hydrothermal* to describe its use. However, there are many other possible solvents. Alcohols, hydrocarbons, pyridine, and many other organic solvents have all been used with varying degrees of success [24].

As with water, these molecular solvents produce significant autogenous pressure at elevated temperatures. The solvents used in solvothermal synthesis vary widely in their properties, from nonpolar and hydrophobic to polar and hydrophilic.

The solvents used in hydrothermal and solvothermal synthesis differ fundamentally from ILs in that they are molecular in nature. The ionic nature of ILs imparts particular properties, including low vapor pressures [25] (and very little, if any, autogenous pressure is produced at a high temperature).

## 3.3
## Ionothermal Aluminophosphate Synthesis

Many ILs used currently often have chemical structures that are very similar to the structures of commonly used SDAs (sometimes also known as *templates*) in the hydrothermal synthesis of zeolites and other porous materials [26]. This realization led to the first attempts to prepare zeotype frameworks using ILs as both the solvent and the template provider at the same time. The potential advantage of this approach is that the competition between the solvent and template for interaction with any growing solid is removed when both the solvent and the template are the same species. In principle, this may lead to improved templating of the growing zeolite crystal structure. The first work in this area, published in 2004, used 1-ethyl-3-methyl imidazolium bromide (EMIM Br) and urea/choline chloride DESs to prepare several different materials depending on the conditions [9].

Since the first breakthroughs in this area, there have been many further attempts to prepare zeotype materials. The ionothermal synthesis of aluminophosphate zeolites has been by far the most successful. Many common ILs are suitable solvents for the preparation of these materials, with both known [27–30] and previously unknown [31] structure types, as well as related low-dimensional materials [32, 33] being synthesized successfully. It is interesting to note that more than simply preparing the base aluminophosphate structure, the ionothermal method is also suitable for incorporating the dopant metal atoms that give the frameworks their chemical activity. Silicon (to make so-called SAPOs) [34] and many different tetrahedral metals (Co, Mg, etc.) can all be incorporated into the ionothermally prepared aluminophosphate zeolites, and aspects such as their catalytic activity [35] and the use of additional templates [36] show some very promising results. A discussion of some of the unusual concepts seen in AlPO synthesis is discussed in the remaining sections of this chapter.

Figure 3.1 illustrates several of the SIZ-*n*(ST.Andrews Ionothermal Zeolite) materials that can be prepared from one particular IL–EMIM Br. Several of these materials have known frameworks but several others were previously unknown. The structure of SIZ-1 consists of hexagonal prismatic units known as *double six rings* joined to form layers that are linked into a three-dimensional framework by units containing four tetrahedral centers (two phosphorus and two aluminum) known as *single four rings*. The formula of the material is $Al_8(PO_4)_{10}H_3{\cdot}3C_6H_{11}N_2$ but the Al–O–P alternation is maintained. The framework is therefore interrupted

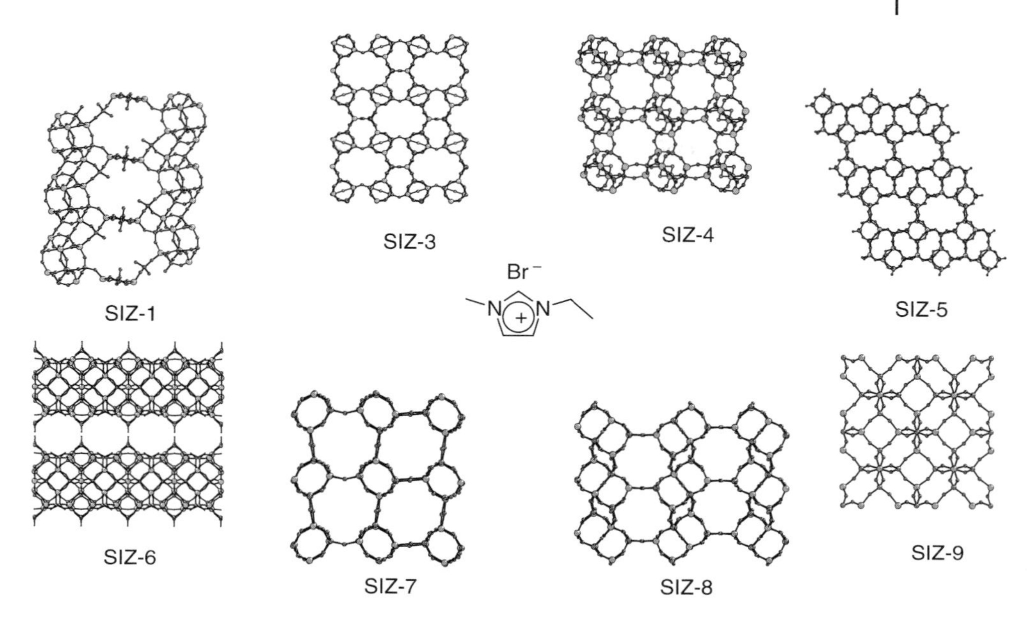


**Figure 3.1** Representatives of the SIZ-*n* series of aluminophosphate zeotype structures prepared using 1-ethyl, 3-methyl imidazolium bromide ionic liquids as both the solvent and structure directing agent.

with some unusual intraframework hydrogen bonding. The negative charge present on the framework (caused by the existence of terminal P–O bonds) balances the charge on the 1-methyl-3-ethyl imidazolium templates that are present in the pores. The overall structure of SIZ-1 shows a two-dimensional channel system parallel to the *a* and *b* crystallographic axes. SIZ-3, SIZ-4, SIZ-5, SIZ-8, and SIZ-9 all have known framework structures (AEL, chabasite (CHA), AFO, AEI, and sodalite (SOD) frameworks respectively). The structure of SIZ-7 is also a novel cobalt aluminophosphate material, given the International Zeolite Association (IZA) code SIV. However, SIZ-7 is a novel framework structure, which joins a family of related zeolites that includes the PHI, GIS, and merlinoite (MER) structure types. This family can be described as consisting of the double-crankshaft chain.

In SIZ-7, these chains run parallel to the crystallographic *a* axis in the structure and are connected to form a one-dimensional small-pore zeolite structure with windows into the pores delineated by rings containing eight tetrahedral atoms (known as *eight-ring windows*). The repeat unit in the *a* direction is 10.2959 (4) Å and equals one repeat unit of the double-crankshaft chain. These chains are linked via four rings in both the *b* and *c* directions to form the eight-ring windows. The relative orientation of neighboring chains means that there are two types of eight-ring channels. The two different windows are of similar size (3.66 × 3.26 Å and 3.40 × 3.52 Å) but are different in shape. In the *b* direction, the same type of eight-ring channel is repeated, leading to a repeat unit in this direction of 14.3715

(5) Å, while in the *c* direction the two types of channel alternate, leading to an approximate doubling of the unit cell dimension in this direction to 28.599 (1) Å.

The overall structure of SIZ-6 is also shown in Figure 3.1. This is a very unusual material comprising 13.5-Å thick anionic aluminophosphate layers of chemical composition $Al_4(OH)(PO_4)_3(HPO_4)(H_2PO_4)^-$. The layers themselves consist of rings containing four, six, and eight nodes (aluminum or phosphorus atoms). The eight-ring windows are large enough to make the layers potentially porous to small molecules. The layers are held together via some relatively strong hydrogen bonding. This occurs because two $H_2PO_4$ groups, one each from two adjacent layers, forming dimeric units with O–O distances across the hydrogen bond of 2.441 Å. In addition, the negative charge on the layers is compensated for by one 1-ethyl-3-methylimidazolium (EMIM) cation, which occupies the interlayer space.

## 3.4
## Ionothermal Synthesis of Silica-Based Zeolites

The ionothermal synthesis of AlPOs is relatively straightforward. Silicon-based zeolites have, however, been much more of a challenge for ionothermal synthesis, although there has been more success in the synthesis of mesostructured silica using ILs [37]. The problem with zeolite synthesis is primarily the solubility of silica starting materials in the commonly used ILs, which is not sufficiently good to allow silicate and aluminosilicate materials to be prepared. Before 2009, there was only one report of a silica polymorph being prepared from an IL [38] and one report of the synthesis of a sodalite [39]. Successful synthesis of zeolites requires the preparation of ILs more suited to silicate dissolution. Recently, in our laboratory, we were successful in preparing ILs comprising mixed halide and hydroxide anions that are suitable solvents for the preparation of purely siliceous and aluminosilicate zeolites. The presence of hydroxide increases the solubility of the silicate starting materials and allows the zeolites to crystallize on a suitable timescale (Figure 3.2, Wheatley and Morris, manuscript in preparation). However, despite this proof of concept work, there is still much to be done to more fully understand the chemistry of silica in ILs, and it is likely that task-specific ILs will need to be developed before silica zeolites can be prepared routinely using ionothermal synthesis.

## 3.5
## Ionothermal Synthesis of Metal Organic Frameworks and Coordination Polymers

Similar to the synthesis of zeolites, ILs can be used as solvents and templates to prepare many other types of solids. One of the most interesting and important class of materials that has been recently developed is that of metal organic frameworks (also known as *coordination polymers*) [40, 41]. These materials offer great promise for many different applications, particularly in gas storage [42–45]. Normally these materials are prepared using solvothermal reactions, with organic solvents such as

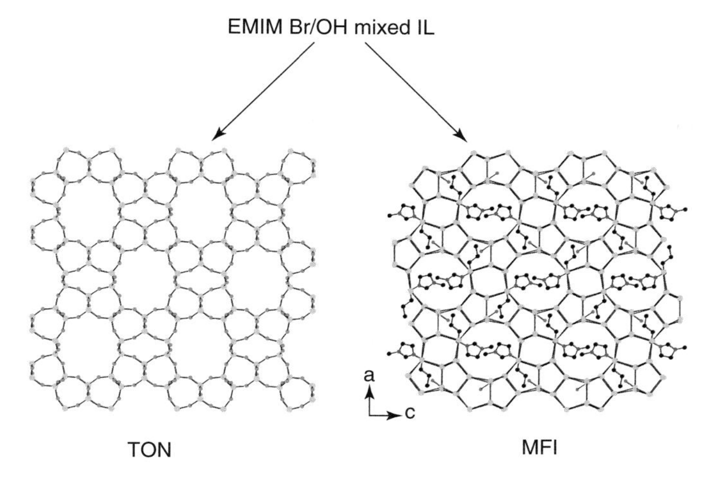


**Figure 3.2** The ionothermal synthesis of pure silica zeolites (TON and MFI) using 1-butyl, 3-methyl imidazolium–based ionic liquids with mixed bromide-hydroxide counteranions. The BMIM cation can be clearly seen from the single-crystal X-ray diffraction structure of MFI.

alcohols and dimethyl formamide. Ionothermal synthesis has been used extensively over the last few years to prepare these types of solid, and there are now many examples in the literature [46–55].

Unlike zeolites, however, the lower thermal stability of coordination polymers leads to several issues regarding removal of ionic templates from the materials to leave porous materials. Often removing the IL cation is not possible without collapsing the structure. However, it is possible to prepare porous materials using DESs, and Bu has recently proven this very elegantly [56].

A great many of the materials prepared ionothermally are relatively low-dimensional solids, and this is clearly a very productive method for the preparation of such materials. It is very clear that in these systems changing the chemistry of the solvent to ionothermal leads to great possibilities in this area.

## 3.6 Ambient Pressure Ionothermal Synthesis

Perhaps the most striking feature of ILs is their very low vapor pressure. This means that, unlike molecular solvents such as water, the ILs can be heated to relatively high temperatures without the production of autogenous pressure. High-temperature reactions therefore do not have to be completed inside pressure vessels such as Teflon-lined steel autoclaves but can be undertaken in simple containers such as round-bottomed flasks. The absence of autogenous pressure at high temperature also makes microwave heating a safer prospect as hot spots in the liquid should not cause excessive increases in pressure with their associated risk of explosion, assuming of course that the IL is stable and does not breakdown into smaller components during heating [57, 58]. Figure 3.3 shows the measured pressure during the synthesis of an aluminophosphate molecular sieve (SIZ-4) using a microwave heating experiment [59]. Figure 3.3a is the pure IL solvent, and it is clear that no autogenous pressure is produced. Figure 3.3b, however, shows that, even when only modest amounts of water are added to the system, significant pressures are evolved.

One of the most interesting potential uses of ambient pressure synthesis of zeolite coatings is for anticorrosion applications. Yushan Yan has shown that ionothermally prepared zeolite films make excellent anticorrosion coatings for several different types of alloys [60, 61]. Given that current coatings technology is based on the use of environmentally unfriendly chromium, there is interest in finding more acceptable alternatives. Sealed zeolites are one such option. However, hydrothermal synthesis of zeolites inside sealed vessels is impractical for

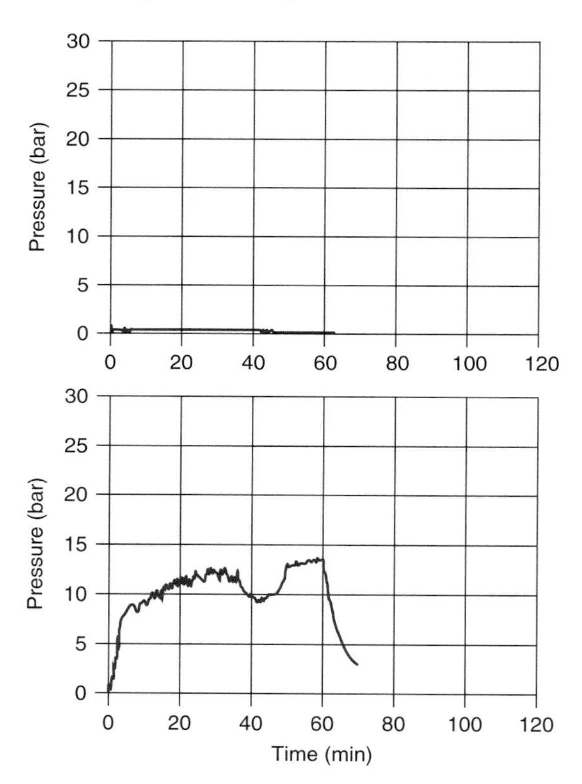


**Figure 3.3** The evolution of pressure (in bar) in the microwave synthesis of aluminophosphate SIZ-4 from (a) a pure ionic liquid solvent with no water added and (b) the same solvent system with 0.018 ml of water added. The maximum temperature is 200 °C and the duration of heating is 60 minutes. There is almost no pressure evolution in the pure ionic liquid.

large, oddly shaped, and cut pieces of metal. Yan contends that ambient pressure ionothermal synthesis eliminates the need for unwieldy sealed vessels, and, given the excellent coatings that can be prepared using this approach, offers an interesting and potentially important alternative technology.

## 3.7 The Role of Cation-Templating, Co-Templating, or No Templating

The original concept behind ionothermal synthesis was to simplify the templating process that occurs in traditional zeolite hydrothermal synthesis by making the solvent and the template the same species. The template molecules normally involved in zeolite synthesis are usually cationic as the resultant framework has a negative charge. The commonly used templating cations are very similar in chemistry to IL cations. It is not surprising therefore that the IL cations are often occluded into the final structures of the materials, in exactly the same way as in traditional zeolite synthesis [9].

In an exactly analogous fashion, metal organic frameworks can also be synthesized using the ILs as both the solvent and the template. Most solvothermally prepared MOFs have neutral frameworks, but when the template is a cation the framework must, for charge balance, have a negatively charged framework, in exactly the same way as zeolites. Of course, the overall goal of all templating-based synthesis is to have control over the architecture of the final material by changing the size of the templating cation. It is well known, however, that apart from rough correlations with the size of the cation, the templating interaction is not really specific enough to yield very precise control over the reaction. Figure 3.4 shows that the same general features hold for ionothermal synthesis. In this work, changing the size of the IL cation does have some effect on the final structure – the larger cations form more open frameworks with the extra space needed to accommodate the large template. However, this is not particularly specific in this type of MOF synthesis, indicating that templating is more likely to be by simple "space filling" rather than any more specific or directed template–framework interactions (Lin and Morris, Unpublished work).

In hydrothermal synthesis, there is also the possibility of adding alternative cations to act as templates. Of course, the situation is exactly analogous in ionothermal synthesis, and added templates offer equally great opportunities. Recently, Xing *et al.* [62] have shown that methylimidazolium (MIA), when added to an EMIM Br IL leads to a cooperative templating effect, occluding both MIA and EMIM in the same solid. The intriguing feature of this solid is that it seems, at least on first inspection, that the material is made of two distinct layers. The MIA is located close to one layer and the EMIM close to the other – perhaps indicating that each cation plays a specific role in directing the structure of each part of the material. It is, of course, impossible to say this for certain until the full mechanism of synthesis is elucidated, something that is very difficult in practice. However, further circumstantial evidence for this maybe the fact that the previously prepared

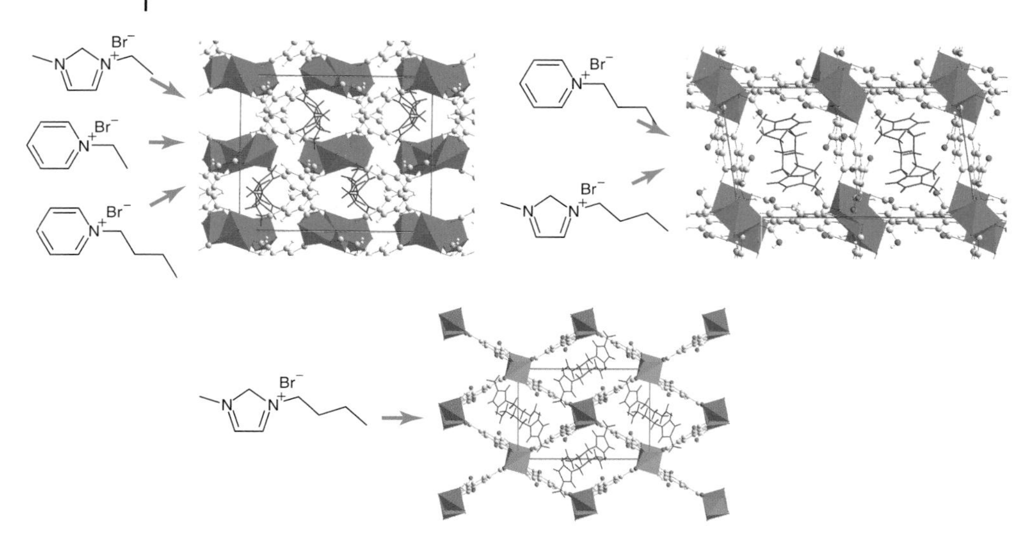


**Figure 3.4** The effect of changing the size of the IL cation on the resulting metal organic framework structure. The materials prepared in this study are Ni (blue) or cobalt (purple) terephthalate MOFs.

EMIM-templated materials SIZ-1 and SIZ-4 have closely related structures to the EMIM-"templated" layer in this material.

Up to now, the cation in the IL has only acted as a template in the synthesis. However, like any other solvent, including water, there is also the possibility of bonding interactions with the frameworks. Most of the ILs that are based on di-alkylated imidazolium cations have no obvious sites through which to coordinate to the metal sites in the way that water does. However, some ILs, under specific conditions, can breakdown to leave the monoalkylated imidazole species that can coordinate to metals [63]. As in hydrothermal synthesis, controlling how the solvent interacts with the framework materials is therefore important in determining the exact nature of the final material. A similar example where the solvent can coordinate to the metal in a metal organic framework comes when using choline chloride/urea-based DES ILs. Normally, this type of solvent is regarded as being relatively unstable, especially the urea portion which can break up and deliver smaller templates into the reaction. However, under conditions where the urea is stable, it is possible to keep this intact, and in the case of ionothermally prepared lanthanum-based MOFs the urea coordinates to the metal [64].

In addition to the templating cations, ILs also contain an anion, and these turn out to be extremely important in controlling the properties of the solvents (Section 3.8). The anions can, in certain circumstances, also be occluded in the structure as a template, most often in combination with the IL cation. Bu *et al.* recently showed that in a series of MOFs (called *ALF-n*) the IL displayed several different types of behavior, including templating by only the cation and templating by both the cation and anion simultaneously, illustrating the multiple functions that ILs can play even in the same systems [56].

Finally, of course, the ILs can only play the role as solvent and not be occluded in the final structure at all. For species like aluminophosphate zeolites and MOFs where the chemistry of the cation is similar to that of commonly used templates one would expect them to be occluded in the final structure. However, there are certain situations where this does not happen. Perhaps the most striking of these is when a very hydrophobic IL is used. In the case of aluminophosphate and MOF synthesis, the more hydrophobic the IL used the less likely the IL cation is to be occluded [32]. Of course, as the chemistry of the system is changed (e.g., by trying to make different types of inorganic material), the balance between the solvent and templating actions of ILs also changes.

## 3.8 The Role of the Anion – Structure Induction

As we have seen above, the common organic cations in ILs are chemically very similar to zeolite templates. However, ILs also contain an anion, and the nature of the anion plays an extremely important part in controlling the nature of the IL. Figure 3.5 demonstrates this dependence of property on anion very clearly. Two low melting ILs that are solid at room temperature can be prepared from the same cation (EMIM) but with two different anions – bromide and triflimide ($NTf_2$). The two ILs have very different properties, especially when it comes to their interaction with water. Figure 3.6 shows what happens when the two compounds are left out in the air for 20 minutes. EMIM $NTf_2$ is a relatively hydrophobic material and there is no change in its properties on exposure to the moisture in the air. EMIM Br, on the other hand, is highly hygroscopic and turns liquid on reaction with moisture in the air.

Clearly, this change in IL chemistry on alteration of the IL anion is bound to have a significant effect on the products of any reaction carried out in such solvents. One example of this is given in Section 3.7, where in the synthesis of aluminophosphates EMIM Br solvents lead to incorporation of the EMIM cation to form zeotype materials, whereas the use of the EMIM $NTf_2$ IL leads to no occlusion of the IL cation [32]. More interesting, however, and potentially extremely useful, is the possibility of mixing the two types of liquid to form solvents with different chemistries from the end member liquids. Figure 3.6 illustrates this for the synthesis of cobalt bezenetricarboxylate MOFs [65]. The two end member ILs, EMIM Br and EMIM $NTf_2$, form two different types of material, while a 50 : 50 mixture of the two ILs, which are miscible, forms a third structure type. This type of result opens up the possibility of mixing ILs to form solvents whose chemistry is different from the end members, giving rise to much more control over the properties of the solvent. In a similar example, a mixed anion IL (50% bromide 50% triflimide) leads to the formation of coordination polymers containing fluorinated

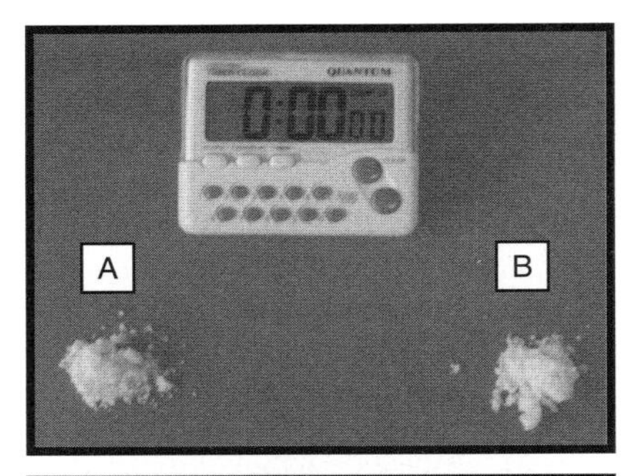

**Figure 3.5** The effect of moist air on hydrophobic EMIM triflimide (sample A) and hydrophilic EMIM Br (sample B). After 20 minutes exposure to normal air at room temperature, the EMIM Br has absorbed enough moisture from the atmosphere to turn from a solid into a liquid.

ligands when ILs containing only one anion (either bromide or triflimide) does not produce any crystalline solid [66].

It is clearly the nature of the anion that determines the final material in these examples. However, the anions themselves are not generally occluded into the structure, and so this is an induction effect rather than a templating of the structure directing effect. It is perhaps not too surprising that changing the chemistry of the solvent will change the type of product in such a manner. In the example illustrated in Figure 3.6, there is no obvious correspondence between the nature of the anions and the nature of the final material. However, in 2007, we published an example of an anion induction using a chiral anion as part of an IL to induce a chiral coordination polymer that contains only achiral building blocks [67] (Figure 3.7). In this example, a chiral IL prepared from the butyl methyl imidazolium (BMIM) cation in combination with L-aspartate as the anion, when used to prepare a cobalt bezenetricarboxylate MOF produced a chiral structure, with all indications that the bulk solid produced was homochiral. Where some specific property of the IL anion manifests itself in the resulting material, despite the fact that it is not actually occluded, the potential for "designer" structure induction becomes very attractive, and one would hope that such properties of ionothermal synthesis will be explored and exploited more thoroughly in the near future.

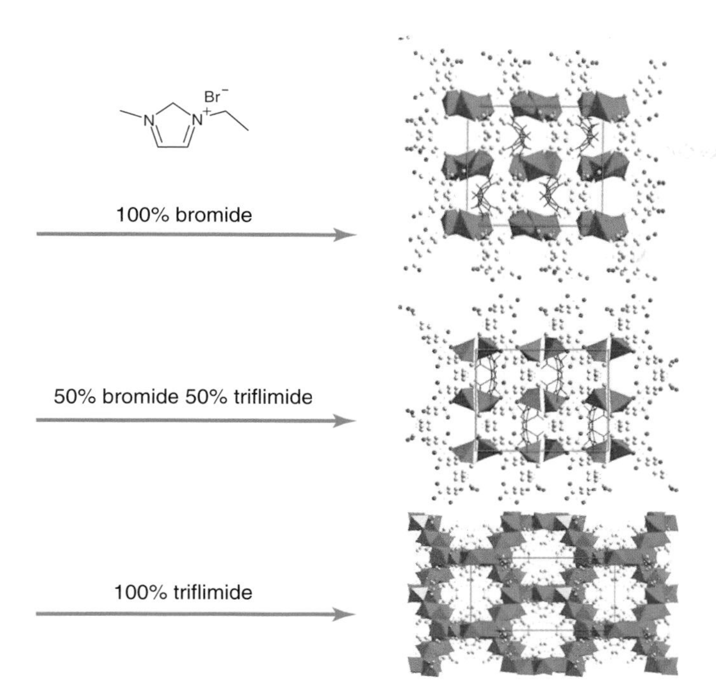


**Figure 3.6** The effect of the anion on the final structure of the material produced in an ionothermal synthesis. The top reaction shows uses EMIM Br as the solvent, and produces one particular cobalt-trimesic acid MOF. A 50 : 50 mixture of EMIM Br and EMIM triflimide produces a different MOF, while using only the EMIM triflimide produces yet another material.

## 3.9
## The Role of Water and Other Mineralizers

One of the very first questions asked about ionothermal synthesis was whether the ILs used were sufficient in their own right to promote the synthesis of zeolites and other inorganic materials, particularly those oxides where water might catalyze the condensation reactions needed to form the required bonds. One of the first things noted about ionothermal synthesis was that too much water was detrimental to the formation of zeolites. At low concentrations of water, zeolites were the main products, but as more water was added to the IL solvents so that they were about equimolar in concentration only dense phases could be prepared. Wragg and coworkers studied this effect in more detail and confirmed through several hundred high-throughput reactions that larger amounts of water did indeed lead to dense phases [63]. The origin of this effect is still under investigation but it is known that the microstructure of water in ILs changes with concentration. At low concentrations, the water is hydrogen bonded relatively strongly to the anion, and exists either as isolated water molecules or as very small clusters [68]. However, as the concentration of water increases, larger clusters and eventually

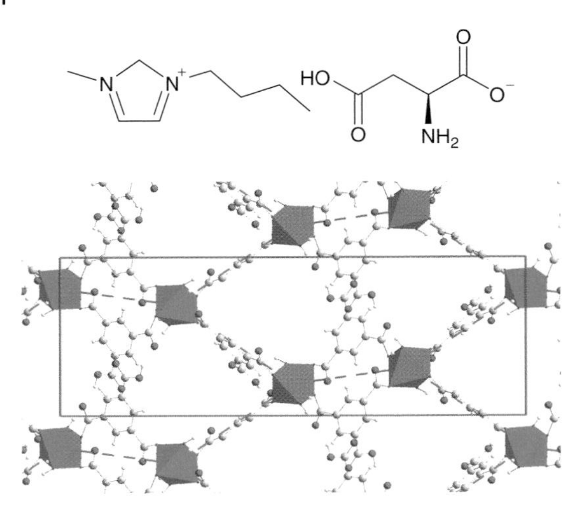


**Figure 3.7** The use of an ionic liquid with a chiral anion induces a chiral MOF structure. Use of an achiral anion produces achiral structures.

hydrogen-bonded networks start to appear, which change the properties of the liquid markedly. Eventually, of course, as more and more water is added, it becomes the dominant chemical component (and therefore the solvent) and the system becomes hydrothermal rather than ionothermal.

The strong binding of isolated water molecules in ILs leads to another interesting effect that can be used in ionothermal synthesis – so-called water deactivation. At low concentrations of water, this strong hydrogen bonding leads to water being less reactive than similar amounts in other solvents. This effect is so strong that highly hydrolytically sensitive compounds such as $PCl_3$ can be stored for relatively long periods, whereas they react quickly, and often violently, in other "wet" solvents [69]. Such water deactivation is probably the reason why some of the materials prepared using ILs can have unusual features. For instance SIZ-13, a cobalt aluminophosphate material, has a layered structure that is closely related to a zeolite, but has Co–Cl bonds. Normally such bonds are hydrolytically unstable and, under hydrothermal conditions, it is unlikely that this material would be stable [27].

In zeolite (and other) synthetic procedures, mineralizers, such as fluoride or hydroxide ions, added to the reaction mixtures in the correct quantities are often vital for crystallization of the desired molecular sieve products. Fluoride in particular has recently been an extremely useful mineralizer for aluminophosphate [70] and silicate [71, 72] synthesis. In addition to helping solubilize the starting materials under the reaction conditions, there is evidence that fluoride itself can play a structure directing role [73] and is intimately involved in template ordering in certain materials [74, 75]. In ionothermal synthesis, the addition of fluoride also seems to be important in determining the phase selectivity of the reaction [9]. It may also help catalyze the bond-forming reactions in zeolite synthesis, as suggested by Camblor and coworkers [71]. For instance, in the synthesis of aluminophosphates, the addition of fluoride leads to the formation of SIZ-4 and SIZ-3, which are both fully four-connected zeolite frameworks, SIZ-1, which is an interrupted structure with some unconnected P–OH bonds.

Tian and coworkers have recently completed an extremely useful kinetic study of the effect of both water and fluoride added to ionothermal systems in zeolite synthesis [76]. It is clear from their results that both small amounts of water, and particularly fluoride, increase the crystallization rate. If the reactions are carried out carefully to exclude as much water as possible, the crystallization of the zeolites becomes very slow indeed, suggesting that for all practical purposes a small reactant amount of water (probably in the IL) is vital if ionothermal synthesis is to be successful.

## 3.10 Unstable Ionic Liquids

In many publications, one often sees the mention of the high thermal and chemical stability of ILs. Bearing in mind of course that it is difficult to generalize across all the possible ILs, this is true under many conditions. However, under ionothermal conditions, some quite common ILs can breakdown. Even some that are often relatively stable such as BMIM bromide can breakdown, especially in the presence of fluoride ions [77]. One possible reaction is the transalkylation reaction that swaps the alkyl groups, leading to the formation of dimethylimidazolium cations, which then templates a zeolite structure [77].

DES ILs based on choline chloride/urea mixtures are also unstable under ionothermal conditions. The urea portion of the IL breaks up to release ammonium ions into the mixture, which then templates the SIZ-2 aluminophosphate material. This type of instability in the ILs is actually extremely repeatable. Deep eutectic ILs made from functionized ureas all break down in the same way to produce the expected functionalized ammonium or diammonium cations that then go on to template many different structures [78]. Such reproduction ability in the reactions of these ILs opens up interesting possibilities for the delivery of small amounts of template to the reaction mixture, as opposed to having the whole IL made up of the template.

## 3.11 Summary and Outlook

Normally, ILs are classed as "green" chemicals because they are most often used to replace volatile organic solvents. However, when preparing the materials, this perspective has been discussed and, in particular, all inorganic framework solids such as zeolites and ILs are more often than not replacing water. In these situations, ionothermal synthesis cannot be called a *green technology* compared to that which it replaces. When replacing organic solvents in, for example, the synthesis of metal organic frameworks there is more justification for using the "green" tag.

However, even in these syntheses, the success of the methodology has to rely on the ILs introducing new chemistry into the system that is not possible using other systems. Fortunately, over recent years, ionothermal synthesis has been recognized as a highly flexible methodology that does indeed bring new chemistry to the system. Features like water deactivation and chiral induction offer many possibilities for the preparation of materials that are unlikely or even impossible to make in other solvents.

One of the most interesting features of ILs for ionothermal synthesis is the sheer number of possible liquids available. There are an estimated 1 million binary ILs available, compared to only a few hundred molecular solvents. The wide range of accessible properties of the liquids provides huge opportunities for matching the chemistry of the solvent system to that of the reactants. However, this also presents huge challenges – it is at the moment extremely difficult to predict *a priori* the properties of the solvent and how they will behave in combination with the reactants. Up to now, only a few of the easily available ILs have been studied, leaving many potentially interesting solvents completely unexplored. One particularly interesting feature of ionothermal synthesis is the use of mixed ILs to tailor the solvent toward a particular reaction chemistry by mixing two different miscible ILs to produce a new solvent with different properties (Section 3.8). Once again the issue of predicting the properties of the mixed ILs is a problem. However, this type of approach is particularly suited to high-throughput methodologies because new solvents can be prepared simply by mixing two ILs in various amounts, and the "brute force" approach afforded by high-throughput instrumentation can at least identify areas of interest in the compositional fields.

The use of ILs in the synthesis of solids has, of course, not been limited to new hybrid and inorganic framework solids. Work in the nanomaterials area and increasingly in other areas, such as the organic solid state, has increased steadily over the last few years. However, there is still much scope to develop the synthesis methodology further.

In the field of zeolite science, the challenges are clear, particularly for the synthesis of silica-based zeolites. Here the plethora of possible ILs is both a blessing and a challenge as we really need to understand more fully the speciation of silicate ions in particular when they are dissolved in ILs. It is clear that the change from molecular to ionic solvents significantly affects the chemistry, and that new zeolite-type structures will inevitably arise from ionothermal preparations. We hope that as we discover ever more about the interesting properties of ILs the field of ionothermal synthesis will develop into an even more useful addition to the armory of synthetic zeolite chemists.

## References

1. Davis, M.E. (2002) *Nature*, **417**, 813–821.
2. Paillaud, J.L., Harbuzaru, B., Patarin, J., and Bats, N. (2004) *Science*, **304**, 990–992.
3. Lee, H., Zones, S.I., and Davis, M.E. (2003) *Nature*, **425**, 385–388.
4. Caullet, P., Paillaud, J.L., Simon-Masseron, A., Soulard, M.,

and Patarin, J. (2005) *C. R. Chim*, **8**, 245–266.
5. Villaescusa, L.A., Lightfoot, P., and Morris, R.E. (2002) *Chem. Commun*, 2220–2221.
6. Villaescusa, L.A., Wheatley, P.S., Bull, I., Lightfoot, P., and Morris, R.E. (2001) *J. Am. Chem. Soc*, **123**, 8797–8805.
7. Blackwell, C.S., Broach, R.W., Gatter, M.G., Holmgren, J.S., Jan, D.Y., Lewis, G.J., Mezza, B.J., Mezza, T.M., Miller, A.M., Moscoso, J.G., Patton, R.L., Rohde, L.M., Schoonover, M.W., Sinkler, W., Wilson, B.A., and Wilson, S.T. (2003) *Angew. Chem*, **42**, 1737–1740.
8. Corma, A., Diaz-Cabanas, M.J., Jorda, J.L., Martinez, C., and Moliner, M. (2006) *Nature*, **443**, 842–845.
9. Cooper, E.R., Andrews, C.D., Wheatley, P.S., Webb, P.B., Wormald, P., and Morris, R.E. (2004) *Nature*, **430**, 1012–1016.
10. Rogers, R.D. and Seddon, K.R. (2003) *Science*, **302**, 792–793.
11. Blanchard, L.A., Hancu, D., Beckman, E.J., and Brennecke, J.F. (1999) *Nature*, **399**, 28–29.
12. Cole-Hamilton, D.J. (2003) *Science*, **299**, 1702–1706.
13. Earle, M.J., Esperanca, J., Gilea, M.A., Lopes, J.N.C., Rebelo, L.P.N., Magee, J.W., Seddon, K.R., and Widegren, J.A. (2006) *Nature*, **439**, 831–834.
14. Chou, S.L., Wang, J.Z., Sun, J.Z., Wexler, D., Forsyth, M., Liu, H.K., MacFarlane, D.R., and Dou, S.X. (2008) *Chem. Mater*, **20**, 7044–7051.
15. Abbott, A.P. and McKenzie, K.J. (2006) *Phys. Chem. Chem. Phys*, **8**, 4265–4279.
16. Miao, W.S. and Chan, T.H. (2006) *Acc. Chem Res*, **39**, 897–908.
17. Mugavero, S.J., Bharathy, M., McAlum, J., and zur Loye, H.C. (2008) *Solid State Sci*, **10**, 370–376.
18. Wasserscheid, P. and Welton, T. (2003) *Ionic Liquids in Synthesis*, Wiley-VCH Verlag GmbH, Weinheim.
19. Reichert, W.M., Holbrey, J.D., Vigour, K.B., Morgan, T.D., Broker, G.A., and Rogers, R.D. (2006) *Chem. Commun*, 4767–4779.
20. Nockemann, P., Thijs, B., Pittois, S., Thoen, J., Glorieux, C., Van Hecke, K., Van Meervelt, L., Kirchner, B., and Binnemans, K. (2006) *J. Phys. Chem. B*, **110**, 20978–20992.
21. Abbott, A.P., Capper, G., Davies, D.L., Rasheed, R.K., and Tambyrajah, V. (2003) *Chem. Commun*, 70–71.
22. Abbott, A.P., Boothby, D., Capper, G., Davies, D.L., and Rasheed, R.K. (2004) *J. Am. Chem. Soc*, **126**, 9142–9147.
23. Cundy, C.S. and Cox, P.A. (2003) *Chem. Rev*, **103**, 663–701.
24. Morris, R.E. and Weigel, S.J. (1997) *Chem. Soc. Rev*, **26**, 309–317.
25. Luo, H.M., Baker, G.A., and Dai, S. (2008) *J. Phys. Chem. B*, **112**, 10077–10081.
26. Lobo, R.F., Zones, S.I., and Davis, M.E. (1995) *J. Mol. Incl. Phen. Mol. Rec. Chem*, **21**, 47.
27. Drylie, E.A., Wragg, D.S., Parnham, E.R., Wheatley, P.S., Slawin, A.M.Z., Warren, J.E., and Morris, R.E. (2007) *Angew. Chem. Int. Ed*, **46**, 7839–7843.
28. Liu, L., Kong, Y., Xu, H., Li, J.P., Dong, J.X., and Lin, Z. (2008) *Microporous Mesoporous Mater*, **115**, 624–628.
29. Han, L.J., Wang, Y.B., Li, C.X., Zhang, S.J., Lu, X.M., and Cao, M.J. (2008) *AIChE*, **54**, 280–288.
30. Hu, Y., Liu, Y.J., Yu, J.Y., Xu, Y.P., Tian, Z.J., and Lin, L.W. (2006) *Chin. J. Inorg. Chem*, **22**, 753–756.
31. Parnham, E.R. and Morris, R.E. (2006) *J. Am. Chem. Soc*, **128**, 2204–2205.
32. Parnham, E.R. and Morris, R.E. (2006) *J. Mater. Chem*, **16**, 3682–3684.
33. Parnham, E.R., Wheatley, P.S., and Morris, R.E. (2006) *Chem. Commun*, 380–382.
34. Xu, Y.P., Tian, Z.J., Xu, Z.S., Wang, B.C., Li, P., Wang, S.J., Hu, Y., Ma, Y.C., Li, K.L., Liu, Y.J., Yu, J.Y., and Lin, L.W. (2005) *Chin. J. Catal*, **26**, 446–448.
35. Wang, L., Xu, Y.P., Wang, B.C., Wang, S.J., Yu, J.Y., Tian, Z.J., and Lin, L.W. (2008) *Chemistry*, **14**, 10551–10555.
36. Wang, L., Xu, Y.P., Wei, Y., Duan, J.C., Chen, A.B., Wang, B.C., Ma, H.J., Tian, Z.J., and Lin, L.W. (2006) *J. Am. Chem. Soc*, **128**, 7432–7433.
37. Wang, T.W., Kaper, H., Antonietti, M., and Smarsly, B. (2007) *Langmuir*, **23**, 1489–1495.
38. Parnham, E.R. and Morris, R.E. (2007) *Acc. Chem. Res*, **40**, 1005–1013.

39. Ma, Y.C., Xu, Y.P., Wang, S.J., Wang, B.C., Tian, Z.J., Yu, J.Y., and Lin, L.W. (2006) *Chem. J. Chin. Univ*, **27**, 739–741.
40. Kitagawa, S., Kitaura, R., and Noro, S. (2004) *Angew. Chem. Int. Ed*, **43**, 2334–2375.
41. Ferey, G. (2008) *Chem. Soc. Rev*, **37**, 191–214.
42. Morris, R.E. and Wheatley, P.S. (2008) *Angew. Chem. Int. Ed*, **47**, 4966–4981.
43. Rosi, N.L., Eckert, J., Eddaoudi, M., Vodak, D.T., Kim, J., O'Keeffe, M., and Yaghi, O.M. (2003) *Science*, **300**, 1127–1129.
44. Banerjee, R., Phan, A., Wang, B., Knobler, C., Furukawa, H., O'Keeffe, M., and Yaghi, O.M. (2008) *Science*, **319**, 939–943.
45. Xiao, B., Wheatley, P.S., Zhao, X.B., Fletcher, A.J., Fox, S., Rossi, A.G., Megson, I.L., Bordiga, S., Regli, L., Thomas, K.M., and Morris, R.E. (2007) *J. Am. Chem. Soc*, **129**, 1203–1209.
46. Chen, S.M., Zhang, J., and Bu, X.H. (2008) *Inorg. Chem*, **47**, 5567–5569.
47. Hogben, T., Douthwaite, R.E., Gillie, L.J., and Whitwood, A.C. (2006) *CrystEngComm*, **8**, 866–868.
48. Ji, W.J., Zhai, Q.G., Hu, M.C., Li, S.N., Jiang, Y.C., and Wang, Y. (2008) *Inorg. Chem. Commun*, **11**, 1455–1458.
49. Liao, J.H. and Huang, W.C. (2006) *Inorg. Chem. Commun*, **9**, 1227–1231.
50. Liao, J.H., Wu, P.C., and Bai, Y.H. (2005) *Inorg. Chem. Commun*, **8**, 390–392.
51. Liao, J.H., Wu, P.C., and Huang, W.C. (2006) *Cryst. Growth Des*, **6**, 1062–1063.
52. Lin, Z.J., Li, Y., Slawin, A.M.Z., and Morris, R.E. (2008) *Dalton Trans*, 3989–3994.
53. Shi, F.N., Trindade, T., Rocha, J., and Paz, F.A.A. (2008) *Cryst. Growth Des*, **8**, 3917–3920.
54. Xu, L., Choi, E.Y., and Kwon, Y.U. (2007) *Inorg. Chem*, **46**, 10670–10680.
55. Zhang, J., Chen, S.M., and Bu, X.H. (2008) *Angew. Chem. Int. Ed*, **47**, 5434–5437.
56. Zhang, J., Wu, T., Chen, S.M., Feng, P., and Bu, X.H. (2009) *Angew. Chem. Int. Ed.*, **48**, 3486–3490.
57. Lin, Z.J., Wragg, D.S., and Morris, R.E. (2006) *Chem. Commun*, 2021–2023.
58. Xu, Y.P., Tian, Z.J., Wang, S.J., Hu, Y., Wang, L., Wang, B.C., Ma, Y.C., Hou, L., Yu, J.Y., and Lin, L.W. (2006) *Angew. Chem. Int. Ed*, **45**, 3965–3970.
59. Wragg, D.S. and Morris, R.E. *Solid State Sci*, in press.
60. Cai, R., Sun, M.W., Chen, Z.W., Munoz, R., O'Neill, C., Beving, D.E., and Yan, Y.S. (2008) *Angew. Chem. Int. Ed*, **47**, 525–528.
61. Morris, R.E. (2008) *Angew. Chem. Int. Ed*, **47**, 442–444.
62. Xing, H.Z., Li, J.Y., Yan, W.F., Chen, P., Jin, Z., Yu, J.H., Dai, S., and Xu, R.R. (2008) *Chem. Mater*, **20**, 4179–4181.
63. Byrne, P.J., Wragg, D.S., Warren, J.E., and Morris, R.E. (2009) *Dalton Trans*, 795.
64. Himeur, F., Wragg, D.S., Stein, I., and Morris, R.E. *Solid State Sci.*, doi: 10.1016/j.solidstatesciences.2009.05.023.
65. Lin, Z., Wragg, D.S., Warren, J.E., and Morris, R.E. (2007) *J. Am. Chem. Soc*, **129**, 10334.
66. Hulvey, Z., Wragg, D.S., Lin, Z., Morris, R.E., and Cheetham, A.K. (2009) *Dalton Trans*, 1131.
67. Lin, Z., Slawin, A.M.Z., and Morris, R.E. (2007) *J. Am. Chem. Soc*, **129**, 4880.
68. Hanke, C.G. and Lyndon-Bell, R.M. (2003) *J. Phys. Chem. B*, **107**, 10873.
69. Amigues, E., Hardacre, C., Keane, G., Migaud, M., and O'Neill, M. (2006) *Chem. Commun*, 72.
70. Morris, R.E., Burton, A., Bull, L.M., and Zones, S.I. (2004) *Chem. Mater*, **16**, 2844–2851.
71. Camblor, M.A., Villaescusa, L.A., and Diaz-Cabanas, M.J. (1999) *Top. Catal*, **9**, 59–76.
72. Zones, S.I., Darton, R.J., Morris, R., and Hwang, S.J. (2005) *J. Phys. Chem. B*, **109**, 652–661.
73. Villaescusa, L.A., Lightfoot, P., and Morris, R.E. (2002) *Chem. Commun*, 2220–2221.
74. Bull, I., Villaescusa, L.A., Teat, S.J., Camblor, M.A., Wright, P.A., Lightfoot, P., and Morris, R.E. (2000) *J. Am. Chem. Soc*, **122**, 7128–7129.

75. Villaescusa, L.A., Wheatley, P.S., Bull, I., Lightfoot, P., and Morris, R.E. (2001) *J. Am. Chem. Soc*, **123**, 8797–8805.
76. Ma, H.J., Tian, Z.J., Xu, R.S., Wang, B.C., Wei, Y., Wang, L., Xu, Y.P., Zhang, W.P., and Lin, L.W. (2008) *J. Am. Chem. Soc*, **130**, 8120.
77. Parnham, E.R. and Morris, R.E. (2006) *Chem. Mater*, **18**, 4882.
78. Parnham, E.R., Drylie, E.A., Wheatley, P.S., Slawin, A.M.Z., and Morris, R.E. (2006) *Angew. Chem. Int. Ed*, **45**, 4962.

# 4
# Co-Templates in Synthesis of Zeolites

*Joaquin Pérez-Pariente, Raquel García, Luis Gómez-Hortigüela, and Ana Belén Pinar*

## 4.1
## Introduction

Since the pioneering work by Barrer, the stabilizing role that guest molecules occluded inside zeolite cavities play in the crystallization of these materials has been widely recognized [1]. Apart from inorganic cations, whose influence in the zeolite crystallization is not discussed here, water and different types of organic molecules contribute to the stabilization of the otherwise intrinsically unstable low-density zeolite frameworks, at the expense of the more stable denser crystalline phases.

In the particular case of organic molecules, the template concept was first used by Aiello and Barrer [2], making reference to the mechanism that allows the trapping of tetramethylammonium (TMA) cations inside the gmelinite cages during the synthesis of zeolites offretite and omega. Since then, the concept of "templating" [3] has been extremely fruitful for the synthesis of a large variety of new zeolite structures [4, 5]. Although in most of the experimental approaches in this field, just only one type of organic molecule is present in the synthesis gel as a structure-directing agent (SDA), there are also a number of structures that require the simultaneous presence of at least two different templates to crystallize. This chapter is devoted to the analysis of this strategy for the synthesis of zeolite-type materials, including both silica-based and aluminophosphate-based structures.

Besides the structure-directing role of organic molecules, experimental evidence accumulated over the years shows the determining influence that inorganic guest species other than cations, such as water and fluoride anions, exert in directing the crystallization of specific zeolitic structures, in combination with the organic molecules. In addition, this structure-directing concept could eventually be extended to the inorganic atoms of the framework other than Si, Al, or P that are incorporated in the microporous networks, such as germanium, for there are several Ge-containing materials which either do not have pure Si counterparts or crystallize more easily when Ge is present in the synthesis gel [6, 7]. However, we discuss the effect of the organic compounds on the synthesis of microporous

*Zeolites and Catalysis, Synthesis, Reactions and Applications. Vol. 1.*
Edited by Jiří Čejka, Avelino Corma, and Stacey Zones

ISBN: 978-3-527-32514-6

materials in this chapter; inorganic species are taken into account only in those cases where they are chemically associated to the organic compound to promote zeolite crystallization.

## 4.2 Templating of Dual-Void Structures

Use of organic molecules in the synthesis of zeolites received a boost at the beginning of the 1980s, when many novel high-silica materials were claimed to crystallize from synthesis gels containing many different types of organic molecules as templates [8]. Most of these results were reported in patents, which also described some of the properties of the obtained zeolites. All these documents were comprehensively reviewed and analyzed in a work that still remains an invaluable source to study zeolite materials [9]. Some of these early works already reported the use of mixed-template systems, to which we refer to as co-templates, for the synthesis of high-silica zeolites, and they showed that the co-templating concept was already successfully applied at that time to crystallize zeolitic structures containing cages and channels of different size.

What may be the first example of this synthesis strategy was provided by Mobil researchers who, in 1980, reported the crystallization of zeolites ZSM-39 and ZSM-48 from gels containing TMA cations and propylamine [10–12]. The structure of ZSM-39 was later solved [13] and seems to be similar to that of dodecasil-3C (MTN structure type). It turned out to be composed of packing of two types of cages of different size, a smaller pentagonal dodecahedron and a larger hexadecahedron. Although no location of the organic molecules was found at that time, subsequent X-ray diffraction (XRD) studies of the as-made material synthesized from mixtures of pyridine and propylamine [14] confirmed what was suspected when the structure was first solved – the selective occupation of both cages by the two organic molecules as a function of their molecular size, the bulky pyridine molecule (or the TMA cation used in the earlier works) being occluded within the large cage, whereas the smaller propylamine molecule is accommodated inside the small one. Hence, crystallization of this material requires two different templates with different size to stabilize two different-sized cages. Synthesis of ZSM-39 was later optimized by using mixtures of TMABr (instead of TMACl) and ethylamine [15].

Similarly, ZSM-48 was obtained in the presence of TMA and *n*-propylamine by Chu in 1980 [16], although no experimental evidence of the occlusion of these organic molecules was provided. An improved synthesis of this structure involves the use of TMABr and octylamine [17].

The structure of this zeolite, ZSM-48, was initially explained using a model based on ferrierite sheets. An intergrowth of two of the structures resulting from the different configurations of the sheets was proposed as a model for the real disordered structure [18]. More recently, ZSM-48 has been described as a family of disordered materials assembled from silicate tubes made up of 10 T-atoms (where

T refers to tetrahedral atom sites). Connection occurs either via four-rings or via zig-zag crankshaft chains, forming a stacking of sheets of directly connected tubes. This building scheme gives rise to several ordered polytypes, two of which are equivalent to the models based on ferrierite sheets proposed earlier [19].

The ferrierite framework has a two-dimensional channel structure, formed by 10-ring channels parallel to the *c*-axis and 8-ring channels parallel to the *b*-axis. The intersection of the eight-ring channels with the smaller six-ring channels, which run along the *c*-axis, forms the ferrierite cavity, accessible through eight-ring windows. Then, the ferrierite structure is appropriate to probe a dual-templating concept, where the small cage and the 10-ring channel would be stabilized by suitable organic guests.

An early example of the synthesis of ferrierite in the presence of two templates was reported in 1981 [20]. In this patent a synthetic ferrierite, FU-9, was prepared employing several combinations of TMA with different tertiary amines (Table 4.1). In the synthesis with TMA + trimethylamine (Si/Al = 10.5) the C/N ratio was found to be 3.6, which is between the C/N ratios of the free molecules, thus suggesting the incorporation of both SDAs in the zeolite. Replacement of triethylamine with tributylamine or triethanolamine resulted also in the crystallization of FU-9.

The templating role of different organic molecules present in the synthesis gels leading to ferrierite crystals has been elucidated by XRD techniques. Fully connected

**Table 4.1** Synthesis of FER and MWW-type materials from co-template systems.

| SDA1 | SDA2 | Phase | $F^-$ | T (°) | References |
|---|---|---|---|---|---|
| FER-type materials | | | | | |
| Propylamine | TMA | ZSM-48 | – | – | [16] |
| Triethanolamine | TMA | FU-9 | – | – | [20] |
| Tributylamine | TMA | FU-9 | – | – | [20] |
| Trimethylamine | TMA | FU-9 | – | – | [20] |
| Propylamine | Pyridine | FER | + | – | [14] |
| bmp | TMA | FER | + | 135, 150 | [21] |
| bmp | Quinuclidine hydrochloride | FER layered | + | 135 | [22] |
| bmp | TEA | FER layered | + | 135, 150 | – |
| bmpm | TMA | FER related | + | 135 | – |
| bmp | Quinuclidine hydrochloride | FER + MWW | + | 150 | [22] |
| bmpm | TMA | FER + MWW | + | 150 | – |
| Quinuclidine | TMA | MCM-65(CDO) | – | 180 | [23] |
| MWW-type materials | | | | | |
| $TMAda^+$ | Dipropylamine | ITQ-1 | – | 150 | [24] |
| $TMAda^+$ | Hexamethyleneimine | ITQ-1 | – | 150 | [24] |
| $TMAda^+$ | Isobutylamine | SSZ-25 | – | 170 | [25] |
| $TMAda^+$ | Piperidine | SSZ-25 | – | 170 | [25] |
| Polycyclic amines | Isobutylamine | SSZ-25 | – | 170 | [25] |

"+" in the $F^-$ column indicates the presence of this anion in the gel.

ferrierite zeolite was prepared with pyridine and propylamine, in HF/pyridine solvent [14]. Single crystal XRD studies showed that pyridine is located in both the ferrierite cavity and the 10-ring channel, while a small amount of propylamine is located exclusively in the channels. Propylamine thus seems to play a nucleating role in the synthesis of ferrierite, according to the small amount of this molecule found in the final product and the decrease of the crystals size when increasing the amount of propylamine in the gel. Instead, ZSM-39 (MTN) was obtained in the absence of propylamine, which remarks the key role of these molecules in the crystallization of ferrierite under these synthesis conditions. This competitive phase is also obtained in the presence of both SDAs at higher temperatures or longer heating times. In this case, propylamine was found in the small cavity, and pyridine, which was too bulky to occupy the small cage, was located in the larger one.

Furthermore, it has been shown that the simultaneous use of the bulky 1-benzyl-1-methylpyrrolidinium (bmp, Figure 4.1) and the smaller TMA cation leads to the crystallization of this zeolite structure [21]. When the size of the co-SDA increases to those of quinuclidine and tetraethylammonium (TEA) while the molecule remains the same, that is, bulky, the outcome of the synthesis is layered ferrierite materials. The structure of such phases is made up by the stacking of the individual ferrierite layers, with the organic molecules located in the interlayer region. As discussed later, for the MWW structure, there is a family of layered materials related to ferrierite, to which many different members belong, such as MCM-47, the borosilicate ERS-12, or the UZM series of materials [26–28]. In the synthesis using quinuclidine, the interlayer distance of these ferrierite layered phases decreased with the crystallization time, which was attributed to the exchange of bmp by quinuclidine as the crystallization proceeded [22]. When a bulkier organic molecule such as TEA was used as a co-SDA, related layered materials were obtained though, for this co-SDA, the interlayer distance did not vary with the crystallization time.

Molecular mechanics calculations revealed a tendency of each of the SDAs used in these preparations to occupy specific sites within the ferrierite structure, evidencing a co-structure-directing effect and allowing to understand the experimental observations: the bulky bmp molecules accommodated along the 10-ring channels and the smaller co-SDA within the ferrierite cages (Figure 4.2). These calculations show that while the TMA cation has a strong preference for accommodation within the ferrierite cages, quinuclidine is too large to fit properly in the ferrierite cavities and it would be better for it to be accommodated in a somewhat larger cavity. For TEA, its location within the ferrierite cavities would lead to an even more unstable situation, which explains that the layered FER structure obtained with TEA is not able to evolve to the fully condensed FER structure, in contrast to the one obtained with quinuclidine. Moreover, XRD studies show that the TMA cations are indeed located in the ferrierite cage, while bmp occupies the 10-ring channel (Pinar *et al.*, manuscript in preparation).

Replacement of the bulky SDA in these preparations, bmp, by a related chiral cation, 2-hydroxymethyl-1-benzyl-1-methylpyrrolidinium (bmpm, see Figure 4.1)

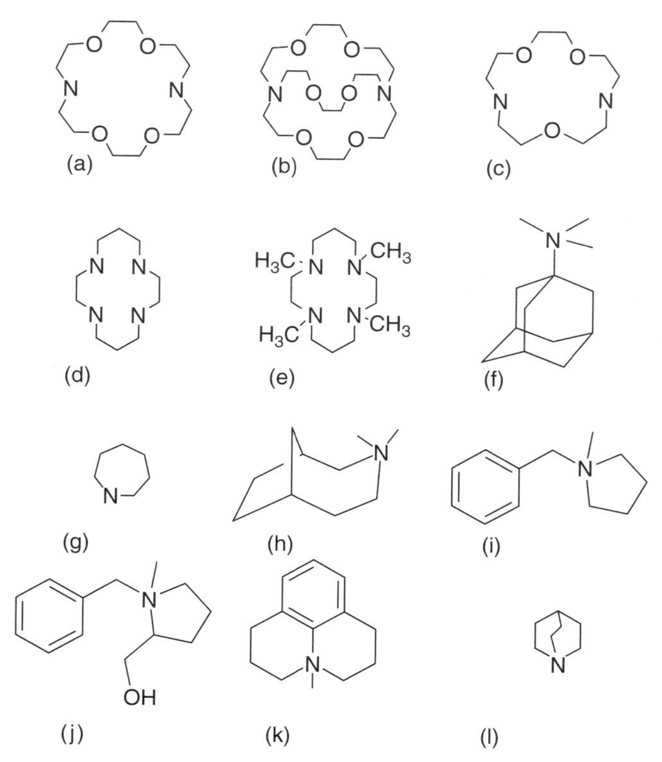


**Figure 4.1** Structures of some of the organic molecules used as co-structure-directing agents in the synthesis of zeolites: (a) Kryptofix22, (b) Kryptofix222, (c) Kryptofix21, (d) 1,4,8,11-tetraazacyclotetradecane (cyclam), (e) 1,4,8,11-tetramethyl-1,4,8,11-tetraaza-cyclotetradecane (Tetramethylcyclam), (f) *N*, *N*, *N*-trimethyl-1-adamantanammonium ($TMAda^+$), (g) hexamethyleneimine (HMI), (h) *N*, *N*-dimethyl-3-azoniabicyclo [4.2.1] nonane, (i) 1-benzyl-1-methylpyrrolidinium (bmp), (j) 2-hydroxymethyl-1-benzyl-1-methyl-pyrrolidinium (bmpm), (k) 4-methyl-2,3,6,7-tetrahydro-1H,5H-pyrido [3.2.1-ij] quinolinium, and (l) quinuclidine.

was shown to direct the synthesis to ferrierite-related phases by using TMA as a co-SDA, evidencing again the high tendency of this cation to produce ferrierite. It is worth noting that in many of the above preparations that yielded ferrierite-related phases, the co-crystallization of a phase of the MWW family was observed at higher synthesis temperatures (150 °C, see Table 4.1).

Another material containing ferrierite layers, MCM-65, can also be synthesized by the combination of quinuclidine and TMA. The ferrierite layers of this precursor are stacked in such a manner that they render a new cage structure upon calcination. The structure of the calcined MCM-65, the CDO structure type, comprises a two-dimensional network of eight-ring channels [23]. Both SDAs used in its synthesis (quinuclidine and TMA) were found to be intact and rigidly held in the voids of the MCM-65 precursor.

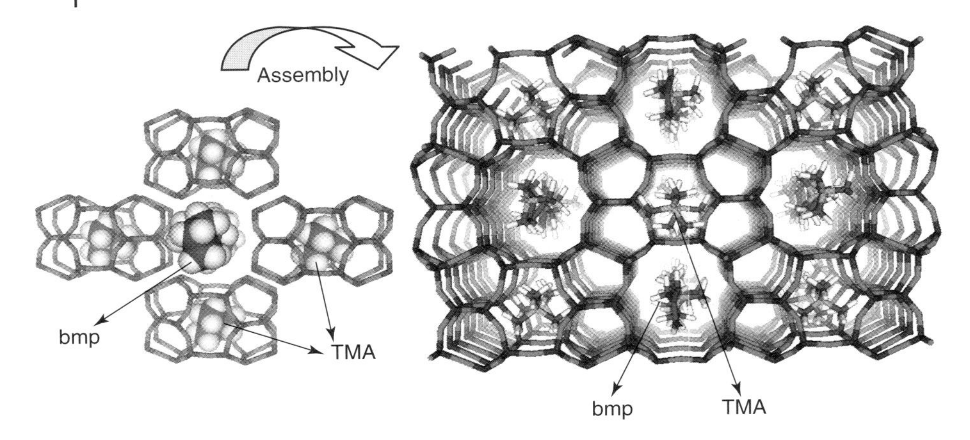


**Figure 4.2** Scheme of the self-assembly of TMA-filled cavities around bmp molecules to give the final ferrierite structure.

One of the zeolite structures frequently obtained in the presence of two SDAs is MWW, which is moreover often found to crystallize together with ferrierite in many preparations. The structure of this zeolite is composed of two independent channel systems accessible through 10-ring pores, one formed by large cages delimited by 10-ring apertures and the other one by two-dimensional sinusoidal channels that run perpendicular to the long dimension (18 Å) of the cages. The three-dimensional MWW structure is usually obtained via a layered precursor MCM-22 (P) that forms the characteristic three-dimensional zeolite when calcined. There is a family of materials with a structure related to MWW, such as MCM-22, MCM-56, and MCM-49, where the differences between them are mainly due to the packing degree of the layers [29]; MCM-49 presents the same framework topology as calcined MCM-22, and MCM-56 maintains the layered structure on calcination. Related materials are the aluminosilicates SSZ-25 [30], the boron-containing analog ERB-1 [31], and the pure silica polymorph ITQ-1 [32].

Camblor *et al.* reported the synthesis of the pure silica zeolite ITQ-1 using *N*, *N*, *N*-trimethyl-1-adamantanammonium (TMAda$^+$, Figure 4.1) as the only SDA [32]. However the synthesis was hardly reproducible. They found that adding a second organic component to the system, hexamethyleneimine (HMI) or dipropylamine (DPA), allowed a faster and more reproducible synthesis [24]. They proposed that TMAda$^+$, since it is too large to fit in the sinusoidal channel system, stabilizes the 12-ring cages, whereas the amine additives help in crystallizing ITQ-1 by filling the 10-ring sinusoidal channel system. Indeed, using a bulkier amine, diisobutylamine (DIBA), under the same conditions as HMI and DPA, did not result in the crystallization of ITQ-1. In addition, the amines seemed to exert some effect on the pH of crystallization, since the use of HMI as the hydrochloride salt yielded the zeolite MCM-35 (MTF structure type) instead of ITQ-1, where only the HMI molecules remained occluded within the cages of the structure [33].

SSZ-25, a MWW-related material, was also synthesized in a two-component organic system, using the quaternary ammonium cation $TMAda^+$ and different amines as pore fillers. Zones *et al.* studied the crystallization of SSZ-25, replacing both components [25]. The system showed a high flexibility and other adamantane derivatives, such as the free amine or the alcohol, could be used instead of $TMAda^+$ at low concentrations to produce SSZ-25. In addition, under the same conditions, other large polycyclic hydrocarbon cations were also shown to yield SSZ-25, combined mostly with isobutylamine and piperidine. All these molecules possess a size that prevents them from fitting in the 10-ring sinusoidal channels and, as commented above, they most likely occupy the large cages in the structure. It was shown that the adamantyl component can be decreased in the synthesis gel, which results in a larger uptake of the isobutylamine. However, a certain amount of this component is required to crystallize the zeolite, since in its absence, other zeolites are produced.

Mixed-template systems involving simple and commonly used tetralkylammonium cations still have a large potential for discovering new zeolite structures that require a unique combination of those SDAs under suitable synthesis conditions. A few years ago, a research team of the UOP company reported the synthesis of two open-framework materials with $Si/Al < 10$, from TMA/TEA mixed systems: the zeolites UZM-4 and UZM-5 [34]. The first one is a 12-ring channel structure with the BPH framework topology, initially synthesized as a beryllophosphate, and the second one is a novel 8-ring structure related to that of zeolite A. The gel rendering UZM-4 also contains lithium, and both organic molecules are incorporated into the structure, the TEA/TMA ratio being much higher in UZM-5.

On some occasions, the mixed-template approach results in failure [35]. Zones *et al.* studied the structure-directing effect of several long symmetric diquaternary ammonium cations of varying chain length under different reaction conditions. One of the synthesis variables of these preparations was the use of a second SDA, the small TMA cation, together with a bulky diquaternary compound. However, in the presence of TMA cations, only clathrasil phases were obtained, probably templated by the TMA cation.

## 4.3 Crystallization of Aluminophosphate-Type Materials

The co-templating strategy has also been applied to the synthesis of different aluminophosphate structures. One of the first examples was the crystallization of silicoaluminophosphate (SAPO)-37 from gels containing mixtures of TMA and tetrapropylammonium (TPA) hydroxides [36–38]. SAPO-37 possesses the faujasite structure, which provides a topological explanation for the role of each template: the TMA cations are located within the sodalite cages, while the bulkier TPA cations occupy the supercages. This material crystallizes in a relatively narrow range of TMA/TPA ratios in the synthesis gel. Unbalance of this ratio produces SAPO-20 (SOD structure-type) if an excess of TMA is used while, on the other side, predominance of TPA favors the crystallization of the large-pore SAPO-5, which possesses unidirectional 12-ring channels but no cavities.

This example illustrates nicely what is commonly observed by using mixed-template systems for the crystallization of zeolite-type materials, namely, a delicate balance between both templates is required for successful synthesis, since both molecules cooperate to stabilize the dual-void structures in preference to the ones favored by each template separately.

One of the zeolite structures frequently obtained with a variety of combinations of SDAs is LTA. This structure type is built up by the connection of double four-rings creating a framework with large $\alpha$-cavities and smaller sodalite cages (Figure 4.3). The macrocycle, azaoxacryptand 4,7,13,16,21,24-hexaoxa-1,10-diazabicyclo[8.8.8] hexacosane (K222) (Figure 4.1), was shown to direct the formation of the aluminophosphate version of the LTA structure and its substituted derivatives when it was used as the only SDA [39]. In the presence of TMA cations and fluoride anions, a rhombohedral variant of the LTA structure was obtained. The use of other Kryptofix-n macrocycles led to the crystallization of the LTA topology in the presence of fluoride and/or TMA cations. In all cases, the Kryptofix-n molecules were most likely located within the $\alpha$-cages of the structure, the fluoride anions in the D4Rs and the TMA cations in the smaller sodalite cages [40], thus showing the cooperation of all the species in the construction of the cavities of the zeolite structure. A similar scenario was found with other combinations of SDAs. A copper–cyclam complex also produced LTA from aluminophosphate gels containing fluoride and trimethylamine [41]. The Cu–cyclam complex was located within the $\alpha$-cages, with trimethylamine in the sodalite cage and $F^-$ anions in the D4R units. Likewise, combination of diethanolamine, fluoride anions, and TMA cations yielded the same framework structure, with each of them structure-directing one of the three polyhedra present [42]. Diethanolamine was located in the $\alpha$-cage, with the nitrogen atom in the middle of the eight-ring of the cavity and the two ethanol

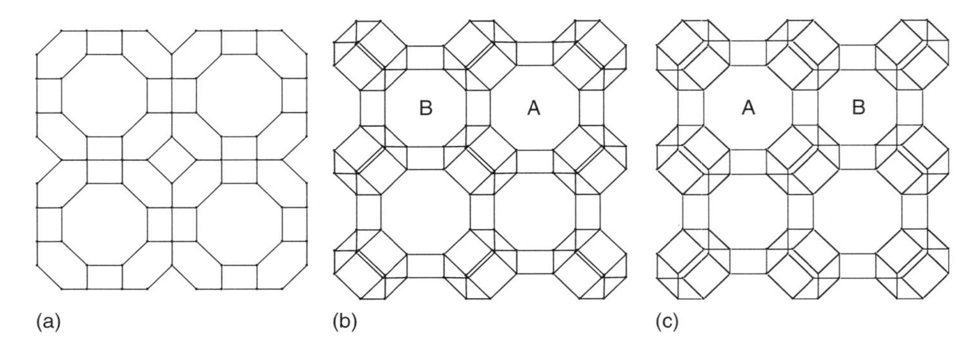


**Figure 4.3** (a) The structure of LTA, showing the smaller sodalite cages at the center of four large $\alpha$-cages; (b) the SAV framework showing the bigger cages (labeled A) and the smaller cages (labeled B) of the structure; and (c) the structure of KFI showing the $\alpha$-cages (A), as those in LTA, and the smaller MER cages (B).

groups pointing to the contiguous sodalite cavities, while $F^-$ and TMA were located in the D4R unit and in the sodalite cage, respectively. Interestingly, eight water molecules were additionally located in the $\alpha$-cage close to the six-rings, probably linked to the terminal OH groups of the amine through hydrogen bonds, and so they were also considered as templates. In this system, the replacement of the small diethanolamine molecule by the bulkier TPA cations led to the crystallization of materials having larger voids accessible through 12-ring apertures, possessing either FAU or AFR structures depending on specific synthesis conditions [43].

The aluminosilicate version of the LTA structure can also be prepared by using a mixture of templates. The cation 4-methyl-2,3,6,7-tetrahydro-1H,5H-pyrido [3.2.1-ij] quinolinium iodide (Figure 4.1) was able to direct the formation of this structure in germanosilicate preparations in fluoride medium; this cation was shown to form self-assembled dimers within the $\alpha$-cages of the structure. Interestingly, the addition of TMA to the synthesis gels, together with the quinolinium derivative, allowed for obtaining the pure silica version of the material, and both cations remained occluded in the structure [44].

The macrocycle tetramethylcyclam (1,4,8,11-tetramethyl-1,4,8,11-tetraazacyclotetradecane) was known to direct the synthesis of two metalloaluminophosphates, STA-6 (SAS structure type) for metals such as Mg, Mn, and Si, and STA-7 (SAV structure type) in preparations with Zn and Co [45]. Both are cage-like structures based on D6R as secondary building units, although the cavities are different in both materials. In particular, the framework of STA-7 possesses two three-dimensional channel systems of eight-ring openings and two cages of different size. The large cage of the SAV structure is stacked along the $c$-axis through planar eight-membered rings, creating one of the channel systems. The cage also possesses four elliptical eight-ring openings to the second channel system that is made up of smaller cages (Figure 4.3).

Addition of TEA cations to the synthesis gels, which in their absence yield STA-6, directs the synthesis toward the formation of STA-[46]. Single crystal diffraction showed that the macrocycle resides in the large cages of the structure while the TEA cations are located in the smaller cages. Furthermore, together with TEA cations, tetramethylcyclam can be replaced with the related and less expensive cyclam (Figure 4.1) to direct the formation of STA-7, reducing the cost of production of this material.

On the basis of the above studies with macrocycles, the co-templating approach was also used to prepare the aluminophosphate version of the aluminosilicate zeolite ZK-5 (KFI structure type) [46]. This structure possesses two types of cages: the large $\alpha$-cage (the same found in the structure of LTA) and a smaller cage also found in the structure of zeolite merlinoite (MER). Computer modeling suggested the TEA cation as a good SDA for the MER cages. Therefore, the KFI structure could be obtained instead of LTA by using the strong structure-directing action of K222 for the $\alpha$-cage and in the presence of TEA as a co-SDA. Crystal diffraction showed that TEA cations are present in the MER cages of the KFI structure, as was predicted by modeling [46].

Application of the co-templating concept is by no means restricted to hydrothermal crystallization. The use of ionic liquids for the synthesis of zeolite-type materials, aluminophosphates in particular, provides opportunities for new advances in this area, as the organic cations forming part of the ionic liquid may themselves act as templates. The first example of co-templating ionothermal synthesis of a new open-framework aluminophosphate denoted by JIS-1 has been recently reported by using the aromatic amine 1-methyl-imidazole (MIA) and the ionic liquid 1-methyl-3-ethylimidazolium bromide (EMIMBr) [47]. The structure consists of an anionic open framework with an Al/P ratio of 6/7 that possesses three-dimensional intersecting 10-ring, 10-ring, and 8-ring channels along the three crystallographic axes, where the protonated amine and cations of the ionic liquid are simultaneously found in the intersection of the channels.

## 4.4
## Combined Use of Templating and Pore-Filling Agents

A usually high cost of SDAs and their difficulty to be removed from the microporous networks led to the search for new zeolite synthesis systems, in which the amount of the SDAs required to crystallize the zeolites is reduced. Here, we find another interesting application for the use of combinations of organic molecules as SDAs in the synthesis of zeolites [48, 49]. Zones and Hwang proposed [48] a new zeolite synthesis system, in which a minor amount of the SDA is used to selectively specify the nucleation product, and then a larger amount of a cheaper, less selective, and smaller molecule is used to provide both pore-filling and basicity capacities in the zeolite synthesis as the crystal continues to grow. This new synthesis system provides a successful route toward the cost-effective making of a number of aluminosilicate materials and, indeed, new zeolite topologies have also been discovered. In addition to reducing the amount of the expensive SDAs required in the synthesis, these systems also accelerate crystallization rates. Interestingly, there is a reagent flexibility of the pore-filling agent to be used. Furthermore, the smaller size of the pore-filling agent makes it easier to be removed from the zeolite network, which can be achieved by simple extraction. As an example, crystallization of SSZ-25 can be accomplished by adding minor amounts of *N, N, N*-trimethyladamantyl ammonium, which directs the formation of the zeolite, and major amounts of isobutyl amine, which are occluded during the crystallization as a pore-filling agent. In addition, a new intergrowth material (SSZ-47) related to the structural family NES/EUO/NON was obtained by using a large amount of isobutylamine together with a smaller amount of a group of bicycloorgano-cations [49], such as *N, N*-dimethyl-3-azoniabicyclo [4.2.1] nonane (Figure 4.1), providing evidence that this approach can also aid in the search for new zeolite materials.

A particular use of these two-component systems is when degradable organic molecules are used as SDAs [50–52]. SDA molecules have to be removed from the zeolite frameworks before their use in adsorption and catalytic applications. Owing to their usual large size, removal of these encapsulated species normally requires high temperature combustion that destroys them, and the associated energy released in combination with the formed water can be extremely detrimental to the zeolite structure. In order to avoid this, Lee *et al.* proposed [50–52] a new class of organic SDAs that have the potential to be degraded into fragments within the pores so that they can be readily extracted from the zeolite under mild conditions, thus avoiding the necessity of the high-temperature calcination process. Furthermore, the extraction of the molecular fragments would allow reuse (recycle) of the SDAs by re-assembly, which might decrease the cost of the zeolite synthesis, as usually SDA species are the most expensive components of the synthesis gel. One class of these degradable organic SDA molecules is ketal-containing species that are very stable at high pH but can be cleaved into ketones and diols at low pHs. These ketal molecules will be intact during the zeolite synthesis that typically occurs at high pH, and fragmented into pieces by lowering the pH (hydrolysis). Under these conditions, access of $H_2O$ and $H^+$ to the ketal group of the SDA is essential for the hydrolysis reaction to occur. If this access is prohibited, the SDAs will remain intact despite the chemical treatment. This is where the use of the second organic molecule, the pore-filling agent, plays its crucial role: the inclusion of a second organic molecule, smaller than the main SDA, facilitates its removal by extraction, and then the void space created is used by $H^+$ and $H_2O$ species to access the SDA for the cleavage reaction to occur. The degradable SDA would remain intact after the acidic treatment without the previous extraction of the pore-filling agent, which demonstrates that the extraction of these species, and thus the use of a two-component system, is essential for the successful application of degradable SDAs.

## 4.5
## Cooperative Structure-Directing Effects of Organic Molecules and Mineralizing Anions

In some of the examples described above the fluoride anions are found to be located in the structure, where they play an additional stabilizing role, showing preference for very small cages, such as the D4R in LTA structures. However, we pay attention in this section to a somewhat different situation, where mineralizing species such as $F^-$ and $OH^-$ anions interact with the organic guest species to provide the actual templating chemical entity. Such a cooperative structure-directing effect has been experimentally observed in the crystallization in fluoride medium of the all-silica EUO structure, which contains 10-ring one-dimensional channels with side-pockets, by using a fluorine-containing molecule as an SDA, *o*-fluorobenzyl-benzyl-dimethylammonium [53]. A combination of XRD studies and computer simulation of the as-made material reveals the development of a strong interaction between the fluoride anions occluded in the structure and the F-containing molecules located in the channel, in such a way that the actual arrangement of both species tends to maximize the electrostatic attraction between the positively charged N atom of the molecule and the $F^-$ anions while reducing

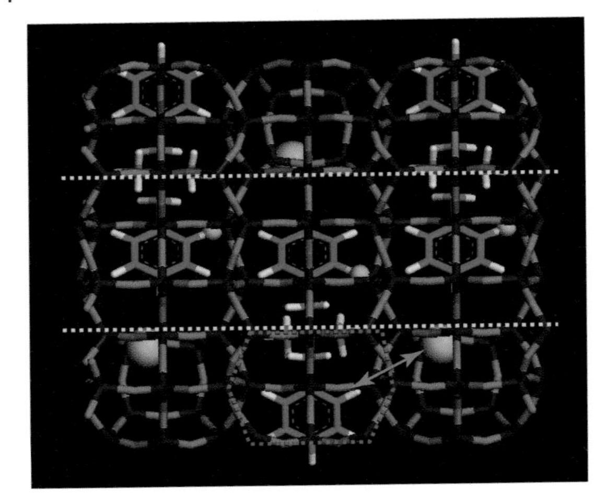

**Figure 4.4** Location of *o*-fluorobenzyl-benzyl-dimethylammonium in the EUO structures; fluorinated rings are invariably located in the 10-ring channel (highlighted between dashed yellow lines). Fluoride anions are displayed as large pink balls, while fluorine organic atoms are displayed as small pink balls. The side-pockets where nonfluorinated aromatic rings are located are shown by dashed green lines. The repulsion between fluoride and *ortho*-F located in the side-pockets that prevents this arrangement is shown by blue arrows.

the $F^-\cdots F$(SDA) repulsion, leading to a well-defined location of the fluorinated aromatic rings in the channel rather than in the cavities (Figure 4.4). This determines the cooperative effect, between both chemical species in the stabilizing EUO crystals, that if no F is present in the SDA, the competing zeolite Beta is obtained, which also crystallizes if both the aromatic rings contain one F atom each. In this last case, strong F---F repulsion cannot be avoided when the fluoride anions and the fluorine-containing SDA are located inside EUO pores, and this makes the nucleation of this zeolite less favorable, leaving a chance for zeolite Beta to crystallize [54].

In the absence of fluoride and low-valent heteroatoms, the necessity for compensating the positive charge of the organic SDA molecules implies the formation of negatively charged structural defects; however, in certain cases, it has been observed that the hydroxide anions are incorporated in the frameworks as charge-compensating species [55], leading to the effective incorporation of neutral $SDA^+\cdots OH^-$ adducts in the microporous framework. In a similar manner as previously described for *o*-fluorobenzyl-benzyl-dimethylammonium in the synthesis of EUO, the presence of active groups (F atoms) in the organic SDA molecules, bis(*o*-fluorobenzyl)-dimethylammonium, used for the synthesis of the AFI structure leads to the development of strong interactions between the organic molecule and other species present in the synthesis gels, such as the hydroxide anions, giving room to the formation of strongly bonded supramolecular organo-inorganic entities. These units are stabilized by electrostatic interactions between the hydroxide

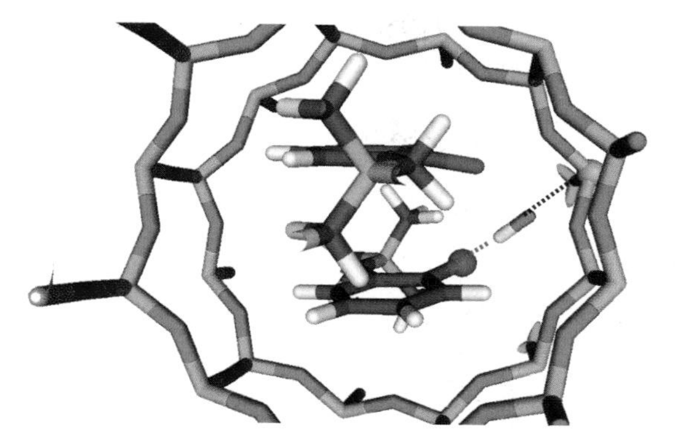

**Figure 4.5** A model of the proposed orientation of hydroxide anions within the 12MR channels, bridging the *ortho*-fluorinated organic SDA and the framework. Dashed green line indicates the H-bond and dashed blue line indicates the coordination bond with Al. Fluorine atom (blue) and pentacoordinated aluminum atom are displayed as balls.

anions and the positive charge localized around the N atom of the organic molecule, and especially by the formation of H-bonds between the neighboring F atoms (in the ortho position, i.e., spatially close to the ammonium group) and the hydroxide anion. Such a stable supramolecular arrangement is the actual SDA of the AFI structure, in which both species are finally incorporated in the framework, with the hydroxide anions also coordinatively bonded to Al framework ions, thus bridging the organic molecule and the framework (Figure 4.5). This mode of cooperative structure direction, where the organic molecules template the formation of the microporous structure in addition to contributing to the stabilization by developing nonbonded interactions, while the hydroxide anions act as charge-compensating units, carries the benefit of incorporating a neutral entity, thus preventing the formation of negatively charged framework defects that destabilize the microporous structure.

## 4.6
## Cooperative Structure-Directing Effect of Organic Molecules and Water

So far in this chapter, we have dealt only with co-templating effects where both co-templating agents were organic molecules and/or anions (fluoride or hydroxide). However, there is also another component in the synthesis gels of zeolite-type materials that could, in principle, play a role in the structure direction of these materials, namely, the water molecules. Traditionally, the hydrothermal synthesis of microporous materials has involved the use of water as the solvent (although new synthesis trends propose the use of other solvents like alcohols or ionic liquids). Nevertheless, recent studies have demonstrated that water can play an important role in structure direction together with the usual organic molecules, apart from its role as a solvent [56–58]. This is especially true in the case of hydrophilic zeolite-type materials, which are susceptible to strong interactions with water molecules.

High-silica zeolites are hydrophobic in nature, and therefore, no strong interaction and so no adsorption of water within these materials is usually found. However, this is not the case for AlPO-based frameworks, where the molecular-ionic nature of the network, which comprises $Al^{3+}$ and $PO_4^{3-}$ units ionically bonded, and the strict alternation of Al and P ions provide these materials with a high hydrophilic character. Therefore, hydrophilic AlPO frameworks are more susceptible to strong interactions with water.

A recent work by our group has evidenced a clear cooperation between water and organic molecules when directing the synthesis of the large-pore AlPO-5 aluminophosphate (AFI structure type) with different organic molecules. Water itself can template the formation of small secondary building units, like the 6-MR channels in the AFI structure, where a clear host–guest correlation between their shape and symmetry can be appreciated (Figure 4.6a). However, water molecules

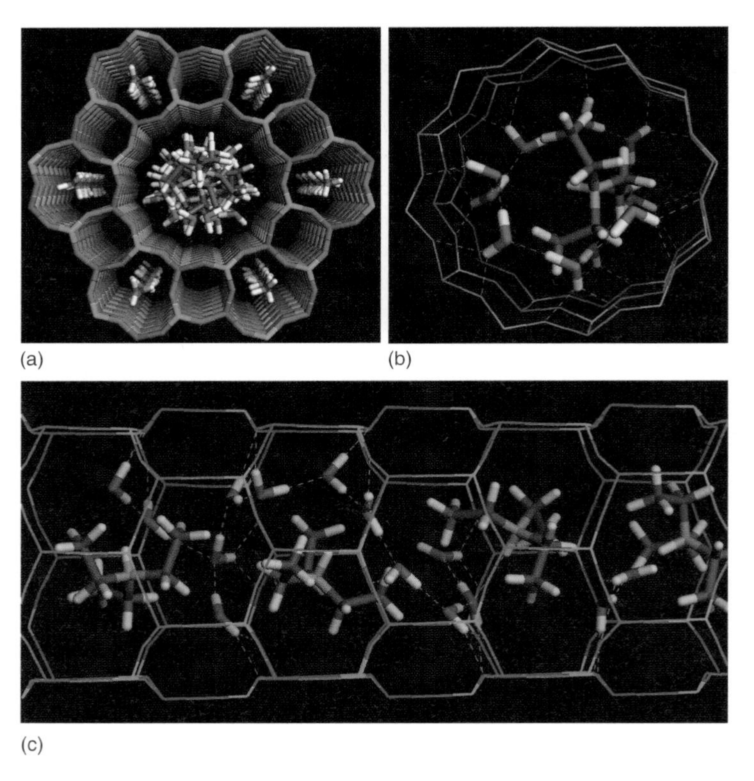


**Figure 4.6** Cooperative structure-directing effect of triethylamine and water molecules in the synthesis of the AFI structure. (a) Occlusion of guest species in the AFI framework. (b,c) Two views of the H-bonded chains of water around triethylamine molecules within the AFI channel.

themselves cannot template the formation of large-pore topological items, such as the 12-ring channels, since the H-bonding network between the water molecules is not strong enough as to hold the large-pore architecture at the high synthesis temperatures usually required. In this regard, the larger organic molecules are inevitably required to template the synthesis of large-pore microporous materials. Nonetheless, water can play a cooperative role in directing the synthesis of hydrophilic microporous aluminophosphates, depending on the structure-directing efficiency and the hydrophilicity of the organic molecule employed as the main SDA. In cases where the interaction of the organic SDA molecules with the microporous framework is not very strong, water can compete for being occluded within the frameworks during crystallization, thus strongly contributing to the final stabilization of the microporous structure by the development of H-bond interactions with the oxygen atoms of the framework. In this case, water molecules have a cooperative structure-directing effect on the organic molecules, where neither water nor the organic molecules themselves are the true SDAs, but a water–organic molecule aggregate in which both species cooperatively act as the SDA entity. Such a cooperativeness involves a synergetic effect of the two species: the large size of the organic SDA molecules provides the templating role of the large-pore architecture, allowing the development of nonbonded interactions with the framework, while the strong dipole of water molecules provides further strong interactions with the structure through the formation of H-bonds with the atoms of the oxygen framework, thus leading to a strong stabilization of the structure. Such a mode of structure direction could not be achieved by the occlusion of the isolated species: despite the high interaction of water molecules, they cannot hold the large-pore architecture of the microporous structures, while the weaker interaction of certain SDA organic molecules is not strong enough as to stabilize the structure to make the crystallization pathway viable. This cooperative effect is nicely illustrated by the crystallization of AlPO-5 in the presence of triethylamine [56] (Figure 4.6), where it can be clearly observed that water molecules arrange as H-bonded chains surrounding the triethylamine organic molecules, locating always close to the channel walls and thus developing strong H-bonds with the framework oxygen atoms. It seems that this cooperative water–organic structure-directing effect occurs for low-interacting organic molecules, where such low interaction needs to be compensated by the simultaneous occlusion of strongly interacting water molecules. Similar cooperative effects of water have been found for benzylpyrrolidine molecules when directing the synthesis of the AFI and the SAO structures, or methylamine in the synthesis of IST-1 and IST-2 [58]. Indeed, another nice example of this cooperative role of water was long time ago observed in the synthesis of Si and metal-substituted $AlPO_4$ materials with the LTA structure, where water molecules are hydrogen bonded to the OH groups of diethanolamine, as has been previously mentioned in this chapter [42]. Instead, molecules that interact more strongly with the framework, like a benzylpyrrolidine analog with an attached methanol group, and (*S*)-*N*-benzyl-pyrrolidine-2-methanol, tend to direct the crystallization of the frameworks by themselves, preventing the simultaneous occlusion of water.

## 4.7
## Control of Crystal Size and Morphology

In the examples discussed above, the use of co-template systems severely alters the crystallization pathway of the synthesis gel to make the synthesis of zeolite structures feasible that would not otherwise be obtained in the absence of such specific combination of co-templating agents. However, there are also examples where such modifications of the nucleation and crystal growth are less severe, and although they do not lead to new structures, they nevertheless affect the crystal size and shape of the resulting zeolite, as well as its chemical composition, in a noticeable way.

Crown ethers, 18-crown-6 and 15-crown-5, have been used for the synthesis of the cubic zeolite faujasite and its hexagonal analog EMT by using ethylene glycol and 1,3,5-trioxane as co-templates [59]. The reason for adding the latter two organic compounds is that they allow the crystallization of high-silica sodalite. As sodalite units are also present in both FAU and EMT structures, the association of these neutral organics with the crown ethers was supposed to provide good candidates to enhance further the Si/Al ratio of these two large-pore materials. Pure EMT and FAU zeolites were obtained from the large and the small crown ethers, respectively; in both cases, the crystal size of the obtained materials increased up to 3–5 μm, which was attributed to a decrease in the supersaturation of the system, since the aluminosilicate species were less soluble in the synthesis media containing the co-templates, evidencing a new effect of the presence of co-templates on the crystal size. However, despite some ethylene glycol was found in the EMT crystals, the Si/Al ratios of both materials were similar to those obtained by using the crown ethers alone, showing that in this case no control over the Si/Al ratio was provided by the use of the co-templates.

Very large crystals of zeolite Y (Si/Al = 1.7) with diameters up to 245 μm were obtained by using bis(2-hydroxyethyl)dimethylammonium chloride (TCl) as a co-template together with triethanolamine. In this case, the amine is used as a chelating agent for the $Al^{3+}$ ions in the gel in order to retard the nucleation of faujasite, thus leading to larger crystals. However, zeolite P was formed as an impurity [60]. Attempts to improve this synthetic procedure by adding seeds of zeolite Y did not completely eliminate the formation of zeolite P [61]. Nonetheless, these works provide a new example of the usefulness of using co-templates for influencing the crystal growth of the microporous materials.

Finally, a dual-template approach has been proposed for the synthesis of nano-sized edingtonite-type (EDI topology) zeolite from gels containing a mixture of TMA and copper amine complexes $[Cu(NH_3)_4]^{2+}$[62]. Moreover, if only copper cations or ammonia are present, pure nanosized FAU zeolite is obtained. Both TMA and the copper complex are present in the crystals, and in some cases 82% of the channel intersections are filled by them, which points to a true co-structure-directing role of the molecules, apart from their effect on the crystal size.

## 4.8
## Membrane Systems

The control of the crystal size and shape provided by the co-templating zeolite synthesis strategy is particularly useful for the production of high-performing zeolite membranes. Mixed-template systems have also been employed to change the morphology of MFI crystals in order to improve the performance of membranes made up of this zeolite in separation processes [63]. The target here is to obtain zeolite crystals with *b*-axis as the preferential orientation, which are the most effective for that purpose [64]. Mixed systems containing *N*-ethyl-hexamethylenetetrammonium bromide as one template, and *n*-propylamine, *n*-butylamine, or ethylamine as the second template were used to crystallize pure silica MFI. It was found that *n*-propylamine is the most effective one in promoting anisotropic growth of MFI crystals. Both the amine and the ammonium amine, and the ammonium cation are found to be present in the zeolite crystals, playing a co-templating role in the formation and growth of the *b*-oriented crystals.

SAPO-34 (CHA structure type) membranes can be conveniently used for the separation of $CO_2/CH_4$ mixtures, and it has been reported that by using a combination of SDAs in the preparation of SAPO-34 membranes by seeding on porous stainless steel supports, higher fluxes and selectivities are obtained [65]. TEAOH was used as the main template, and di-*n*-propylamine and cyclohexylamine as the secondary ones. The addition of both amines to the synthesis gel decreases the crystal size, but the former is more effective. In this case, cubic crystals with a narrow size distribution $(0.7 \pm 0.06)$ μm were obtained, resulting in more effective membranes for the separation of $CO_2$ and $CH_4$: smaller crystals with narrow size distributions pack better than large crystals, which leads to a decrease in the size of the intercrystalline pores and so to higher $CO_2/CH_4$ selectivities.

## 4.9
## Use of Co-Templates for Tailoring the Catalytic Activity of Microporous Materials

A key issue in zeolite catalysis refers to the possibility of controlling the distribution of aluminum atoms in the framework, and hence that of their associated acid sites. The preference for different SDAs to be located within specific and different void architectures in a given zeolite structure gives unexpected perspectives for gaining an effective control over Al, for these molecules would drive aluminum toward different T-sites according to their location. This hypothesis has been recently experimentally proven in the case of zeolite ferrierite [66–68]. For this purpose, crystals of zeolite ferrierite $(Si/Al = 15)$ were synthesized in fluoride media in the absence of alkali cations but in presence of different SDAs, pyrrolidine, pyrrolidine + TMA, and TMA + bmp. X-ray refinement of the crystals showed that pyrrolidine is occluded in both the cage and the 10-ring channels and, if pyrrolidine TMA is also present in the synthesis gel, this cation is located in some of the FER cages as well, but not in the 10-ring channel. If a TMA + bmp mixed-template system is used, TMA is found in the cages, while bmp is located in the 10-ring channel. It has been shown that the accessibility of a basic probe molecule such as pyridine to the active sites increases with the population of the ferrierite cages by TMA, with the sample containing only pyrrolidine exhibiting the lowest accessibility. Pyridine is too bulky to enter the cage through the eight-ring windows, so these results evidenced that the population of Brönsted acid sites within the cage and in positions accessible through the channel varies according to the combination of SDAs used in the synthesis. Catalytic activity of these materials in *m*-xylene isomerization and *n*-butene skeletal isomerization showed an increase with the acid sites accessibility.

In line with the previous argument for ferrierite zeolite, a similar approach can be used for tailoring the catalytic activity of SAPO materials. In this case, the use of different co-templates is envisaged to control the incorporation mechanisms whereby heteroatoms are incorporated in the network of microporous materials, rather than to control their spatial distribution as in the previous case. Silicon can be incorporated in $AlPO_4$ networks through different substitution mechanisms: SM2, which leads to the formation of isolated $Si(OAl)_4$, leading to one Brönsted site per Si ion with a low acid strength, or SM2 + SM3, giving room to the formation of stronger acid sites associated with silicon islands, but in a lower concentration. There is experimental evidence that different organic SDA molecules direct the incorporation of Si heteroatoms in $AlPO_4$ networks through the different substitution mechanisms, thus altering the acidity and catalytic activity of the SAPO catalysts obtained. For instance, in SAPO-5, triethylamine favors the incorporation through SM2, leading to more abundant but weaker acid sites, while benzylpyrrolidine favors the insertion through the formation of Si islands, with a lower concentration of acid sites but of higher acidity. This different behavior of the SDA molecules is additive, that is, the use of combinations of the two molecules as SDAs leads to SAPO materials with Si distributions, and so with acid and catalytic properties related to the ratio of the two SDAs employed in the synthesis. Therefore, the catalytic activity of these materials in reactions that require different acid strengths can be easily tailored as needed by using combinations of organic molecules that direct the incorporation of heteroatoms through different replacement mechanisms in the desired ratio [69].

Other examples of the influence of mixed-template systems over the Si distribution in SAPO materials have been reported. The use of mixtures of diethylamine (DEA) and di-*iso*-propylamine in the synthesis of SAPO-11 led to a better dispersion of Si by favoring the substitution mechanism SM2 as compared with that occurring when both templates are used separately, and they are more active and selective to isomerization products in the hydroisomerization of *n*-tetradecane [70]. A similar effect has been found by adding a small amount of methylamine as a co-template in the synthesis of SAPO-11 and SAPO-34 (CHA structure) from di-*n*-propylamine (DPA) and TEAOH containing gels, respectively. The effect in this case is attributed to the fact that methylamine prevents silica polymerization in the gel, favoring the insertion of small Si oligomers in the framework, thus increasing the number of negative charges in the framework which are compensated by the incorporation of small amounts of methylammonium cations [71]. In the synthesis of the one-dimensional 12-ring channel SAPO-41, a mixture of two secondary amines with different chain lengths, DPA and DEA, led to materials with smaller crystal size and stronger acidity, where both templates were occluded inside the channels. In this case, the addition of DEA increases the average size of the Si islands. The obtained catalysts were also more active, yet less selective, in the hydroisomerization of *n*-octane [72].

As a conclusion, we have tried to highlight in this chapter the important influence that cooperative structure-directing effects between different species present in the synthesis gels, including those between different types of organic molecules and between these and mineralizing agents or water, exerts in molecular sieve science, ranging from the production of complex topological structures to the control of the crystalline growth as well as the tailoring of the catalytic activity of the materials. This synthesis strategy thus opens up new possibilities in the synthesis and applications of zeolite-type materials that would be otherwise unviable in traditional synthesis systems involving only one type of SDA.

## 4.10 Summary and Outlook

A comprehensive review of the zeolite materials obtained by the combined use of two different organic SDAs in the synthesis gel has revealed that this strategy is most suitable for the efficient crystallization of structures built up by assembling cages and channels of different sizes and shapes. Following this approach, each template is generally accommodated in one specific site, dictated by geometrical correspondences between the guest template and the host void volume.

While the above model on the action mechanism of dual-template systems can be recognized in many cases, progress in the characterization of actual crystals obtained by using either two or even only one template evidenced a more complex picture, yet even more challenging, in which inorganic species other than cations, namely, water molecules and anions such as hydroxyl and fluoride, also cooperate with the organic guest molecules to stabilize the zeolite crystals. All these are but ingredients of a rich "chemical soup" that nurtures the nascent zeolite nuclei, which absorb from the medium whatever elements serve them in their continuous growth. In some cases, these small chemical entities serve to stabilize zeolite cages which are too small to be occupied by organic molecules, contributing in this way not in a small quantity to the overall stability of the structure, while in others they are actually involved in chemical interactions with the organic species to create supramolecular assemblies which behave as the actual SDAs for the nucleation of the corresponding zeolite structure. The additional stabilization energy provided by such chemical cooperativeness is often reflected in nucleation and crystal growth-depending properties of the resulting crystals, such as their size and/or shape, both being most relevant for catalysis and adsorption applications.

The structure co-directing role of different templates, each of which would in principle be located in different zeolite cavities, would also open unexpected perspectives for influencing and eventually controlling the location of heteroatoms, such as aluminum, for example, in the framework. The assumption here is that each template would affect the location of the aluminum atoms in its vicinity in a nearly independent way; hence, the aluminum sitting could be modified by using different combinations of template molecules. As long as different templates would be occluded in different cavities, this strategy would result in crystals where the aluminum would also be unevenly distributed among different cavities. Moreover, if those cavities could accommodate more than one type of organic molecule, changes in the chemical nature of the corresponding template would further affect the location of aluminum, for it would be possible in this way to affect the template–aluminum interaction independently in each cavity.

This strategy for influencing the aluminum's location in the framework has already been proven to be valid for zeolite ferrierite, but it could be extended to many other structures as well, which contain cages and/or channels of different size.

Taking all these considerations into account, it can be concluded that co-templating synthesis strategies entail specific characteristics that make them suitable for developing zeolite materials with new properties. Although this approach can be dated back to the earliest times of the discovery of high-silica zeolites and aluminophosphate materials, when it was used as a valuable tool for the crystallization of new structures, it has however been scarcely applied since then in the exploration of the synthesis of new zeolite topologies. Several interesting structures have been discovered in recent years by using different synthesis strategies, all of which involve however just one type of template. It would not be without interest to revisit such strategies by using co-SDAs that would be able to stabilize small cages or channels that, in combination with the void topologies of actual structures, would eventually render related low-density materials with higher complexity, or even totally new structures. This would however require a careful exploration of the synthesis parameters to overcome the crystallization of the competing cage-like structures that are usually formed by using the small templates alone. This work would benefit by the use of advanced modeling techniques that can help in selecting the most appropriate template combinations.

Control of heteroatom sitting would be feasible by a wise choice of SDAs, and this strategy can be extended to a range of zeolite-type materials, from zeolites to metal-containing aluminophosphates, and to heteroatoms ranging from aluminum in zeolites (or other trivalent elements or tetravalent ones like germanium, tin, or titanium) to silicon, cobalt, tin, zinc, or magnesium, to name a few, in aluminophosphates.

The chemical phenomena encompassed under the heading of "co-templating" are adding new elements to the already complex field of zeolite crystallization, where they contribute to reshape the classical scenario toward a more colorful one, one that would provide new resources for creative chemistry.

## Acknowledgments

The authors acknowledge the Spanish Ministry of Science and Education for financial support (project CTQ2006-06282). A. B. P. and L. G. H. are grateful to the Spanish Ministry of Science and Innovation for a predoctoral grant and a postdoctoral fellowship, respectively. R. G. acknowledges CSIC for the J.A.E. contract.

## References

1. Barrer, R.M. (1982) *Hydrothermal Chemistry of Zeolites*, Academic Press, New York.
2. Aiello, R. and Barrer, R.M. (1970) *J. Chem. Soc. A*, 1470.
3. Flanigen, E.M. (1973) *Adv. Chem. Ser*, **121**, 119.
4. Davis, M.E. and Lobo, R.F. (1992) *Chem. Mater*, **4**, 756.
5. Pérez-Pariente, J. and Gómez-Hortigüela, L. (2008) *Zeolites: From Model Materials to Industrial Catalysts*, Chapter 3 Transworld Research Network, pp. 33–62.
6. Corma, A., Navaro, M.T., Rey, F., and Valencia, S. (2001) *Chem. Commun*, 1486.
7. Corma, A., Díaz-Cabañas, M.J., Jorda, J.L., Martinez, C., and Moliner, M. (2006) *Nature*, **443**, 842.
8. Lok, B.M., Cannan, T.R., and Messina, C.A. (1983) *Zeolites*, **3**, 282.
9. Jacobs, P.A. and Martens, J.A. (1987) *Synthesis of High-silica Aluminosilicate Zeolites*, Elsevier, Amsterdam.
10. Pelrine, B.P. (1981) US Patent 4,259, 306.
11. Dwyer, F.G. and Jenkins E.E. (1981) US Patent 4,287,166.
12. Casci, J.L. Lowe, B.M., and Whittam, T.V. (1981) UK Patent Application GB 2 077 709.
13. Schlenker, J.L., Dwyer, F.G., Jenkins, E.E., Rohrbaugh, W.J., and Kokotailo, G.T. (1981) *Nature*, **294**, 340.
14. Weigel, S.J., Gabriel, J.C., Gutiérrez-Puebla, E., Monge-Bravo, A., Henson, N.J., Bull, L.M., and Cheetham, A.K. (1996) *J. Am. Chem. Soc*, **118**, 2427.
15. Ref 9, recipe 3.b, p. 11.
16. Chu, P. (1980) EPA 0023089.
17. Ref 9, p. 22.
18. Schlenker, J.L., Rohrbaugh, W.J., Chu, P., Valyocsik, E.W., and Kokotailo, G.T. (1985) *Zeolites*, **5**, 355.
19. Kirschhock, C.E.A., Liang, D., Van Tendeloo, G., Fécant, A., Hastoye, G., Vanbutsele, G., Bats, N., Guillon, E., and Martens, J.A. (2009) *Chem. Mater*, **21**, 371.
20. Seddon, D. and Whittam, T.V. (1981) EPA 55,529.
21. Pinar, A.B., Gómez-Hortigüela, L., and Pérez-Pariente, J. (2007) *Chem. Mater*, **19**, 5617.
22. García, R., Gómez-Hortigüela, L., Díaz, I., Sastre, E., and Pérez-Pariente, J. (2008) *Chem. Mater*, **20**, 1099.
23. Dorset, D.L. and Kennedy, G.J. (2004) *J. Phys. Chem. B*, **108**, 15216.
24. Camblor, M.A., Corma, A., Díaz-Cabañas, M.J., and Baerlocher, C. (1998) *J. Phys. Chem. B*, **102**, 44.
25. Zones, S.I., Hwang, S.-J., and Davis, M.E. (2001) *Chem. Eur. J*, **7** (9), 1990.
26. Burton, A., Accardi, R.J., Lobo, R.F., Falcioni, M., and Deem, M.W. (2000) *Chem. Mater*, **12**, 2936.
27. Millini, R., Carluccio, L.C., Carati, A., Bellussi, G., Perego, C., Cruciani, G., and Zanardi, S. (2004) *Microporous Mesoporous Mater*, **74**, 59.
28. Knight, L.M., Miller, M.A., Koster, S.C., Gatter, M.G., Benin, A.I., Willis, R.R., Lewis, G.J., and Broach, R.W. (2007) *Stud. Surf. Sci. Catal*, **170**, 338.
29. Roth, W.J. (2005) *Stud. Surf. Sci. Catal*, **158**, 19.
30. Zones, S.I. (1987) US Patent 4 665 110.
31. Millini, R., Perego, G., Parker, W.O., Bellussi, G., and Carluccio, L. (1995) *Microporous Mesoporous Mater*, **4**, 221.

32. Camblor, M.A., Corell, C., Corma, A., Diaz-Cabañas, M.J., Nicolopoulos, S., Gonzalez-Calbet, J.M., and Vallet-Regi, M. (1996) *Chem. Mater*, **8**, 2415.
33. Barrett, P.A., Diaz-Cabañas, M.J., and Camblor, M.A. (1999) *Chem. Mater*, **11** (10), 2919.
34. Blackwell, C.S. *et al.* (2003) *Angew. Chem. Int. Ed*, **42**, 1737.
35. Jackowski, A., Zones, S.I., Hwang, S.-J., and Burton, A. (2009) *J. Am. Chem. Soc*, **131**, 1092.
36. Lok, P.M., Messina, C.A., Patton, R.L., Gajek, R.T., Cannan, T.R., and Flanigen, E.M. (1984) US Patent 4 440 871.
37. Edwards, G.C., Gilson, P.J., and Mc Daniel, V. (1987) US Patent 4 681 864.
38. de Saldarriaga, L.S., Saldarriaga, C., and Davis, M.E. (1987) *J. Am. Chem. Soc*, **109**, 2686.
39. Schreyeck, L., D'agosto, F., Stumbe, J., Caullet, P., and Mougenel, J.C. (1997) *Chem. Commun*, 1241.
40. Paillaud, J.-L., Caullet, P., Schreyeck, L., and Marler, B. (2001) *Microporous Mesoporous Mater*, **42**, 177.
41. Wheatley, P.S. and Morris, R.E. (2002) *J. Solid State Chem*, **167**, 267.
42. Sierra, L., Deroche, C., Gies, H., and Guth, J.L. (1994) *Microporous Mesoporous Mater*, **3**, 29.
43. Sierra, L., Patarin, J., Deroche, C., Gies, H., and Guth, J.L. (1994) *Stud. Surf. Sci. Catal*, **84**, 2237.
44. Corma, A., Rey, F., Rius, J., Savater, M.J., and Valencia, S. (2004) *Nature*, **431**, 287.
45. Wright, P.A., Maple, M.J., Slawin, A.M.Z., Patinec, V., Aitken, R.A., Welsh, S., and Cox, P.A. (2000) *J. Chem. Soc., Dalton Trans*, **8**, 1243.
46. Castro, M., Garcia, R., Warrender, S.J., Slawin, A.M.Z., Wright, P.A., Cox, P.A., Fecant, A., Mellot-Draznieks, C., and Bats, N. (2007) *Chem. Commun*, 3470.
47. Xing, H., Li, J., Yan, W., Chen, P., Jin, Z., You, J., Dai, S., and Xu, R. (2008) *Chem. Mater*, **20**, 4179.
48. Zones, S.I. and Hwang, S.-J. (2002) *Chem. Mater*, **14** (1), 313.
49. Lee, G.S., Nakagawa, Y., and Zones, S.I. (2000) US Patent 6,156,290.
50. Lee, H., Zones, S.I., and Davis, M.E. (2003) *Nature*, **425**, 385.
51. Lee, H., Zones, S.I., and Davis, M.E. (2005) *J. Phys. Chem. B*, **109**, 2187.
52. Lee, H., Zones, S.I., and Davis, M.E. (2006) *Microporous Mesoporous Mater*, **88**, 266.
53. Arranz, M., Pe'rez-Pariente, J., Wright, P.A., Slawin, A.M.Z., Blasco, T., Gómez-Hortigüela, L., and Cora, F. (2005) *Chem. Mater*, **17**, 4374.
54. Arranz, M., García, R., and Pérez-Pariente, J. (2004) *Stud. Surf. Sci. Catal*, **154**, 257.
55. Gómez-Hortigüela, L., Corà, F., Márquez-Álvarez, C., and Pérez-Pariente, J. (2008) *Chem. Mater*, **20**, 987.
56. Gómez-Hortigüela, L., Pérez-Pariente, J., and Corà, F. (2009) *Chem. Eur. J*, **15**, 1478.
57. Gómez-Hortigüela, L., López-Arbeloa, F., Corà, F., and Pérez-Pariente, J. (2008) *J. Am. Chem. Soc*, **130**, 13274.
58. Fernandes, A., Ribeiro, M.F., Borges, C., Lourenço, J.P., Rocha, J., and Gabelica, Z. (2006) *Microporous Mesoporous Mater*, **90**, 112.
59. Chatelaine, T., Patarin, J., Soulard, M., and Guth, J.L. (1995) *Zeolites*, **15**, 90.
60. Ferchiche, S., Valcheva-Traykova, M., Vaughan, D.E.W., Warzywoda, J., and Sacco, A. Jr. (2001) *J. Cryst. Growth*, **222**, 801.
61. Berger, C., Gläser, R., Rakoczy, R.A., and Weitkamp, J. (2005) *Microporous Mesoporous Mater*, **83**, 333.
62. Kecht, J., Mintova, S., and Bein, T. (2008) *Microporous Mesoporous Mater*, **116**, 258.
63. Yu, H., Wang, X.-Q., and Long, Y.-C. (2006) *Microporous Mesoporous Mater*, **95**, 234.
64. Lai, Z.P., Bonilla, G., Díaz, I., Nery, J.G., and Tsapatsis, M. (2003) *Science*, **300**, 456.
65. Carreon, M.A., Li, S., Falconer, J.L., and Noble, R.D. (2008) *Adv. Mater*, **20**, 729.
66. Pinar, A.B., Pérez-Pariente, J., and Gómez-Hortigüela, L. (2008) WO2008116958A1.
67. Pinar, A.B., Márquez-Álvarez, C., Grande-Casas, M., and Pérez-Pariente, J. (2009) *J. Catal*, **263**, 258.

68. Márquez-Alvarez, C., Pinar, A.B., García, R., Grande-Casas, M., and Pérez-Pariente, J. (2009) *Top. Catal*, **52**, 1281.
69. Gómez-Hortigüela, L., Márquez-Álvarez, C., Grande-Casas, M., García, R., and Pérez-Pariente, J. (2009) *Microporous Mesoporous Mater*, **121**, 129.
70. Liu, P., Rien, J., and Sun, Y. (2008) *Microporous Mesoporous Mater*, **114**, 365.
71. Fernandes, A., Ribeiro, F., Lourenço, J.P., and Gabelica, Z. (2008) *Stud. Surf. Sci. Catal*, **174A**, 281.
72. Li, L. and Zhang, F. (2007) *Stud. Surf. Sci. Catal*, **170A**, 397.

# 5
# Morphological Synthesis of Zeolites

*Sang-Eon Park and Nanzhe Jiang*

## 5.1
## Introduction

Zeolites are crystalline inorganic materials with unique, characteristic micropores. These micropores in molecular scale, which have one to three dimensions, are due to interconnected cavities or channels. The types of assembly of cavities or channels control the size of micropores, orientation of microporous channels, and the morphology of zeolite particles. Undoubtedly, the morphology is originally related to the framework type and also closely related to the micropore size, crystal size, and shape, and directly affects the physicochemical properties of zeolites.

Zeolites are extensively used in the fields of heterogeneous catalysis, separations, ion exchange, chemical separation, adsorption, host/guest chemistry, microelectronic devices, optics, and membranes [1–4]. Such applications of zeolites are strongly affected by the pore size, type of channels, and morphologies [5–9]. Morphological synthesis of zeolites is particularly important in catalytic applications where the particle shape can have a dramatic effect on the product distribution due to differences in rates of transport/diffusion and reaction. Recent research interests have also been directed toward the crystallography of single zeolite crystals by applying nanotechnology. Thus, there has been a great interest in developing synthetic approaches to control crystal size and morphology of zeolites [9–11].

Fine-tuning of zeolite crystal size and shape is usually accomplished by systematic variation of the composition of the precursor mixture (including the use of additives such as salts) [12]. But through the development of new technology and equipment, synthetic parameters such as temperature, pressure, and stirring rate or even gravity can be controlled.

This chapter provides a brief introduction to the morphological synthesis of zeolites and related fabrication methods such as as microwave. Examples in the preparation of well-shaped single crystals, fine control of common particles, and microwave driven fabrication are discussed.

*Zeolites and Catalysis, Synthesis, Reactions and Applications. Vol. 1.*
Edited by Jiří Čejka, Avelino Corma, and Stacey Zones

ISBN: 978-3-527-32514-6

## 5.2
## Morphology of Large Zeolite Crystals

The synthesis of large zeolite single crystals is of a great interest for a large number of requirements, including single-crystal structure analysis; fine-structure analysis; study of crystal growth mechanisms; study of adsorption and diffusion; and the determination of anisotropic electrical, magnetic, or optical properties [13–16]. For example, although the high surface areas of small zeolite particles might initially appear advantageous for heterogeneous catalysis, in the case of shape-selective catalysis by zeolites it is well established that a higher selectivity is offered by larger crystallites, because the internal surface area accounts for a much greater proportion of the total zeolite surface area [17].

Zeolites have distinct framework types that can be built in a periodic pattern by linking the basic building units (BBUs), called *tetrahedrons.* These BBUs can be continuously linked together to form more complex composite building units (CBUs), such as rings, which in turn will further lead to the next level, that is, cages (Figure 5.1). The type of building units or the way to connect them is the intrinsic factor deciding morphology, especially morphology of single crystals. And another important implication is to understand the crystallization of zeolites, that is, the ability to engineer crystal morphology and crystal size.

### 5.2.1
### Large Crystals of Natural Zeolites

In nature, large zeolite crystals are often found in volcanogenic sedimentary rocks, which are believed to have formed via dissolution of volcanic glass [18]. Figure 5.2 and Table 5.1 show some well-known crystalline natural zeolites. Typical morphologies are the rhombic-shaped polyhedron or needle-shaped fibrous zeolites of several centimeters in size. In nature, zeolites may grow under hydrothermal conditions in the earth's crust, which are termed *closed hydrologic conditions* [19]. And natural zeolites usually occur in combined form with other silicate minerals

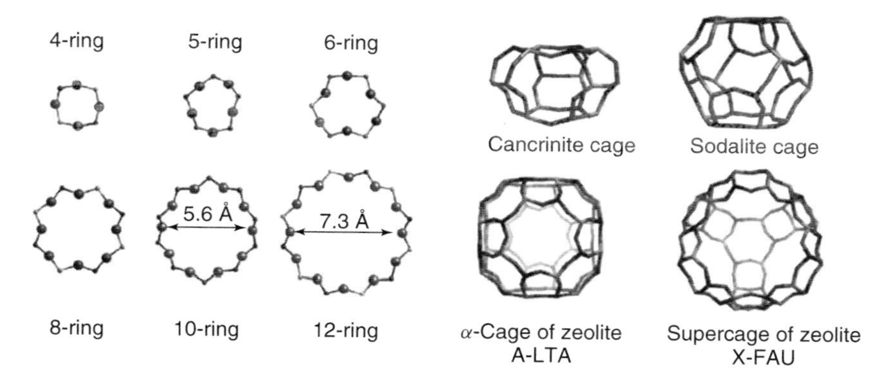


**Figure 5.1** Rings and cages frequently found in zeolite structures.

**Figure 5.2** Some natural zeolite crystals: (a) ammonioleucite (Fujioka, Japan); (b) analcime (Quebec, Canada); (c) erionite (Oregon, USA); (d) natrolite (California, USA); and (e) mordenite (Washington, USA) (*http://www.iza-online.org*).

such as clays and dense forms of silica, and their chemical composition varies from one location to another.

### 5.2.2
### Synthesis of Large Zeolite Crystals

Synthesis of large zeolite single crystals has been achieved by several excellent means [20]. A detailed study of the crystal structure, crystallization mechanism, and relation with morphology has become possible. Therefore, the synthesis of large, single zeolite crystals is highly desirable from the point of view of structural determination or discovering the crystallization mechanism. Although recent advances in X-ray powder diffractometry combined with sensitive spectroscopic methods such as multinuclear NMR methods have made structural determination from polycrystalline powders amenable (indeed most zeolites are structurally characterized by powder data) [20], diffraction data from a single crystal often remain the most accurate means of unambiguous structure elucidation, including the location of both the framework and extraframework atoms in a zeolite [21–28]. Hence it is possible to synthesize submicrometer-sized zeolite crystals [29–33] as well as very large crystals [34–37] by reproducible synthesis. The former relies on the syntheses where nucleation rates are very high, often involving monomeric silica and alumina sources, high alkalinity, and low temperatures. The latter relies on syntheses employing low-solubility silicon sources, fluoride as the mineralizing

**Table 5.1** Material data for some nature zeolites (data from *http://www.iza-online.org*).

| Zeolites | Composition | Framework type | CBUs | Morphology |
|---|---|---|---|---|
| Ammonioleucite | $\|NH_4K)\|[AlSi_2O_6]$ | ANA | 6-2 or 6 or 4-[1,1] or 1-4-1 or 4 (SBUs) | Tetragonal |
| Analcime | $\|Na\ (H_2O)\|[AlSi_2O_6]$ | ANA | | Isometric or pseudoisometric single crystals are trapezohedra in sizes ranging from millimeters to several centimeters |
| Erionite | Ca $\|K_2(Ca_{0.5},Na)_8(H_2O)_{30}\|[Al_{10}Si_{26}O_{72}]$ | ERI | d6r, can | Hexagonal, 6/*m*2/*m*2/*m* single crystals are hexagonal prisms terminated by a pinacoid with sizes under 3 mm, fibrous, and wool-like |
| Natrolite | Natrolite $\|Na_2(H_2O)_2\|[Al_2Si_3O_{10}]$ | NAT | nat | Orthorhombic *mm2* single crystals are pseudotetragonal prisms terminated by a pyramid sizes range from a few millimeters to several centimeters |
| Mordenite | $\|Na_2,\ Ca,K_2)_4(H_2O)_{28}\|[Al_8Si_{40}O_{96}]$ | MOR | mor | Orthorhombic *mmm* or *mm2* single crystals are thin fibers, 0.1–10 mm long |

agent, and high temperatures. These methods mainly focus on lowering the speed of nucleation or crystallization steps extremely by (i) adding chemical reagents or (ii) applying novel synthetic conditions. Here we will show some typical examples.

Various nucleation suppressing agents are applied to prepare large zeolite crystals with well-shaped morphology by extremely slow and controlled nucleation and crystallization steps. The most common LTA and FAU zeolites were synthesized into big crystals (Figure 5.3) by applying a nucleation suppression agent such as tertiary alkanolamine. The morphology of big crystals has the same shape symmetry as the framework type [38]. The unit cell and crystal structure would be understandable easily through such morphology of single crystals.

The most interesting agent is the fluoride anion. By applying the fluoride anion as the mineralizing agent instead of hydroxide, large zeolite crystals with the framework types MFI (Figure 5.4), FER, MTT, MTN, and TON have been prepared.

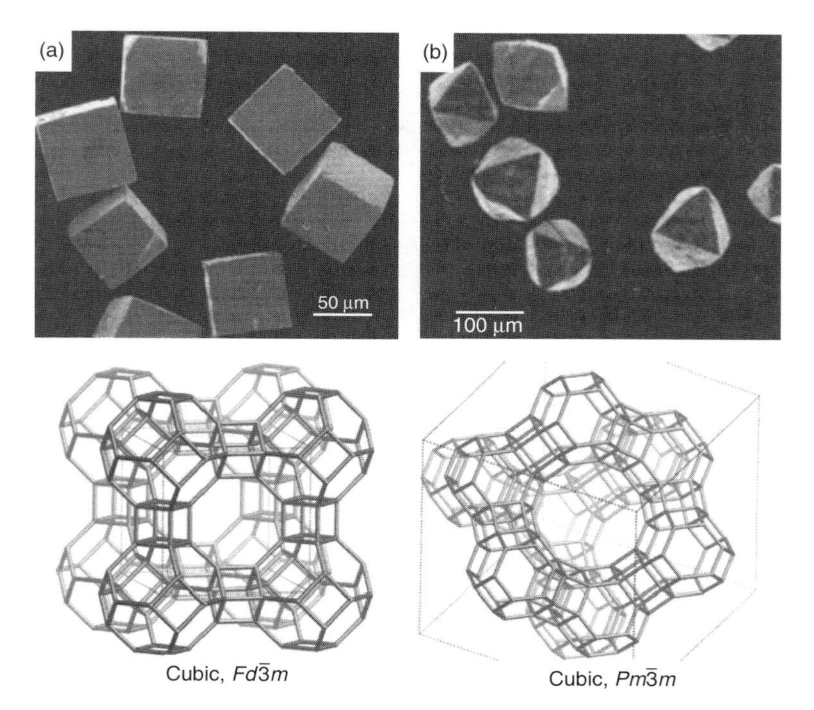


**Figure 5.3** Large single crystals of (a) LTA and (b) FAU [38].

This method has allowed not only the formation of large zeolite crystals but also the growth of crystals with a high silica content and free of defects compared with those prepared by conventional routes [39, 40]. Even though $F^-$ has a tendency to form complexes with the initial reactant species such as silicon, these complexes slowly hydrolyze to release less fluorinated silicon, which gradually supplies the nutrients for crystal growth, permitting large crystallites to be formed [41–43]. Under these conditions of low supersaturation, growth is favored at the expense of nucleation and thus a small number of large zeolite crystals are ultimately formed. The shape and size of large crystals are controllable by varying the chemical compositions (Figure 5.4b).

Single crystals of Si-MFI (all-silica MFI) zeolite with different crystalline sizes ranging from $9 \times 3 \times 2\,\mu m$ to $165 \times 30 \times 30\,\mu m$ were obtained by adding benzene-1,2-diol as a complexing agent. Crystals synthesized in the presence of benzene-1,2-diol have a much larger size than those synthesized in its absence, and their size and shape were largely influenced by the content of benzene-1,2-diol in the reaction system (Figure 5.5). It is interesting to note that the aspect of ratio increased as a function of the content of benzene-1,2-diol, and the crystal size increased along the length rather than the width of the crystal [46].

Very interestingly, through a novel crystallization technique (the bulk-material dissolution (BMD) technique), giant zeolite crystals with size of several millimeters were synthesized by controlling the release and solubility of reactive solution

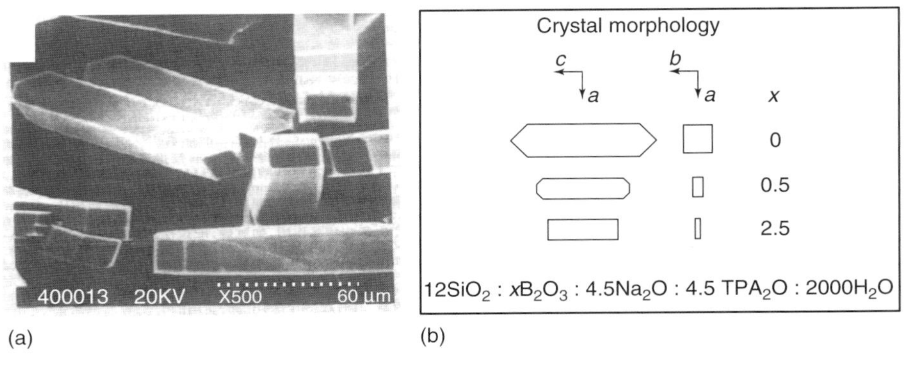


**Figure 5.4** (a) (B, Al)-MFI (B-ZSM-5) [44]. (b) Modification of silicalite morphology with addition of boric acid to conventional hydrothermal synthesis [45].

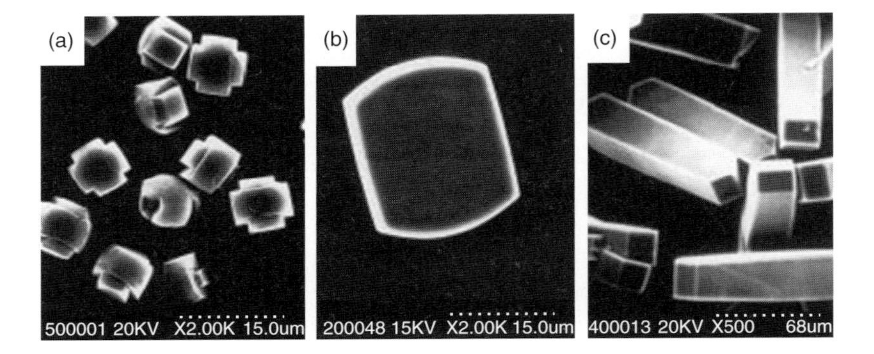


**Figure 5.5** Crystal size of Si-MFI zeolite synthesized at 180 °C with the molar composition: $SiO_2$-0.2TPABr-$x$R-0.5NaOH-30$H_2O$ (R: benzene-1,2-diol) (a) $x = 0$; (b) $x = 0.2$; (c) $x = 0.4$ [46].

species in organothermal systems [47]. Si-MFI zeolite, Analcime (ANA), and JBM crystals with sizes of about 3 mm were successfully synthesized by using bulk materials (quartz tube) as the silica and alumina sources (Figure 5.6). Well-shaped giant crystals with different morphologies obviously prove that they originated from the zeolite framework type.

Instead of above novel method, hydrothermal synthesis was intensively investigated for synthesizing single crystals of Al-MFI zeolite (ZSM-5) with varying chemical compositions and conditions to facilitate detailed structural studies of the MFI framework. Single crystals of Al-MFI zeolite were synthesized in systems containing $Na^+$-tetrapropylammonium (TPA), $Li^+$-TPA, and $NH_4^+$-TPA. The samples consisted of a fully crystalline and pure zeolitic phase with good homogeneity with crystal sizes up to 420 µm [48]. In the alkaline-free $NH_4^+$-TPA system, homogeneous and pure single crystals of Al-MFI zeolite were prepared of lengths up to 350 µm [49]. The crystal sizes and yields were found to depend on

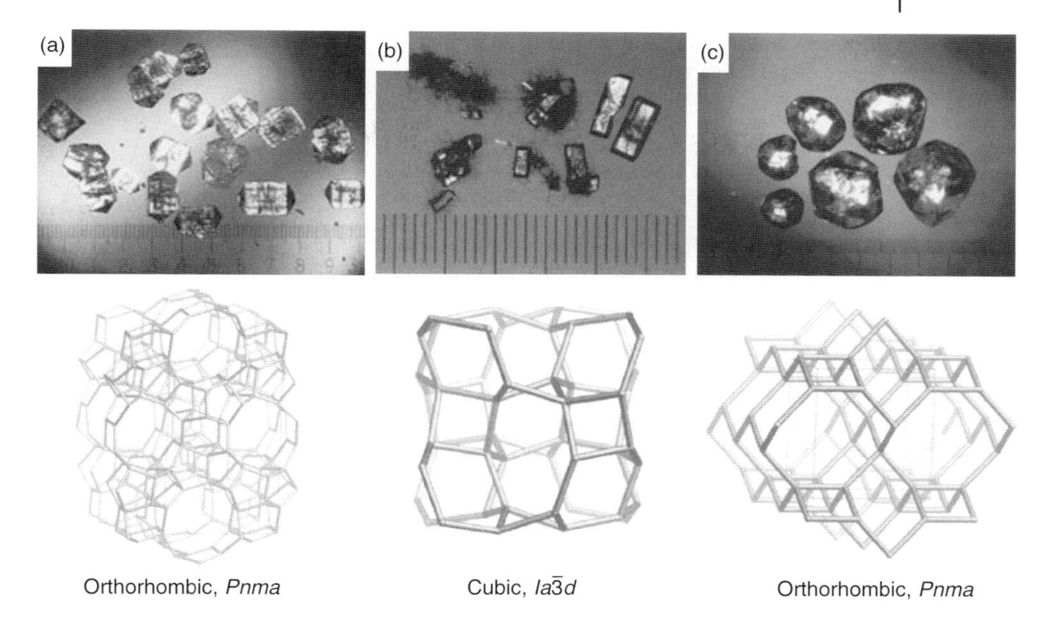


**Figure 5.6** (a) Giant crystals of Si-MFI zeolite; (b) Giant crystals of ANA zeolite; and (c) Giant crystals of JBW zeolite [47].

the water content of the starting reaction mixture and on the type of aluminum source. Large single crystals of Si-MFI zeolite have been synthesized with the choline cation, 1,4-diazabicyclo [2.2.2]octane, and tetramethylammonium (TMA) cations from an alkaline-free medium [50]. At high temperature (300 °C) and high pressure (100 MPa), the structure-directing organic template of $TPA^+$ is stabilized, and hydrolyzed tetraethylortosilicate (TEOS) grow to millimeter-sized big crystals of Si-MFI zeolite [51]. The elevated temperature and pressure favor the formation of crystals with improved quality. Prismatic Si-MFI zeolite crystals with a uniform size of about $0.7 \times 0.2 \times 0.2$ mm have been obtained by heating a gel prepared from TMA-silicate solution, TPABr, and sodium hexafluorosilicate at 250 °C under a pressure of 80 MPa (Figure 5.7). The influence of synthesis conditions on the crystal size has been studied systematically by changing the temperature, pressure, and gel compositions. Under the specific conditions of 250 °C and 80 MPa, a strong correlation between the crystal size and the F/Si mole ratio of the starting gel was found, which enabled the preparation of uniform crystals of Si-MFI zeolite with preset dimensions [51].

Generally, crystallization of large, single zeolite crystals can be achieved by controlling the nucleation and crystallization steps. But the various reagents applied would suppress the nucleation and slow down the crystal growth. At such low crystallization rates, it is difficult to separate nucleation and crystallization clearly. So reagents such as $F^-$, tetraethylammonium (TEA), or various amines and alcohols affect both processes. The synthesis conditions were varied by controlling

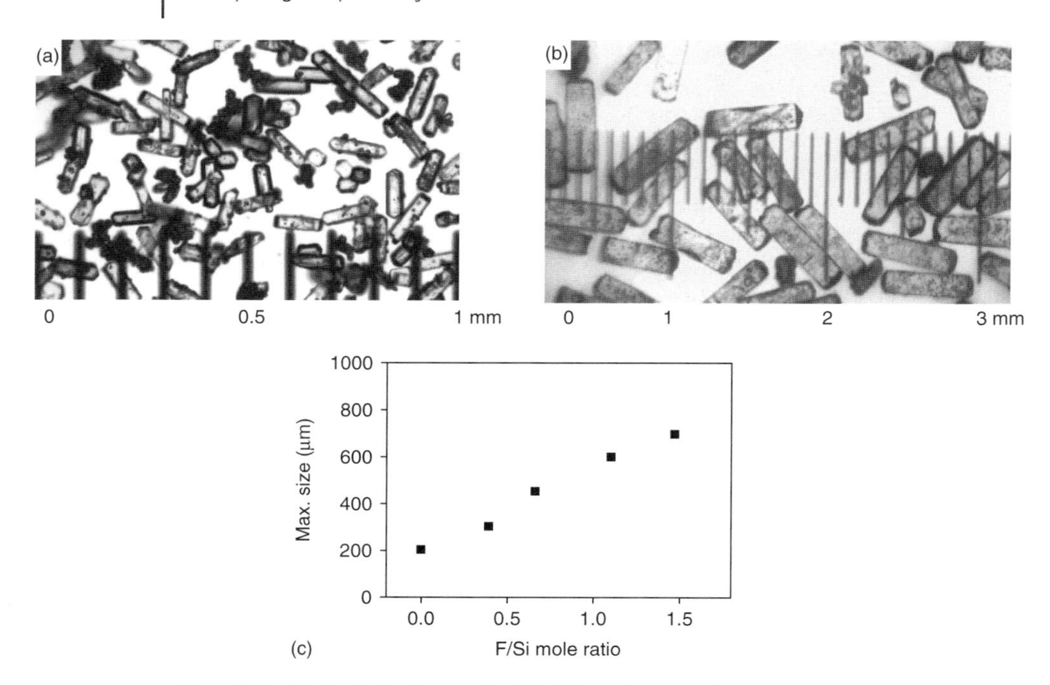


**Figure 5.7** (a) Optical micrograph of Si-MFI zeolite crystals synthesized at 300 °C and 100 MPa; (b) Optical micrograph of Si-MFI zeolite crystals synthesized at 250 °C and 80 MPa; (c) Correlation between Si-MFI zeolite crystal size and the reagent F/Si ratio adjusted by using $Na_2SiF_6$ for syntheses at 250 °C and 80 MPa [51].

pH of the synthetic precursor and/or by applying high pressure, temperature, and even gravity. Also, in space perfect single crystals of zeolites A and X have been synthesized [12]. All synthesized single crystals exactly reflect their own bulk symmetry from their unit cells in nanoscale.

## 5.3 Morphology Control of MFI Zeolite Particles (of Size Less than 100 μm)

Depending on the crystal morphology, pore openings of a particular channel system can be present at the crystal surface to different extents. As a consequence, the access to the intracrystalline volume may be facilitated or hindered. The crystal morphology and size define also the diffusion paths via the channels, which often have a great impact on their applications such as reaction kinetics. The diffusivity of guest molecules in zeolite crystals is closely related to the pore size, but they also depend on the crystal morphology. Especially as shape catalysts, the shape selectivity of zeolites is affected by their morphology (shape and size): for example, large crystals give higher selectivity but their long pathways decrease the efficiency.

So the selectivity and effectiveness need to be compromised by optimization of the crystal size and shape [37, 52].

Even though more than 130 different zeolites have been discovered so far, only a few are available as industrial catalysts. The crystal morphology of the materials possessing a three-dimensional channel system, for example, FAU- and LTA-type zeolites, is not expected to have a great impact on their properties. In contrast, the performance of materials with mono- or bidimensional channel systems might be strongly affected by the morphology of the crystals. Among them, we prefer to introduce morphological synthesis of zeolites having MFI-type framework.

MFI-type zeolite possesses an anisotropic framework with two intersecting 10-ring channels. The straight channels are parallel to *b* axis and the zigzag channels with an estimated pore opening of 0.51 nm × 0.55 nm are parallel to the *a* axis (Figure 5.8). The *b* channels and *a* channels are interconnected with each other, so diffusion along the *c* direction is also possible. It has been known that MFI-type zeolites with various crystal shapes can be prepared, such as spherical, hexagonal twined disk, rodlike, and so on [53–68]. In the synthesis of Al-MFI, the various factors influencing the dimensions along each axis of the crystal have been investigated systematically [69]. The dependence of metal cations, structure-directing agents, chemical source, and compositions has been summarized by Singh and Dutta [70].

## 5.3.1 Dependence of Structure-Directing Agents (SDAs)

The typical structure-directing agent (SDA) for MFI-type zeolite is the TPA cation. Instead of TPA, synthesis of MFI zeolite in the presence of dC6 (Figure 5.9) has been reported in several studies [72, 73]. The characteristic crystal shape of TPA-Si-MFI

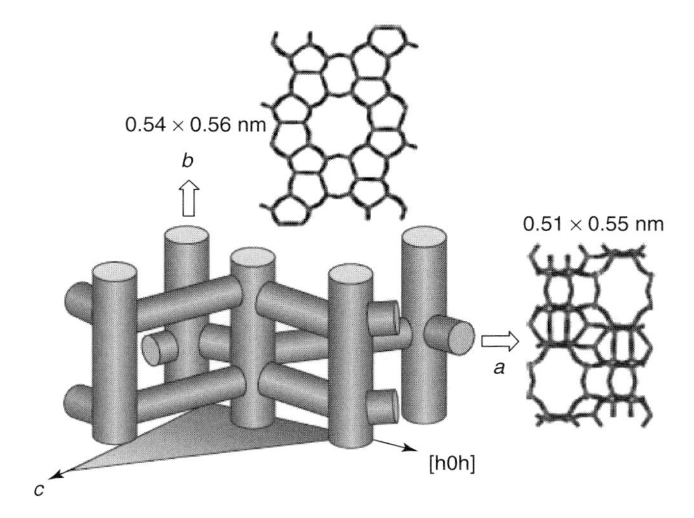


**Figure 5.8** Pore structure of the AL-MFI [71].

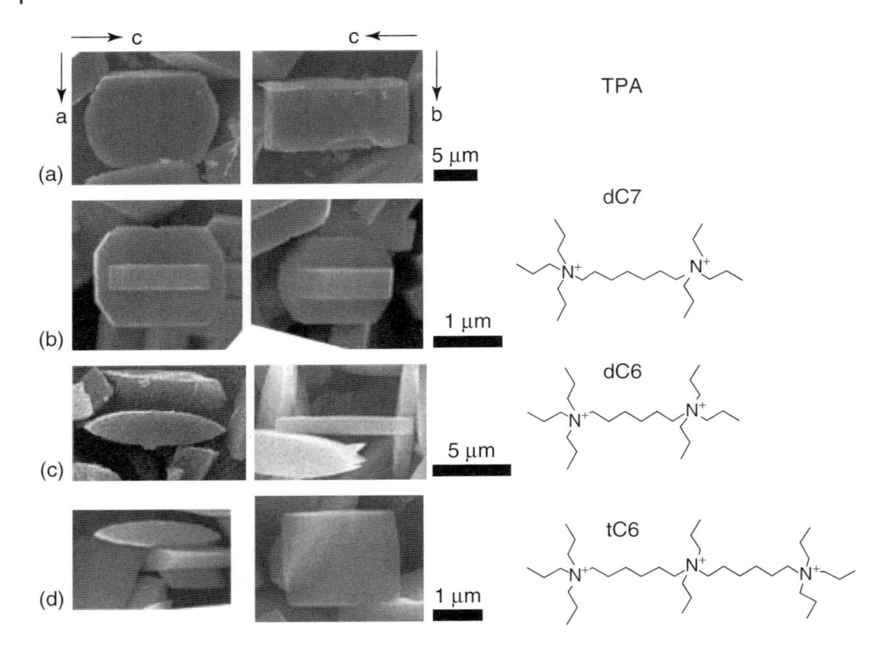


**Figure 5.9** SEM images of Si-MFI: (a) pill- or coffin-shaped crystals using TPA; (b) octagonal shaped crystals with twin intergrowths using dC7; (c) leaf-shaped crystals from dC6; (d) *b*-elongated leaf-shaped (or platelike) crystals from tC6 [74].

is hexagonal prismatic, more commonly referred to as a *coffin shaped*, with the order of crystal dimensions $L_c > L_a > L_b$ (where $L_i$ indicates crystal size along $i$ axis). Tsapatsis *et al.* controlled the order of crystal dimensions to $L_c > L_a = L_b$ with dimer of TPA (dC6, Figure 5.9) and to $L_c > L_b > L_a$ with trimer tC6 [74].

The morphological changes for Si-MFI zeolites with amine additives and TPABr have been reported [75]. The TPABr-containing crystals are rather elongated ($a \times b \times c = 80 \times 40 \times 20\ \mu m^3$), whereas the crystals containing TPA and DPA were smaller ($30 \times 25 \times 20\ \mu m^3$ and $6 \times 5 \times 4\ \mu m^3$, respectively) and isometric in shape (Figure 5.10).

Al-MFI crystals have been synthesized into cubic crystals in pyrrolidine-containing hydrous gels with uneven size and the diameter ranging from 0.5 to 4 μm [76].

By varying the TPABr content, Si-MFIs have been grown in a rodlike shape. Because fewer nuclei are formed at lower TPABr concentrations, the volume of the individual crystallites was inversely proportional to the initial TPABr concentration [77].

Si/TPA ratios have varied with values of 10, 24, and 48 [78]. With a ratio of 10, tablet-shaped crystals were formed with knobs at the top and bottom; for a ratio of 24, the crystals had a similar shape with sharp corners and were significantly larger. The larger size was a reflection of lower TPA content and reduced rate of

**Figure 5.10** SEM images of MFI-type zeolites prepared with structure directing agents: (a) TPABr; (b) tripropyl amine; and (c) dipropyl amine.

nucleation. With a Si/TPA ratio of 48, the size and shape remained the same as in 24, but there appeared to be a solid phase growing on the surface of the crystals.

### 5.3.2 Dependence on Alkali-Metal Cations

The morphology of AL-MFI was found to be dependent on the presence of alkali-metal ions [79]. Li and Na zeolites consisted of spheroidal 2–5 and 8–15 μm crystal aggregates of very small platelet-like units, respectively. K, Rb, and Cs zeolites consisted of twins of rounded (K, Rb) or sharp-edged crystals (Cs). $(NH_4)$-Al-MFI consisted of large lath-shaped, well-developed, and double-terminated single crystals.

The (Li, Na)-, Na-, and (Na, K)-Al-MFI zeolites have spherical or egg-shaped polycrystallites [80], and similar morphology of the Al-MFI zeolites for Na and K have been observed [81].

Morphology of Al-MFI synthesized from Na, K-TPA depended on the relative ratio of the alkali-metal cations. With both Na and K cations present at a ratio of $K/(K + Na) = 0.75$, large crystal aggregates were obtained in the range of 5–10 μm [82, 83].

Here, some of the batch compositions were studied, that is, $x Na_2O/8TPABr/100SiO_2/1000H_2O$ and $x TPA_2O/(8-2x)TPABr/100SiO_2/1000H_2O$, where $x$ varies from 0.5 to 4.0. As the alkalinity of the reaction mixture was reduced from $x = 4$ to $x = 0.5$, the aspect ratio (length/width) of the crystals increased from 0.9 to 6.7. Both nucleation and crystallization occurred more rapidly in the presence of $Na^+$. Synthesis of Al-MFI in glycerol solvent has been reported, and the morphology of the crystals was found to be hexagonal columns [84].

Addition of $Li_2O$ in the synthesis of zeolite TPA-Al-MFI with $(NH_4)_2O/Al_2O_3 = 38$ produces unusually uniform, large, lath-shaped crystals of Al-MFI about $140 \pm 10$ μm in length [85].

## 5.4
## Morphological Synthesis by MW

Zeolites can be effectively and rapidly synthesized by using microwave heating techniques [86]. The advantages in the microwave synthesis of zeolites are homogeneous nucleation, fast synthesis by rapid heat-up time, selective activation of the reaction mixture by microwaves, phase selective synthesis by fine-tuning of synthesis conditions, uniform particle, size-facile morphology control, fabrication of small crystallites and enhancement of crystallinity, and so on [87].

### 5.4.1
### Examples of MW Dependency

AFI-type molecular sieves Aluminophosphate five such as AlPO-5 and SAPO-5 have one-dimensional channels with micropores (0.73 nm) and they have been synthesized with various morphologies under microwave irradiation [88–91]. The morphologies were controlled by varying the reaction conditions and addition of extra components such as fluoride and silica (Figure 5.11). Rodlike crystals (with aspect ratio of about 40) were obtained with the addition of fluoride and increase of template and water concentrations. The platelike crystals (with an aspect ratio of about 0.2) were synthesized in an alkaline condition with the addition of an appropriate concentration of silica sol. In this case, it was supposed that silica might hinder the crystal growth in the *c* direction and the fluoride ion might retard the nucleation rate.

Most recently, Xu *et al.* controlled the morphology Si-MFI crystals (Figure 5.12) from a microwave-assisted solvothermal synthesis system in the presence of diols, that is, ethylene glycol (EG), diethylene glycol (DEG), triethylene glycol (TEG), and tetraethylene glycol (tEG) [92]. Under microwave radiation, the Si-MFI crystals with tunable sizes, shapes, and aspect ratios were crystallized.

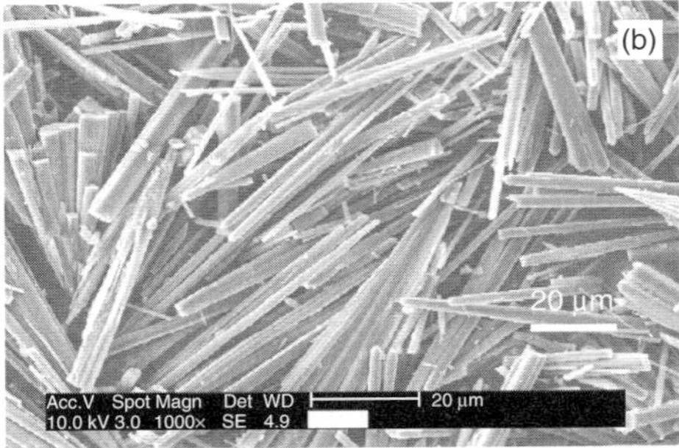


**Figure 5.11** SEM images of typical AFI molecular sieves: (a) platelike crystal and (b) rodlike crystal. White scale bar corresponds to 10 µm [87].

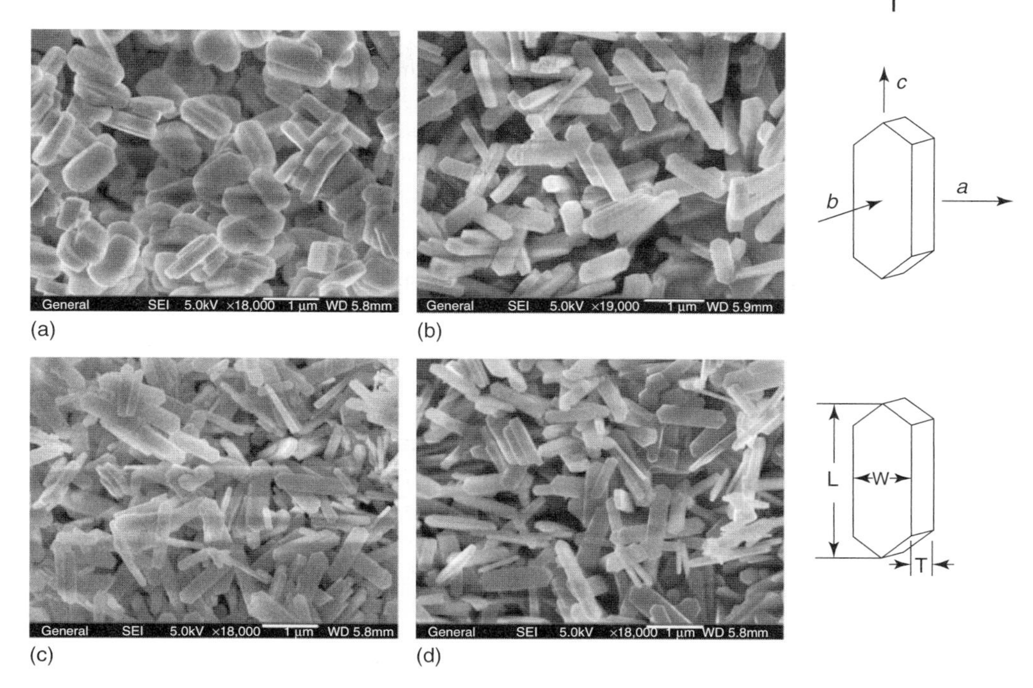


**Figure 5.12** The SEM images of the Si-MFI crystals crystallized from the microwave-assisted solvothermal synthesis system in the presence of the diols: (a) EG, (b) DEG, (c) TEG, and (d) tEG. A schematic identifying the crystal faces is shown on the right part of the figure. Gel composition (in molar ratio), SiO : TPAOH : EtOH : diols : $H_2O$ = 1 : 0.357 : 4.0 : 7 : 21.55 [92].

## 5.4.2 Morphological Fabrication by MW

Fabrication of nanostructured zeolites has attracted much attention in order to (i) optimize zeolite performance (no pore blocking and zeolite diluting binding additives are present), easy handling, and attrition resistance; (ii) minimize diffusion limitations with the secondary larger pores; and (iii) apply to nonconventional applications, such as guest encapsulation, bioseparation, enzyme immobilization, and so on [33]. By applying various templates or nanotechniques, zeolites could be fabricated into membranes and films, biomimic or hierarchical structures, and micro-/mesoporous materials [33]. Among them, fabrication of nanoporous materials using a chemical glue can be used for implementing nanoscopic or microscopic arrays of these materials. So far, there have been a few reports on the utilization of chemical glues such as inorganic glue [93], nano-glue [94], and organic covalent linkers [95]. Recently, we reported microwave fabrication of zeolites directly from the synthetic solution and proposed the incorporation of a transition metal as nano-glue [96].

In the microwave synthesis of Ti-MFI zeolite (TS-1), the surface titanol groups behave as inorganic glue and will stack Ti-MFI crystals to a fibrous morphology. This technique could be expanded to the fabrication of zeolite films and zeolite coatings [96]. Pure and metal-incorporated MFI crystals were synthesized by microwave heating. These samples will be denoted as M-MFI-MW, where M stands for the incorporated metal (Ti, Fe, Zr, and Sn) and MW for the microwave condition; metal-free MFI by the microwave synthesis will be denoted as Si-MFI-MW [96]. The microwave induces a dramatic change in the morphology depending on the composition. Si-MFI-MW and Ti-MFI-CH show the characteristic hockey-puck-like crystals of submicrometer sizes with well-developed, large (010) faces (Figure 5.13a and b). The microwave syntheses of metal (Ti, Sn) incorporated systems produced similar primary crystals, but in this case the crystals were all stacked on top of each other along their (010) direction (*b* axis) to form a wormlike or fibrous morphology (Figure 5.13c and d). This stacking is sufficiently robust so as not to be destroyed by a sonication treatment for more than 1 h, indicating that this morphology is not a result of simple aggregation but of strong chemical bonding between the crystals. This fibrous morphology persists as long as there is incorporated Ti, whose concentration is kept in the range of Si/Ti = 70–230. When the concentration of Ti is increased (Si/Ti $\leq$ 50), the product is composed of isolated ellipsoidal crystals

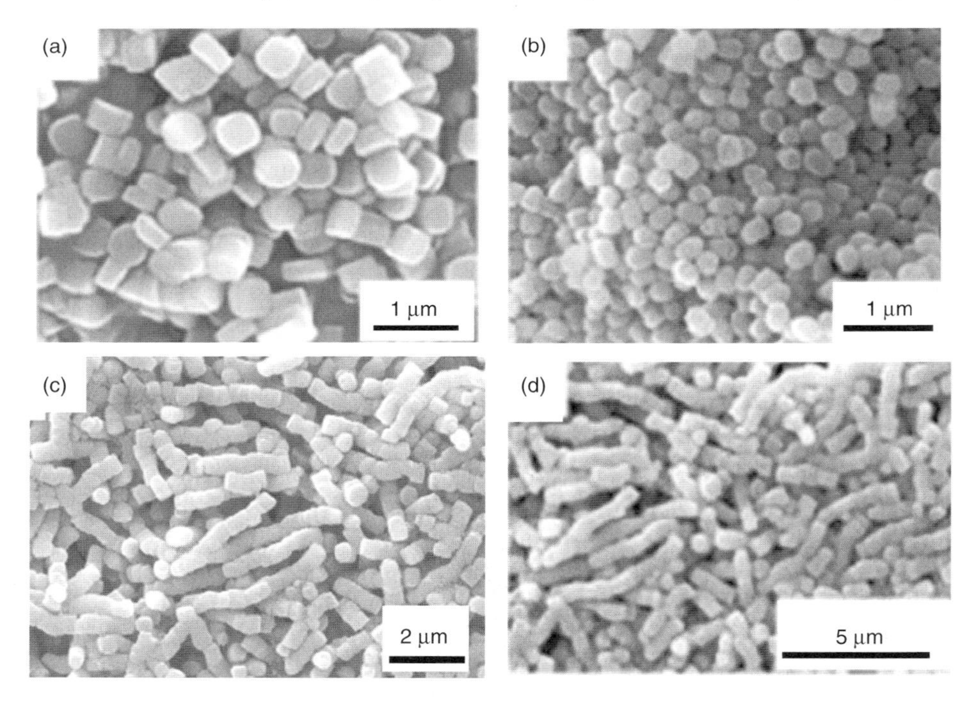


**Figure 5.13** SEM images of (a) Si-MFI-MW, (b) Ti-MFI-CH (Si/Ti = 70), (c) Ti-MFI-MW(Si/Ti = 70), and (d) Sn-MFI-MW(Si/Sn = 70) [96].

(not shown). This is probably because the high concentration of Ti interferes with the crystal growth mechanism and the lack of flat surfaces does not allow crystal stacking. The crystals with other incorporated metals (Fe, Zr, and Sn) also show the fibrous morphology.

High-resolution TEM image and ED patterns were used to observe the closely connected boundary between crystals (Figure 5.14). The connected parts have well-crystallized structures. We assume that these crystallized parts are mostly connected with the straight channels of each other except some edge parts, which allows the formation of mesopores. Although it is not clear how the incorporated metals induce the stacking of crystals under microwave conditions, it appears to be related to the magnitude of the local dipole moment of the M–O bonds that might be originated from the differences in the electronegativities. Electrically insulating materials absorb microwave energy through the oscillation of dipoles, and the magnitude of absorption increases with the increase of the dipole moment. The magnitude of a dipole moment is mainly determined by the difference in the electronegativities ($\Delta\chi$) of the two bonded atoms. Therefore, the Ti–O bond ($\Delta\chi = 2.18$ according to the Allred–Rochow scheme) [91] is a better microwave absorber than the Si–O bond ($\Delta\chi = 1.76$). The Ti–O bonds on the surface are strongly activated by microwave absorption and can undergo condensation reactions to form Ti–O–Ti and/or Ti–O–Si bonds between crystals. The same explanation applies to the Fe-MFI-MW and Zr-MFI-MW zeolites because of the large $\Delta\chi$ values for Fe–O and Zr–O bonds. In the case of the Sn-MFI-MW, the $\Delta\chi$ value of the Sn–O bond (1.78) is rather small, close to that of Si–O bond. However, because Sn is large in size, the valence electron density of the Sn–O bond is shifted to the O side, making this bond more polar than estimated by the $\Delta\chi$ value alone, namely, the homopolar contribution to the dipole moment [96], and the above explanation of microwave absorption by polar bonds can be applied to the Sn-MFI case. Further work is needed to fully understand the role of the incorporated metals in the stacking of crystals under the microwave condition.

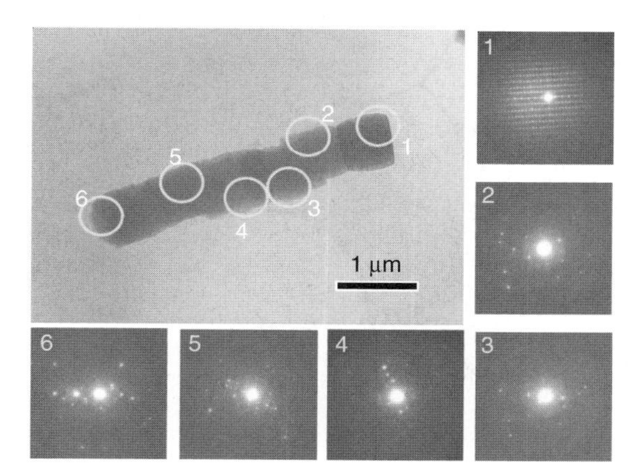


**Figure 5.14** HRTEM images and ED patterns of Ti-MFI-MW [87].

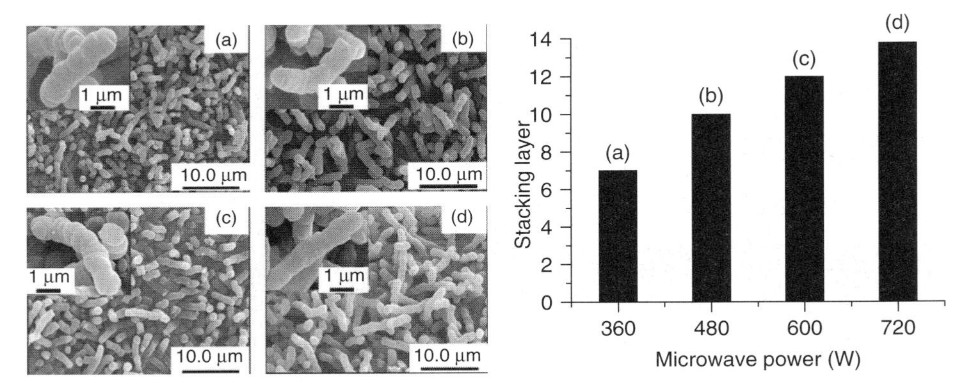


**Figure 5.15** SEM images and average number of stacking layers of stacked Ti-MFI-MW synthesized under different microwave powers: (a) 360 W, (b) 480 W, (c) 600 W, and (d) 720 W [87].

Microwave power was controlled and varied for figuring out the formation of nanostacked Ti-MFI zeolite [97]. FE-SEM images of Ti-MFI zeolites prepared at different powers are given in Figure 5.15. All the samples show stacked morphologies. The crystals are all stacked on top of each other along their (010) direction to form a wormlike or fibrous morphology. And the average number of stacking layers increased from 7 to 13 as the microwave power was increased from 360 to 720 W. From the above discussion, we see that microwave can strongly affect the nano stacking process. This is because a higher power will give more condensation for the dehydration between hydroxyl groups on the crystal surface.

### 5.4.3
### Formation Scheme of Stacked Morphology

In this study, the formation of the stacked morphology was observed more clearly through the SEM images of Ti-MFI depending on different microwave irradiation times [96]. At the first stage of the microwave irradiation (Figure 5.16a–c), small zeolite seeds grew up to uniform crystals; after 40 min they started to form the stacked morphology and the number of stacks kept increasing till 60 min (Figure 5.16d–f). In the microwave synthesis, the silica precursor would be crystallized to uniform, hockey-puck-shaped morphology during the first synthesis step. Under prolonged irradiation, those small crystals adhered together along the *b* orientation to form a stacked morphology. The surface Ti-OH groups seemed to be activated by microwaves and accelerate the condensation reaction between the OH groups on the surface of the crystals (Figure 5.17).

MW irradiation induces a three-dimensional stacking of zeolite particles with opal-like morphology through bimetal incorporation [98]. Microwave synthesis of Al- and Ti-bimetal incorporated MFI zeolite ((Al, Ti)-MFI) gives both fibrous and arrayed morphologies. Uniform zeolite crystals were stacked to form long lines and

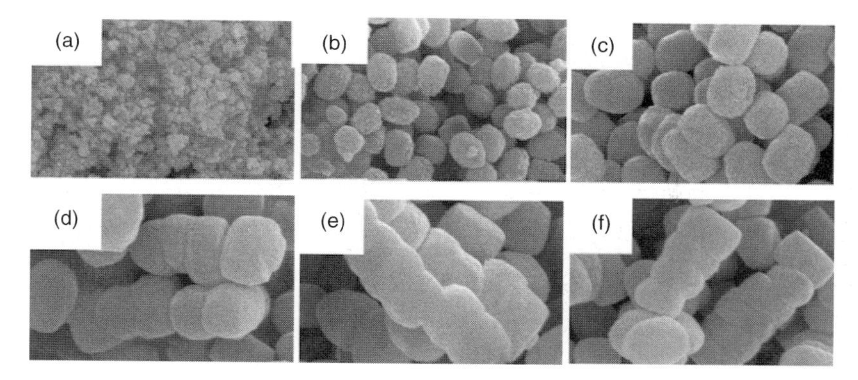


**Figure 5.16** SEM of Ti-MFI with different microwave irradiation times: (a) 10 min, (b) 20 min, (c) 30 min, (d) 40 min, (e) 50 min, and (f) 60 min [87].

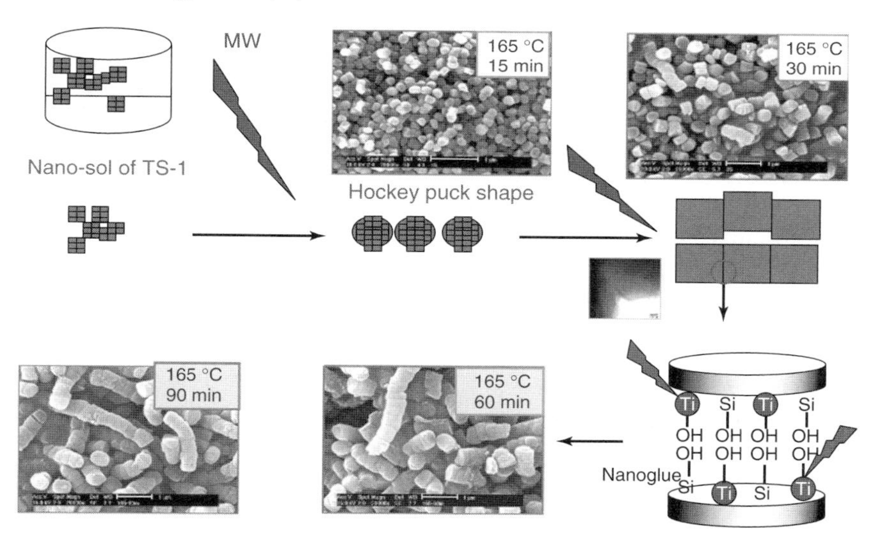


**Figure 5.17** Formation scheme of stacked Ti-MFI [87].

those lines were arrayed to form the three-dimensional opal structure (Figure 5.18). The submicrometer-sized pucklike crystals also were stacked face to face of the (010) plane (Figure 5.18c). Microwave synthesis of bimetal-incorporated systems produced the primary crystals, which were stacked on top of each other along their (010) direction to form long lines (Figure 5.18a–c). The void spaces in the nanoarrayed materials were observed through the SEM image of carbon replicas (Figure 5.18d).

Xu and coworkers synthesized silicalite-1 (Si-MFI) crystals by applying microwave-assisted solvothermal heating (Figure 5.19). Even without the metal as a nano-glue, the zeolite particles could be stacked into fibrous morphology

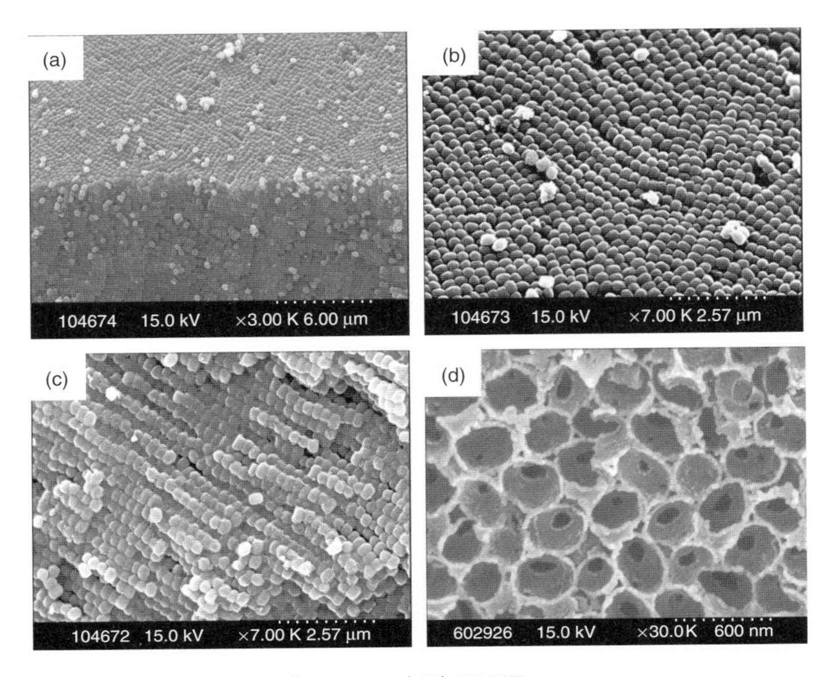


**Figure 5.18** SEM images of nanoarrayed (Al, Ti)-MFI: (a) cross view, (b) top, (c) side, and (d) carbon replica.

**Figure 5.19** SEM images of Si-MFI crystals crystallized using different alcohol cosolvents under microwave radiation conditions: (a) ethylene glycol ($x = 37$), (b) methanol ($x = 32.6$), (c) ethanol ($x = 24.3$), (d) 1-propanol ($x = 20.1$), (e) isopropanol ($x = 18.3$), (f) *n*-butanol ($x = 17.8$), and (g) hexanol ($x = 13.3$). $x$ is the dielectric constant [99].

by controlling the dipolar cosolvents [99]. The low polarity (dielectric constant) of the cosolvent may favor the formation of abundant Si-OH groups in the precursor species and, on the surface of the nanocrystals formed at the early stage of the crystallization, might undergo further condensation to form self-stacked crystals due to the rapid crystallization under microwave conditions. The crystals are stacked on top of each other along their *b* direction to form a self-stacked morphology. The "fiber" cannot be destroyed even by long-time and strong ultrasonication, which indicates that strong chemical bonds might exist between the individual crystals.

## 5.5 Summary and Outlook

The representative techniques for the growth of large single crystals of zeolites are reviewed, especially for typical zeolites such as FAU, LTA, or MFI. To prepare perfect zeolite particles, we should understand the mechanism of nuclei formation and the crystal growth process properly in order to critically control the process. Morphology of large zeolite crystals reflects the expanded shape of their building units.

For fine-tuning the growth, the morphology of small zeolite particles was studied properly. Various additives, for example, alkali cations, alcohols, and amines, were investigated. Different alkali-metal cations result in distinct morphologies; a mixed cationic system will provide uniform, large, lath-shaped crystals (about 140 μm). The use of different structure-directing molecules changes the crystal morphology, as noted with the results in the case of various amines. Hydroxide ion content can alter the morphology significantly. Even for a particular ion, such as TPA, the amount used can alter the morphology. The crystallites tend to be larger at lower template concentrations, presumably because of the formation of fewer nuclei. Morphologies of crystals from mixed solvents or nonaqueous media are distinct from comparable compositions in an aqueous medium.

Microwaves can control the morphology of zeolites by utilizing the precursor solutions including lossy components such as metals and various organic and inorganic additives. They have played as the absorption sites of microwave energy, which were termed nano-glues, for the fabrication of oriented fiber morphology. Such preferred orientations were found to be helpful in the selective adsorption, transportation, or diffusion of longer molecules and could provide advanced shape-selective catalysis.

Morphological synthesis of zeolite crystals is still an expanding area, and the preparation of high-quality zeolite crystals will continue to be of great importance. In order to maximize the performance of zeolite catalysts, it is important to understand the crystallographic organization within zeolite crystallites, particularly regarding access to the variously structured pores. Special attention is paid to the study of well-shaped zeolite crystals by utilizing *in situ* microspectroscopic techniques [100]. Unique insight into diffusion, intergrowth structure, and catalysis can be achieved by applying newly developed technology. It is the way to understand and design appropriate zeolite catalysts for a specific reaction or study the specific reaction mechanism over zeolite catalysts [101].

Additionally, the catalytic performance of zeolites often depends on the morphology of primary zeolite particles, especially in the case of anisotropic zeolite crystals, where the required property is characteristic for a particular crystallographic direction of the zeolites. The crystal morphology and size define also the diffusion paths via the channels, which often have a great impact on the reaction kinetics [102]. Thus, the close control of zeolite crystal morphology is highly desirable for zeolite catalysis.

Moreover, the morphological control of a given zeolite structure is desirable for the fabrication of zeolites into oriented membranes, sensor, or electric devices, especially to the hierarchical nanoporous structured materials.

Finally, morphological control is one of the important keys to understand the mechanism of zeolite crystallization and their performance in various applications. Finding new and simple methods to control morphology is one of the challenges in the field of academic study or industrial application of zoelites. Such studies make the future of morphological synthesis of zeolites crystals bright.

## Acknowledgments

This work was supported by the Korea Science and Engineering Foundation grants (National Research Laboratory Program), BK21 and Nano Center for Fine Chemicals Fusion Technology.

## References

1. Davis, M.E. (1991) *Ind. Eng. Chem. Res.*, **30**, 1675.
2. Davis, M.E. (2002) *Nature*, **417**, 813.
3. Corma, A. (1995) *Chem. Rev.*, **95**, 559.
4. Yu, J. and Xu, R. (2006) *Chem. Soc. Rev.*, **35**, 593.
5. Csicsery, S.M. (1984) *Zeolites*, **4**, 202.
6. Weitkamp, J. and Puppe, L. (1999) *Catalysis and Zeolites: Fundamentals and Applications*, 1st edn, Springer, Berlin.
7. Wojciechowski, B.W. and Corma, A. (1986) *Catalytic Cracking: Catalysis, Chemistry, and Kinetics*, Dekker, New York.
8. Lai, Z., Bonilla, G., Diaz, I., Nery, J.G., Sujaoti, K., Amat, M.A., Kokkoli, E., Terasaki, O., Thompson, R.W., Tsapatsis, M., and Vlachos, D.G. (2003) *Science*, **300**, 456.
9. Kuperman, A., Nadimi, S., Oliver, S., Ozin, G.A., Garces, J.M., and Olken, M.M. (1993) *Nature*, **365**, 239.
10. Feng, S. and Bein, T. (1994) *Science*, **265**, 1839.
11. Shi, F., Chen, X., Wang, L., Niu, J., Yu, J., Wang, Z., and Zhang, X. (2005) *Chem. Mater.*, **17**, 6177.
12. (a) Singh, R. and Dutta, P.K. (2003) in *Handbook of Zeolite Science and Technology* (eds S.M.Auerbach, K.A. Carrado, and P.K. Dutta), Marcel Dekker, New York, p. 21; (b) Cundy, C.S. (2005) in *Zeolites and Ordered Mesoporous Materials: Progress and Prospects*, Studies in Surface Science and Catalysis, Vol. 157 (eds J. Čejka and H. van Bekkum), Elsevier, Prague, The 1st FEZA School on Zeolites, p. 101.
13. (a) Beschmann, K., Kokotailo, G.T., and Reikert, L. (1988) *Stud. Surf. Sci. Catal.*,

39, 355; (b) Muller, U. and Unger, K.K. (1988) *Stud. Surf. Sci. Catal.*, **39**, 101.

14. Terasaki, O., Yamazaki, K., Thomas, J.M., Ohsuna, T., Watanabe, D., Saunders, J.V., and Barry, J.C. (1987) *Nature*, **330**, 58.
15. Cox, S.D., Gier, T.E., Stucky, G.D., and Bierleein, J. (1988) *J. Am. Chem. Soc.*, **110**, 2987.
16. Qiu, S., Yu, J., Zhu, G., Terasaki, O., Nozue, Y., Pang, W., and Xu, R. (1998) *Microporous Mesoporous Mater.*, **21**, 245.
17. (a) DiRenzo, F. (1998) *Catal. Today*, **41**, 37; (b) Coker, E.N., Jansen, J.C., DiRenzo, F., Fajula, F., Martens, J.A., Jacobs, P.A., and Sacco, A. (2001) *Microporous Mesoporous Mater.*, **46**, 223.
18. Mumpton, F.A. (1991) *Proc. Natl. Acad. Sci. U.S.A.*, **96**, 3463.
19. Langella, A., Cappelletti, P., and de Gennaro, M. (2001) in *Hydrologic Natural Zeolite Growth*, Reviews in Mineralogy and Geology 45 (eds D.L.Bish and D.W. Wing), Geoscience-word p. 235.
20. Lethbridge, Z.A.D., Williams, J.J., Walton, R.I., Evans, K.E., and Smith, C.W. (2005) *Microporous Mesoporous Mater.*, **79**, 339.
21. Stöcker, M. (1993) *Acta Chem. Scand.*, **47**, 935.
22. Terskikh, V.V., Moudrakovski, I.L., Du, H.B., Ratcliffe, C.I., and Ripmeester, J.A. (2001) *J. Am. Chem. Soc.*, **123**, 10399.
23. Binder, G., Scandella, L., Kritzenberger, J., Gobrecht, F., Koegler, J.H., and Prins, R. (1997) *J. Phys. Chem. B*, **101**, 483.
24. Megelski, S., Lieb, A., Pauchard, M., Drechsler, A., Glaus, S., Debus, C., Meixner, A.J., and Calzaferri, G. (2001) *J. Phys. Chem. B*, **105**, 25.
25. Jackson, K.T. and Howe, R.F. (1994) *Zeolites and Microporous Crystals*, Studies in Surface Science and Catalysis, Vol. 83 (eds H.T.Hattori and T. Yashima), Elsevier p. 187.
26. Heink, W., Kärger, J., Pfeifer, H., Salverda, P., Datema, K.P., and Nowak, A. (1992) *J. Chem. Soc., Faraday Trans.*, **88**, 515.
27. Snurr, R.Q., Hagen, A., Ernst, H., Schwarz, H.B., Ernst, S., Weitkamp, J., and Kärger, J. (1996) *J. Catal.*, **163**, 130.
28. Schwarz, H.B., Ernst, S., Kärger, J., Knorr, B., Seiffert, G., Snurr, R.Q., Staudte, B., and Weitkamp, J. (1997) *J. Catal.*, **167**, 248.
29. Karge, H.G. and Weitkamp, J. (1998) *Zeolite Synthesis*, Vol. 1, Springer-Verlag, Berlin.
30. Cundy, C.S., Lowe, B.M., and Sinclair, D.M. (1990) *J. Cryst. Growth*, **100**, 189.
31. Persson, A.E., Schoeman, B.J., Sterte, J., and Ottesstedt, J.E. (1994) *Zeolites*, **14**, 557.
32. Mintova, S., Olson, N.H., Valtchev, V., and Bein, T. (1999) *Science*, **283**, 958.
33. Tosheva, L. and Valtchev, V.P. (2005) *Chem. Mater.*, **17**, 2494, and references therein.
34. Charnell, J.F. (1971) *J. Cryst. Growth*, **8**, 291.
35. Sun, Y.Y., Song, T., Qiu, S., Pang, W., Shen, J., Jiang, D., and Yue, Y. (1995) *Zeolites*, **15**, 745.
36. Shimizu, S. and Hamada, H. (2001) *Microporous Mesoporous Mater.*, **48**, 39.
37. DiRenzo, F. (1998) *Catal. Today*, **41**, 37.
38. Qiu, S., Yu, J., Zhu, G., Tarasaki, O., Nozue, Y., Pang, W., and Xu, R. (1998) *Microporous Mesoporous Mater.*, **21**, 245.
39. Camblor, M.A., Villaescusa, L.A., and Diaz-Cabanas, M. (1999) *Top. Catal*, **9**, 59.
40. Axon, S.A. and Klinowski, J. (1992) *Appl. Catal. A*, **81**, 27.
41. Qiu, S., Pang, W., and Xu, R. (1997) *Stud. Surf. Sci. Catal.*, **105**, 301.
42. Qui, S., Yu, J., Zhu, G., Terasaki, O., Nozue, Y., Pang, W., and Xu, R. (1998) *Microporous Mesoporous Mater.*, **21**, 245.
43. Guth, J.L., Kessler, H., Higel, J.M., Lamblin, J.M., Patarin, J., Seive, A., Chezeau, J.M., and Wey, R. (1989) *ACS Symp. Ser.*, **398**, 176.
44. Qiu, S., Pang, W., and Yao, S. (1989) *Stud. Surf. Sci. Catal.*, **49**, 133.

45. Jansen, J.C., Engelen, C.W.R., and van Beckkum, H. (1989) *ACS Symp. Ser.*, **398**, 257.
46. Shao, C., Li, X., Qiu, S., Xiao, F.-S., and Terasaki, O. (2000) *Microporous Mesoporous Mater.*, **39**, 117.
47. Shimizu, S. and Hamada, H. (1999) *Angew. Chem. Int. Ed. Engl.*, **38**, 2725.
48. Kornatowski, J. (1988) *Zeolites*, **8**, 77.
49. Mueller, U. and Unger, K.K. (1988) *Zeolites*, **8**, 154.
50. Zhang, D., Qiu, S., and Pang, W. (1990) *J. Chem. Soc. Chem. Commun.*, 1313.
51. Wang, X. and Jacobson, A.J. (2001) *Mat. Res. Soc. Symp.*, **658**, GG8.1.
52. Coker, E.N., Jansen, J.C., DiRenzo, F., Fajula, F., Martens, J.A., Jacobs, P.A., and Sacco, A. (2001) *Microporous Mesoporous Mater.*, **46**, 223.
53. Ghamami, M. and Sand, L.B. (1983) *Zeolites*, **3**, 155.
54. Tuel, A. (1997) *Stud. Surf. Sci. Catal.*, **105**, 261.
55. Iwasaki, A., Sano, T., and Kiyozumi, Y. (1998) *Microporous Mesoporous Mater.*, **25**, 119.
56. Persson, A.E., Schoeman, B.J., Sterte, J., and Otterstedt, J.-E. (1994) *Zeolites*, **14**, 557.
57. Kalipcilar, H. and Culfaz, A. (2000) *Cryst. Res. Technol.*, **35**, 933.
58. Cizmek, A., Subotica, B., Kralj, D., Babil-Ivancic, V., and Tonejc, A. (1997) *Microporous Mesoporous Mater.*, **12**, 267.
59. Gao, F., Zhu, G., Li, X., Li, B., Terasaki, O., and Qiu, S. (2001) *J. Phys. Chem. B*, **105**, 12704.
60. Franklin, K.R. and Lowe, B.M. (1988) *Zeolites*, **8**, 501.
61. Burchart, E.V., Janse, J.C., van der Graaf, B., and van Bekkum, H. (1993) *Zeolites*, **13**, 216.
62. Beck, L.W. and Davis, M.E. (1998) *Microporous Mesoporous Mater.*, **22**, 107.
63. de Moor, P.-P.E.A., Beelen, T.P.M., van Santen, R.A., Beck, L.W., and Davis, M.E. (2000) *J. Phys. Chem. B*, **104**, 7600.
64. Hussein, M.Z., Zainal, Z., and Masdan, S.A. (2001) *Res. J. Chem. Environ.*, **5**, 21.
65. Dwyer, J. and Zhao, J. (1992) *J. Mater. Chem.*, **2**, 235.
66. Ke, J.-A. and Wang, I. (2001) *Mater. Chem. Phys.*, **68**, 157.
67. Aiello, R., Crea, F., Nigro, E., Testa, F., Mostowicz, R., Fonseca, A., and Nagy, J.B. (1999) *Microporous Mesoporous Mater.*, **28**, 241.
68. Derouane, E.G. and Gabelica, Z. (1986) *J. Solid State Chem.*, **64**, 296.
69. Ban, T., Mitaku, H., Suzuki, C., Matsuba, J., Ohya, Y., and Takahashi, Y. (2005) *J. Cryst. Growth*, **274**, 594 –602.
70. Auerbach, S.M., Carrado, K.A., and Dutta, P.K. (2003) *Handbook of Zeolite Science and Technology*, Marcel Dekker Inc., New York, p. 43.
71. Lai, Z., Tsapatsis, M., and Nicolich, J.P. (2004) *Adv. Funct. Mater.*, **14**, 716.
72. Beck, L.W. and Davis, M.E. (1998) *Microporous Mesoporous Mater.*, **22**, 107.
73. de Moor, P., Beelen, T.P.M., van Santen, R.A., Beck, L.W., and Davis, M.E.J. (2000) *Phys. Chem. B*, **104**, 7600.
74. Bonilla, G., Diaz, I., Tsapatsis, M., Jeong, H.K., Lee, Y., and Vlachos, D.G. (2004) *Chem. Mater.*, **16**, 5697.
75. Patarin, J., Soulard, M., Kessler, H., Guth, J.-L., and Baron, J. (1989) *Zeolites*, **9**, 397.
76. Suzuki, K., Kiyozumi, Y., Shin, S., Fujisawa, K., Watanabe, H., Saito, K., and Noguchi, K. (1986) *Zeolites*, **6**, 290.
77. Crea, F., Nastro, A., Nagy, J.B., and Aiello, R. (1988) *Zeolites*, **8**, 262.
78. Ahmed, S., El-Faer, M.Z., Abdillahi, M.M., Siddiqui, M.A.B., and Barri, S.A.I. (1996) *Zeolites*, **17**, 373.
79. Gabelica, Z., Blom, N., and Derouane, E.G. (1983) *Appl. Catal.*, **5**, 227.
80. Crea, F., Aiello, R., Nastro, A., and Nagy, J.B. (1991) *Zeolites*, **11**, 521.
81. Aiello, R., Crea, F., Nastro, A., and Pellegrino, C. (1987) *Zeolites*, **7**, 549.
82. Erdem, A. and Sand, L.B. (1979) *J. Catal.*, **60**, 241.
83. Lowe, B.M., Nee, J.R.D., and Casci, J.L. (1994) *Zeolites*, **14**, 610.
84. Kanno, N., Miyake, M., and Sato, M. (1994) *Zeolites*, **14**, 625.

85. Nastro, A. and Sand, L.B. (1983) *Zeolites*, **3**, 57.
86. (a) Cundy, C.S. (1998) *Collect. Czech. Chem. Commun.*, **63**, 1699; (b) Cundy, C.S., Plaisted, R.J., and Zhao, J.P. (1998) *Chem. Commun.*, 1465.
87. Park, S.-E. and Jiang, N. (2008) in *Zeolites: From Model Materials to Industrial Catalysis* (eds J. Čejka, J. Perez-Pariente, and W.J. Roth), Transworld Research Network, Kerala, p. 81.
88. Carmona, J.G., Clemente, R.R., and Morales, J.G. (1997) *Zeolites*, **18**, 340.
89. (a) Yates, M.Z., Ott, K.C., Birnbaum, E.R., and McCleeskey, T.M. (2002) *Angew. Chem. Int. Ed. Engl.*, **41**, 476; (b) Ganschow, M., Schulz-Ekloff, G., Wark, M., Wendschuh-Josties, M., and Wöhrle, D. (2001) *J. Mater. Chem.*, **11**, 1823.
90. Mintova, S., Mo, S., and Bein, T. (1998) *Chem. Mater.*, **10**, 4030.
91. Jacobsen, C.J.H., Madsen, C., Houzvicka, J., Schmidt, I., and Carlsson, A. (2000) *J. Am. Chem. Soc.*, **122**, 7116.
92. Chen, X., Yan, W., Cao, X., Yu, J., and Xu, R. (2009) *Microporous Mesoporous Mater.*, **119**, 217.
93. Che, M., Masure, D., and Chaquin, P. (1993) *J. Phys. Chem.*, **97**, 9022.
94. Morris, C.A., Anderson, M.L., Stroud, R.M., Merzbacher, C.I., and Rolison, D.R. (1999) *Science*, **284**, 622.
95. Kulak, A., Park, Y.S., Lee, Y.J., Chun, Y.S., Ha, K., and Yoon, K.B. (2000) *J. Am. Chem. Soc.*, **122**, 9308.
96. Hwang, Y.K., Chang, J.-S., Park, S.-E., Kim, D.S., Kwon, Y.-U., Jhung, S.H., Hwang, J.-S., and Park, M.-S. (2005) *Angew. Chem. Int. Ed. Engl.*, **44**, 446.
97. Jin, H., Jiang, N., and Park, S.-E. (2008) *J. Phys. Chem. Solids*, **69**, 1136.
98. Choi, K.-M., David, R.B., Han, S.-C., and Park, S.-E. (2007) *Solid State Phenom.*, **119**, 167.
99. Chen, X., Yan, W., Shen, W., Yu, J., Cao, X., and Xu, R. (2007) *Microporous Mesoporous Mater.*, **104**, 296.
100. Roeffaers, M.B.J., Ameloot, R., Baruah, M., Uji-I, H., Bulut, M., De Cremer, G., Müller, U., Jacobs, P.A., Hofkens, J., Sels, B.F., and De Vos, D.E. (2008) *J. Am. Chem. Soc.*, **130**, 5763.
101. Schoonheydt, R.A. (2008) *Angew. Chem. Int. Ed. Engl.*, **47**, 9188.
102. Larlus, O. and Valtchev, V.P. (2004) *Chem. Mater.*, **16**, 3381.

# 6
# Post-synthetic Treatment and Modification of Zeolites

*Cong-Yan Chen and Stacey I. Zones*

## 6.1
## Introduction

Zeolites and other molecular sieves find widespread applications in catalysis, adsorption, and ion exchange due to their unique shape selective properties and high internal surface areas [1, 2]. Their use in these applications depends largely on both the crystalline structures (e.g., 8-, 10-, 12-ring, or even larger pore openings and one-, two-, or three-dimensional channel systems) and framework compositions (e.g., Si/Al ratios and the associated acidity and hydrophobicity/hydrophilicity). According to the Structure Commission of the International Zeolite Association, 191 framework structure codes have been assigned to zeolites and zeotype materials as of July 2009 [3–5]. Various theoretical studies demonstrate that this number represents only a small fraction of the structures possible for zeolites and other molecular sieves [6–11]. The major challenge in tailoring and utilizing zeolites for specific applications remains the development of synthesis methods to produce desirable structures with desirable framework compositions. In principle, there are two routes to achieve this goal: (i) direct synthesis and (ii) post-synthetic treatment and modification. In this chapter, we will first briefly discuss direct synthesis. Then the emphasis will be placed on two selected methods of the post-synthetic treatment and modification of zeolites, namely, aluminum reinsertion into zeolite framework using aqueous $Al(NO_3)_3$ solution under acidic conditions and preparation of pure-silica zeolites via hydrothermal treatment with acetic acid.

## 6.2
## Direct Synthesis of Zeolites

The direct synthesis is the primary route of the synthesis of zeolites [12–28]. The major variables that have a predominant influence on the zeolite structure crystallized include the composition of synthesis mixture, synthesis temperature and time, as well as the preparation procedure of the synthesis mixture such as

*Zeolites and Catalysis, Synthesis, Reactions and Applications. Vol. 1.*
Edited by Jiří Čejka, Avelino Corma, and Stacey Zones

ISBN: 978-3-527-32514-6

aging and seeding. Furthermore, the selection of the structure directing agents (SDAs) plays a critical role in the selective formation of zeolites, as reported in the above references.

Depending on the nature of the zeolites involved and the chemistry of their formation, some zeolites can be synthesized in a broad spectrum of framework compositions, as exemplified by IFR-type zeolites which are characterized by an undulating, one-dimensional, 12-ring channel system. IFR is the framework type code of the isostructural zeolites ITQ-4, SSZ-42, and MCM-58, according to the Structure Commission of the International Zeolite Association [3, 4]. As reported in the literature, IFR-type zeolites that can be synthesized via direct synthesis route include ITQ-4 (pure-silica synthesized via fluoride route) [29, 30], Si-SSZ-42 (pure-silica synthesized via conventional hydroxide route) [31], B-SSZ-42 (borosilicate) [22, 32–34], Al-MCM-58 (aluminosilicate) [22, 34–37], Fe-MCM-58 (ferrosilicate) [37, 38], and Zn-SSZ-42 (zincosilicate) [31].

In contrast to IFR-type zeolites, the synthesis of some other structures succeeds only if certain heteroatom X (X = B, Ga, Ge, or Al, for example, or X = none of these elements for pure-silica zeolites) is present in the synthesis mixture and, in turn, incorporated into the framework. In many cases, certain zeolite structures containing specific heteroatoms can be synthesized only within a limited range of Si/X ratio or in the presence of certain specific structure directing agents. For example, borosilicate zeolite SSZ-33 (which contains intersecting 10-/12-/12-ring channels with CON topology [3, 4]) can be synthesized with *N,N,N*-trimethyl-8-tricyclo-[5.2.1.0$^{2,6}$]decane ammonium cations as the structure directing agent [39, 40]. The direct synthesis of aluminosilicate SSZ-33 using this structure directing agent is so far not successful. On the other hand, aluminosilicate zeolite SSZ-26 (which has a very similar crystalline structure to that of SSZ-33) can be synthesized using *N,N,N,N′,N′,N′*-hexamethyl[4.3.3.0]propellane-8,11-diammonium cations as the structure directing agent [39, 41]. However, this structure directing agent is more difficult to synthesize and more expensive than *N,N,N*-trimethyl-8-tricyclo[5.2.1.0$^{2,6}$]decane ammonium which is used for the synthesis of B-SSZ-33.

Furthermore, Table 6.1 shows the aluminum, boron, and zinc effects on the synthesis of some zeolites when using different structure directing agents. Here we have different molar ratios of silicon to heteroatoms, Si/X, in the synthesis mixtures where X is Al, B, or Zn. These results present here some additional examples which demonstrate the complicated relationship between zeolite structures, framework compositions, and structure directing agents in the direct synthesis.

In short, the direct synthesis route often does not lead to the formation of zeolites with the properties desirable for the target applications. As will be discussed in the next section, the post-synthetic treatment and modification of zeolites constitute a powerful technique to reach this goal.

**Table 6.1** Aluminum, boron, and zinc effects on the synthesis of some zeolites when using different structure directing agents.

| | Molar ratio in synthesis mixture | | | |
|---|---|---|---|---|
| **SDA** | **$SiO_2$ (pure silica)** | **$SiO_2/Al_2O_3$ <50** | **$SiO_2/B_2O_3$ <30** | **$SiO_2/ZnO$ <100** |
| 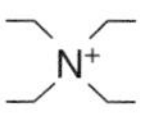 $N^+$ | ZSM-12 | Beta | Beta | VPI-8 |
| 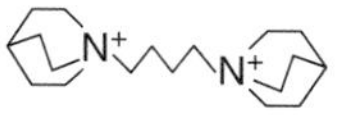 $N^+$ $N^+$ | ZSM-12 | Beta | Beta | VPI-8 |
| 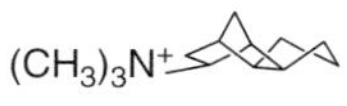 $(CH_3)_3N^+$ | ZSM-12 | Mordenite | Beta | Layered |
| 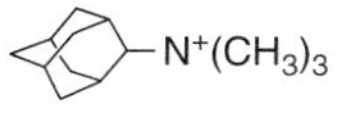 $N^+(CH_3)_3$ | SSZ-24 | SSZ-25 | SSZ-33 | – |
| 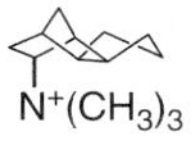 $N^+(CH_3)_3$ | SSZ-31 | Mordenite | SSZ-33 | VPI-8 |
| 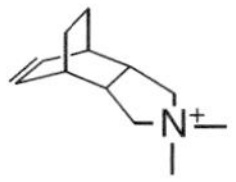 $N^+$ | SSZ-31 | SSZ-37 | SSZ-33 | – |

## 6.3 Post-synthetic Treatment and Modification of Zeolites

In addition to the direct synthesis method, post-synthetic treatment often provides a more practical route to modify the zeolites to acquire desirable framework compositions and other properties. Kühl [42] and Szostak [43] have published two comprehensive review papers about the modification of zeolites. These review papers cover many important aspects of post-synthetic treatment and modification of zeolites such as ion exchange, preparation of metal-supported zeolites, dealumination, reinsertion of heteroatoms into zeolite framework, and other modification methods. The readers are referred to these two review papers and the references therein for the details on the various post-synthetic treatment methods. In this chapter, we will review two newer methods of the post-synthetic treatment and modification of zeolites for the lattice substitution, both of which are based on the same principle: the desired atoms such as Al or Si are inserted into lattice sites previously occupied by other T-atoms such as B. These two methods are given below:

1) Aluminum reinsertion into zeolite framework using aqueous $Al(NO_3)_3$ solution under acidic conditions [44, 45];
2) Synthesis of pure-silica zeolites via hydrothermal treatment with acetic acid [46].

### 6.3.1 Aluminum Reinsertion into Zeolite Framework Using Aqueous $Al(NO_3)_3$ Solution under Acidic Conditions

As discussed in Section 6.2, many borosilicate zeolites have been successfully synthesized via the direct synthesis route. However, borosilicate zeolites do not possess sufficiently high acidity required for many hydrocarbon reactions. The preparation of the catalytically more active aluminosilicate counterparts of these borosilicate zeolites relies, therefore, on the post-synthetic treatment.

Various methods have been used to reinsert aluminum into the framework of zeolites [42, 43]. Two of these methods are succinctly described below:

1) Aqueous solution of sodium aluminate is used to insert aluminum into the tetrahedral vacancies and/or substitute aluminum for framework silicon [47]. It is noteworthy is that, due to the chemical nature of sodium aluminate, this method functions under basic conditions which leads to some dissolution of silicon species from the framework of zeolites.
2) Aluminum is inserted into the framework by treating zeolites with $AlCl_3$ vapor in nitrogen at temperatures above 773 K [48–50]. This is basically a reversal of the dealumination reaction using $SiCl_4$ vapor [51–55].

There is a newer post-synthetic treatment method to reinsert aluminum into the framework of zeolites, which operates with an aqueous solution of $Al(NO_3)_3$ under acidic conditions [44, 45], in contrast to the method with sodium aluminate described above. At low pH, this technique minimizes the dissolution of silicon species from the framework of zeolites. This method will be reviewed next.

#### 6.3.1.1 Experimental Procedures

This post-synthetic technique consists of two elementary stages: (i) deboronation of borosilicate zeolites under acidic conditions and (ii) reinsertion of aluminum in the lattices of deboronated zeolites using an aqueous $Al(NO_3)_3$ solution. Basically, there are two ways to perform the experiments [44, 45]:

1) **Two-step method:** These two steps are carried out sequentially. First, the calcined borosilicate zeolite is deboronated at room temperature in 0.01 M aqueous HCl solution for about 24 hours. The zeolite to HCl solution ratio of 1 : 40 by weight is usually used. After washing and air-drying, the deboronated zeolites are then combined with certain amount of aqueous $Al(NO_3)_3$ solution (e.g., 1 g zeolite with 100 g of 1 M $Al(NO_3)_3$ solution) and treated under various conditions (e.g., reflux for 100 hours). Typically, the temperature ranges between 363 and 443 K. The final zeolite products are washed, filtered, and air-dried. The resulting aluminosilicate zeolites are already in the H-form.
2) **One-step method:** Here the deboronation and aluminum reinsertion are conducted in the same single step. The experiments start directly with the parent borosilicate zeolites and $Al(NO_3)_3$ solution. The deboronation is carried out by the $Al(NO_3)_3$ solution which has a low pH ($<3.5$). Other experimental conditions are similar to those of the two-step method.

Important experimental parameters for this aluminum reinsertion technique include $Al(NO_3)_3$/zeolite ratio, pH of the slurry, reaction time, and temperature (see above).

#### 6.3.1.2 One-Step Method versus Two-Step Method

The two-step method separates the deboronation and aluminum reinsertion steps, similar to many organic syntheses with a series of steps. It demonstrates how these two individual steps proceed and helps our scientific understanding of the chemistry involved. The one-step method has the advantage of combining these two steps together and providing a more efficient way to reinsert aluminum into the framework of zeolites.

Analytical results from X-ray diffraction (XRD), $N_2$/hydrocarbon adsorption, elemental analysis, $^{11}B$ magic angle spinning nuclear magnetic resonance (MAS NMR) with B-SSZ-33 and other zeolites reveal that the deboronated samples from the deboronation of two-step method are essentially boron-free, and their crystalline structures and micropore volumes remain unchanged [44, 45]. Likewise, the final products of the one-step method have similar properties in terms of deboronation, namely, they are also boron-free. These results indicate that aqueous $Al(NO_3)_3$ solutions in the one-step method have low enough pH ($<3.5$) to provide an acidic environment for efficient deboronation, as does the aqueous HCl solution with the two-step method.

Table 6.2 compares two Al-SSZ-33 zeolites prepared from B-SSZ-33 via two-step and one-step method, respectively [44, 45]. The molar $(Si/B)_{bulk}$ and $(Si/Al)_{bulk}$ ratios were measured via elemental analyses, while the molar $(Si/Al)_{framework}$ ratios for the framework compositions were determined via $^{27}Al$ MAS NMR.

**Table 6.2** Comparison of Al-SSZ-33 prepared from B-SSZ-33 via two-step and one-step methods.

| Method | | Two-step | One-step |
|---|---|---|---|
| Starting material: B-SSZ-33 | $(Si/B)_{bulk}$ | 18.1 | 18.1 |
| | Micropore volume (milliliters per gram) | 0.19 | 0.19 |
| Intermediate material: deboronated SSZ-33 | $(Si/B)_{bulk}$ | $>200$ (below detection limit) | – |
| | Micropore volume (milliliters per gram) | 0.18 | – |
| Final product: Al-SSZ-33 | $(Si/Al)_{bulk}$ | 13.1 | 12.9 |
| | $(Si/Al)_{framework}$ | 16.6 | 16.3 |
| | Micropore volume (milliliters per gram) | 0.21 | 0.21 |

$^{27}$Al MAS NMR results showed that it is possible to exceed the boron lattice substitution in the resulting aluminum contents, but not all the aluminum is in the framework sites. These and additional $^{27}$Al MAS NMR results (Chen, Zones, and Wilson, unpublished results) confirm that most of aluminum in the resulting aluminosilicate zeolites is incorporated into the framework and these aluminum species are already placed into the tetrahedral positions upon the post-synthetic treatment described here. Initially we speculated that some "intermediate Al-species" were first attached to the silanol nests (or defects) created by deboronation, and subsequently a calcination step at a high temperature (e.g., >623 K) might be needed to finally convert these "intermediate Al-species" into tetrahedral aluminum positioned in the framework of zeolites. But such "intermediate Al-species" were not observed in our experiments. Therefore, there is no need to add any additional steps such as calcination to reinsert aluminum into the framework of zeolites. The high micropore volumes of ~0.2 ml/g (determined with cyclohexane adsorption at $P/Po = 0.3$ and room temperature) reveal that there is no pore blocking in the channels of the starting material B-SSZ-33, deboronated SSZ-33, and resulting Al-SSZ-33 samples. Supported by the results from XRD, $N_2$/hydrocarbon adsorption, and elemental analysis, both two-step and one-step methods essentially result in the same aluminosilicate products. Once the chemical roles of the $Al(NO_3)_3$ and HCl solutions for deboronation are understood, from the practical point of view, the deboronation step with aqueous HCl solution can be omitted in practice and the one-step method is the preferred one to use due to its feature of simplicity.

#### 6.3.1.3 **Effects of the Ratio of $Al(NO_3)_3$ to Zeolite**

In Section 6.3.1 we mentioned that the reinsertion of Al into frameworks can be accomplished with sodium aluminate under basic conditions [47]. It is also well known that framework aluminum of zeolites can be extracted via dealumination under acidic conditions [42, 43]. For example, we treated an Al-SSZ-33 sample ($(Si/Al)_{bulk} = 13.1$) in a HCl solution (pH ~0.8) under reflux for 32 hours. After this treatment in the acidic environment, the crystalline structure of the resulting product remained intact based on XRD and $N_2$ adsorption results but its $(Si/Al)_{bulk}$ increased to 36.9. It reveals that the equilibrium does not favor aluminum reinsertion when we use aqueous solution of $Al(NO_3)_3$ for aluminum reinsertion under acidic conditions. To shift the equilibrium toward aluminum reinsertion under acidic conditions, we have to increase the concentration of $Al(NO_3)_3$ in the reaction slurry to force the conversion in an excess of Al cations. This trend and relationship between aluminum reinsertion and $Al(NO_3)_3$ to zeolite ratio are demonstrated with the results shown in Table 6.3 [45]. When the weight ratio of $Al(NO_3)_3$ solution to zeolite increases, the Si/Al ratio of the products decreases, indicating that more aluminum is reinserted. Although a significant amount of $Al(NO_3)_3$ is employed in this technique, it is noteworthy that the excess of $Al(NO_3)_3$ can be always recycled and reused.

**Table 6.3** Preparation of Al-SSZ-33 from B-SSZ-33 at various ratios of $Al(NO_3)_3$ solution to zeolite via two-step method.

| Zeolite | $(Si/B)_{bulk}$ | $(Si/Al)_{bulk}$ | Remarks |
|---|---|---|---|
| B-SSZ-33 | 18.1 | – | Starting material |
| Deboronated SSZ-33 | >200 (below detection limit) | – | Prepared via deboronation of B-SSZ-33 in 0.01 M HCl |
| Al-SSZ-33 | – | 24.7 | S/Z = 16 : 1[a] |
| Al-SSZ-33 | – | 20.1 | S/Z = 25 : 1 |
| Al-SSZ-33 | – | 17.0 | S/Z = 50 : 1 |
| Al-SSZ-33 | – | 13.1 | S/Z = 100 : 1 |

[a] S/Z stands for the weight ratio of 1 M $Al(NO_3)_3$ solution to deboronated SSZ-33.

#### 6.3.1.4 Effects of pH, Time, Temperature, and Other Factors

The solubility of aluminum salt becomes lower at higher pH, enhancing the precipitation of aluminum species. When using 1 M $Al(NO_3)_3$ solution for aluminum reinsertion, the final pH value of the zeolite/$Al(NO_3)_3$ slurries is below 1. We extraneously raised the pH of the zeolite/$Al(NO_3)_3$ slurries in two preparations by adding some ammonium acetate and water [44, 45]. Their final pH was 3.34 and 4.10, whereas the resulting aluminosilicate products had a very low molar bulk Si/Al ratio of 5.5 and 7.5, respectively. These two samples with very high aluminum contents were, however, considerably less active than other Al-SSZ-33 materials prepared at lower pH. Apparently, the precipitation of aluminum species at high pH is the reason for it. Therefore, it is necessary to keep the pH of the zeolite/$Al(NO_3)_3$ slurry below 3.5 to enhance the efficiency of aluminum incorporation.

Aluminum reinsertion does not occur instantaneously. Shorter reaction time is accompanied by a less efficient aluminum reinsertion. We usually run the experiments for at least 24 hours to reach the reaction equilibrium under the corresponding conditions [44, 45]. The reaction temperature stretches from about 363 to 443 K, which is also the typical temperature range for zeolite synthesis. A higher reaction temperature leads to a quicker reach for the equilibrium. So, the same synthesis devices can be used for both direct synthesis and post-synthetic treatment with $Al(NO_3)_3$. The experiments can be also easily carried out in heated glass flasks under reflux. Although it is simple and convenient to carry out the reactions under static conditions (e.g., in a simple Teflon-lined autoclave), stirring, tumbling, or refluxing is preferred in order to enhance the mass and heat transfer.

#### 6.3.1.5 Applicable to Medium Pore Zeolite?

Hydrated aluminum cations are too bulky to penetrate through the pores of medium pore zeolites. Therefore, this technique is not applicable to medium pore zeolites. This finding is supported by the following two parallel experiments conducted in our lab [44, 45]. B-ZSM-11 and Al-ZSM-11 zeolites were synthesized. B-ZSM-11 was then post-synthetically treated with $Al(NO_3)_3$ solution in an attempt to convert it to its aluminosilicate counterpart. The resulting material is denoted here as B-(Al)-ZSM-11. When comparing B-(Al)-ZSM-11 with Al-ZSM-11 for *n*-hexane/3-methylpentane cracking, B-(Al)-ZSM-11 had a much lower activity than Al-ZSM-11. Results from elemental analyses indicated that B-(Al)-ZSM-11 took very little aluminum. Since both B-(Al)-ZSM-11 and Al-ZSM-11 exhibit similar XRD patterns characteristic of ZSM-11 and had the same micropore volume (0.17 ml/g) as determined by $N_2$ adsorption, the inactivity of the B-(Al)-ZSM-11 is not due to any pore plugging. We conclude that the real reason is related to the mismatch between the medium pore zeolites (e.g., ZSM-11) with relatively small pore sizes and the bulky hydrated aluminum cations.

In addition, based on this feature of medium pore zeolite in relation to the post-synthetic treatment with $Al(NO_3)_3$, we can use this technique as a tool, together with other physicochemical and catalytic methods, to distinguish medium pore zeolites from large or extra-large pore ones when we have a new zeolite with unknown structure.

### 6.3.2 Synthesis of Hydrophobic Zeolites by Hydrothermal Treatment with Acetic Acid

Hydrophobic pure-silica zeolites are useful materials for the separations of hydrophobic from hydrophilic compounds [56]. Defect-free or defect-deficient pure-silica zeolites are particularly desirable for the characterization and determination of zeolite structures [57]. Various techniques have been developed to make such materials [42, 43]. Reactions with $SiCl_4$ vapor [51–55] and hexafluorosilicate [54, 58, 59] as well as steaming [60] are three well-known techniques for the dealumination of zeolites and preparation of pure-silica zeolites.

As discussed in Section 6.3.1.2, boron in borosilicate zeolites can be easily hydrated and removed from the framework. Similar to the aluminum reinsertion discussed above, the silanol nests (also called *vacancies* or *defects*) created by deboronation can be repopulated with silicon. Jones *et al.* developed a new method to synthesize hydrophobic pure-silica zeolites via post-synthetic treatment of borosilicate zeolites with aqueous acetic acid [46]. With this single-step method, boron is expelled from borosilicate zeolites under acidic conditions and the defects created by the boron removal are subsequently healed with silicon dissolved from other parts of the crystal. By use of this procedure, highly crystalline, hydrophobic pure-silica CIT-1 and SSZ-33 (CON topology) zeolites were synthesized for the first time. In the following sections, the features of this technique will be reviewed and some experimental results from our lab will be discussed.

#### 6.3.2.1 Experimental Procedures

The acetic acid treatment experiments are typically carried out between 373 and 458 K for six days. The calcined borosilicate zeolite is treated in Teflon-lined autoclaves in an oven with rotation at 60 rpm. In a typical experiment, 0.2 g of zeolite is added to a 45 ml autoclave filled with 25 g of water and 10 g of glacial acetic acid (pH ~1.65). After six days of heating, the solid products are washed extensively with water and acetone and recovered by filtration. So, the experiments are conducted in a way similar to the direct synthesis of zeolites.

#### 6.3.2.2 **Highly Crystalline Pure-Silica Zeolites Prepared via This Technique**

Jones *et al.* [46] reported successful synthesis of highly crystalline, hydrophobic pure-silica zeolites CIT-1 and SSZ-33 (both CON topology [3, 4]) as well as ERB-1 (MWW topology) and beta (*BEA topology, originally synthesized in fluoride media) under appropriate conditions. They used XRD, $^{29}Si$ BD NMR, elemental analysis, as well as $N_2$ and $H_2O$ adsorption to investigate the parent zeolites and the products from acetic acid treatment. The acetic acid treated samples remain crystalline as revealed by the XRD and $N_2$ adsorption results. $^{29}Si$ BD NMR results show impressive differences between the parent and acetic acid treated zeolites. Significant sharpening of the $Q^4$ peaks in the $^{29}Si$ BD NMR spectra allows crystallographically distinct T-sites to be distinguished. The enhanced hydrophobicity of the acetic acid treated zeolites is evidenced by the reduced amount of $Q^3$ silicon species and the decreased $H_2O$ adsorption capacity.

This method is effective on zeolites CIT-1, SSZ-33, and ERB-1, all containing 10- and 12-ring channels. But it is noteworthy that the 12-ring pore openings of ERB-1 do not open to the exterior of the crystal. This method appears to be less efficient on ZSM-5, likely suggesting that the small pore apertures of this zeolite could limit migration of soluble silica species. Jones *et al.* [46] also reported that some beta zeolite samples are either less sensitive (e.g., no significant change in hydrophobicity) or oversensitive (e.g., loss of porosity or collapse of the crystalline structure) to this technique within the condition ranges of their investigation. Apparently, further investigations are needed to address these issues.

#### 6.3.2.3 **Effects of Type of Acid, pH, Temperature, and Other Factors**

Jones *et al.* [46] compared acetic acid with various aqueous mineral acids (HCl, $HNO_3$, and $H_2SO_4$) for this technique and found that treatment with mineral acids results in materials with more significant porosity losses than those from treatment with acetic acid. It is speculated that the mineral acids more readily solubilize silica than does the organic acid. Therefore, acetic acid is preferred for this technique.

The treatment appears to be most effective near the isoelectric point of silica (pH 0–2). At these conditions, the dissolution of zeolite is slow enough to prevent significant loss of microstructure while at the same time allowing sufficient dissolution of silica to provide soluble silicon species for healing of the defects without an external silicon source.

Temperature typically ranges between 373 and 458 K and is another important factor. For zeolites that contain no $F^-$, such as CIT-1, SSZ-33, and ERB-1 (synthesized in hydroxide media), higher temperatures (433–458 K) give better results with limited loss in porosity due to structural degradation. In contrast, zeolites synthesized in the presence of $F^-$ require lower temperatures for effective healing of defects with silicon, likely due to structural degradation by traces of residual $F^-$ when at higher temperature.

Jones *et al.* [46] also reported that additional aluminum or gallium source can be added to the acetic acid solution, and the silanol defects that are not filled with aluminum or gallium are repopulated with silicon during the treatment. As a result, it is possible to produce a more hydrophobic metallosilicate zeolite than the one containing internal silanol defects.

#### 6.3.2.4 Experimental Results from Our Lab

We prepared the following three SSZ-33 samples to investigate their physicochemical and catalytic properties [61]:

1) A borosilicate SSZ-33 was synthesized using *N,N,N*-trimethyl-8-ammoniumtricycle-[5.2.1.0$^{2,6}$]decane hydroxide as structure directing agent [40]. It was calcined following the standard procedure to remove the occluded structure directing agent molecules. The calcined sample is denoted as B-SSZ-33.
2) A deboronated SSZ-33 was prepared by stirring B-SSZ-33 in an aqueous HCl solution (0.01 M) at room temperature and a zeolite-to-solution weight ratio of 1 : 40 for 24 hours. The deboronation creates SiOH nests in place of boron T-atoms and the resulting sample is denoted as SiOH-SSZ-33.
3) A pure-silica SSZ-33 was prepared by tumbling B-SSZ-33 in acetic acid (28.6 wt%) at 458 K and a zeolite-to-solution weight ratio of 1 : 25 for six days according to Jones *et al.* [46]. The resulting pure-silica zeolite is denoted as Si-SSZ-33.

Each of B-SSZ-33, SiOH-SSZ-33, and Si-SSZ-33 was impregnated with 0.5 wt% Pt. The resulting Pt-loaded catalysts are denoted as Pt/B-SSZ-33, Pt/SiOH-SSZ-33, and Pt/Si-SSZ-33, respectively. These were investigated with XRD, elemental analysis, thermogravimetric analysis (TGA), $N_2$ physisorption, $H_2$ chemisorption if applicable, $^{11}$B MAS NMR, $^{29}$Si BD NMR, hydrocracking of *n*-octane, and other techniques.

All samples (B-SSZ-33, SiOH-SSZ-33, and Si-SSZ-33 as well as Pt/B-SSZ-33, Pt/SiOH-SSZ-33, and Pt/Si-SSZ-33) possess XRD peaks characteristic of SSZ-33 and are highly crystalline. They all have a micropore volume of about 0.2 cc/g as determined by $N_2$ adsorption. These results indicate that the SSZ-33 structure is well retained after deboronation with HCl, treatment with acetic acid, and Pt-loading. Results from elemental analysis and $^{11}$B MAS NMR indicate that boron is essentially completely removed from the zeolite in SiOH-SSZ-33 and Si-SSZ-33.

Figure 6.1 shows the $^{29}$Si BD NMR spectra of B-SSZ-33, SiOH-SSZ-33, and Si-SSZ-33. Typical of borosilicate zeolites, the B-SSZ-33 sample shows a broad resonance of $Q^4$ silicon T-atoms at about −110 ppm, indicating that the seven crystallographically distinct T-atoms of SSZ-33 are not distinguishable as a result of the distribution of chemical shifts caused by the presence of lattice boron atoms and framework defects. Consistent with this is a signal at −100 ppm from $Q^3$ silicon atoms. SiOH-SSZ-33 exhibits an increased amount of $Q^3$ silicon atoms that are created as SiOH nests in the framework upon the deboronation with HCl. The treatment with acetic acid expels boron from the zeolite and the defects created by

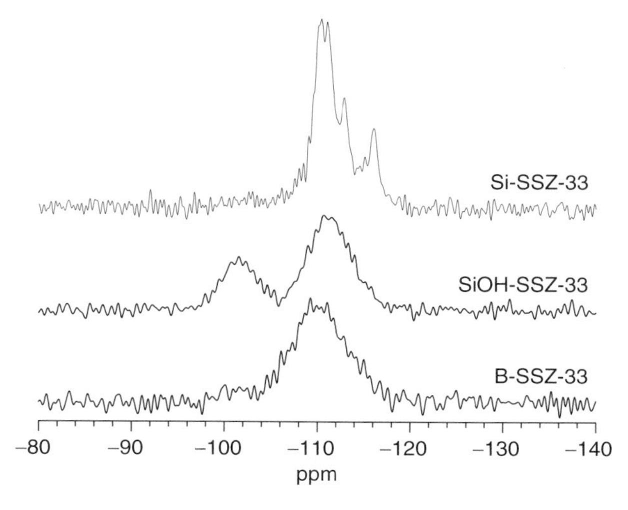


**Figure 6.1** $^{29}$Si BD NMR spectra of B-SSZ-33, SiOH-SSZ-33, and Si-SSZ-33 [61].

deboronation are subsequently healed with silicon dissolved from other parts of the crystal [46]. With Si-SSZ-33, the $Q^3$ silicon groups are reduced via acetic acid treatment, rendering a more defined $^{29}$Si BD NMR spectrum, which is indicative of fewer internal defects in Si-SSZ-33.

TGA results also provide useful information on the hydrophobicity, dehydration, and dehydroxylation of B-SSZ-33, SiOH-SSZ-33, and Si-SSZ-33. As shown in Figure 6.2, these samples exhibit two distinct stages of weight loss: the first weight loss (~298–373 K) is due to dehydration and the second stage (above ~573 K) is related to dehydroxylation. The lower weight loss of SiOH-SSZ-33 versus B-SSZ-33 between ~298 and 373 K reflects some difference in hydrophobicity between B-SSZ-33 and SiOH-SSZ-33. When compared to B-SSZ-33, SiOH-SSZ-33 loses more weight above ~573 K. This weight loss is attributed to dehydroxylation of the

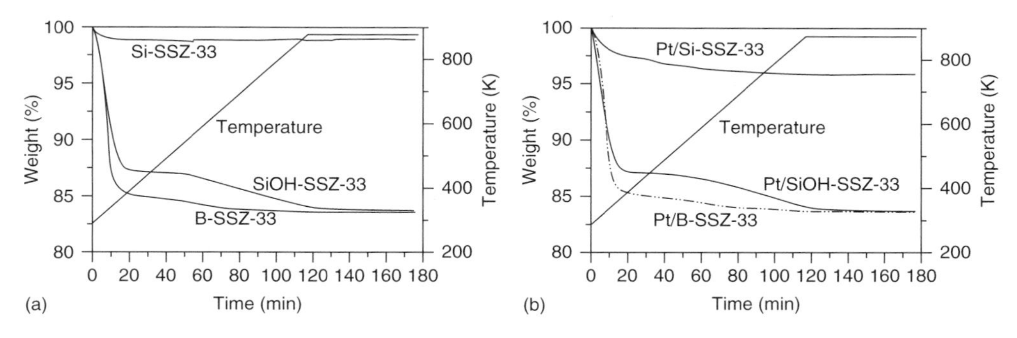


**Figure 6.2** TGA of (a) B-SSZ-33, SiOH-SSZ-33, and Si-SSZ-33 and (b) Pt/B-SSZ-33, Pt/SiOH-SSZ-33, and Pt/Si-SSZ-33. All samples were saturated with moisture in ambient atmosphere [61].

silanol groups present in abundance in SiOH-SSZ-33. Si-SSZ-33 clearly exhibits a higher hydrophobicity (reduced dehydration) and lower content of structural defects (reduced dehydroxylation). The TGA profiles and weight losses of Pt/B-SSZ-33 and Pt/SiOH-SSZ-33 catalysts are essentially the same as those of their parent zeolites B-SSZ-33 and SiOH-SSZ-33. Pt/Si-SSZ-33 sample contains some more water than Si-SSZ-33, maybe due to the presence of Pt particle.

The Pt dispersion in Pt/B-SSZ-33, Pt/SiOH-SSZ-33, and Pt/Si-SSZ-33 catalysts was determined via $H_2$ chemisorption. As reported earlier [61], Pt/B-SSZ-33 has the best Pt dispersion, most likely due to the stronger Pt–B interaction versus Pt–SiOH interaction for Pt/SiOH-SSZ-33. The very low Pt dispersion of Pt/Si-SSZ-33 indicates that structural defects are needed to anchor and disperse Pt in zeolites for catalytic applications, although, by contrast, defect-free zeolite samples are always desirable for structure characterization and determination [57].

It is known that iso-alkanes (e.g., iso-butane) are preferably yielded by hydrocracking of normal alkanes (e.g., *n*-octane). With decreasing acid strength of the catalyst, the yield of *n*-butane from *n*-octane hydrocracking becomes more predominant. Sulfiding minimizes the hydrogenolysis activity of the Pt-loaded catalysts. We measure the weak acid strength of the catalysts with the so-called Acidity Index [61, 62] which is defined as the yield ratio of iso-butane to *n*-butane produced at 95% C5+ yield of *n*-octane (containing 20 ppm sulfur) hydrocracking over the sulfided catalysts. A lower value of the Acidity Index corresponds to a weaker acid strength. Pt/B-SSZ-33 and Pt/SiOH-SSZ-33 have an Acidity Index of 0.049 and 0.010, respectively. Pt/SiOH-SSZ-33 has a lower catalytic activity than Pt/B-SSZ-33, as evidenced by a difference in reaction temperature of 661 K for Pt/B-SSZ-33 versus 672 K for Pt/SiOH-SSZ-33 to reach 95% C5+ yield of *n*-octane hydrocracking. As a reference, Pt/mordenite has an Acidity Index above 3, as expected from the higher acid strength of this aluminosilicate zeolite. With its higher Pt dispersion, higher Acidity Index, and higher catalytic activity as compared to Pt/SiOH-SSZ-33, Pt/B-SSZ-33 appears to have a stronger ability of anchoring and dispersing Pt particles inside the channel system of SSZ-33 and exhibits a higher acid strength of the acid sites. Both factors are attributed to the framework boron sites. Because of the lack of acid function as well as Pt aggregation and its consequent low catalytic activity, the Acidity Index of Pt/Si-SSZ-33 is difficult to be determined meaningfully.

## 6.4
## Summary and Outlook

Kühl and Szostak have published two comprehensive review papers on the post-synthetic treatment and modification of zeolites [42, 43]. These review papers cover many important aspects of post-synthetic treatment and modification of zeolites such as ion exchange, preparation of metal-supported zeolites, dealumination, reinsertion of heteroatoms into zeolite framework, and other modification methods. Complementary to these two papers, here we have reviewed two newer methods for post-synthetic lattice substitution. These two methods are (i) aluminum reinsertion into zeolite framework using aqueous $Al(NO_3)_3$ solution under acidic conditions and (ii) synthesis of pure-silica zeolites via hydrothermal treatment with acetic acid. Both methods use borosilicate zeolites as starting materials and operate on the basis of the same principle, namely, the desired atoms such as aluminum or silicon are inserted into lattice sites previously occupied by boron atoms. They are carried out under conditions typical of direct synthesis of zeolites.

Aluminum reinsertion into zeolite framework using aqueous $Al(NO_3)_3$ solution operates at pH below 3.5. It is known, however, that the reaction equilibrium favors dealumination over aluminum reinsertion under acidic conditions. Therefore, a high ratio of $Al(NO_3)_3$ to zeolite in an excess of aluminum cations is required to force the equilibrium to shift toward aluminum reinsertion under acidic conditions. Although a significant amount of $Al(NO_3)_3$ is employed in this technique, it is noteworthy that the excess of $Al(NO_3)_3$ can be always recycled and reused. In addition, the dissolution of silicon species from the framework is minimized under acidic conditions, overcoming the inherent disadvantage of the aluminum reinsertion method using sodium aluminate under basic conditions. Aluminum reinsertion method using aqueous $Al(NO_3)_3$ solution is not applicable to medium pore zeolites because hydrated aluminum cations are too bulky to penetrate through the pores of medium pore zeolites. As demonstrated by converting B-SSZ-33, B-UTD-1, and many other borosilicate zeolites to their catalytically more active aluminosilicate counterparts, this method has proved to be a useful way to synthesize large or extra-large pore aluminosilicate zeolites which are difficult or impossible to be prepared via direct synthesis. In the future, it will continue to serve as a practical and efficient tool to synthesize these aluminosilicate zeolites before a more economic method of direct synthesis is found.

Hydrothermal treatment of calcined borosilicate zeolites with aqueous acetic acid near the isoelectric point of silica (pH 0–2) effectively produces hydrophobic pure-silica zeolites with few structural defects. Here, the boron is expelled from the zeolites and the defects created by the deboronation are subsequently healed with silicon dissolved from other parts of the crystal. By use of this technique, highly crystalline, hydrophobic, pure-silica CIT-1 and SSZ-33 zeolites were synthesized for the first time. It appears to be a powerful alternative to other post-synthetic treatment methods such as steaming as well as reactions with $SiCl_4$ vapor and hexafluorosilicate. This technique provides a new route to synthesize high-quality zeolites for the characterization and determination of zeolite structures.

## Acknowledgments

We thank Chevron Energy Technology Company for support of our zeolite research, especially Dr. C.R. Wilson and Dr G.L. Scheuerman.

## References

1. Weitkamp, J. and Puppe, L. (eds) (1999) *Catalysis and Zeolites – Fundamentals and Applications*, Springer, p. 564.
2. Čejka, J., van Bekkum, H., Corma, A., and Schüth, F. (eds) (2007) *Introduction to Zeolite Science and Practice*, Studies in

Surface Science and Catalysis, Vol. 168, 3rd revised edn, Elsevier, p. 1058.

3. *http://www.iza-structure.org/* (last accessed February 03, 2010).
4. Baerlocher, Ch., McCusker, L.B., and Olson, D.H. (2007) *Atlas of Zeolite Framework Types*, 6th revised edn, Elsevier, p. 398.
5. McCusker, L.B. and Baerlocher, Ch. (2007) in *Introduction to Zeolite Science and Practice*, Studies in Surface Science and Catalysis, Vol. 168, 3rd revised edn (eds J. Čejka, H. van Bekkum, A. Corma, and F. Schüth), Elsevier, pp. 13–37.
6. Treacy, M.M.J., Rao, S., and Rivin, I. (1993) in *Proceedings of the 9th International Zeolite Conference, Montreal, 1992* (eds R. von Ballmoos, J.B. Higgins, and M.M.J. Treacy), Butterworth-Heinemann, pp. 381–388.
7. Treacy, M.M.J., Randall, K.H., Rao, S., Perry, J.A., and Chadi, D.J. (1997) *Z. Krist.*, **212**, 768–791.
8. Treacy, M.M.J., Randall, K.H., and Rao, S. (1999) in *Proceedings of the 12th International Zeolite Conference, Baltimore, 1998* (eds M.M.J.Treacy, B.K. Marcus, M.E. Bisher, and J.B. Higgins), Materials Research Society, pp. 517–532.
9. Falcioni, M. and Deem, M.W. (1999) *J. Chem. Phys.*, **110**, 1754–1766.
10. Le Bail, A. (2005) *J. Appl. Cryst.*, **38**, 389–395.
11. Foster, M.D. and Treacy, M.M.J. A Database of Hypothetical Zeolite Structures, *http://www.hypotheticalzeolites.net* (last accessed February 03, 2010).
12. Zones, S.I. and Nakagawa, Y. (1994) *Microporous Mesoporous Mater.*, **2**, 543–555.
13. Lobo, R.F., Zones, S.I., and Davis, M.E. (1995) *J. Inclusion Phenom. Mol. Recognit. Chem.*, **21**, 47–78.
14. Zones, S.I. and Davis, M.E. (1996) *Curr. Opin. Solid State Mater. Sci.*, **1**, 107–117.
15. Zones, S.I. and Maxwell, I.E. (1997) *Curr. Opin. Solid State Mater. Sci.*, **2**, 55–56.
16. Davis, M.E. and Zones, S.I. (1997) in *Synthesis of Porous Materials: Zeolites, Clays and Nanostructures*, (eds M.L. Occelli and H. Kessler), Marcel Dekker, pp. 1–34.
17. Nakagawa, Y., Lee, G.S., Harris, T.V., Yuen, L.T., and Zones, S.I. (1998) *Microporous Mesoporous Mater.*, **22**, 69–85.
18. Zones, S.I., Nakagawa, Y., Lee, G.S., Chen, C.Y., and Yuen, L.T. (1998) *Microporous Mesoporous Mater.*, **21**, 199–211.
19. Millini, R., Perego, G., and Bellussi, G. (1999) *Top. Catal.*, **9**, 13–34.
20. Camblor, M.A., Villaescusa, L.A., and Díaz-Cabañas, M.J. (1999) *Top. Catal.*, **9**, 59–76.
21. Lee, G.S., Nakagawa, Y., Hwang, S.J., Davis, M.E., Wagner, P., Beck, L.W., and Zones, S.I. (2002) *J. Am. Chem. Soc.*, **124**, 7024–7034.
22. Zones, S.I. and Hwang, S.J. (2003) *Microporous Mesoporous Mater.*, **58**, 263–277.
23. Corma, A. and Davis, M.E. (2004) *Chem. Phys. Chem.*, **5**, 304–313.
24. Corma, A. (2004) in *Book of Abstracts of the 14th International Zeolite Conference, Cape Town, 2004* (eds E. van Steen, L.H. Callanan, and M. Claeys), pp. 25–41. Also available on CD.
25. Burton, A.W., Zones, S.I., and Elomari, S.A. (2005) *Curr. Opin. Colloid Interface Sci.*, **10**, 211–219.
26. Burton, A.W. and Zones, S.I. (2007) in *Introduction to Zeolite Science and Practice*, Studies in Surface Science and Catalysis, Vol. 168, 3rd revised edn (eds J. Čejka, H. van Bekkum, A. Corma, and F. Schüth), Elsevier, pp. 137–179.
27. Wilson, S.T. (2007) in *From Zeolites to Porous MOF Materials, Proceedings of the 15th International Zeolite Conference, Beijing, 2007*, Studies in Surface Science and Catalysis, Vol. 170 (eds R. Xu, Z. Gao, J. Chen, and W. Yan), Elsevier, pp. 3–18.
28. Robson, H. and Lillerud, K.P. (2001) *Verified Syntheses of Zeolitic Materials*, 2nd revised edn, Elsevier on behalf of the Synthesis Commission of the International Zeolite Association, p. 266.
29. Camblor, M.A., Corma, A., and Villaescusa, L.A. (1997) *J. Chem. Soc., Chem. Commun.*, 749–750.

30. Barrett, P.A., Camblor, M.A., Corma, A., Jones, R.H., and Villaescusa, L.A. (1997) *Chem. Mater.*, **9**, 1713–1715.
31. Chen, C.Y., Zones, S.I., Hwang, S.J., Burton, A.W., and Liang, A.J. (2007) in *From Zeolites to Porous MOF Materials, Proceedings of the 15th International Zeolite Conference, Beijing, 2007*, Studies in Surface Science and Catalysis, Vol. 170 (eds R. Xu, Z. Gao, J. Chen, and W. Yan), Elsevier, pp. 206–213.
32. Chen, C.Y., Finger, L.W., Medrud, R.C., Crozier, P.A., Chan, I.Y., Harris, T.V., and Zones, S.I. (1997) *J. Chem. Soc., Chem. Commun.*, 1775–1776.
33. Chen, C.Y., Finger, L.W., Medrud, R.C., Kibby, C.L., Crozier, P.A., Chan, I.Y., Harris, T.V., Beck, L.W., and Zones, S.I. (1998) *Chem. Eur. J.*, **4**, 1312–1323.
34. Chen, C.Y., Zones, S.I., Yuen, L.T., Harris, T.V., and Elomari, S.A. (1999) in *Proceedings of the 12th International Zeolite Conference, Baltimore, 1998* (eds M.M.J. Treacy, B.K. Marcus, M.E. Bisher, and J.B. Higgins), Materials Research Society, pp. 1945–1952.
35. Valyocsik, E.W. (1995) US Patent 5,441,721.
36. Ernst, S., Hunger, M., and Weitkamp, J. (1997) *Chem. Ing. Tech.*, **68**, 77–79.
37. Košová, G., Ernst, S., Hartmann, M., and Čejka, J. (2005) *Eur. J. Inorg. Chem.*, 1154–1161.
38. Košová, G., Ernst, S., Hartmann, M., and Čejka, J. (2004) in *Book of Abstracts of the 14th International Zeolite Conference Cape Town*, 2004, (eds E. van Steen, L.H. Callanan, and M. Claeys), pp. 362–363. Also available on CD.
39. Lobo, R.F., Pan, M., Chan, I.Y., Medrud, R.C., Zones, S.I., Crozier, P.A., and Davis, M.E. (1994) *J. Phys. Chem.*, **98**, 12040–12052.
40. Zones, S.I., Holtermann, D.L., Santilli, D.S., Jossens, L.W., and Kennedy, J.V. (1990) US Patent 4,963,337.
41. Zones, S.I., Santilli, D.S., Ziemer, J.N., Holtermann, D.L., and Pecoraro, T.A. (1990) US Patent 4,910,006.
42. Kühl, G.H. (1999) in *Catalysis and Zeolites – Fundamentals and Applications* (eds J. Weitkamp and L. Puppe), Springer, pp. 81–197.
43. Szostak, R. (2001) in *Introduction to Zeolite Science and Practice*, Studies in Surface Science and Catalysis, Vol. 137, 2nd completely revised and expanded edn (eds H. van Bekkum, P.A. Jacobs, E.M. Flanigen, and J.C. Jansen), Elsevier, pp. 261–297.
44. Chen, C.Y. and Zones, S.I. (2002) US Patent 6,468,501.
45. Chen, C.Y. and Zones, S.I. (2001) in *Zeolites and Mesoporous Materials at the Dawn of the 21st Century, Proceedings of the 13th International Zeolite Conference, Montpellier, 2001*, Studies in Surface Science and Catalysis, Vol. 135 (eds A. Galarneau, F. Di Renzo, F. Fajula, and J. Vedrine), Elsevier, pp. 211–218.
46. Jones, C.W., Hwang, S.J., Okubo, T., and Davis, M.E. (2001) *Chem. Mater.*, **13**, 1041–1050.
47. Sulikowski, B., Rakoczy, J., Hamdan, H., and Klinowski, J. (1987) *J. Chem. Soc., Chem. Commun.*, 1542–1543.
48. Jacobs, P.A., Tielen, M., Nagy, J.B., Debras, G., Derouane, E.G., and Gabelica, Z. (1983) in Proceedings of the 6th International Zeolite Conference, Reno, 1983 (eds D.H.Olson and A. Bisio), Butterworth, pp. 783–792.
49. Anderson, M.W., Klinowski, J., and Liu, X. (1984) *J. Chem. Soc., Chem. Commun.*, 1596–1597.
50. Dessau, R.M. and Kerr, G.T. (1984) *Zeolites*, **4**, 315–318.
51. Beyer, H.K., Belenykaya, I. (1980) in *Catalysis by Zeolites*, Studies in Surface Science and Catalysis, Vol. 5 (eds B. Imelik *et al.*), Elsevier, pp. 203–210.
52. Beyer, H.K., Belenykaya, I., Hange, F., Tielen, M., Grobet, P.J., and Jocobs, P.A. (1985) *J. Chem. Soc., Faraday Trans. I*, **81**, 2889–2901.
53. Grobet, P.J., Jacobs, P.A., and Beyer, H.K. (1986) *Zeolites*, **6**, 47–50.
54. Weitkamp, J., Sakuth, M., Chen, C.Y., and Ernst, S. (1989) *J. Chem. Soc., Chem. Commun.*, 1908–1910.
55. Li, H.X., Annen, M.J., Chen, C.Y., Arhancet, J.P., and Davis, M.E. (1991) *J. Mater. Chem.*, **1**, 79–85.
56. Weitkamp, J., Kleinschmit, P., Kiss, A., and Breke, C.H. (1993) in *Proceedings of the 9th International Zeolite Conference, Montreal, 1992*, Vol. 2 (eds

R. von Ballmoos, J.B. Higgins, and M.M.J. Treacy), Butterworth-Heinemann, pp. 79–87.

57. Fyfe, C.A., Gies, H., Kokotailo, G.T., Marler, B., and Cox, D.E. (1990) *J. Phys. Chem.*, **94**, 3718–3721.

58. Skeels, G.W. and Breck, D.W. (1983) in *Proceedings of the 6th International Zeolite Conference, Reno, 1983* (eds D.H.Olson and A. Bisio), Butterworth, pp. 87–96.

59. Garralon, G., Fornes, V., and Corma, A. (1988) *Zeolites*, **8**, 268–272.

60. McDanial, C.V. and Maher, P.K. (1968) in *Molecular Sieves* (ed. R.M.Barrer), Society of Chemical Industry, p. 186.

61. Chen, C.Y., Zones, S.I., Hwang, S.J., and Bull, L.M. (2004) in *Book of Abstracts of the 14th International Zeolite Conference, Cape Town, 2004* (eds E. van Steen, L.H. Callanan, and M. Claeys), pp. 1547–1554. Also available on CD.

62. Chen, C.Y., Rainis, A., and Zones, S.I. (1997) in *Proceedings of Materials Research Society Symposium on "Advanced Catalytic Materials – 1996"*, vol. 454 (eds P.W. Lednor, M.J. Ledoux, D.A. Nagaki, and L.T. Thompson), Materials Research Society, pp. 205–215.

# 7
# Structural Chemistry of Zeolites

*Paul A. Wright and Gordon M. Pearce*

## 7.1
## Introduction

Zeolites are tetrahedrally connected framework solids, based on silica, with intricate structures that possess channels and cages large enough to contain extra-framework cations and to permit the uptake and desorption of molecules varying from hydrogen to complex organics up to 1 nm in size. Their crystalline structure directly controls their properties and consequently their performance in applications such as ion exchange, separation, and catalysis, and is therefore of great interest to academics and technologists alike. A "ball and stick" representation of the most widely used zeolite A, with tetrahedral Al and Si atoms linked by O atoms and with charge-balancing $Na^+$ cations, is given in Figure 7.1.

Originally discovered as aluminosilicate minerals, synthetic zeolites with a range of compositions are now widely prepared and subsequently modified for a wide range of applications. Many excellent articles, reviews, and books describe the structures of these solids, often in well-illustrated texts [1–3] and online resources [4]. Here we start by summarizing their structural chemistry, beginning with the key features of the best known and most widely used zeolites A and Y. Besides covering the periodic structures of these solids, other features such as secondary mesoporosity are also important to their performance and are discussed. This is followed by a summary of the structural chemistry of some of the important zeolite types prepared using inorganic and simple organic cations prior to the 1990s.

Over the last 20 years there has been a major international effort to prepare new zeolitic materials, in the search for improved adsorbents and catalysts. Much of this has focused on the exploration of the use of complex organic alkylammonium ions, synthesized specifically as potential "templates," giving high-silica-content zeolites or pure silica polytypes. The diversity of structures has also been increased by the inclusion of elements other than Al for Si in the framework, which may be either aliovalent (2+ or 3+) or isovalent (4+), and the search for new materials is encouraged by the tantalizing arrays of hypothetical structures that have been shown to be energetically feasible [5, 6]. The remarkable products of this ongoing odyssey continue to show new structural features that are both intriguing and of practical importance or potential: increased crystallographic complexity leading to structures with novel

*Zeolites and Catalysis, Synthesis, Reactions and Applications. Vol. 1.*
Edited by Jiří Čejka, Avelino Corma, and Stacey Zones

ISBN: 978-3-527-32514-6

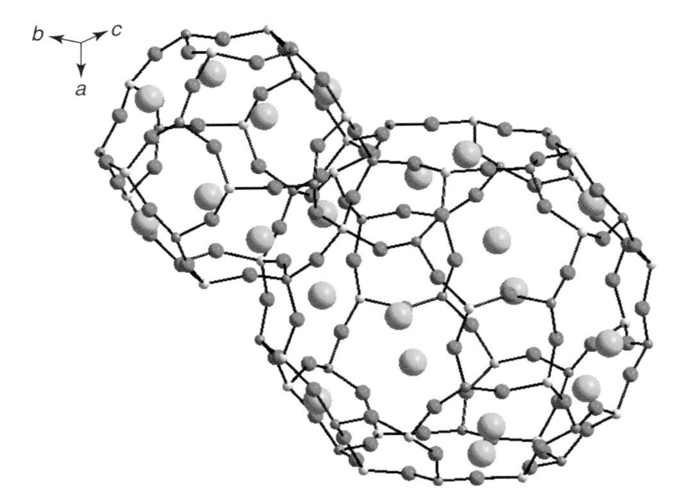


**Figure 7.1** "Ball and stick" representation of part of the structure of the sodium form of zeolite A (Na-A), showing ordered Si and Al framework atoms (dark and light grey spheres) linked by O atoms (red spheres). Extra-framework $Na^+$ cations (orange spheres) occupy sites in the six-membered and eight-membered rings (where this refers to the number of tetrahedrally coordinated cations in the ring).

combinations of pore size, geometry and connectivity; coordination of framework cations above fourfold; chirality and mesoporosity; ordered defects; and layered zeolitic precursors. The second main section of this chapter therefore concentrates on some of the latest structural chemistry to be discovered and its significance.

The structures and structural chemistry described in this chapter have been established by a combination of diffraction methods. Synthetic zeolites rarely crystallize as single crystals of sufficient size or quality for single crystal diffraction, so that the careful and often inspired analysis of powder diffraction profiles, often in combination with electron microscopy, has been essential. This is described elsewhere in this book. Similarly, a full picture of how these structures arise can only be achieved through an understanding of the conditions and mechanism of their crystallization. The first steps in unraveling this process are underway and are also described in this volume.

## 7.2 Zeolite Structure Types Exemplified by Those Based on the Sodalite Cage

### 7.2.1 Introduction

Strictly defined, zeolites are aluminosilicates with tetrahedrally connected framework structures based on corner-sharing aluminate ($AlO_4$) and silicate ($SiO_4$)

tetrahedra. Conceptually, they are based on pure silica frameworks with Si substituted by Al. This aliovalent ($Al^{3+} \leftrightarrow Si^{4+}$) substitution imparts an overall negative charge to the framework. This is balanced by the presence of extra-framework charge-balancing cations within the pore space, which is also able to take in neutral atoms and molecules small enough to enter via the pore windows.

A simplified empirical formula for an aluminosilicate zeolite is

$$M^{n+}_{x/n}Al_xSi_{1-x}O_2 \cdot yX \tag{7.1}$$

where $M^{n+}$ represents inorganic or organic cations and X are included or adsorbed species.

The building blocks of the aluminosilicate framework are the tetrahedra. Typically, Al–O and Si–O bond distances are 1.73 and 1.61 Å, with OTO angles (where T is the tetrahedral cation) close to the tetrahedral angle, 109.4° [7]. There is more variation in the TOT bond angles between tetrahedra, where the average angle is 145° but there is considerable spread. (A recent review of the crystallographic data on pure silica zeolites gives the following distribution: range 133.6–180°, mode 148°, mean 154(±9)° [8].) The variation in TOT angles is in large part responsible for the great structural diversity they exhibit. Values close to 180° are often reported, although this can commonly be attributed to fractional atomic coordinates of oxygen atoms averaged over different positions. A recent electron microscopic and diffraction study of SSZ-24 [9], a pure silica polymorph, sheds some light on the 180° TOT bond angle observed in this solid. In addition to the usual crystallographic repeat along the channel axis (parallel to the linear bond) an incommensurate structural repeat was observed, with a repeat unit of 2.63 × the c repeat. This indicates that there is a long-range modulation in the deviation of this TOT bond angle from 180°, with the O atom moving off its average position.

Al–O–Al linkages are not observed in hydrothermally synthesized aluminosilicate zeolites, because of the unfavorable interaction of adjacent negative charges associated with aluminate tetrahedra – an observation expressed as Löwenstein's rule. A similar rule applies for Ga–O–Ga linkages in gallosilicates. At higher Al (or Ga) contents, Löwenstein's rule results in short-range ordering (confirmed by solid-state NMR) and ultimately long-range ordering. When Si/Al approaches 1, in zeolite A, for example, there is strict alternation between Si and Al.

Several elements other than Al are able to substitute for Si in tetrahedral positions in the framework. These include the divalent cations $Be^{2+}$ and $Zn^{2+}$, other trivalent cations such as $B^{3+}$, $Ga^{3+}$, and $Fe^{3+}$, and tetravalent cations such as $Ti^{4+}$ and $Ge^{4+}$. These can substitute either at low levels, where they may not affect the zeolite structure to form, or at higher concentrations, where $Zn^{2+}$, $Be^{2+}$, and $Ga^{3+}$, for example, tend to give rise to novel structures. This can be ascribed in part to their different ionic radii [10] ($Be^{2+}$, 0.27 Å; $Zn^{2+}$, 0.60 Å; $B^{3+}$, 0.11 Å; $Al^{3+}$, 0.39 Å; $Ga^{3+}$, 0.47 Å; $Fe^{3+}$, 0.49 Å; $Ti^{4+}$, 0.42 Å; $Ge^{4+}$, 0.39 Å) compared to $Si^{4+}$, 0.26 Å, which in some cases favor particular structural units. The incorporation of Ge in silicate frameworks, for example, favors the formation of D4Rs [11].

Topological examination of the possible ways of assembling tetrahedra within all possible space group symmetries indicates that the number is practically limitless,

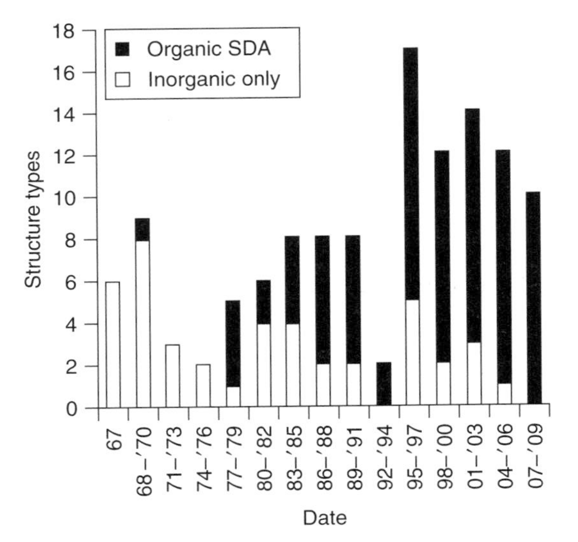


**Figure 7.2** Chart showing the history of the first preparations of synthetic silicates with tetrahedral open frameworks, as reported in the "Atlas of Zeolite Structures" [4]. A distinction is made between those prepared by including only inorganic cations in extra-framework positions and those that include organic species as structure-directing agents, or SDAs.

even taking into account the constraints of energetic feasibility. Considering only silicates, 144 framework topologies have so far been observed, either as natural minerals or synthetic materials [4]. Figure 7.2 shows the trend in the first observation of different synthetic zeolitic silicate structures, making a distinction between materials prepared with inorganic or organic cations acting as structure-directing agents (SDAs). This figure underlines the increasing importance of organic species, typically alkylammonium cations, controlling the synthesis of novel structures. Because these are typically bulky molecular cations, their inclusion in the pore space requires the silicate framework to have lower negative charge density than those formed in the presence of metal cations, so that in aluminosilicates, for example, these organically "templated" zeolites have high framework Si/Al ratios.

A full description of all observed zeolite structure types is available on the web, courtesy of the Structure Commission of the International Zeolite Association [4]. Each observed framework topology is given a three letter code (e.g., FAU for faujasite, MFI for ZSM-5, etc.), and for the first-observed type structure, full details (symmetry, atomic coordinates, secondary building units (SBUs), coordination sequences of tetrahedral nodes, pore connectivity, pore dimensions, etc.) are given. Typically, zeolites with windows defined by 8-rings (made up of eight tetrahedrally coordinated cations and eight bridging O atoms) with pore sizes of around 4 Å are described as small-pore zeolites, those with 10-ring windows as medium-pore zeolites (typically about 5–5.5 Å), those with 12-ring windows as large-pore zeolites

(7–8 Å) and those with windows made up by larger rings (14-ring or 18-ring, for example) are described as extra-large-pore solids. An account is also kept of the different chemical compositions of reported structures which have each topology type, along with references. This "Atlas of Zeolite Structures" includes all porous solids which possess fully four-connected nets, regardless of chemical composition, so that there are many phosphates, germanates, nitrides, and even sulfides that fulfill these geometric requirements. This compilation, which is rigorously checked and kept up-to-date and contains many of the most relevant structural references, is therefore the essential source of information for zeolite structures.

### 7.2.2
### The Framework: Secondary Building Units in Zeolite Structural Chemistry

Silicate (and aluminate, borate, gallate, germanate, titanate, etc.) tetrahedra are the primary building units of zeolites, but consideration of the observed framework types leads naturally to their description in terms of SBUs which are characteristic arrangements of tetrahedra. These are assembled, either on their own or in combination with others, to give the periodic structures. Whether these units exist independently, for example, in solution, remains an open question. Examples of SBUs are given in Figure 7.3, where only the topology of the linked tetrahedral

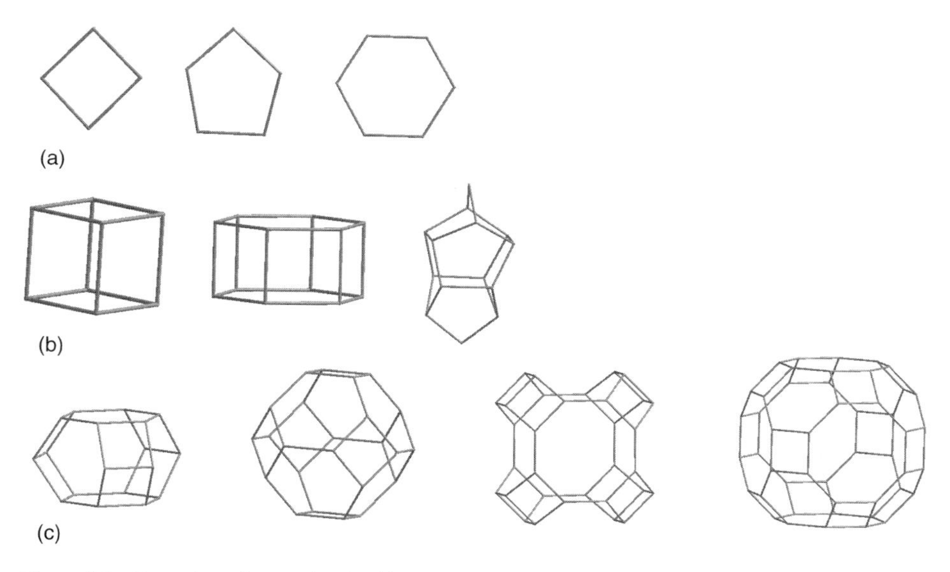


**Figure 7.3** Examples of secondary building units (SBUs) found in zeolite structures, represented by single lines joining the tetrahedral atom (T-atom) positions. (a) Rings; (b) double four-membered rings (D4Rs), D6Rs, and pentasil units; and (c, left to right) cancrinite, sodalite, paulingite, and $\alpha$-cages.

cations is given – in reality, these will be linked by O atoms. The SBUs include rings with different numbers of tetrahedral cations (three-rings, four-rings, etc.), double four-rings, D4Rs, (which contain two rings of four tetrahedral cations), D6Rs, and an array of polyhedral units and cages containing faces with even number of edges, including the cancrinite cage, the paulingite cage, the sodalite or $\beta$-cage and the $\alpha$-cage. One way to represent these cages is in terms of their faces (or rather the number of edges on each face). In this way, a D4R is represented as $[4^6]$, a D6R as $[4^6 6^2]$, and a sodalite cage as $[6^8 4^6]$. Besides these polyhedral units, many high silica zeolites have SBUs containing five-rings (including the so-called pentasil zeolites ZSM-5 and ZSM-11). In these latter solids the characteristic pentasil unit is observed. Besides these polyhedral units, characteristic chains are observed in zeolite structures, such as the zig-zag, sawtooth, crankshaft, natrolite 4=1, and pentasil chains (Figure 7.4). The crankshaft chain is a key building unit in zeolites, in both single and double chains. Adjacent crankshaft chains can be related either by a mirror plane (giving the double crankshaft chain) or by a center of inversion (giving the narsarsukite chain, observed in zeolites but particularly common in aluminophosphates). The important natural zeolites of the natrolite family contain the 4=1 chain and zeolite ZSM-5 the pentasil chain.

Among all the SBUs, the sodalite cage, or $\beta$-cage, is of great importance, because it is a key SBU in two of the most important zeolites, A and Y. We will therefore use the class of zeolites containing sodalite cages to illustrate some of the key features of zeolite structural chemistry.

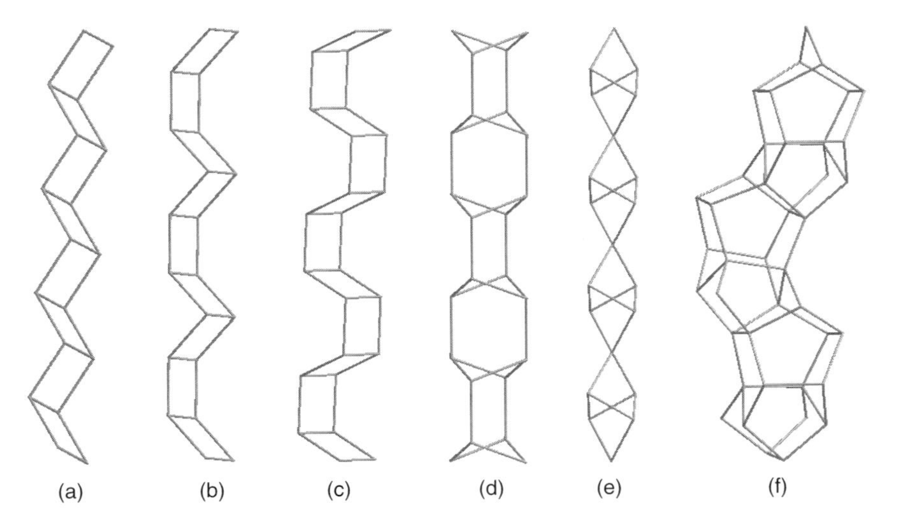


**Figure 7.4** Chains found in zeolite structures: (a) zig-zag, (b) sawtooth, (c) double crankshaft, (d) narsarsukite, (e) natrolite 4=1, and (f) pentasil chains. For clarity, in this and most of the following framework representations, single lines link the tetrahedral atom (T-atom) positions and O atoms are omitted.

### 7.2.3
### Assembling Sodalite Cages: Sodalite, A, Faujasites X and Y, and EMC-2

The mineral sodalite is a so-called "feldspathoid," the semi-open structure of which consists entirely of face-sharing sodalite cages ($\beta$-cages) (Figure 7.5a). The sodalite cages are connected through six-rings which permit the passage of water but no other molecules [4]. Nevertheless, the ability of the sodalite cage to host brightly colored sulfide radicals gives rise to important blue and purple pigments and semi-precious stones (ultramarine, Lapis lazuli, etc.) [3]. If the sodalite cages

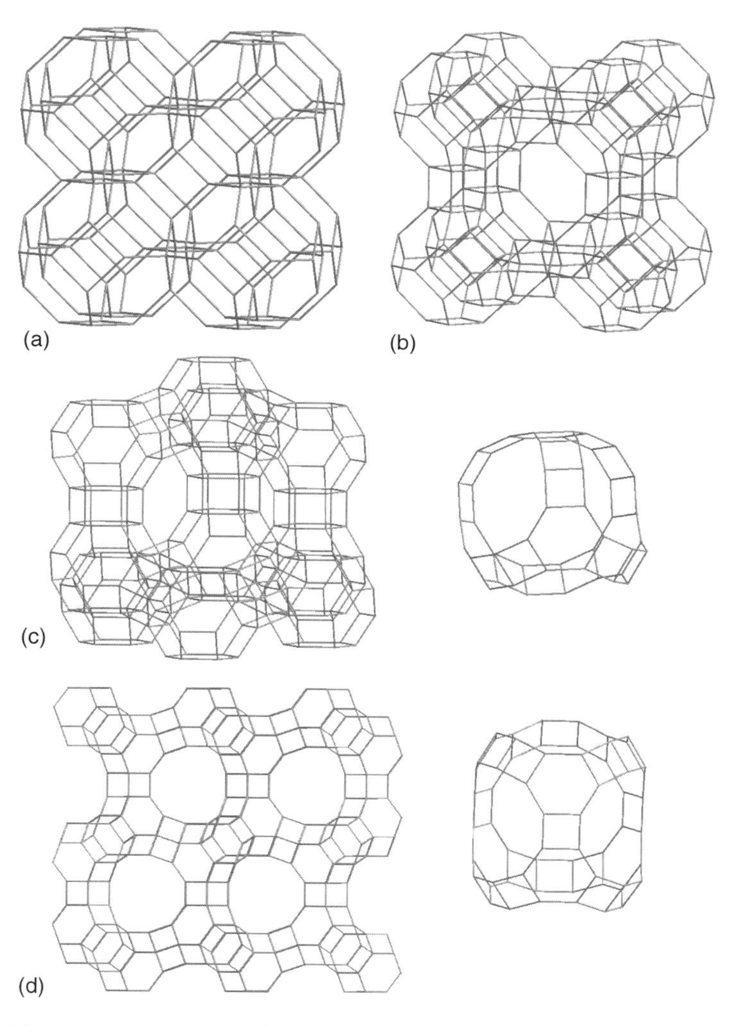


**Figure 7.5** Arrangement of sodalite cages in (a) sodalite, (b) zeolite A, (c) faujasite (with supercage, right), (d) EMC-2 (with supercage, right).

are linked through their four-rings via D4R units, the zeolite A framework results (topology type LTA, Figure 7.5b). Besides D4Rs and $\beta$-cages, the structure contains $\alpha$-cages, which share eight-ring openings that can allow the uptake of small molecules with molecular sizes up to 4 Å. In the aluminum-rich zeolite A (Si/Al = 1) many extra-framework cations are required for charge balance, and these are distributed between the $\beta$- and $\alpha$-cages. Those in the $\alpha$-cage can influence the pore size if they are close to the window. For Na–A the effective pore size is around 4 Å ("zeolite 4A"), whereas for the potassium form, K–A, the larger cations restrict the effective window size ("zeolite 3A"). If $Na^+$ is exchanged by half the number of $Ca^{2+}$ cations, the effective pore size is increased ("zeolite 5A").

Sodalite cages can also be assembled into frameworks by being linked through D6Rs on their six-ring faces (Figures 7.5c,d). Whereas there is only one way of arranging the cages for sodalite and zeolite A, there are different ways of linking layers of sodalite cages through D6Rs. The two end member variants of stacking of layers of sodalite cages are cubic zeolites with the FAU structure type [12], where FAU refers to the mineral form of this material, faujasite and its hexagonal polytype EMC-2 (structure type EMT) [4]. In faujasitic materials the Si/Al ratio can vary from 1.1 to infinity (for X, Si/Al = 1.1–1.8, for Y, Si/Al = 1.8–$\infty$). Disordered stacking sequences are also observed [13]. Which polytype crystallizes is determined by the cationic and other templating species included in the synthesis gel. Besides the D6R and $\beta$-cages, larger cavities, or supercages, are formed.

### 7.2.4
### Faujasitic Zeolites X and Y as Typical Examples

The faujasitic zeolites X and Y provide excellent examples of many of the important types of zeolite structural chemistry. They can be prepared directly with Si/Al ratios from 1.1 to 5, and the Si/Al ratio can be increased by post-synthetic treatments, even to pure silica solids. In Al-containing X and Y the siting of extra-framework cations introduced during synthesis or subsequently by cation exchange is very important in the determination of properties, and has been extensively studied. Some of the sites observed in these studies are shown in Figure 7.6. These include sites in the D6Rs, in the sodalite cages, and in the supercages. If the deammoniation of the $NH_4$-exchanged Y is performed under carefully controlled conditions, the protonic form of the zeolite is prepared. Neutron diffraction studies of this solid have shown that the protons are located on bridging oxygen atoms, Si–OH–Al, giving strong Bronsted acid sites [14].

One of the most important uses of zeolite Y is as a solid acid catalyst for oil cracking. However, the need for both hydrothermal stability and acidity in working catalysts requires very high Si/Al framework ratios. This can be achieved by the process of ultrastabilization, in which the ammonium form is heated in an atmosphere containing steam. The result is deammoniation and the preparation of Bronsted acidic bridging protons on the framework. Under these conditions, Al is removed from the framework to give extra-framework aluminum in the pores and this leaves vacant tetrahedral sites. Simultaneously, Si can leave lattice sites and migrate to fill

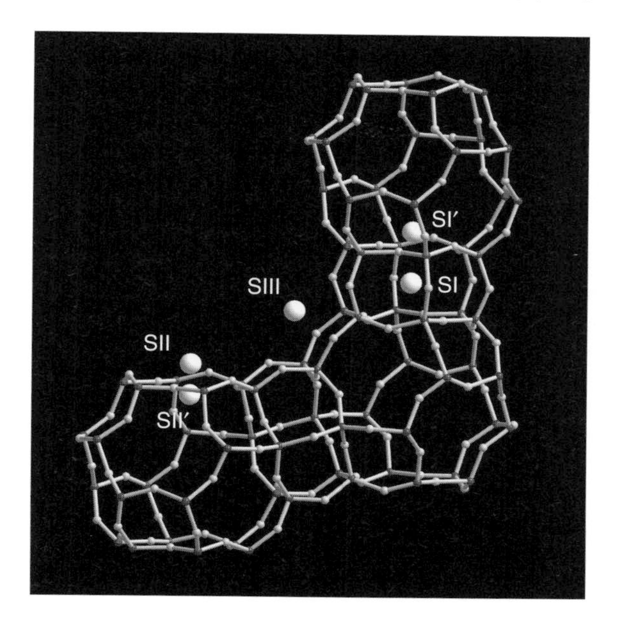


**Figure 7.6** Cation sites (SI, SI′, SII, SII′, and SIII) observed in zeolites X and Y. (Reproduced with permission from [3]).

these vacant sites, resulting in a more hydrothermally stable framework with an increased Si/Al ratio and secondary mesopores [3]. A similar process is observed for some other zeolites (such as zeolite Rho), but for zeolites with very high aluminum contents there is insufficient Si to fill the vacancies and the structure collapses. Where secondary mesoporosity is observed, it improves access of molecules to the interior of the crystallites and has an important beneficial effect in catalytic reactions.

### 7.2.5
### Key Inorganic Cation-Only Zeolites Pre-1990

Once the pioneers of synthetic zeolite chemistry had shown that it was possible to prepare aluminosilicate zeolites such as A in the laboratory, the way was open for an exhaustive investigation of the use of inorganic cations as SDAs in these syntheses, varying cation ratios, Si/Al ratios, temperature, and time. Of the synthetic zeolites produced in this work, mordenite, zeolite L, chabazite and related zeolites, and ZK-5 and Rho are important examples [4].

Mordenite is typically prepared with a Si/Al ratio around 5–10 from $Na^+$-containing gels. It has a framework that contains both 4-rings and 5-rings, giving a system of one-dimensional 12-ring channels connected via "side-pockets" with staggered 8-ring openings to adjacent 12-ring channels (Figure 7.7a). Extra-framework cations are mainly located in these side-pockets. Mordenite is hydrothermally very stable and secondary mesoporosity can be introduced, improving transport between the large channels. These properties, together with

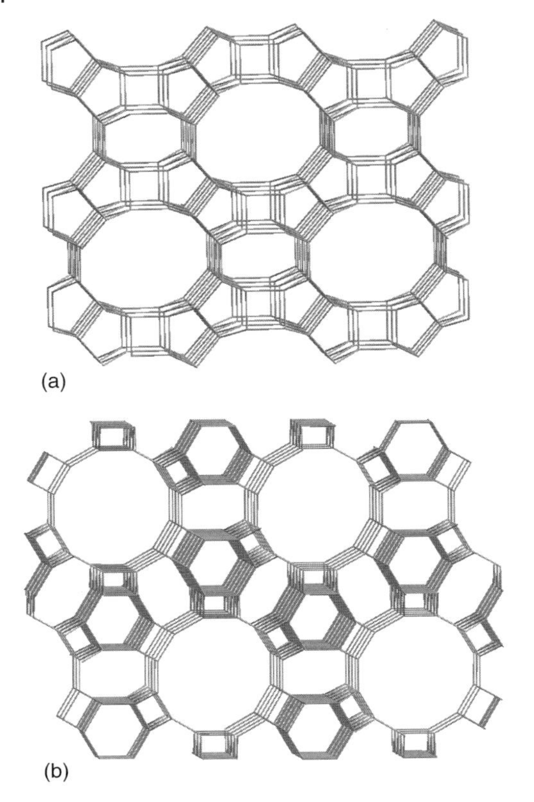


**Figure 7.7** Framework topologies of (a) mordenite and (b) zeolite L, in each case viewed down the 12MR channel.

strong solid acidity when in the protonic form, have led to its application in petrochemical conversion processes.

Zeolite L is a typical product of syntheses in the presence of $K^+$ cations, with a moderate Si/Al ratio of about 3 (Figure 7.7b). The key structural elements of the framework are columns of cancrinite cages alternating with D6Rs that link to give a hexagonal zeolite with one-dimensional 12-ring channels. In between the 12-ring openings, the channels open out to a larger diameter. $K^+$ cations adopt sites in the cancrinite cages (suggesting a strong templating role), in the intercage regions and in the large channels.

The zeolite chabazite can also be prepared from $K^+$-containing gels, and in the $Li^+$-exchanged form it has important properties in the noncryogenic separation of oxygen and nitrogen in air. Structurally, chabazite is one of a large family of synthetic and natural zeolites that can be thought to be built up entirely from six-rings, stacked in well-ordered sequences according to their position (A, B, or C) at (0,0), $(\frac{2}{3}, \frac{1}{3})$, or $(\frac{1}{3}, \frac{2}{3})$ in the $xy$ plane of a hexagonal cell (Figure 7.8a). In chabazite, all six-rings are part of D6Rs: the stacking sequence being AABBCC. All D6Rs have the same orientation, and link to other D6Rs to give a structure that contains

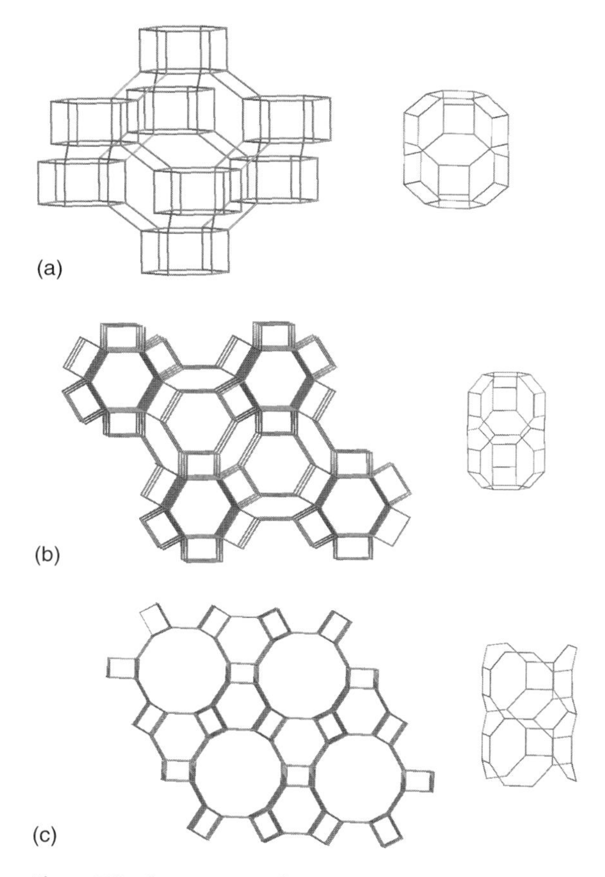


**Figure 7.8** Arrangement of 6MR SBUs giving (a) chabazite, (b) erionite, and (c) offretite topologies, with associated cages and channel (right).

the chabazite cage. Each chabazite cage is linked to six others via eight-ring windows giving a highly porous, three-dimensionally connected eight-ring structure. Variation in the six-ring stacking sequence gives structures with cages and channels of different sizes: the cage and channel of erionite and offretite, respectively, two other zeolites of this family that are prepared with $K^+$ and $Na^+$ cations that have the stacking sequences AABAAC and AAB, are shown in Figure 7.8b,c.

The zeolites ZK-5 and Rho are two cubic, small-pore zeolites with high pore volume, which contain $\alpha$-cages (Figure 7.9). Both are prepared with alkali-metal cations, including $K^+$ and $Cs^+$. In ZK-5, as in chabazite, the structure can be thought to be built entirely from D6Rs, giving rise to $\alpha$- and *pau* cages that alternate along the crystallographic axes, giving a cage structure connected by planar eight-ring windows. In zeolite Rho the structure can be envisaged as $\alpha$-cages linked via D8Rs, giving two interpenetrating pore volumes. Each pore volume is itself connected via eight-rings, but the two different pore systems are not connected.

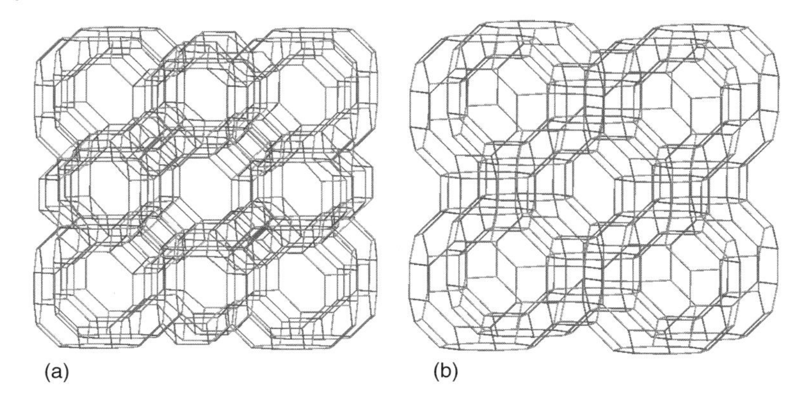


**Figure 7.9** Arrangement of α-cages in the cubic zeolites (a) ZK-5 and (b) Rho.

There is no inaccessible pore space in the zeolite Rho structure and so it has a high pore volume. When in the $Cs^+$-form, it shows structural flexibility upon dehydration, as the D8Rs distort to allow better coordination to the $Cs^+$ cations [15].

## 7.2.6
## Structures Templated by Simple Alkylammonium Ions

Once the success of using alkali-metal cations as SDAs was apparent, it was not long before organic alkylammonium cations were investigated for the same purpose. In the period before the 1990s, these were mainly commercially available ions, especially tetraalkylammonium ions $NR_4{}^+$ (R = $CH_3$, $C_2H_5$, $C_3H_7$, and $C_4H_9$: $TMA^+$, $TEA^+$, $TPA^+$, and $TBA^+$, respectively). Using these, often in combination with alkali-metal cations, a new series of zeolites with medium to high Si/Al ratios was obtained [4].

Using $TMA^+$ gives, in addition to the structures with the LTA topology with Si/Al > 1 (ZK-4), the important structures EAB, gmelinite, and ZSM-4 (which has the structure of the mineral mazzite), each of which contains cages. The cages in these structures (gmelinite cages are present in both gmelinite and ZSM-4) are shown in Figure 7.10. Among the most important new structure types prepared with $TEA^+$ were zeolite Beta and ZSM-12. Although it was one of the first zeolites to be prepared, the structure of zeolite Beta defied solution until 1988 due to its disorder [16]. The structure is built up from layers with tetragonal symmetry that can be connected via stacking offsets of $\pm a/3$ and $\pm b/3$. In zeolite Beta, these offsets occur with a high degree of disorder, but this does not block the straight 12-ring channels parallel to the layers (shown in Figure 7.11), or the undulating and interconnected 12-ring channels that run perpendicular to these. The high silica content of Beta infers to it a high degree of stability, and by virtue of its 3D connectivity and ease of synthesis it is one of the most important zeolites used in catalysis. ZSM-12 is another zeolite that crystallizes in the presence of $TEA^+$. It has a 1D 12-ring channel system, the projection of which is similar to that of the straight channel in zeolite Beta.

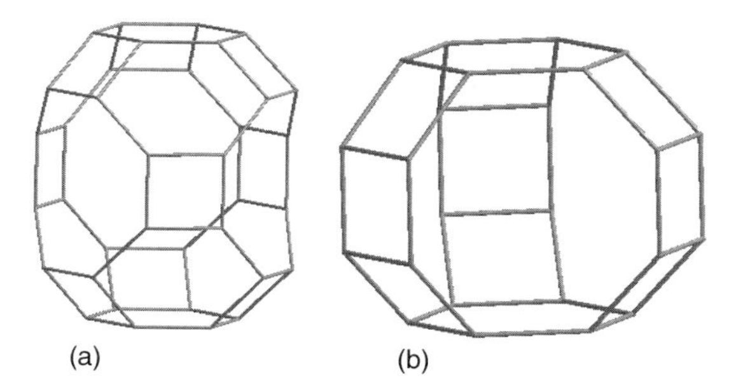


**Figure 7.10** Cages found in (a) EAB and (b) gmelinite (and ZSM-4 or mazzite).

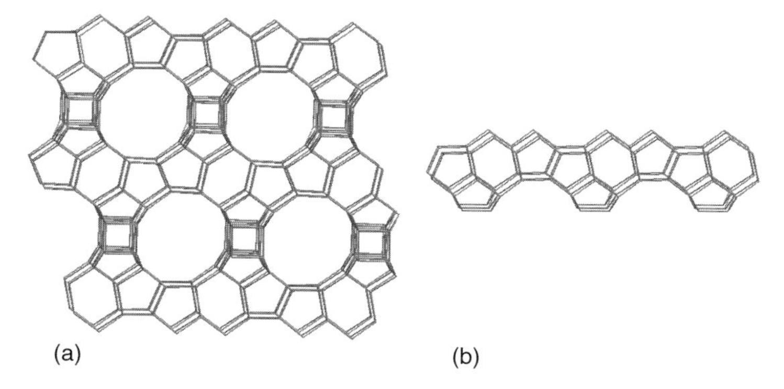


**Figure 7.11** The topology of zeolite Beta, viewed along one set of its straight 12MR channels (a). A single layer of the Beta structure is shown (b).

One of the major breakthroughs in zeolite chemistry came with the synthesis of ZSM-5 using $TPA^+$ cations as SDAs. This solid possesses a three-dimensionally connected 10MR pore system consisting of intersecting straight and undulating medium-pore channels [17]. The pore structure, together with the high stability and strong acid strength that result from its high Si/Al ratio (reaching infinity for its pure silica polymorph, silicalite-1), makes it a highly active and shape-selective catalyst, particularly in conversions of monoaromatics. The structure of ZSM-5 is built up from the pentasil units shown in Figure 7.12. These link to form chains, which in turn link to form sheets. The ZSM-5 structure results when these sheets are linked across a center of inversion, as shown in Figure 7.12, with the $TPA^+$ cations located at the channel intersections. Straight 10-ring channels run parallel to the sheets, and are connected by undulating channels that lie in the plane parallel to them. Although there are only two sets of channels, they are connected so that any part of the pore space in a crystal is accessible to any other. If the same sheets are stacked so that they are related to adjacent sheets by mirror planes, a different structure, ZSM-11, is formed, also with a 3D 10-ring channel system. ZSM-11 is

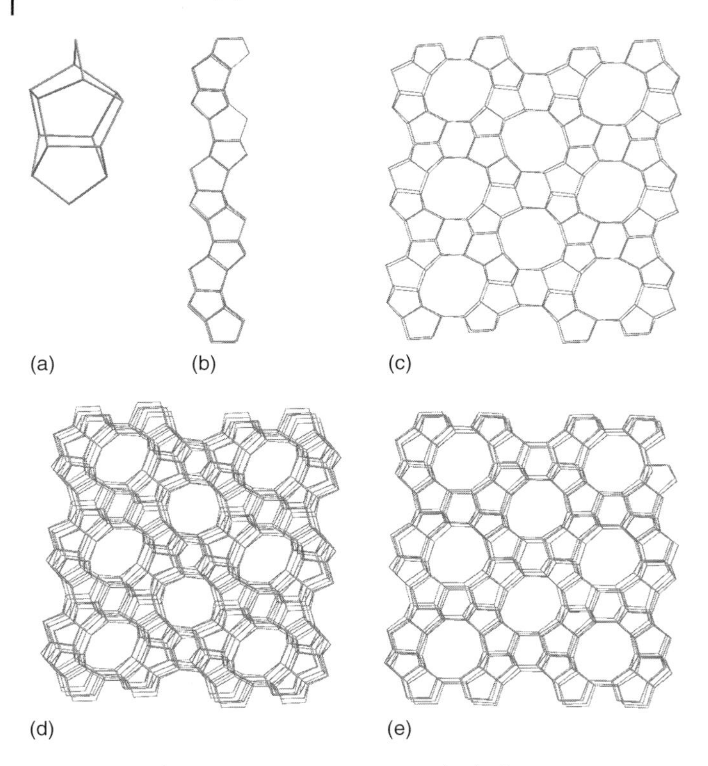


**Figure 7.12** The ZSM-5 (MFI) structure can be built up successively: (a) the pentasil unit; (b) chains of pentasil units; (c) layers of these chains; and (d) layers linked across inversion centers. If the layers are linked across mirror planes, (e), ZSM-11 results.

prepared in the presence of $TBA^+$ cations, among other SDAs. The observation that different structure types can often be formed from the same sheets, but with different stacking sequences, is a common one and the "Atlas" lists several families of this type [4].

In this first period of zeolite structure synthesis in the presence of organic SDAs, initial efforts were made to use more complex alkylammonium ions. Casci used a series of diquaternary cations of the form $[(H_3C)_3N(CH_2)_nN(CH_3)_3]^{2+}$ (where $n = 3-10$) and discovered the high silica 1D 10-ring zeolite EU-1 [18], which was templated by hexamethonium ions ($n = 6$). This was one of the first examples of the use of more complex SDAs.

### 7.2.7
### Lessons from Nature

From the elucidation of the structure of the natural mineral sodalite in 1930 to the present day, natural zeolites have provided an inspiration to zeolite chemists. Many of the commercially important zeolites have been observed as minerals, either before or after they have been synthesized in the laboratory, and frequently as crystals suitable for single-crystal diffraction. The structures of ferrierite, mordenite, faujasite, and mazzite, for example, were all determined from the minerals, whereas for mutinaite (MFI type), tschernichite (*BEA), and direnzoite (ECR-1, EON) the crystal structures were first solved from synthetic samples before their natural counterparts were unearthed.

For zeolite chemists, the observation of mineral zeolite structures with attractive features but no synthetic counterpart acts as a spur, because it indicates that their synthesis should be feasible, and furthermore that this should be possible without the use of organic templates, which are not thought to be involved under geological conditions. The mineral boggsite, for example, first found and solved in 1990, has a 2D channel system in which 12-ring and 10-ring channel systems intersect. The possible advantages such a pore system could possess in terms of shape selectivity make it an attractive target. Subsequent synthetic studies have realized this 12-ring × 10-ring feature of pore systems in several novel solids. More recently, discovery of the aluminosilicate mineral direnzoite, an ordered intergrowth of mordenite and mazzite sheets [19], strongly suggests that this should be preparable without the complex organic that was initially used for its synthetic counterpart ECR-1 [20]. Recent reports suggest this to be the case [21].

Finally, some zeolite mineral structures remain without synthetic counterparts. Examples include boggsite, terranovaite (which has a 2D 10-ring channel system), and tschörtnerite (TSC). TSC, which like CHA and KFI is composed entirely of D6Rs (Figure 7.13) [4], has a remarkable interconnected pore system containing $\beta$-cages, $\alpha$-cages, and a larger supercage, and its synthesis remains an attractive target.

## 7.3 The Expanding Library of Zeolite Structures: Novel Structures, Novel Features

### 7.3.1 Introduction

The discovery that mono, di, and triquaternary alkylammonium cations, some prepared specifically, could direct the crystallization of new zeolite structures prompted several research groups to prepare a wide variety of such species with novel geometries and to screen them as SDAs in a variety of gel compositions. This approach has since been extended by the use of phosphonium ions [22]. The groups of Zones (Chevron) and Corma (ITQ, Valencia) pioneered these studies and realized that expert synthetic organic chemistry was an integral part of them. Other academic and industrial research groups have also made important and sustained contributions in this area, including those of Davis (Caltech), Hong (Taejon), and Xou (Stockholm) and groups at ExxonMobil, UOP, Mulhouse and the Institut Francais du Petrole. Examples of their work are referenced in the following text

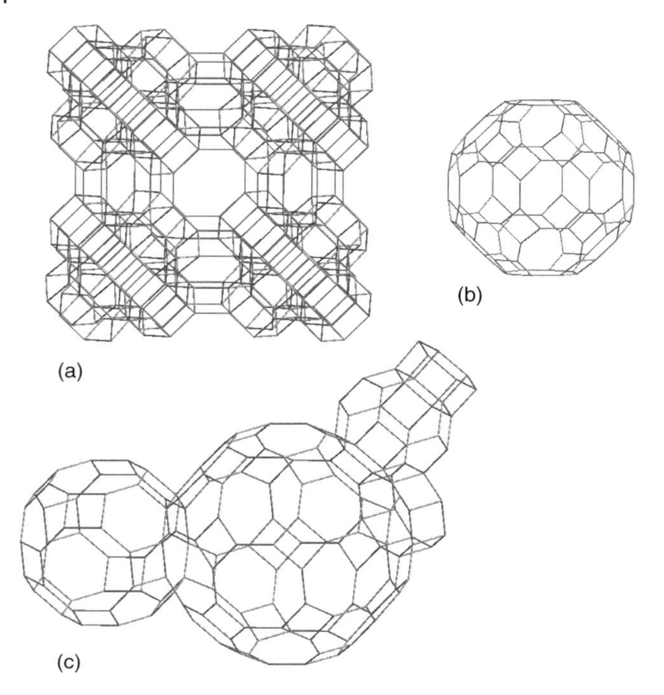


**Figure 7.13** The tschörtnerite mineral's framework structure (a) contains the tschörtnerite supercage (b), a combination of D6Rs, $\beta$-cages, $\alpha$-cages, and the supercage (c).

alongside some of the structures they have produced. In parallel with this, strategies of using combinations of organic bases as potential co-template mixtures have also had important results in the synthesis of zeolites and related solids, particularly as shown via recent high-throughput studies by Blackwell *et al.* at UOP [23].

In addition to the use of designed potential SDAs, modification of the inorganic components of the gel has been found to play a key role in phase selectivity. The work of Camblor at the ITQ on the introduction of fluoride ions into low water content synthesis gels gave many new porous silica polymorphs of zeolites with large pore volumes, where the fluoride ions have a dual role of assisting silicate condensation and crystallization and balancing the positive charge on the framework [24]. Early examples of the success of this model for new structures include the highly porous small-pore zeolite ITQ-3 (with 2D 8-ring pore system), the large-pore 1D channel structure ITQ-4 (there are now many 1D 12-ring channel structures known), and the 3D 12-ring ITQ-7 [25]. This approach has subsequently been used very productively by many researchers in the field. The structural role of the fluoride is discussed in Section 7.3.3.

Changing the composition and elemental ratios of framework-forming cations has also been found to exert a large influence on the phase to form. Zones has investigated the effect of variation of Si/Al and Si/B ratios in the gel as additional parameters in the syntheses with new potential SDAs [26], and Corma (and the Mulhouse and Stockholm groups) found that the inclusion of germanium had a strong structure-directing influence, because of its propensity to favor the formation of D4Rs [11, 27, 28] . Furthermore, the addition of inorganic cations and variation of alkalinity ($OH^-/T$) have been shown to have an important influence on the zeolite structure to form [29].

A combination of these innovative synthetic strategies has been responsible for the upsurge in reported novel synthetic zeolite types since 1990 (Figure 7.2). The need for state-of-the-art structural characterization, combined with specialist organic synthesis and high-throughput screening of potential SDAs and mixtures of SDAs under a very wide range of synthetic variables, means that the synthesis of ever more complex structures is a specialized enterprise. This approach has brought many significant structural highlights in the last decade or so, including larger pore sizes, new pore connectivities, increased structural complexity, a broadening of the compositional range of known structure types, and chiral structures. In the second main part of this chapter we summarize some of the most important developments.

### 7.3.2
### Novel Structures and Pore Geometries

One of the most obvious advances in new structural chemistry has been the preparation of zeolites with extra-large pores, larger than the 12-rings found in faujasites and Beta (Table 7.1). The first 14-ring zeolite, UTD-1 [30] (University of Texas, Dallas), was prepared using the permethylated cobalticenium ion (Figure 7.14). Other 14-ring window pure silicate and germanosilicates were prepared at Caltech (CIT-5) [31], at Chevron, (SSZ-53 and SSZ-59) [32], and at Mulhouse (IM-12) [33], and a beryllosilicate (OSB-1) with 14-ring openings in the framework has also been synthesized [34]. Furthermore, two silicates with channels bounded by 18-rings have been reported, the gallosilicate ECR-34 [35] and most recently the germanosilicate ITQ-33 [36]. The dimensions of the pore openings as defined crystallographically are given in Table 7.1. The large-pore nature of ECR-34 has been demonstrated by the adsorption of large hydrocarbons such as perfluorotri-n-butylamine. The gallium and germanium contents have important roles in directing these structures, for example, in favoring D4Rs in ITQ-33, and tend to reduce their overall hydrolytic stability. Nevertheless, these structures signpost the way to thermally regenerable extra-large-pore acid catalysts. Finally, a germanosilicate ITQ-37 with channels linked by highly noncircular 30-rings has recently been reported [37]. This is described in more detail in Section 7.3.5.

Many new structure types with 3D pore connectivity have resulted from these structures, including those with connectivity in all dimensions via openings of 10-ring or greater (Table 7.2). These are of great interest as adsorbents and catalysts because of their enhanced molecular transport properties and their resistance to blocking in catalytic reactions. TNU-9 [38] and SSZ-74 [39], for example, add to the important class of zeolites with 3D 10-ring channel systems, previously

**Table 7.1** Window sizes of extra-large pore silicates.

| Zeolite (code) | Framework composition | Connectivity (MRs) | Window dimensions of the largest pore (Å) | Reference (year) |
|---|---|---|---|---|
| UTD-1 (DON) | $SiO_2$ | 14 | (8.2 × 8.1) | [30] (1999) |
| CIT-5 (CFI) | $SiO_2$ | 14 | (7.5 × 7.2) | [31] (1998) |
| SSZ-53 (SFH) | $Si_{0.97}B_{0.03}O_2$ | 14 | (8.7 × 6.4) | [32] (2003) |
| SSZ-59 (SFN) | $Si_{0.98}B_{0.02}O_2$ | 14 | (8.5 × 6.2) | [32] (2003) |
| IM-12 (UTL) | $Si_{0.82}Ge_{0.18}O_2$ | 14 × 12 | (9.5 × 7.1) × (8.5 × 5.5) | [33] (2004) |
| OSB-1 (OSO) | $Si_{0.66}Be_{0.33}O_2$ | 14 × 8 × 8 | (7.3 × 5.4) × (3.3 × 2.8) × (3.3 × 2.8) | [34] (2001) |
| ECR-34 (ETR) | $Si_{0.75}Ga_{0.24}Al_{0.01}O_2$ | 18 × 8 × 8 | (10.1) × (6.0 × 2.5) | [35] (2003) |
| ITQ-33 | $Si_{0.66}Al_{0.04}Ge_{0.30}O_2$ | 18 × 10 × 10 | (12.2) × (6.1 × 4.3) × (6.1 × 4.3) | [36] (2006) |
| ITQ-37 | $Si_{0.58}Ge_{0.42}O_2$ | 30 × 30 × 30 | (19.3 × 4.9) × (19.3 × 4.9) × (19.3 × 4.9) | [37] (2009) |

exemplified by ZSM-5 and ZSM-11. They are the most complex structures yet observed and are discussed further in Section 7.3.4, along with the 2D 10-ring zeolite IM-5 [40]. In addition, new 3D-connected 12-ring silicates ITQ-17 [25] and germanosilicates ITQ-17 [41], ITQ-21 [42], and ITQ-26 [43] add to the previously known structures with this connectivity, the faujasite and Beta structures (see, for example, Figure 7.15). ITQ-17 is related to the disordered Beta structures originally prepared, having the same framework layers, but these are stacked differently, in an ordered tetragonal arrangement and including D4Rs. In this structure, originally hypothesized and named Beta polymorph C (Beta C), the three perpendicular 12-ring channel systems intersect at the same place. Originally prepared as a germanate FOS-5 [44], this has more recently been prepared as the germanosilicate ITQ-17. ITQ-21 and -26 also possess 3D 12-ring channel systems, and here again the D4Rs typical of germanium-containing silicates are a crucial structural element.

One of the most important novel classes of 3D-interconnected channel structures that has been prepared contains 12-ring channels intersecting with 10-ring channels. The zeolites CIT-1 [45], ITQ-24 [46], and MCM-68 [47] are examples of this type of framework, each prepared with a complex SDA. There is considerable interest in investigating possible novel shape-selective catalytic performances in this type of structure. Other novel solids with 3D connectivity include ITQ-33 (18 × 10 × 10). In addition to these new structures with 3D-connected porosity,

**Table 7.2** Recent (post-1990) tetrahedrally connected zeolite structures with 3D connectivity via at least 10MR openings, compared with ZSM-5, Y, and Beta zeolites.

| Zeolite (code) | Framework composition | Space group | Pore system (MR) | Dimensions (Å) | Reference (year) |
|---|---|---|---|---|---|
| Faujasite (FAU) | $Si_{1-x}Al_xO_2$ ($x = 0–0.4$) | *Fd3m* | 12 × 12 × 12 | (7.4) × (7.4) × (7.4) | [12] (1958) |
| Beta (*BEA) | $Si_{1-x}Al_xO_2$ ($x = 0–0.05$) | *P4122* | 12 × 12 × 12 | (6.7 × 6.6) × (6.7 × 6.6) × (5.6 × 5.6) | [16] (1988) |
| ZSM-5 (MFI) | $Si_{1-x}Al_xO_2$ ($x = 0–0.05$) | *Pnma* | 10 × 10* | (5.1 × 5.5) × (5.3 × 5.6) | [17] (1978) |
| ITQ-7 (ISV) | $SiO_2$ | $P4_2/mmc$ | 12 × 12 × 12 | (6.5 × 6.1) × (6.5 × 6.1) × (6.6 × 5.9) | [25] (1999) |
| CIT-1 (CON) | $Si_{0.96}B_{0.04}O_2$ | $C2/n$ | 12 × 12 × 10 | (7.0 × 6.4) × (7.0 × 5.9) × (5.1 × 4.5) | [45] (1995) |
| ITQ-17 (BEC) | $Si_{0.64}Ge_{0.36}O_2$ | $P4_2/mmc$ | 12 × 12 × 12 | (6.7 × 6.6) × (6.7 × 6.6) × (5.6 × 5.6) | [41] (2001) |
| ITQ-21 (no code) | $Si_{0.66}Ge_{0.34}O_2$ | *Fm-3c* | 12 × 12 × 12 | (7.5) × (7.5) × (7.5) | [42] (2002) |
| ITQ-24 (IWR) | $Si_{0.89}Ge_{0.09}Al_{0.06}O_2$ | *Cmmm* | 12 × 10 × 10 | (5.8 × 6.8) × (4.6 × 5.3) × (4.6 × 5.3) | [46] (2003) |
| MCM-68 (MSE) | $Si_{0.90}Al_{0.10}O_2$ | $P4_2/mnm$ | 12 × 10 × 10 | (6.8 × 6.4) × (5.8 × 5.2) × (5.2 × 5.2) | [47] (2006) |
| ITQ-33 (no code) | $Si_{0.66}Al_{0.04}Ge_{0.30}O_2$ | $P6/mmm$ | 18 × 10 × 10 | (12.2) × (6.1 × 4.3) × (6.1 × 4.3) | [36] (2006) |
| TNU-9 (TUN) | $Si_{0.95}Al_{0.05}O_2$ | $C2/m$ | $10^2$ × 10* | (5.6 × 5.5),(5.5 × 5.1) × (5.5 × 5.4) | [38] (2006) |
| SSZ-74 (-SVR) | $Si_{0.96}\square_{0.04}O_2$ | *Cc* | 10 × 10* | (5.9 × 5.5) × (5.6 × 5.6) | [39] (2008) |
| ITQ-26 (IWS) | $Si_{0.8}Ge_{0.2}O_2$ | $I4/mmm$ | 12 × 12 × 12 | (7.05) × (7.3 × 7.0) × (7.3 × 7.0) | [43] (2008) |
| ITQ-37 (no code) | $Si_{0.58}Ge_{0.42}O_2$ | $P4_132$ or $P4_332$ | 30 × 30 × 30 | (19.3 × 4.9) × (19.3 × 4.9) × (19.3 × 4.9) | [37] (2009) |

*indicates that 3D connectivity is achieved via the intersection of two channel systems.
□indicates a tetrahedral cation vacancy.

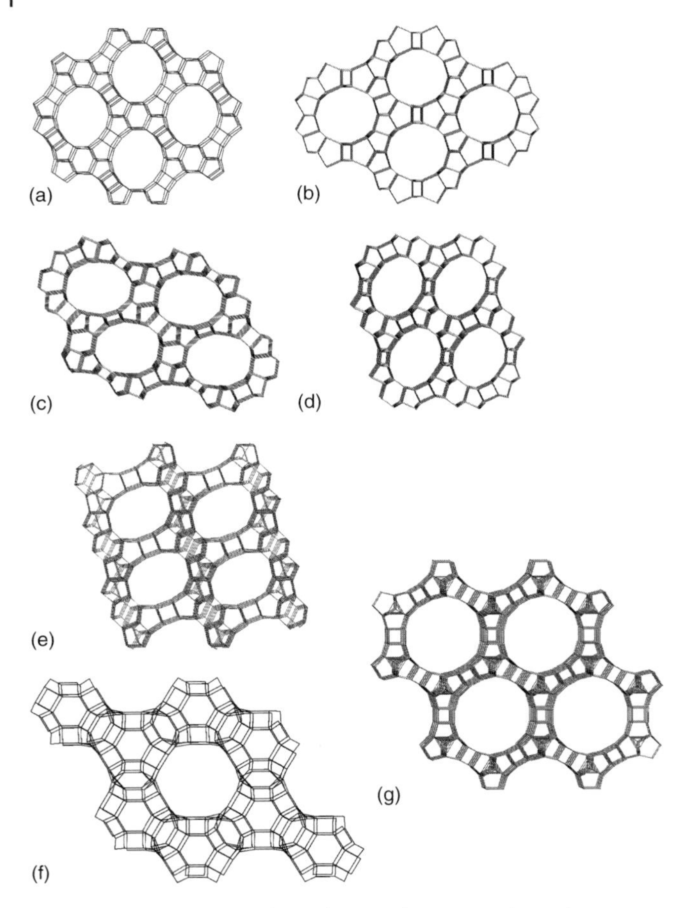


**Figure 7.14** Projections down the extra-large pore channels of the 14MR zeolites (a) UTD-1, (b) CIT-5, (c) SSZ-53, (d) SSZ-59, and (e) IM-12 and of the 18MR zeolites (f) ECR-34, and (g) ITQ-33.

several new 2D-connected materials have been prepared and their structures solved (MCM-22, ITQ-3, ITQ-13, ITQ-22, SSZ-56, etc.) [4]. Besides possessing interesting gas adsorption and catalytic properties in their calcined forms, post-synthetic treatment is a possible route to secondary mesoporosity to increase the dimensionality of molecular transport. Finally, new structures that show chirality have been prepared and are described in Section 7.3.5: the chiral and mesoporous ITQ-37 is a remarkable example.

Another major result that has been achieved by these synthetic studies is a widening of the available compositional range of zeolites with known structures. The pure silica version of zeolite A (ITQ-29) [48] has been prepared by using organic species (with low charge densities) in the syntheses, rather than $Na^+$ cations. This silica shows much higher hydrolytic stabilities than zeolite A. Similar results have

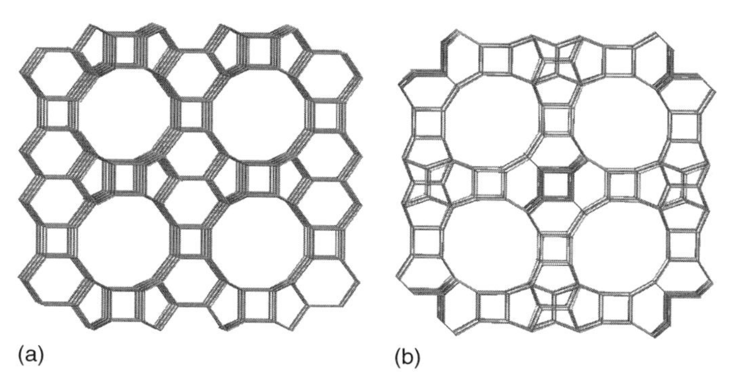


**Figure 7.15** Projections down 12MR channels of (a) ITQ-17 (BEC) and (b) ITQ-26, both of which possess 3D-connected 12MR channel systems.

been achieved in UZM-4 (BPH) [23], which is a higher silica and more stable version (Si/Al > 1.5) of Linde Q (BPH, Si/Al = 1.1) [4]. In a similar way, templating studies have given silicate versions of structures initially prepared as germanates or aluminophosphates. Among the latter, SSZ-16, SSZ-24, SSZ-55, and SSZ-73 are the silica versions of $AlPO_4$-56, $AlPO_4$-5, $AlPO_4$-36, and STA-6, respectively [4]. It is likely that silicate versions of other aluminophosphates without existing zeolite analogs will also be obtained: the silicate versions of the larger pore, 3D-connected DAF-1(DFO) and STA-14(SAO) [4] would be of particular interest.

### 7.3.3
### Expansion of the Coordination Sphere of Framework Atoms

Although zeolites are defined by their tetrahedrally connected frameworks, there are some examples where their framework cations adopt higher coordination. The most important of these are when Si expands its coordination by bonding to fluoride during synthesis and when framework Ti expands its coordination, for example, with water or upon uptake of hydrogen peroxide. Whereas the first observation is important in understanding fluoride synthesis [24], the second has important catalytic consequences [49].

The fluoride ion behaves as an efficient mineralizing agent in the synthesis of pure silica zeolites [24], where it catalyzes the hydrolysis of silica and enables the formation of silicate frameworks at pH values of 7–9, where no reaction would occur in its absence. A series of crystallographic and NMR studies have shown that in silicas prepared in fluoride media, $F^-$ ions are present in the as-prepared solids, coordinated to lattice silicon atoms, where they raise the coordination number of Si to 5 ($SiO_4F$) [50]. The fluoride ion is often found to occur within small cages, distributed over a number of different silicon sites with partial occupancy. For example, it is found in $[4^6]$ cages (LTA, AST) and also in cages in nonasil $[4^15^46^2]$, EU-1 $[4^15^46^2]$, silicalite (MFI $[4^15^26^2]$, ITQ-4 (IFR $[4^26^4]$) SSZ-23 $[4^35^4]$, and so on. This is illustrated in Figure 7.16 for $F^-$ in the $[4^15^46^2]$ cage of EU-1 [51]. In this

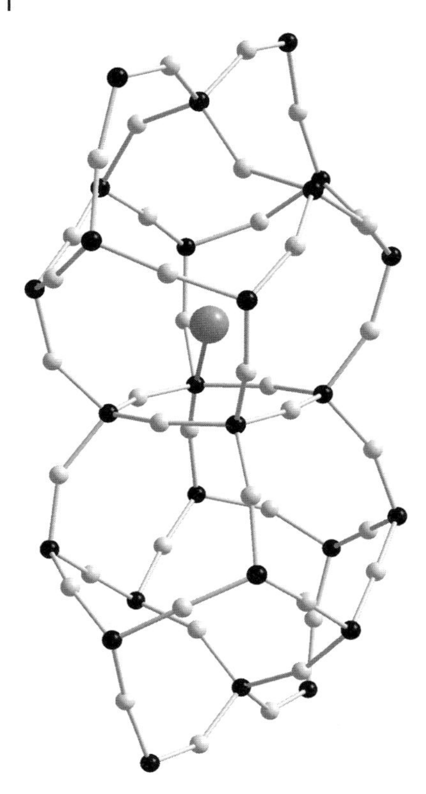

**Figure 7.16** Part of the as-prepared EU-1 structure prepared in fluoride medium. The fluoride ion is located within a small cage and is connected to one of the Si atoms, raising its coordination fivefold. (Reproduced with permission from [50]).

way $F^-$ acts as an inorganic SDA. At the site where it coordinates to the silicon, the tetrahedrally arranged $SiO_4$ group is distorted so that the three OTO angles that are closest to the $F^-$ ion are increased to minimize the O–F repulsion [8]. Once incorporated into the structure, the $F^-$ ion balances the positive charge of the alkylammonium ion template. Calcination results in simultaneous removal of both the organic cation and the $F^-$, leaving a $SiO_2$ framework with very few framework defects. Such materials are hydrophobic and as a direct result have potential applications in adsorption.

Titanium can substitute for Si at low concentrations in pure silica zeolites, and adopts tetrahedral coordination once the as-prepared solid is calcined and dehydrated. Upon exposure to, for example, aqueous hydrogen peroxide [52], it expands its coordination, acting as a Lewis acid, and the local geometry is distorted. In this way, titanosilicate zeolites can act as important oxidation catalysts, especially when they activate hydrogen peroxide [49].

### 7.3.4 The Current Limits of Structural Complexity in Zeolites

The ongoing synthetic efforts described above continue to give rise to structures with increasing complexity, both in terms of their crystallographic description and the diversity of their framework architecture. In this section, we discuss in detail the current limits of structural complexity, as defined by the number of crystallographically distinct tetrahedral cation environments (T-sites) present in the repeat unit of the structure. In terms of crystallographic complexity, the structures of both zeolite A and zeolite Y are very simple, for although there are many tetrahedra within the unit cells of each framework, they are all related by the many symmetry elements present in the two structures, so there is only one unique position in each case. By contrast, the structure of ZSM-5 is far more complex and its framework structure is built from layers with 12 crystallographically distinct T-sites. As described in Section 7.3.1, each layer is linked to symmetrically equivalent layers by centers of symmetry. (In fact, high silica materials with this structure distort to lower symmetry at low temperatures so that they have 24 different sites, but all the sites retain their original coordination sequences and the distortion is a subtle one) [53, 54].

Recently two zeolite structures have been solved which have 24 crystallographically distinct sites, each with a different coordination sequence, TNU-9 [38] and IM-5 [40]. A third reported structure, SSZ-74 [39], is similarly complex, with 23 different T-sites and an ordered vacancy. None forms as single crystals and their complexity proved a major challenge to structure solution. Fortunately, in a tour-de-force of powder crystallography, by combining experimental X-ray diffraction (XRD) and phases determined from high-resolution electron microscopy with crystallographic computing algorithms, the groups of McCusker, Baerlocher, and Terasaki (for TNU-9) and McCusker, Baerlocher, and Zou (for IM-5 and SSZ-74) have successfully elucidated the structures of these materials. For TNU-9, high-resolution powder XRD and HRTEM have been successfully combined with the FOCUS program, designed to search trial electron density maps for tetrahedra. Sufficient correct phase information was obtained from the electron micrographs to make this possible. For IM-5 and SSZ-74, similar types of experimental diffraction and imaging data have been combined with so-called enhanced charge flipping crystallographic algorithms adapted from other application programs. These charge flipping programs can be applied without applying structural constraints to obtain solutions (and so are generally applicable, regardless of framework connectivity).

The structure of TNU-9 has a very similar projection (down [010]) to that observed for ZSM-5, but has two different sets of straight 10-ring channels, labeled A and B in Figure 7.17, rather than one. Perpendicular to [010], channels of type B are linked to each other via short 10-ring channels and to channels of type A via 10-ring windows. It is instructive to investigate how a structure as complex as TNU-9 is built from repeating units, and as a consequence to suggest how it may assemble from solution, in the presence of organic SDAs [55]. The TNU-9 framework can be built up from a single kind of chain, similar to what is seen in ZSM-5. These

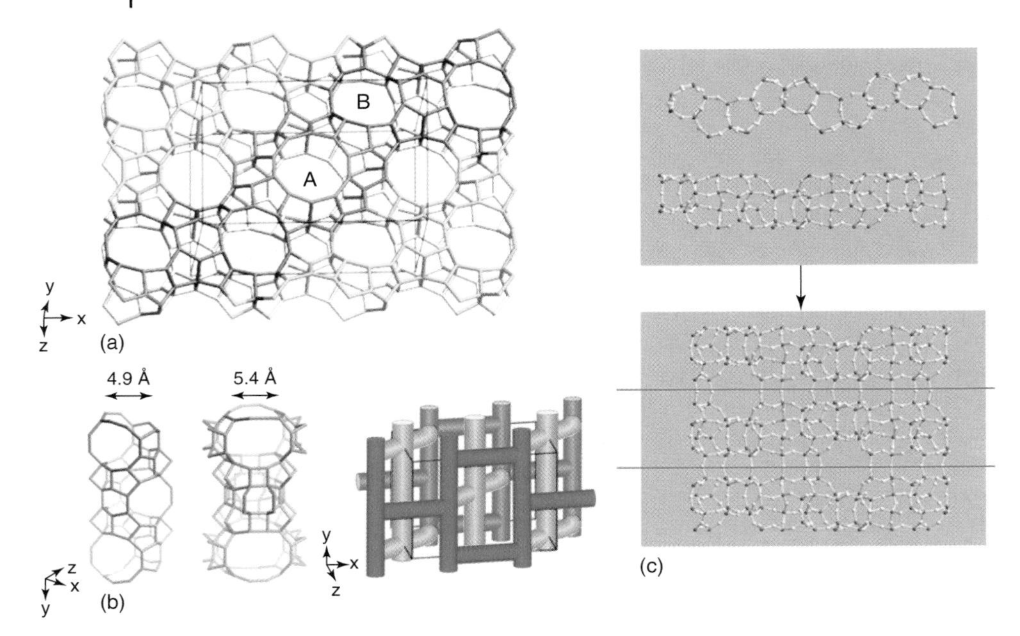


**Figure 7.17** The complex topology of TNU-9 (a) contains two kinds of channel (b), as depicted in pink and blue, which show complex 3D connectivity. The framework itself is built from one kind of chain that links to form layers (c), which join to give the final structure. (Reproduced with permission from [38] and [52]).

chains are connected via mirror planes to give sheets which are asymmetric (i.e., one side is different from the other). These sheets can only be linked to each other via their similar sides, so that there are two different intersheet regions. Modeling indicates that the organic SDA is able to interact favorably in different sites in both intersheet regions, so indicating a mechanism by which stacking of layers can be favored during synthesis [55].

The high silica zeolite IM-5 is, like ZSM-5 and TNU-9, highly thermally stable and shows interesting performance for hydrocarbon processing and selective NO reduction [40]. The structure has a 2D-connected 10-ring channel system. It is, like TNU-9, similar in projection to ZSM-5 but rather than complete 3D connectivity there are slabs (thickness 25 Å) that contain channels that show connectivity in three dimensions, complete with complex channel intersections. Each slab is isolated from adjacent sheets by impermeable silicate sheets. The catalytic performance could only be properly explained once this connectivity was understood. The work also described how, because the charge flipping structure solution methodology did not need to make use of symmetry, it determined the location of 288 silicon atoms and 576 oxygen atoms in the unit cell. This is very encouraging, as ever more complex structures are prepared, and suggests that an inability to grow single crystals should not limit our discovery of their intricate architectures. Both TNU-9 and IM-5 crystallize in the presence of the same organic SDA, bis-N-methylpyrrolidiniumbutane, which has also been shown to favor the crystallization of several other zeolites depending on solution pH, Si/Al ratio, additional cations, and so on [29]. This is a clear indication that the detailed composition of the hydrothermal reaction mixture plays a crucial role in supplying species for the growth of these complex solids.

Finally, a third highly complex zeolite, SSZ-74, prepared using a template very similar to that used for TNU-9 and IM-5 (bis-N-methylpyrrolidiniumheptane) has recently been solved by the same combination of XRD, TEM, and application of the charge flipping algorithm used successfully for IM-5. The structure possesses 23 different T-sites and an ordered tetrahedral vacancy (effectively a 24th T-site that is vacant). The structure contains an undulating 10-ring channel connected by straight 10-ring channels and leading to 3D connectivity limited by 10-rings similar to that for ZSM-5. The implications of the observed vacant site are discussed in more detail in Section 7.3.6.

### 7.3.5 Chirality and Mesoporosity

As described above, considerable progress has been made toward the synthesis of zeolite structures with different connectivities and larger pore sizes, at least up to the 12.2 Å circular pore openings observed in ITQ-33. One of the key remaining challenges is to produce batches of chiral zeolites which consist entirely of crystals of one enantiomer, so that these might find application in enantioselective separation and catalysis. Very few silicate zeolites with potentially chiral porous structures are known, and the most important examples are described below.

Zeolite beta, described in Section 7.3.3, typically exhibits stacking faults and does not crystallize as an ordered polymorph. Theoretically, though, there is a regular stacking sequence that would give a polymorph (Beta-A) that would be chiral, and efforts have been made to obtain this by the use of chiral templates. No fully ordered chiral polymorph A has yet been prepared, and one of the difficulties is to achieve chiral recognition over the long helical pitch of the channel that this polymorph would have, using either single molecules or molecules that could order.

More recently, two zeolitic silicates have been prepared as mixtures of chiral crystals. The silicogermanate SU-32 is one of a family of fully tetrahedrally connected frameworks templated by the achiral ammonium ion $(CH_3)_2CHNH_3^+$ that are built up from chiral layers with 12-ring openings that consist of "4−1" repeating units of tetrahedra, which point up and down alternately (Figure 7.18) [56]. Relating adjacent layers of this type across inversion centers results in the achiral 12-ring structure SU-15, whereas stacking layers with a $\pm 60^\circ$ rotation between adjacent sheets (which enables fortuitous coincidence) results in chiral polymorphs of SU-32 (space groups $P6_122$ or $P6_522$). Remarkably, crystals form in a mixture of pure enantiomorphs, which is different from what is observed for zeolite Beta. The resulting structure comprises only D4Rs and $4^6 5^8 8^2 10^2$ cavities that share 10-ring openings and make up helical channels which are either right or left handed, with pore openings 5 × 5.5 Å along the channels and intersected by eight-ring channels that run parallel to the channel axis (4.7 × 3 Å). The challenge now is to prepare this topology in the form of a more stable silicate or aluminosilicate, and even to prepare only one of the two enantiomeric forms.

The second type of zeolitic solid that crystallizes in chiral form is the silicogermanate ITQ-37 that crystallizes in the chiral space groups $P4_132$ and $P4_332$, with a structure related to the gyroidal (G) periodic minimal surface. This G-surface is exhibited by micelle-templated mesoporous amorphous silicas. In these mesoporous solids two pore systems of interconnected channels of opposite hand are separated by a silica wall located at the G-surface, whereas ITQ-37 can be thought of as having one of these pore systems empty and the other filled by a chiral zeolitic framework. The remarkable structure is beautifully illustrated by Sun *et al.* [37], and the cavity of this structure and the entrance window are illustrated in Figure 7.19. Besides possessing chiral channels, the structure has the lowest framework density observed for a zeolite (10.2 T/1000 Å$^3$) and cavity dimensions in the mesoporous regime (>20 Å). As in other germanosilicate structures, D4Rs are important SBUs. In ITQ-37 these have one or two terminal hydroxy groups, and this interrupted nature of the framework is crucial in enabling the large cages to form.

The generation of a material that is both intrinsically chiral and with mesoporous cavities is a significant step forward, but before chiral zeolites can find application,

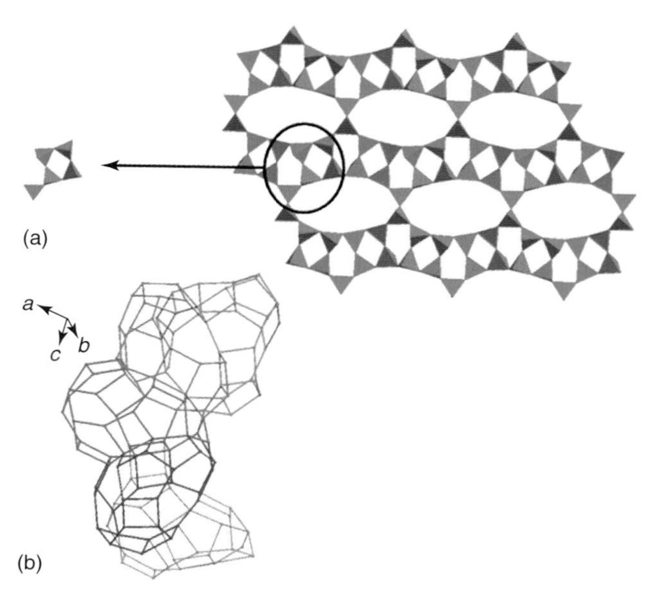


**Figure 7.18** The chiral germanosilicate SU-32 is built from sheets of tetrahedra made up from 4–1 units alternatively pointing in opposite directions (a). These stack to give a chiral structure that possesses helices of cages (b). Adjacent cages are depicted in different colors.

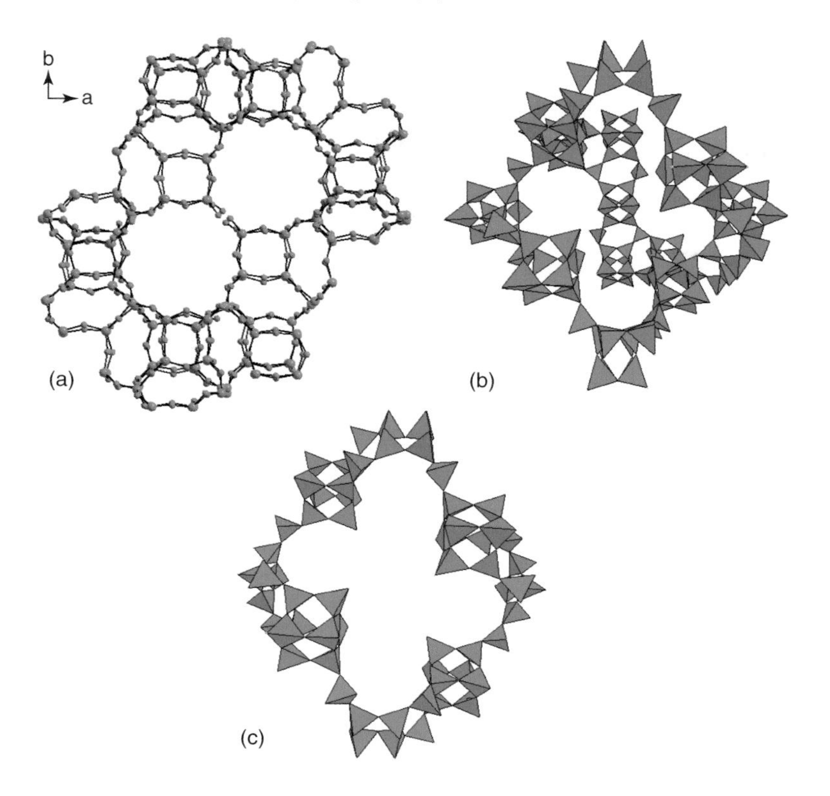


**Figure 7.19** (a) Part of the complex ITQ-37 structure (as "ball and stick") viewed down [100] O atoms red, Si and Ge blue. The structure contains mesoporous cavities (b) accessible through strongly noncircular windows (c). In (b) and (c) the germanosilicate tetrahedra, which can be four-, three-, or two-connected to other tetrahedra, are depicted in blue. Note the predominance of D4Rs in this structure.

they must be prepared as bulk samples of crystals of only one enantiomorph. This might be achieved by a combination of the use of chiral templates or surface-modifying agents coupled with chiral amplification using seeding with chiral crystals of one enantiomorph only.

### 7.3.6 Ordered Vacancies and Growth Defects

Isolated structural defects in the frameworks of zeolites have long been postulated, for example, as intermediates in the process of ultrastabilization and generation of secondary mesoporosity in zeolite Y (Section 7.2.4). Elegant solid-state NMR studies have previously identified the inclusion of defects resulting from T-site vacancies as a method of charge balancing organic cations in pure silica polymorphs synthesized in alkaline media (in the absence of fluoride). Measured compositions suggest that a vacancy includes two silanol groups and two siloxy groups pointing inwards and hydrogen bonded, resulting in a characteristic $^1H$ MASNMR signal observed at 10 ppm [57]. The recent observation of an ordered vacancy in the pure silica polymorph SSZ-74 (the pore structure of which was described in Section 7.3.4) is therefore of great interest. The crystallographic structure of the as-prepared form of this solid gives an atomistic picture of a vacancy and how it interacts with the organic cation (Figure 7.20). The vacant T-site is surrounded by four framework oxygen atoms forming a distorted tetrahedron. Two of these oxygen atoms make closer contacts with the nitrogen atom of the charged template and are taken to be siloxy groups ($Si{-}O^-$) leaving the other two as silanol groups, hydrogen atoms of which are involved in H-bonds with the siloxy oxygens. Although ordering of such defects is rare, it is likely that interactions of this type are a widespread mechanism of charge balancing of pure silica polymorphs templated by organocations when the option of the coordination of fluoride to silicon is not available, so that the vacant T-site in SSZ-74 can be used as a model system. Upon calcination and removal of the organic, it is likely that at least some of these vacancies remain. If this is the case, the intriguing possibility of including additional functionality at this site (for example, a titanium atom) could enable site-selective catalysts to be prepared and permit crystallographic determination of the structure of these catalytically active sites.

High resolution electron microscopy [58] of pure silica zeolite Beta has identified a second type of defect that also has relevance to the crystallization of high silica zeolites. The framework of Beta is made up of layers stacked with a displacement of one-third of the unit cell vector in either direction, as described in Section 7.2.6. In a recent study of zeolite Beta, TEM images show large-pore defects that can only be explained by the nucleation on a given layer of two domains that are stacked with opposite displacements (Figure 7.21). After three layers of growth, these become coincident again and the defect is healed. The implication of this observation is that zeolite Beta crystallizes by layer growth, and this can lead to additional nonperiodic porosity and the presence of extended defects. AFM studies reported elsewhere in this book by Anderson and Cubillas show that layer growth is a general mechanism and can be modeled atomistically.

### 7.3.7
### Zeolites from Layered Precursors

As discussed above and elsewhere, recent microscopic studies have indicated that the growth of zeolites occurs by a layer-by-layer mechanism. A subset of zeolites has also been found which can be prepared via a two-step process that includes the post-synthetic condensation of layered silicate precursors prepared hydrothermally. The initial crystallization gives a layered silicate in which the silicate layers are terminated on each side by Si–OH groups of $(SiO)_3SiOH$, "$Q^3$," silicon atoms and separated from one another by organic SDAs. Upon calcination the organic material is removed and the silanol groups on adjacent layers condense (2 SiOH $\rightarrow$ SiOSi) to give a fully four-connected porous tetrahedral framework. Examples of this include the synthesis of the known zeolites ferrierite (FER) and MCM-22(MWW)

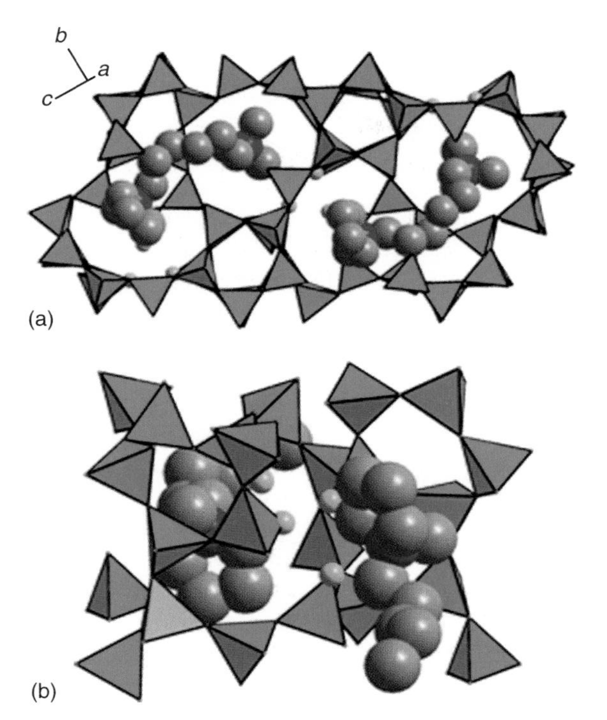


**Figure 7.20** Two views of the ordered framework vacancy in as-prepared SSZ-74. The silicate tetrahedra are shown in blue and the terminal hydroxyl group oxygen atoms of the SiOH/SiO units of the vacancy are represented in orange. Note the proximity to the defect of the charged moieties of the bisalkylammonium template (C atoms as grey spheres) around the quaternary nitrogen atoms (blue).

from the precursor layered silicates PREFER and MCM-22(P), respectively [59, 60]. The latter conversion is represented schematically in Figure 7.22. In addition, the new structure types CDS-1 [61], RUB-24 [62], and RUB-41 [63] have been prepared from the layered silicate precursors PLS-1, RUB-18, and RUB-39, respectively. Such systems, and especially MCM-22(P) – MCM-22, have been chemically manipulated by careful treatment of the laminated phase in order to introduce catalytically active species and prepare pillared microporous solids with novel catalytic features.

### 7.3.8
### Substitution of Framework Oxygen Atoms

As described above, zeolite frameworks show extensive substitution of silicon by other metal cations (divalent Be, Zn; trivalent B, Al, Ga, Fe; tetravalent Ge and Ti). Besides having a strong influence on the structure type that crystallizes, this also has a major effect on the stability and catalytic properties of the zeolite. Much less progress has been made in substituting framework oxygen, but there is a slow

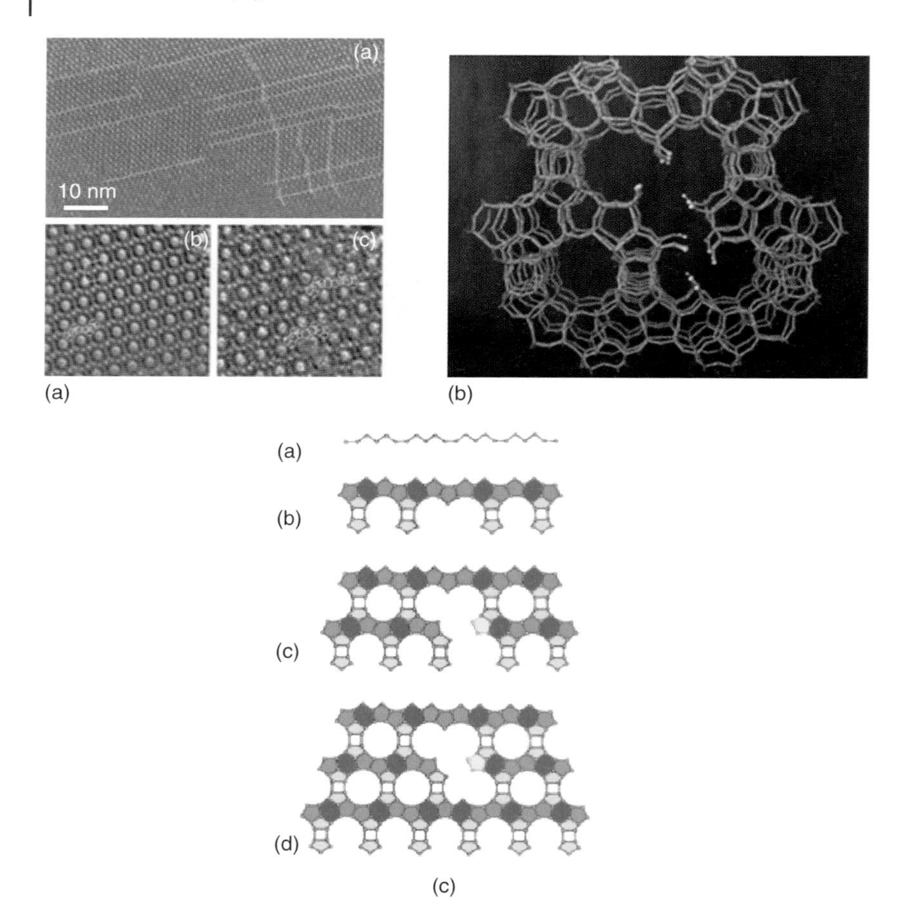


**Figure 7.21** High-resolution electron microscopy of zeolite Beta along one set of its 12MR channels (a) reveals "double pore defects" that can be modeled (b) by layers stacked in two different orientations and coming back into coincidence after three layers. This is consistent with a layer-by-layer growth model where domains with different stacking offsets form on the same layer and converge ((c), (i)–(iv)). (Reproduced with permission from [55]).

accumulation of evidence that partial substitution with retention of structure is possible. Interest has centered on the isoelectronic substitution of O with $CH_2$ or NH groups. Initial attempts to include organic groups into the framework involved the inclusion of aminopropylsiloxanes as well as tetraethoxysilane into the synthesis gel, so that co-crystallization resulted in connectivity defects and pores lined with organic groups [64]. For the inclusion of methylene groups the most promising route is to use bis-(triethoxysilyl)methane, $(EtO)_3Si{-}CH_2{-}Si(OEt)_3$, as a silica precursor within the preparation, so that if complete incorporation into a tetrahedral lattice were achieved, one of the seven framework oxygen atoms would

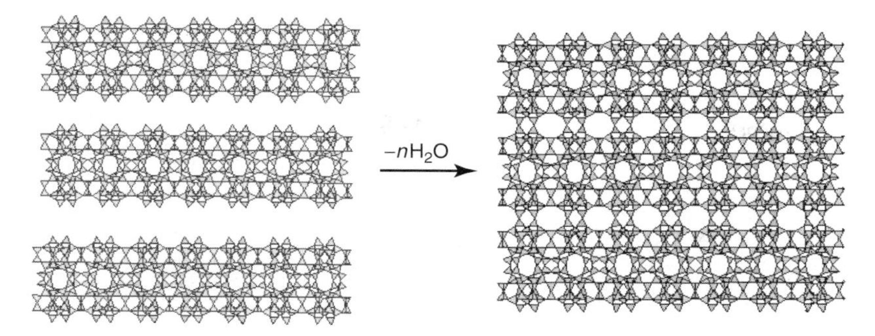


**Figure 7.22** Scheme showing the condensation of the layered zeolitic precursor MCM-22(P) to the fully tetrahedrally connected zeolite MCM-22 upon heating. (Reproduced with permission from [3]).

be replaced by methylene groups. Yamamoto *et al.* [65] observed that analogs of zeolites A, Beta, and ZSM-5 can be prepared using this siloxane precursor, but that the siloxane is partially hydrolyzed to give a mixture of $Si{-}CH_3$ and Si–OH as well as $Si{-}CH_2{-}Si$ in the final framework product.

The inclusion of amine groups into framework positions has also been investigated, with the target of preparing stable shape-selective basic catalysts. The suggested route to the inclusion of NH groups that has received the most attention is via post-synthetic high temperature treatment with ammonia. The inclusion of NH groups into zeolites without loss of crystallinity has been demonstrated, and $^{29}Si$ MASNMR of the product solids gives signals at about −67 and −86 ppm that are attributed on the basis of theoretical calculations to $Al{-}NH_2{-}Si{-}NH_2{-}Al$ and $Al{-}NH_2{-}Si$ species (Si–NH–Si linkages are expected to be less favorable) [66]. If correct, this holds considerable potential in the preparation of solid base catalysts.

## 7.4 Summary and Outlook

### 7.4.1 Summary

From the foregoing discussion, it is clear that the structural features of the zeolites that are most important in applications of ion exchange, adsorption, and catalysis (such as zeolites A, X, Y, chabazite, mordenite, ZSM-5, and Beta) have been studied in great detail because of the direct link between structure and function. The crystallographic structures give time- and space-averaged atomic positions, whereas local features of framework and extra-framework cation disorder have been established by combining diffraction, spectroscopy (NMR, IR, etc.), and computational simulation. The structural stability and activity of zeolites have been established over a wide range of conditions, especially *in situ* under realistic working conditions for adsorption and catalysis. For example, as an early example of the study of the activation of a zeolitic catalyst, $Ni^{2+}$ cation migration within Ni–Y has been followed *in situ* by diffraction and X-ray absorption spectroscopy [67]. In addition, the high temperature limits to structural stability have been probed by both diffraction and inelastic scattering methods, through which a model for the amorphization of zeolite structure has been established [68, 69]. Low-frequency phonon features in zeolites have been found to be responsible for destabilizing and collapsing the zeolite structure and for converting the resulting glass from a low-density to a high-density amorphous phase.

In parallel to the increase in understanding of structural chemistry of important zeolite materials, many tens of new zeolite structures with increasing complexity have been prepared via novel synthetic routes, and their structures solved via advanced crystallographic methods. A simple investigation of the number of hypothetical tetrahedrally connected frameworks with levels of complexity up to those recently observed (24 topologically distinct sites in the unit cell) that could be energetically feasible soon indicates that the known structures are a very small subset of what should be achievable. Many of these new structures have arisen from syntheses that include heteroatoms (other than Al) in the framework: Ge has been of particular interest in this regard, giving striking structures at least partly due to its strong tendency to favor the formation of D4R SBUs. A smaller number of studies have shown that it is also possible to introduce C and N as framework constituents, in the place of linking O atoms.

The range of known zeolite structural chemistry has therefore been widened greatly since 1990, through a combination of ingenious synthesis and structural analysis. However, this has taken place simultaneously with the development of microporous metal organic frameworks (MOFs) [70] and covalent organic frameworks (COFs) [71] and also mesoporous silicas [72]. These new families of porous solids have outstripped the zeolite field in terms of their chemical and structural diversity, and there is no space to describe them in detail here, other than to state that mesoporous silicas, for example, can be prepared with narrow distributions of pore sizes from 4 Å to >100 Å, high porosities and stabilities, and readily functionalizable surfaces, and that MOFs can be prepared, fully crystalline, that possess a much wider chemical range than zeolites, much higher surface areas, great flexibility, inorganic and organic functionalities, and potentially in-built chirality. The outlook for research in porous solids has therefore never been more interesting: the key question here is what role will zeolite structures (compared to that of their more recently discovered relatives) have in future academic and industrial research?

### 7.4.2 Outlook

With great current interest in new types of porous solids, it should be stated that zeolites remain overwhelmingly the most important in applications. While this may be expected due to their being the first materials discovered, synthesized, and investigated, it also derives from key structural and stability advantages they possess over mesoporous silicas and MOFs. These advantages are the relatively low cost of manufacture, at least for (alumino)silicates prepared without exotic templates, high cation exchange capacities (for ion exchange), the much stronger acidity of Bronsted acid sites they possess, a direct consequence of their crystalline structure (for acid catalysis), their ability to include Lewis acidic framework titanium sites (for peroxide activation and selective oxidation), and especially higher thermal and hydrothermal stability of high silica zeolites compared with either mesoporous silicas or MOFs (for most applications, especially catalysis). In addition, they have been shown to be biologically compatible [73]. As a result, developments in zeolite structural chemistry are likely to continue to be of great importance in order to improve their performance as functional materials, especially in their traditional uses (the other materials will find their own fields of application). We also envisage developments in at least two broadly related areas of structural chemistry: New structure types and chemistries, and control of morphology and microstructure.

Current methods in the discovery of novel zeolite structures combine the synthesis of novel organic templates with variation in the inorganic composition of the gel. Whereas the synthesis of organic compounds proceeds via well-established methodologies, inorganic syntheses under hydrothermal conditions proceed via poorly understood processes of aging, nucleation, and growth. Consequently, high-throughput techniques that explore a wide range of composition space and reaction conditions will play a role in these studies. There is also likely to be significant progress in preparing zeolites by the simultaneous use of two or more organic SDAs in the same synthesis. This approach has already been successful for zeotypes with different cages in the structure (STA-7(SAV) and STA-14(KFI)) [74] and for channel and cage sites in ferrierite [75]. It is likely that structurally complex structures will have several potential template sites of different sizes and shapes (as does TNU-9 [38]) that would better suit different SDAs. The trend to preparing novel topologies by the inclusion of elements such as Ge or Ti will also continue to be productive, with the proviso that there is a balance between the higher Ge-content silicate materials having greater structural diversity but lower hydrothermal stability than silicate or aluminosilicate analogs. The novel structure types containing Al that result will generate potential solid acid catalysts and, if Ti is included, selective oxidation catalysts, in each case with new shape selectivities.

In most of the examples of crystalline structure given in this chapter, it is assumed that the zeolite framework extends indefinitely. In reality the particle will have a certain size and shape, and may contain microstructural features such as twin planes and stacking faults. Recent attempts to control these features have been made, in order to tailor the material for particular applications such as enhancing diffusion along one-dimensional channel systems for catalysis, aligning crystals for sensing technologies, or organizing crystal orientations in membranes and thin films [76].

In terms of size, zeolite crystallites from typical hydrothermal preparations are typically of the order of microns, but there has been much recent interest in their preparation as nanoparticles [77] (as precursors for thin film or membrane growth, for example) or as single crystals approaching millimeters in dimensions, which are large enough to enable measurements of anisotropic diffusion, mechanical properties, and so on [78]. The morphology of zeolite crystals, which is a direct consequence of the growth rates on different faces, is found to vary strongly with the gel composition and conditions of crystallization, including the SDA used. The work of the group of Tsapatsis in controlling the crystallographic orientation in silicalite membranes via choice of SDA is an elegant example [79]. Recent advances in understanding the mechanism of crystal growth [80] should ultimately enable morphological control and therefore the control of surfaces that are exposed. The nature and extent of the surface of zeolite crystals is important because it is that part of the crystal that interfaces with the external environment, both during the crystallization and, in the zeolite product, during its application as an adsorbent, ion exchanger, catalyst, or in a medical application. The increased resolution of surface microscopies, coupled with transmission electron microscopy, surface spectroscopies, and measurements of surface charge, are all likely to be important in research in understanding zeolite surface structures as a complement to their bulk crystallographic structure.

## References


1. McCusker, L.B. and Baerlocher, Ch. (2007) in *Introduction to Zeolite Science and Practice*, 3rd edn, Studies in Surface Science and Catalysis, Vol. 168 (eds J. Čejka, H. van Bekkum, A. Corma, and F. Schuth), Elsevier, pp. 13–37.
2. Lobo, R.F. (2004) in *Handbook of Zeolite Science and Technology*(eds S.Auerbach, K. Carrado, and P. Dutta), Marcel Dekker, New York.
3. Wright, P.A. (2007) *Microporous Framework Solids*, RSC Publishing, Cambridge.
4. Database of Zeolite Structures, Structure Commission of the International Zeolite Association, *http://www.iza-structure.org/databases/* (last accessed February 03, 2010).
5. Delgado-Friedrichs, O., Dress, A.W.M., Huson, D.H., Klinowski, J., and Mackay, A.L. (1999) *Nature*, **400**, 644–647.
6. Foster, M.D., Simperler, A., Bell, R.G., Delgado-Friedrichs, O., Paz, F.A.A., and Klinowski, J. (2004) *Nat. Mater.*, **3**, 234–238.
7. Liebau, F. (1985) *Structural Chemistry of Silicates: Structure, Bonding and Classification*, Springer-Verlag, Berlin, pp. 14–30.
8. Wragg, D.S., Morris, R.E., and Burton, A.W. (2008) *Chem. Mater.*, **20**, 1561–1570.
9. Liu, Z., Fujita, N., Terasaki, O., Ohsuna, T., Hiraga, K., Camblor, M.A., Diaz-Cabanas, M.-J., and Cheetham, A.K. (2002) *Chem. Eur. J.*, **8**, 4549–4556.
10. Shannon, R.D. (1976) *Acta Cryst.*, **A32**, 751–767.
11. (a) Li, H. and Yaghi, O.M. (1998) *J. Am. Chem. Soc.*, **120**, 10569–10570; (b) O'Keeffe, M. and Yaghi, O.M. (1999) *Chem. Eur. J.*, **5**, 2796–2801.
12. Baur, W.H. (1964) *Am. Mineral.*, **49**, 697–704.
13. Newsam, J.M., Treacy, M.M.J., Vaughan, D.E.W., Strohmaier, K.G., and Mortier, W.J. (1989) *Chem. Commun.*, 493–495.
14. Czjzek, M., Jobic, H., Fitch, A.N., and Vogt, T. (1992) *J. Phys. Chem.*, **96**, 1535–1540.
15. Parise, J.B., Gier, T.E., Corbin, D.R., and Cox, D.E. (1984) *J. Phys. Chem.*, **88**, 1635–1640.
16. (a) Higgins, J.B., LaPierre, R.B., Schlenker, J.L., Rohrman, A.C., Wood, J.D., Kerr, G.T., and Rohrbaugh, W.J. (1988) *Zeolites*, **8**, 446–452;

(b) Newsam, J.M., Treacy, M.M.J., Koetsier, W.T., and de Gruyter, C.B. (1988) *Proc. R. Soc. Lond. A*, **420**, 375–405.

17. Kokotailo, G.T., Lawton, S.L., Olson, D.H., and Meier, W.M. (1978) *Nature*, **272**, 437–438.
18. Casci, J.L. (1994)in *Zeolites and Related Microporous Materials : State of the Art 1994*, Studies in Surface Science and Catalysis, Vol. 84 (eds J.Weitkamp, H.G. Karge, H. Pfeifer, W. Holderich), Elsevier, pp. 133–140.
19. Galli, E. and Gualtieri, A.F. (2008) *Am. Mineral.*, **93**, 95–102.
20. Chen, C.S.H., Schlenker, J.L., and Wentzek, S.E. (1996) *Zeolites*, **17**, 393–400.
21. Song, J.W., Dai, L., Ji, Y.Y., and Xiao, F.S. (2006) *Chem. Mater.*, **18**, 2775–2777.
22. Corma, A., Diaz-Cabanas, M.J., Jorda, J.L., Rey, F., Sastre, G., and Strohmaier, K.G. (2008) *J. Am. Chem. Soc.*, **130**, 16482–16483.
23. Blackwell, C.S., Broach, R.W., Gatter, M.G., Holmgren, J.S., Jan, D.Y., Lewis, G.J., Mezza, B.J., Mezza, T.M., Miller, M.A., Moscoso, J.G., Patton, R.L., Rohde, L.M., Schoonover, M.W., Sinkler, W., Wilson, B.A., and Wilson, S.T. (2003) *Angew. Chem. Int. Ed.*, **42**, 1737–1740.
24. Camblor, M.A., Villaescusa, L.A., and Diaz-Cabanas, M.-J. (1999) *Top. Catal.*, **9**, 59–76.
25. Villaescusa, L.A., Barrett, P.A., and Camblor, M.A. (1999) *Angew. Chem. Int. Ed.*, **38**, 1997–2000.
26. e.g. Jackowski, A., Zones, S.I., Hwang, S.-J., and Burton, A.W. (2009) *J. Am. Chem. Soc.*, **131**, 1092–1100.
27. Corma, A. (2004)in *Recent Advances in the Science and Technology of Zeolites and Related Materials*, Proceedings of the 14th International Zeolite Conference, Studies in Surface Science and Catalysis, Vol. 154 (eds E. van Steen, M. Claeys, and L.H. Callanan), Elsevier, Amsterdam, pp. 25–40.
28. Sastre, G., Vidal-Moya, J.A., Blasco, T., Rius, J., Jorda, J.L., Navarro, M.T., Rey, F., and Corma, A. (2002) *Angew. Chem. Int. Ed.*, **41**, 4722–4726.
29. Hong, S.B., Lear, E.G., Wright, P.A., Zhou, W., Cox, P.A., Shin, C.-H., Park, J.-H., and Nam, I.-S. (2004) *J. Am. Chem. Soc.*, **126**, 5817–5826.
30. Wessels, T., Baerlocher, Ch., McCusker, L.B., and Creyghton, E.J. (1999) *J. Am. Chem. Soc.*, **121**, 6242–6247.
31. Yoshikawa, M., Wagner, P., Lovallo, M., Tsuji, K., Takewaki, T., Chen, C.Y., Beck, L.W., Jones, C., Tsapatsis, M., Zones, S.I., and Davis, M.E. (1998) *J. Phys. Chem. B*, **102**, 7139–7147.
32. Burton, A.W., Elomari, S., Chen, C.Y., Medrud, R.C., Chan, I.Y., Bull, L.M., Kibby, C., Harris, T.V., Zones, S.I., and Vittoratos, E.S. (2003) *Chem. Eur. J.*, **9**, 5737–5748.
33. Paillaud, J.L., Harbuzaru, B., Patarin, J., and Bats, N. (2004) *Science*, **304**, 990–992.
34. Cheetham, A.K., Fjellvag, H., Gier, T.E., Kongshaug, K.O., Lillerud, K.P., and Stucky, G.D. (2001) *Stud. Surf. Sci. Catal.*, **135**, 158.
35. Strohmaier, K.G. and Vaughan, D.E.W. (2003) *J. Am. Chem. Soc.*, **125**, 16035–16039.
36. Corma, A., Diaz-Cabanas, M.J., Jorda, J.L., Martinez, C., and Moliner, M. (2006) *Nature*, **443**, 842–845.
37. Sun, J., Bonneau, C., Cantin, A., Corma, A., Diaz-Cabanas, M.J., Moliner, M., Zhang, D., Li, M., and Zou, X. (2009) *Nature*, **458**, 1154–1157.
38. Gramm, F., Baerlocher, C., McCusker, L.B., Warrender, S.J., Wright, P.A., Han, B., Hong, S.B., Liu, Z., Ohsuna, T., and Terasaki, O. (2006) *Nature*, **444**, 79–81.
39. Baerlocher, Ch., Xie, D., McCusker, L.B., Hwang, S.-J., Chan, I.Y., Ong, K., Burton, A.W., and Zones, S.I. (2008) *Nat. Mater.*, **7**, 631–635.
40. Baerlocher, Ch., Gramm, F., Massüger, L., McCusker, L.B., He, Z., Hövmuller, S., and Zou, X. (2007) *Science*, **315**, 1113–1116.
41. Corma, A., Navarro, M.T., Rey, F., Rius, J., and Valencia, S. (2001) *Angew. Chem. Int. Ed.*, **40**, 2277–2280.
42. Corma, A., Diaz-Cabanas, M.J., Martinez-Triguero, J., Rey, F., and Rius, J. (2002) *Nature*, **418**, 514–517.
43. Dorset, D.L., Strohmaier, K.G., Kliewer, C.E., Corma, A., Diaz-Cabanas, M.J.,

Rey, F., and Gilmore, C.J. (2008) *Chem. Mater.*, **20**, 5325–5331.
44. Conradsson, T., Dadachov, M.S., and Zou, X.D. (2000) *Microporous Mesoporous Mater.*, **41**, 183–191.
45. Lobo, R.F. and Davis, M.E. (1995) *J. Am. Chem. Soc.*, **117**, 3764–3779.
46. Castaneda, R., Corma, A., Fornes, V., Rey, F., and Rius, J. (2003) *J. Am. Chem. Soc.*, **125**, 7820–7821.
47. Dorset, D.L., Weston, S.C., and Dhingra, S.S. (2006) *J. Phys. Chem. B*, **110**, 2045–2050.
48. Corma, A., Rey, F., Rius, J., Sabater, M.J., and Valencia, S. (2004) *Nature*, **431**, 287–290.
49. Perego, C., Carati, A., Ingallina, P., Mantegazza, M.A., and Bellussi, G. (2001) *Appl. Catal. A*, **221**, 63–72.
50. Koller, H., Wolker, A., Villaescusa, L.A., Diaz-Cabanas, M.J., Valencia, S., and Camblor, M.A. (1999) *J. Am. Chem. Soc.*, **121**, 3368–3376.
51. Arranz, M., Pérez-Pariente, J., Wright, P.A., Slawin, A.M.Z., Blasco, T., Gómez-Hortiguela, L., and Corà, F. (2005) *Chem. Mater.*, **17**, 4374–4385.
52. Bonino, F., Damin, A., Ricchiardi, G., Ricci, M., Spano, G., D'Aloisio, R., Zecchina, A., Lamberti, C., Prestipino, C., and Bordiga, S. (2004) *J. Phys. Chem. B*, **108**, 3573–3583.
53. Fyfe, C.A., Gobbi, G.C., Klinowski, J., Thomas, J.M., and Ramdas, S. (1982) *Nature*, **296**, 530–533.
54. Fyfe, C.A., Kennedy, G.J., Kokotailo, G.T., Lyeria, J.R., and Fleming, W.W. (1985) *J. Chem. Soc., Chem. Commun.*, 740–742.
55. Hong, S.B., Min, H.K., Shin, C.-H., Cox, P.A., Warrender, S.J., and Wright, P.A. (2007) *J. Am. Chem. Soc.*, **129**, 10870–10885.
56. Tang, L., Shi, L., Bonneau, C., Sun, J., Yue, H., Ojuva, A., Lee, B.-L., Kritikos, M., Bell, R.G., Bacsik, Z., Mink, J., and Zou, X. (2008) *Nat. Mater.*, **7**, 381–385.
57. Koller, H., Lobo, R.F., Burkett, S.L., and Davis, M.E. (1995) *J. Phys. Chem.*, **99**, 12588–12596.
58. Wright, P.A., Zhou, W., Perez-Pariente, J., and Arranz, M. (2005) *J. Am. Chem. Soc.*, **127**, 494–495.
59. Schreyeck, L., Caullet, P., Mougenel, J.C., Guth, J.L., and Marler, B. (1996) *Microporous Mater.*, **6**, 259–271.
60. Leonowicz, M.E., Lawton, J.A., Lawton, S.L., and Rubin, M.K. (1994) *Science*, **264**, 1910–1913.
61. Ikeda, T., Akiyama, Y., Oumi, Y., Kawai, A., and Mizukami, F. (2004) *Angew. Chem. Int. Ed.*, **43**, 4892–4896.
62. Marler, B., Ströter, N., and Gies, H. (2005) *Microporous Mesoporous Mater.*, **83**, 201–211.
63. Wang, Y.X., Gies, H., Marler, B., and Müller, U. (2005) *Chem. Mater.*, **17**, 43–49.
64. Tsuji, K., Jones, C.W., and Davis, M.E. (1999) *Microporous Mesoporous Mater.*, **19**, 339–349.
65. Yamamoto, K., Nohara, Y., Domon, Y., Takahashi, Y., Sakata, Y., Plèvert, J., and Tatsumi, T. (2005) *Chem. Mater.*, **17**, 3913–3930.
66. Hammond, K.D., Dogan, F., Tompsett, G.A., Agarwal, V., Conner, W.C., Grey, C.P., and Auerbach, S.M. (2008) *J. Am. Chem. Soc.*, **130**, 14912–14913.
67. Dooryhee, E., Catlow, C.R.A., Couves, J.W., Maddox, P.J., Thomas, J.M., Greaves, G.N., Steel, A.T., and Townsend, R.P. (1991) *J. Phys. Chem.*, **95**, 4514–4521.
68. Greaves, G.N., Meneau, F., and Sankar, G. (2002) *Nucl. Instr. Methods Phys. Res. B*, **199**, 98–105.
69. Greaves, G.N., Meneau, F., Majerus, O., Jones, D.G., and Taylor, J. (2005) *Science*, **308**, 1299–1302.
70. Férey, G. (2007)in *Introduction to Zeolite Science and Practice*, 3rd revised edn, Studies in Surface Science and Catalysis, Vol. 168 (eds J. Čejka, H. van Beckkum, A. Corma, F. Schüth), Elsevier, pp. 327–374.
71. El-Kaderi, H.M., Hunt, J.R., Mendoza-Cortes, J.L., Côté, A.P., Taylor, R.E., O'Keeffe, M., and Yaghi, O.M. (2007) *Science*, **316**, 268–272.
72. Zhao, D. and Wan, Y. (2007)in *Introduction to Zeolite Science and Practice*, 3rd revised edn, Studies in Surface Science and Catalysis, Vol. 168 (eds J. Čejka, H. van Beckkum, A. Corma, and F. Schüth), Elsevier, pp. 241–300.

73. Schainberg, A.P.M., Ozyegin, L.S., Kursuoglu, P., Valerio, P., Goes, A.M., and Leite, M.F. (2005)in *Bioceramics 17*, Key Engineering Materials, Vol. 284–286 (eds P.Li, K. Zhang, and C. W. Colwell), Trans Tech Publications, pp. 561–564.
74. Castro, M., Garcia, R., Warrender, S.J., Wright, P.A., Cox, P.A., Fecant, A., Mellot-Draznieks, C., and Bats, N. (2007) *Chem. Commun.*, 3470–3472.
75. Pinar, A.B., Gomez-Hortiguela, L., and Pérez-Pariente, J. (2007) *Chem. Mater.*, **19**, 5617–5626.
76. Drews, T.O. and Tsapatsis, M. (2005) *Curr. Opin. Colloid Interface Sci.*, **10**, 233–238.
77. Tosheva, L. and Valtchev, V.P. (2005) *Chem. Mater.*, **17**, 2494–2513.
78. Lethbridge, Z., Williams, J.J., Walton, R.I., Evans, K.E., and Smith, C.W. (2005) *Microporous Mesoporous Mater.*, **79**, 339–352.
79. Choi, J., Ghosh, S., Lai, Z.P., and Tsapatsis, M. (2006) *Angew. Chem. Int. Ed.*, **45**, 1154–1158.
80. Brent, R. and Anderson, M.W. (2008) *Angew. Chem. Int. Ed.*, **47**, 5327–5330.

# 8
# Vibrational Spectroscopy and Related *In situ* Studies of Catalytic Reactions Within Molecular Sieves

*Eli Stavitski and Bert M. Weckhuysen*

## 8.1
## Introduction

Scientists in the field of catalysis are in a perpetual pursuit of optimization of existing processes in terms of activity/selectivity as well as development of new ones. In many cases, trial-and-error modus operandi prevails upon the rational approach, even with a level of sophistication and knowledge modern science and technology delivers. In order to improve on this situation, catalyst scientists require better insight into the key stages of the reaction process and, in particular, the catalyst's modus operandi. Armed with a rigorous understanding of this, it could then be possible to prepare the archetypal designer catalyst with the desired superior performance for the reaction in question. However, such information can only be obtained reliably by monitoring a catalyst "in action." In order to do this, it is essential to adapt catalytic reactors and/or spectroscopic/scattering techniques to study the processes in real time – an approach that gives rise to the field of *in situ* or operando spectroscopy [1, 2].

A catalytic cycle typically consists of a sequence of reaction steps that describe the transformation of substrate molecules into a final reaction product at a catalytically active site. Although scientists have been working for decades to decipher such events, only in a limited number of cases has a deeper understanding of the processes involved been achieved. If such information is to be obtained, one requires sufficient detailed information about the catalyst material at each step in its life cycle: that is, synthesis → calcination → activation → reaction → deactivation → regeneration (where possible). Conventional characterization (hereby termed accordingly *ex situ characterization*) focuses on the study of the catalyst materials at these various stages, away from the reactor and is often performed under ambient conditions; that is, at room temperature and atmospheric pressure. Although this approach yields interesting information, it is incapable of providing direct insight into the processes occurring in the catalyst during the course of the reaction. Therefore, the catalyst scientist is forced to develop analytical tools that enable the continuous monitoring of the catalyst "in action." However, we observe that at this stage a drawback of *in situ* methods is that the gas and/or liquid phase

*Zeolites and Catalysis, Synthesis, Reactions and Applications. Vol. 1.*
Edited by Jiří Čejka, Avelino Corma, and Stacey Zones

ISBN: 978-3-527-32514-6

surrounding the catalyst are probed simultaneously, including both the active surface and the inactive bulk of the material in question, making the interpretation prone to ambiguities.

One can trace the origin of *in situ* spectroscopy in catalysis back to 1954 with two seminal papers from the Eischens group [3, 4]. In these reports, the interaction of carbon monoxide with Cu, Pt, Pd, and Ni supported on $SiO_2$ and of ammonia molecules with cracking catalysts were studied with infrared (IR) spectroscopy. It is noteworthy that, from the modern standpoint, it can be argued to what extent the study in question can be considered as *in situ* as the conditions described are a far cry from those actually found in a catalytic process. Nevertheless, they represented an important step forward for the *in situ* approach as for the first time they considered the importance of the dynamics of a catalyst surface in the presence of adsorbates. Indeed, it was most probably the first spectroscopic-reaction cell designed for measuring IR spectra of heterogeneous catalysts. Since these pioneering studies, continuous progress toward the use of *in situ* spectroscopic techniques can be observed. This is illustrated in Figure 8.1, which shows the evolution of the number of *in situ* spectroscopy papers in the zeolite literature over the past decades.

Of all the commonly employed *in situ* techniques, IR spectroscopy has perhaps the longest history and is most often used in the field of zeolite research. In the early stage of the development, IR measurements were made using self-supported wafers (simple transmission/absorption measurements). However, since this time, IR has been developed further in order to obtain better quality spectra more quickly. Such improvements have been brought about by the development of more sensitive detection systems and improved sampling methods and enable the measurement of IR spectra under more relevant reaction conditions. A great

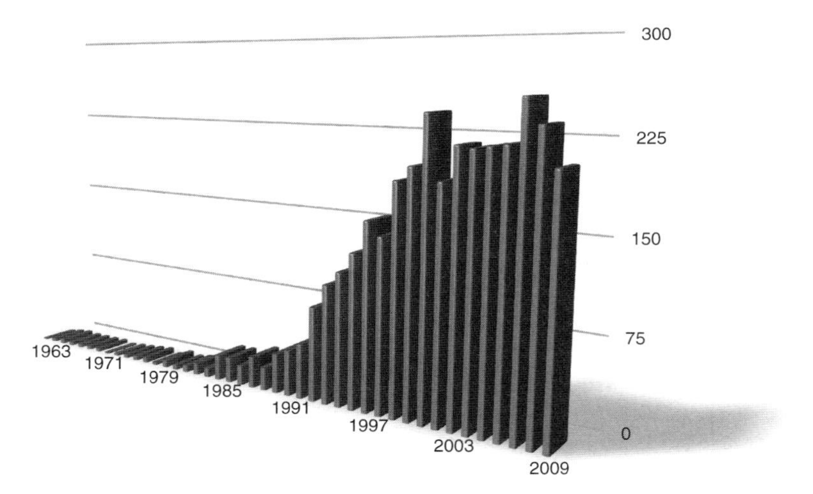


**Figure 8.1** Estimated number of journal publications on *in situ* characterization of catalyst materials (based on the results of an ISI and Chemical Abstracts database search using the terms "*in situ*" and "zeolite").

improvement in the technique was developed in the early 1980s by the introduction of Fourier transform IR (FTIR) instruments, which allowed for short recording times (seconds to minutes) and high resolution (0.5–4 $cm^{-1}$), adjustable to obtain the desired signal-to-noise ratio. Nowadays price-wise, the FT-IR spectrometer is comparatively cheap and, as such, is one of the workhorses in a typical academic and industrial heterogeneous catalysis laboratory.

As most probably known to the reader, IR probes the transitions between vibrational states of the absorber over an energy range of $0.496 > E > 0.0496$ eV (4000–400 $cm^{-1}$ commonly known as *medium/mid infrared* or MIR) and $0.0496 > E > 1.24 \times 10^{-3}$ eV (400–10 $cm^{-1}$ known as *far infrared* or FIR). Thus, the technique is able to probe the chemical and geometric structure of (adsorbed) molecules (by causing changes in the molecular dipolar moment) and solids (changes in lattice vibrations and/or acoustic modes), thus covering the whole gamut of catalyst characterization. In general, there are three commonly employed modes by which IR spectroscopy is used to probe catalytic reactions under *in situ* conditions. First of all, there is transmission IR, which involves the preparation of the catalyst sample in the form of a self-supporting disk and is perhaps the most widely used. A second way of collecting IR data is by diffuse reflectance infrared Fourier transform (DRIFT) spectroscopy, which measures light scattering and absorption phenomena. It is particularly attractive because it does not employ complicated sample preparation approaches and therefore is often very close to realistic reaction conditions. On the other hand, the interpretation of DRIFT spectra is based on the phenomenological theory of Kubelka and Munk and some precautions have to be taken when comparing transmission IR and DRIFT with each other.

The goal of this chapter is to give the reader some background of what currently is possible with the IR technique and related spectroscopic techniques for monitoring physicochemical phenomena during, for example, a catalytic reaction within molecular sieves. The first part of the chapter is dedicated to use of IR to investigate acid–base properties of zeolites by making use of relevant probe molecules. The second part involves the study of zeolite synthesis processes, while the third part discusses two prototypical examples of reactions catalyzed by molecular sieves, namely the catalyst removal of $NO_x$ and the methanol-to-olefin (MTO) process. A final section is devoted to the use of IR microspectroscopy to shed insight into template decomposition phenomena, adsorption, and alignment of probe molecules and catalytic processes taking place within large zeolite crystals. The chapter ends with some concluding remarks.

## 8.2 Acidity Determination with IR Spectroscopy of Probe Molecules

Acid–base reactions within zeolite-based materials are perhaps the most technologically relevant class of heterogeneous catalytic processes. Similar to other solid acids, zeolites possess both Brønsted and Lewis acid sites, which are typically hydroxyl groups and coordinatively unsaturated cations, respectively.

Either Brønsted or Lewis acidity can dominate, depending on the chemical content, crystalline structure, postsynthetic treatments and, last but not the least, the state of hydroxylation under the reaction conditions. In order to draw connections between the acidic properties of zeolites and their catalytic properties, it is essential to obtain quantitative information on the number, nature, location, and strength of the acidic sites.

In general, there are two theories that describe the acid–base properties of the solid materials, that is, Brønsted and Lewis theories. In the former theory, an elementary acid–base reaction is a proton transfer from an acid (AH) to the base (B): $AH + B \rightarrow A + BH$. In this case, *acid strength* may be defined as the tendency to give up a proton, which can be quantified by determining the state of equilibrium, for example, in aqueous solution. In the Lewis approach, any species possessing a vacant orbital can be considered as an acid. The acid–base reaction $A + B \rightarrow A - B$ is then the formation of a bond involving unshared electron pair of B and the vacant orbital of A. However, it is important to emphasize that the determination of acidity based on these concepts is strictly valid only for acidic molecules dissolved in homogeneous media (solvents). It is evident that acid groups located on surfaces of catalytic solids cannot be treated in the same way because the concept of homogeneous medium has lost any validity. This consideration is particularly valid for acid (and basic) groups located in the channels and cavities of microporous catalysts which, by definition, are highly inhomogeneous.

IR has established itself as an essential tool to investigate intermolecular interactions, for example, hydrogen and coordination bondings. It has been shown that the most valuable information on the zeolite acidic properties can be obtained using adsorbed probe molecules interacting with acid sites. Perturbations introduced to the IR spectra of both the surface groups and probe molecules upon the interaction, such as intensity variations and frequency shifts, can be translated into the properties of relevant acid sites. The extensive criteria for a selection of suitable probe molecules have been formulated in great detail (e.g., [5–7]). In general, small and weakly interacting molecules are recommended for probing surface properties of acidic and basic solids. Other criteria include the following: (i) selectivity of interaction with specific surface site; (ii) low reactivity, even at catalytically relevant conditions; (iii) detectable magnitude of spectral response upon interaction; (iv) high and experimentally measurable extinction coefficients; and (v), whenever possible, the reactant itself shall be used as a probe molecule.

IR spectra of activated zeolites typically feature two or more major bands in a hydroxyl spectral region (in some cases, more groups of bands can be observed, as, for example, in the case of SSZ-33 zeolite [8]). The typical values for band positions are summarized in Table 8.1. Firstly, silanol groups give rise to bands centered at 3710–3760 $cm^{-1}$ (Figure 8.2). External and internal silanols (four individual components) can be differentiated in the deconvoluted spectra [9]. In the case of large zeolite crystallites, the individual contribution can be even resolved in the spectra [10]. Secondly, the IR band at $\sim$3600 $cm^{-1}$ is attributed to the O–H stretching mode of the Si(OH)Al bridged group. The latter is the most important chemical entity as these hydroxyl groups are typically strong Brønsted acid sites.

**Table 8.1** IR band positions for hydroxyl groups in various zeolite structures[a].

| Zeolite | Silanol OH groups position ($cm^{-1}$) | Bridged OH group position ($cm^{-1}$) | References |
|---|---|---|---|
| SM-5 | 3747 | 3612 | [12] |
| Mordenite | 3745 | 3605 (3612, 3585) | [13, 14] |
| | 3740 | 3640, 3550 | [15] |
| ZSM-2 | 3714 | 3667 | [16] |
| Ferrierite | 3746 | 3602 (3609, 3601, 3587, 3565) | [17] |
| UZ-4 | 3746 | 3602 (3565, 3592, 3610) | [17] |
| SAPO-34 | 3710 (shoulder) | 3610 (3631, 3617, 3600) | [18, 19] |
| SZ-13 | 3740 | 3616 | [20] |

[a]Only a selection of zeolites, relevant to the scientific content of this chapter are included in this table. For a comprehensive set of data, the reader is referred to [22].

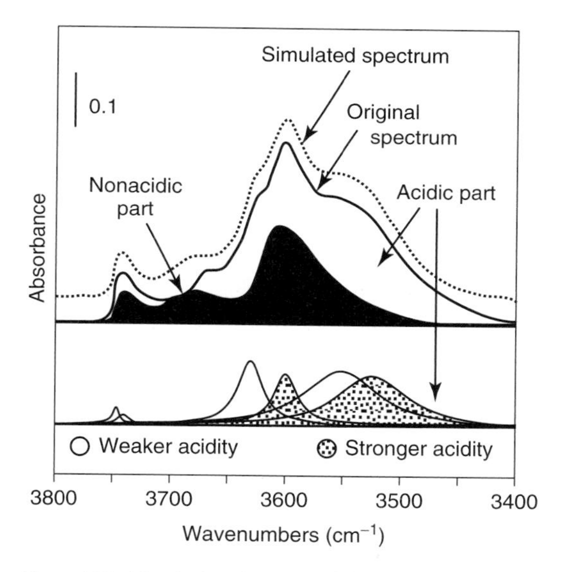


**Figure 8.2** The hydroxyl region of the IR spectrum of ultrastable zeolite Y and its deconvolution into individual components. (Copyright American Chemical Society 1993; Reproduced with permission from [11].)

This band may appear asymmetric due to heterogeneity of the sites located at different crystallographic positions within zeolite unit cell. In the cases when zeolite crystallites are defect rich, a broad band at 3460 $cm^{-1}$, assigned to hydrogen-bonded silanol groups located in the interior of the crystal, appears in the spectrum [10].

Several classes of probe molecules have been successfully used to characterize acidic properties of zeolites, namely, diatomic molecules ($CO, N_2$), nitrogen-containing organic bases (pyridine and its derivatives, nitriles), and hydrocarbons (ethene, benzene). CO forms hydrogen-bonded complexes with hydroxyl groups, which can be identified in the C–O and O–H spectral regions. Upon the H-band formation, O–H stretching band is redshifted and broadened, with the magnitude of shift related to the acidity of the site. At the same time, the C–O band is blueshifted compared to the gas-phase CO frequency. In the case of dinitrogen, the molecule looses its inversion center upon binding, making molecular vibration IR active. Small molecular diameter, chemical inertness, and weak basicity of $N_2$ have made it a very useful probe. Pyridine, first proposed as a reporter molecule in as early as 1963 [21], leads to a similar effect on the O–H bands, whereas ring vibrations in the region 1600–1400 $cm^{-1}$ allow to identify pyridine bound to different surface sites (see Table 8.2 and Figure 8.3). An especially advantageous approach involves using two probes: one is weakly basic, such as CO to access the acidic strength of the sites and a strong base, for example, pyridine, to obtain information of their nature and location within zeolite framework. Several comprehensive reviews covering investigation of zeolite acidic properties with probe molecules were published in the [23]. Therefore, our aim is to highlight some main applications of the IR technique from research work of the last decade in this chapter.

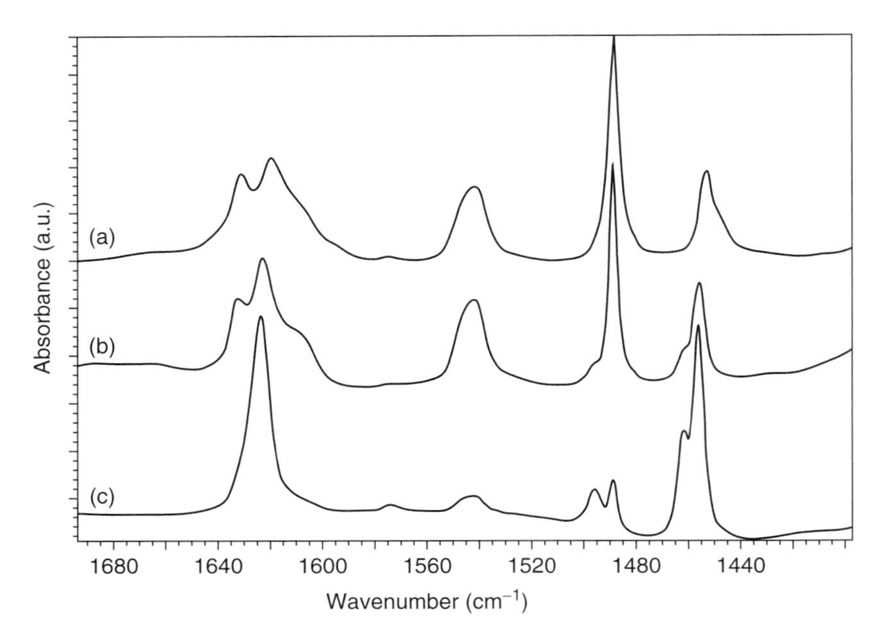


**Figure 8.3** IR spectra taken after pyridine adsorption and subsequent evacuation at (a) 373; (b) 623; and (c) 773 K. (Copyright Royal Society of Chemistry 1996; Reproduced with permission from [24].)

When several acid sites differing in their nature and strength, coexist in a zeolite material, they can be distinguished using one or more molecular probes. For example, two types of Brønsted sites were identified in mordenite, that is, hydroxyl groups located in the main channels and in the side pockets [25, 26]. The former are more acidic, as evidenced by the redshift upon CO adsorption, and accessible to pyridine. Buzzoni *et al.* compared the acidity of several acidic zeolites using pyridine [27]. It was shown that for all the materials studied pyridine reacts quantitatively with the available acid sites to form pyridinium species. Formation of pyridine dimers was also detected. Unlike zeolites, featuring three-dimensional pore system (ZSM-5 and $\beta$), mordenite showed a pore blockage upon pyridine adsorption. Moreover, the acid sites in the small side pockets of the mordenite channels were also shown not to be accessible. This conclusion is confirmed by another study where only the partial protonation of pyridine on mordenite was observed [11]. It was inferred that pyridine could only slightly enter the side pockets, making protonation improbable due to sterical hindrances.

Pyridine derivatives are claimed to be advantageous over pyridine when one probes weak Brønsted acid sites due to their stronger basicity and weaker affinity for Lewis sites, as steric hindrances induced by the methyl groups arise. In addition, several IR bands in the spectra of absorbed methylpyridines are sensitive to the acid site strength. For example, the 2,6-dimethylpyridine spectrum informs of the strength of the Brønsted acid sites not only by the position of the (NH) band (like for pyridinium species) but also more directly by the position of the $\nu_{8a}$ band. Sensible correlations have been established between the spectral shifts upon CO adsorption, the *(NH) as well as the $\nu_{8a}$ band position upon 2,6-DMP adsorption (Table 8.2) [28].

An interesting application of bulky pyridine derivatives is the assessment of acid sites with different accessibility. We illustrate this point with a few examples. One of the important aspects of zeolite catalysis is to overcome the transport limitations, which stems from the exclusively microporous character of these materials by

**Table 8.2** IR band assignments of different pyridine specie detected on solid acids[a].

| Band position ($cm^{-1}$) | Assignment | Acid site |
|---|---|---|
| 1397 | $PyH^+$ | B |
| 1455 | Py | L |
| 1490 | Py + $PyH^+$ | B + L |
| 1545 | $PyH^+$ | B |
| 1576 | Py | L |
| 1621 | Py | L |
| 1635 | $PyH^+$ | B |

From [24].
[a] Py = pyridine; $PyH^+$ = pyridinium ion; B = Brønsted acid site; L = Lewis acid site.

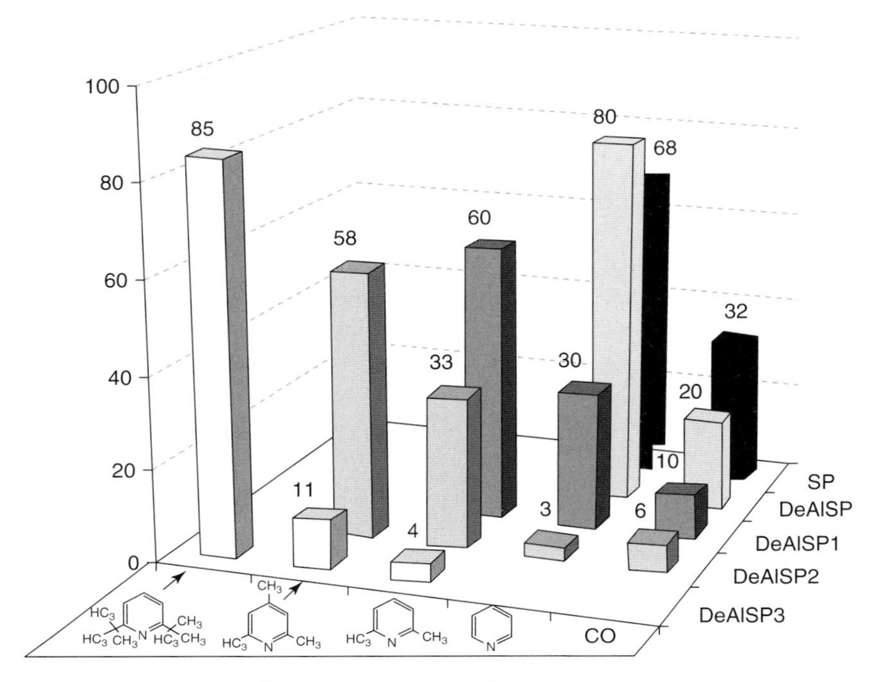


**Figure 8.4** Distribution of Brønsted acid sites of different accessibility over mordenite samples studied. (Copyright Elsevier 2007; Reproduced with permission from [28].)

introducing controlled mesoporosity. Such a treatment would unavoidably affect the acidic properties. A methodology for characterization and quantification of the acidic sites of different accessibility has been developed and applied for a series of dealuminated mordenites. These materials were probed with IR spectroscopy of adsorbed alkylpyridines of increasing molecular dimensions, followed by the characterization of the nonaccessible sites using subsequently adsorbed CO [29], allowing for a step-by-step characterization of the nature and strength of sites with different accessibility. The results of this approach are shown in Figure 8.4. It was shown that upon dealumination of the small-port mordenite samples the acid sites in the side pockets become accessible for pyridine due to the partial destruction of side pockets upon dealumination. Subsequently, the formation of secondary mesoporosity systems due to further dealumination make the zeolite crystals completely accessible to relatively bulky molecules, such as lutidine, collidine, and 2,6-di-*tert*-butylpyridine.

In another related study, desilicated ZSM-5 materials have been probed by applying CO and collidine (2,4,6-trimethylpyridine) [10]. CO adsorption at low temperatures showed no significant alterations in Brønsted acidity after desilication. Collidine, which is too bulky to enter the micropore system of ZSM-5, was used to probe the surface of the mesopores, where Lewis acid sites can be generated from dislodged framework aluminum ions. By first saturating the zeolite with collidine and subsequently adsorbing CO, it was demonstrated that almost all Lewis sites coordinate collidine, whereas the Brønsted acidity was continuously protected in the micropore system. These observations allowed ruling out the higher Brønsted acidity as well as the synergism between two types of acid sites as an explanation for the enhanced catalytic activity of the desilicated material [10].

Thibault-Starzyk *et al.* determined an accessibility of the acid sites in hierarchical zeolites [30]. These materials have been extensively studied in the last decade, as they exhibit an improved catalytic performance compared to their microporous parents due to the integration of the catalytic properties of the native micropores and the facilitated transport via a mesopore network. It was shown that with the postsynthetic desilication up to 40% of the ZSM-5 acid sites could be made available for molecules as bulky as collidine. An accessibility index (ACI), which is defined as the number of acid sites detected by adsorption of the probe molecules divided by the total amount of acid sites in the zeolite based on the measured aluminum content was shown to be a quantitative indicator for evaluating the effectiveness of synthetic strategies for hierarchical zeolites preparation.

Another interesting route toward mesopore introduction in zeolites is partial dissolution in highly alkaline media, followed by recrystallization into mesostructure. The strength of the acid sites of recrystallized Beta zeolite was studied by IR spectroscopy of adsorbed CO [31]. Two types of Brønsted acid sites with different acid strength were identified in recrystallized samples. This was rationalized in terms of highly crystalline areas of zeolite bearing stronger sites, and less crystalline areas featuring weaker sites, due to the partial destruction of Si–O and Al–O bonds. Next, pyridine and 2,6-di-*tert*-butylpyridine were selected as probe molecules to assess the accessibility of Brønsted acid sites. The highest content of the sites accessible for pyridine was observed on parent zeolite, whereas the sample that recrystallizes in mild conditions has shown the highest amount of sites accessible for 2,6-di-*tert*-butylpyridine. More severe conditions of recrystallization lead to acid properties similar to those of mesoporous MCM-41.

An influence of dealumination on acidic properties of zeolite Y was reported by Datka *et al.* [15]. Steaming was shown to lead to the appearance of two new IR bands, assigned to strongly acidic hydroxyls, which interact with extra-framework Al, and Al–OH groups. Two kinds of Al–OH groups can be differentiated, that is, one accessible and another inaccessible to benzene molecules. The dealumination was reverted by KOH treatment, leading to reincorporation of Al ions into the framework and reconstruction of the hydroxyls IR bands of the parent material. Pyridine sorption has shown that in both the realuminated and the nondealuminated samples a part of hydroxyl group is inaccessible to the bulky molecules. Acidic treatment of dealuminated HY was also used to reinsert Al into the zeolite structure: depending on the conditions used, a degree of realumination of 60% can be achieved, with almost all Lewis sites related to extra-framework Al converted into Brønsted ones [32].

Recently developed zeolite structures have also fallen into the scope of acidity measurements. For example, SUZ-4 zeolite, structurally related to ferrierite, has been investigated by Zholobenko *et al.* [17]. The ferrierite family has attracted attention due to the ability to catalyze the selective transformation of olefins. Four major types of hydroxyl groups of different nature could be identified in the OH region of ferrierite spectrum, which were assigned to (i) bridging OH groups located in 10-ring channels; (ii) large cages in eight-ring channels; (iii and iv) in eight- and six rings, respectively. It was suggested that extended rings at the intersections of eight- and six-ring channels are enriched with substituting aluminum, resulting in a high concentration of OH groups in the large cages in eight-ring channels. In the case of SUZ-4 zeolite, the deconvolution of the spectrum gave three IR bands. The intensity of the band, assigned to OH groups in six rings, decreased compared to that in ferrierite, indicating that the nonexchangeable potassium cations are concentrated in small cages or in double six rings. The hydroxyls in 10 rings, which constitute about 50% of Si(OH)Al groups, are accessible to both *n*-hexane and isobutane. The peak at 3592 $cm^{-1}$ (40%) assigned to OH groups in eight-ring channels is affected with *n*-hexane but not with isobutane. A comparison of the OH band spectral shifts to *n*-hexane adsorption has suggested the following order of acidic strength in zeolites: H-ZSM-5 $>$ H-SUZ-4 $\geq$ H-ferrierite. Another zeolite structure with relatively unexplored catalytic applications is ZSM-2, with pore openings of 0.74 nm. Covarrubias *et al.* synthesized and characterized nano-sized ZSM-2 particles [16]. From the pyridine adsorption experiments, both Brønsted and Lewis sites were identified. Steaming of this zeolite led to an increase in the number of Lewis sites at the expense of the B counterpart, possibly due to dealumination.

Alongside zeolites, the acidic properties of silicoaluminophosphates have been investigated by the IR technique. For example, SAPO-34 cannot be probed with pyridine due to small pore openings (4.4 Å). However, with ammonia as a probe molecule, both Lewis and Brønsted sites in SAPO-34 can be accessed [18]. The latter constitute most of the acid sites; they exhibit moderate acidity which, along with the shape selectivity, accounts for high selectivity of the catalyst toward light olefin formation. When probing SAPO-34 with CO and ethene, the nature of the acid sites can be further clarified [19]. Three distinct hydroxyl groups were identified: two major components attributed to different crystallographic positions and the third exhibiting acidic properties comparable with those of zeolites (330 $cm^{-1}$ shift upon CO adsorption) related to species formed at the borders of silica islands or inside aluminosilicate domains. A related material, namely, high-silica chabazite H-SSZ-13 exhibits four types of hydroxyl groups with similar acidic strength but with different accessibilities for CO [20]. Overall, little difference has been found in acidity between this material and SAPO-34.

## 8.3
## Zeolite Synthesis Processes

Zeolites are built from tetrahedral $SiO_4$ and $AlO_4$ units. The manner in which these basic structural units are connected defines the final porous architecture of the channel networks, that is, the pore diameter and interconnection of microporous channels. Even subtle variations in these parameters can significantly affect the chemical reactivity and catalytic performance of the material. As the synthetic conditions directly determine the structure of the resulting zeolites, a number of studies have been directed toward obtaining a detailed understanding of the processes that occur during zeolite preparation. In more detail, nucleation and growth behavior has to be correlated with the changes in the nature of aluminum and silicon source, and synthesis duration, temperature, and pressure. Even though a considerable amount of research has been devoted to the investigation of underlying synthesis factors, a lack of basic insight into the process has led to a rather empirical approach in the attempts of designing new zeolite structures and modifying existing ones. In what follows, we briefly discuss a variety of applications of *in situ* techniques to investigate zeolite synthesis processes.

In one of the pioneering studies, Engelhardt and coworkers used $^{29}Si$ and $^{27}Al$ MAS NMR to investigate the nature of the intermediates during zeolite A synthesis [33]. The initial gel was found to consist of tetrahedral $Si(OAl)_4$ and $Al(OSi)_4$, which form an amorphous network with alternating Al–Si ordering. This structure is converted into a highly crystalline zeolite with time. There was no direct correlation between the initial Si/Al ratio and the intermediate aluminosilicate gels composition. The variations have been explained in terms of different aluminum and silicon sources used [34]. This method has been extended to other systems, such as mordenite and sodalite. However, in all experiments of this kind, the solid and liquid phases were separated. Shi *et al.* demonstrated that the MAS MNR technique could be applied *in situ* to follow the speciation as a function of time [34]. The main distinct species observed during zeolite A synthesis were amorphous $Al(OSi)_4$, and the growing zeolite phase with small fractions of $Si(OH)_4$ and silicalite species. Higher concentration of the gel led to faster crystallization due to a higher number of nuclei formed. The experimental findings were in favor of the crystallization model in which a growth proceeds through a fast deposition of the $Si(OH)_4$ and $Al(OH)_4^-$ on the surface. Small-angle X-ray scattering (SAXS) measurements showed that before the zeolite A crystallization begins uniformly sized precursor particles of about 10 nm are formed [35].

The crystallization of SAPO-34 has been monitored by Vistad *et al.* using *in situ* X-ray diffraction and NMR [36]. Heating rates were shown to significantly influence the course of the process, for example, slow heating produces the precursor phase, which is critical to obtain SAPO-34. The transformation of this layered phase into a chabazite structure occurs through the partial dissolution of the former. As no changes in the diffraction peak widths were observed, other rearrangement mechanisms were ruled out. NMR experiments (Figure 8.5) allowed the identification of four steps in the SAPO-34 crystallization with increasing temperature, namely, dissolution of the initial gel, formation of the four-ring structures, formation of the layered $AlPO_4$-F prephase, and, finally, dissolution of the latter and condensation of the four-ring structures into a triclinic chabazite structure. Incorporation of silicon was proposed to be a limiting step of the process.

The synthesis process of aluminophosphates (AlPOs) has also attracted significant attention. For example, the crystallization of the cobalt-containing $AlPO_4$-5 (CoAPO-5) has been investigated with a combined Raman/SAXS/UV–vis/X-ray

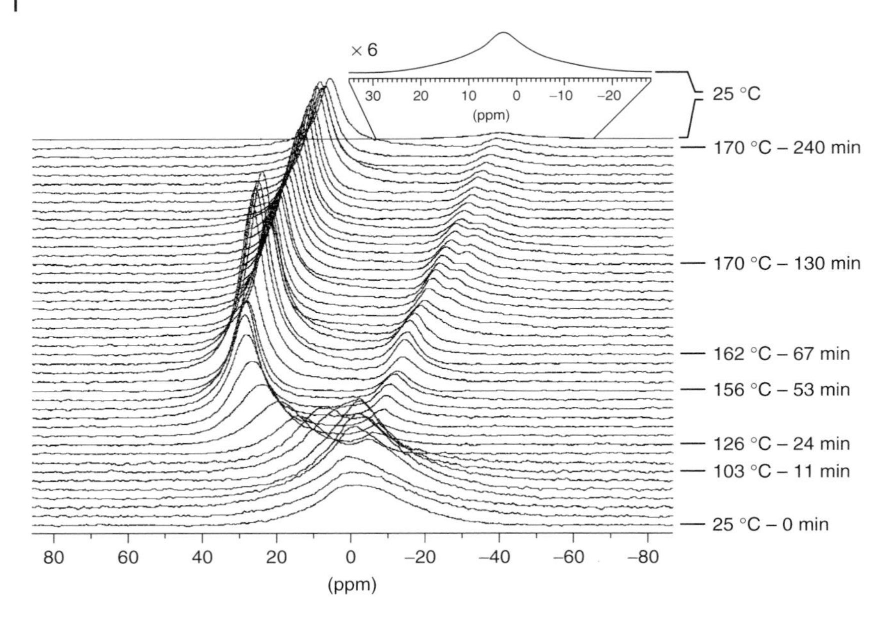


**Figure 8.5** *In situ* $^{27}Al$ NMR spectra recorded during synthesis of triclinic SAPO-34. (Copyright American Chemical Society 2003; Reproduced with permission from [37].)

absorbtion setup [38]. Raman spectroscopy allowed the identification of the instantaneous formation of the Al–O–P bonds in the gel with a wide distribution of particle sizes, as shown by SAXS [35]. Then, the one-dimensional chains with alternating Al–P arrangement are formed, most probably from four-membered units. The chains condense into larger 1D rodlike structures, which undergo rearrangement into 2D and, subsequently 3D network. *In situ* UV–vis data shows the gradual transformation of the octahedral coordination of the Co ions in the precursor gel into a tetrahedral one in the resulting material before the onset and during crystallization. These findings are corroborated by X-ray absorbtion data. Conformation of the structure-directing agent molecules in the synthesis of AlPO-5 and metal-subsituted APO-34 has been assessed with Raman [39]. A strong interaction between the template molecules and inorganic network containing transition metal ions leads to deformation of the template structure; this does not occur in case of the synthesis of a metal-free AlPO-based molecular sieve, as illustrated in Figure 8.6.

Raman spectroscopy often presents challenges when applied for studying zeolite synthesis due to the strong fluorescence of the reaction mixture. This prompts a use of, for example, a UV laser source, which allows avoiding the fluorescence and increasing the sensitivity. Fan *et al.* have successfully applied UV Raman to study hydrothermal crystallization of zeolite X [40]. It was shown that amorphous solid phase initially dissolves to form monomer silicate species in the liquid phase, while amorphous aluminosilicate species composed of predominately four rings

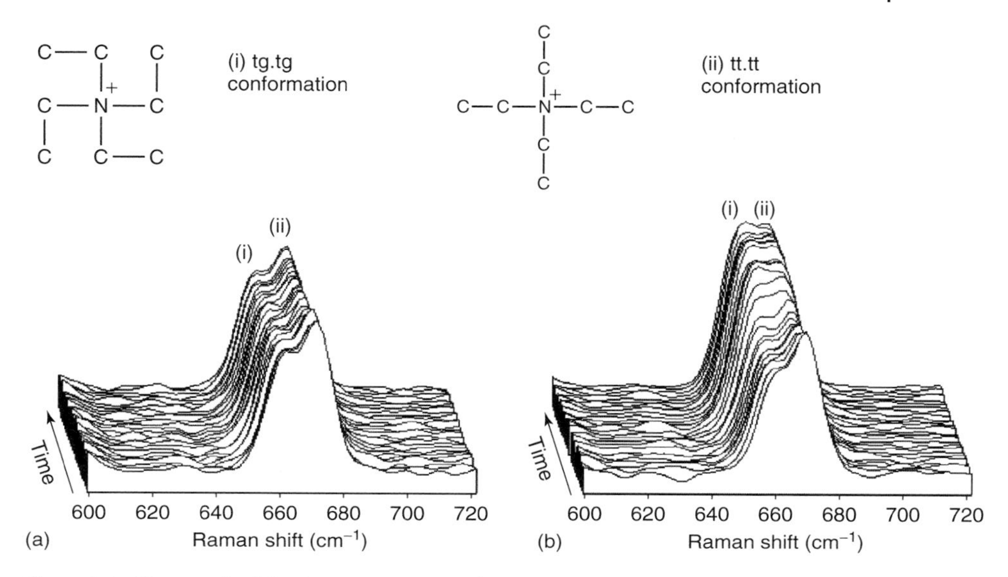


**Figure 8.6** Time-resolved *in situ* Raman spectra for synthesis gels with (a) no framework substitution and (b) 30% substituted $Zn^{2+}$. The structures of the two tetraethylammonium hydroxide template conformations are also indicated. (Copyright American Chemical Society 2006; Reproduced with permission from [39].)

are formed in the solid phase in the early stage of nucleation. These four rings are connected with each other via double six rings together with the monomer silicate species in the liquid phase to form the crystalline zeolite X framework.

Another experimental obstacle is the development of a suitable reaction cell. Successful examples of hydrothermal vessels allowing for laser excitation have been reported [38], including variable focal point design allowing to probe the liquid and solid phase independently [40]. Furthermore, a setup for microwave-assisted zeolite synthesis adapted for simultaneous SAXS/WAXS and Raman measurements has been developed by Tompsett *et al.* [41].

## 8.4 Selection of Zeolite-Based Catalytic Reactions

### 8.4.1 Catalytic Decomposition of Nitric Oxides

Ever-tightening control over nitrogen-containing emissions has prompted an extensive search for effective catalyst materials for nitric oxides abatement, often referred to as *DeNO*$_x$. A number of oxide catalysts have been proposed for this process. Protonated and metal-containing zeolites were shown to be active in various $DeNO_x$ routes, such as direct decomposition and selective catalytic reduction (SCR) by hydrocarbons and ammonia. In what follows, the insights into the mechanism of $DeNO_x$ on zeolites gained by using *in situ* spectroscopy are discussed.

Copper-containing ZSM-5 was found to be active in NO decomposition. However, the nature of the active center is still under discussion despite the large characterization efforts. It was established that during pretreatment and under reaction conditions di- and monovalent copper ions, along with bi- and, possibly, polynuclear complexes, are present in the materials, On the basis of IR investigations of CuHZSM-5 and CuZSM-5 with NO and $NO/O_2$, $Cu^I$ sites were suggested to be catalytically active [42]. In support of this proposal, photoluminescence measurements have demonstrated a correlation between the $Cu^I$ concentration and the activity toward NO decomposition [43, 44]. On the basis of a combination of IR spectroscopy and the molecular modeling, mono-adducts of NO with $Cu^I$ are proposed to be the key intermediates in the reaction, with the N–N bond formed in the interaction of Cu-coordinated and gas-phase NO molecules [45].

In contrast, Kucherov *et al.* reported on the strong interaction of NO with $Cu^{II}$ centers based on EPR (electron paramagnetic resonance) spectroscopy [46]. This technique suggests three different types of paramagnetic $Cu^{II}$ ions being formed upon CuZSM-5 dehydration [47]. Only one species react with adsorbed NO to form $Cu^{II}$–NO, while two other species are inactive. This is illustrated in Figure 8.7. $Cu^I$ is proposed to react with two NO molecules to form the complex, which subsequently transforms into $Cu^{II}$ (NO)$O^-$. Adsorbed species are generated via the interaction of $Cu^{II}O^-$ with NO. Accordingly, a mechanism involving the aforementioned intermediates is proposed.

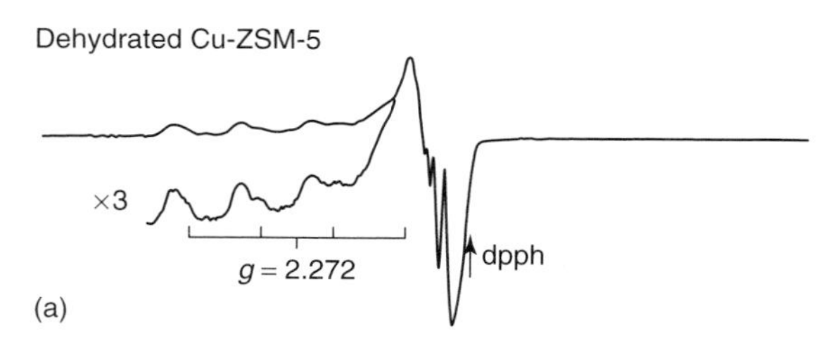


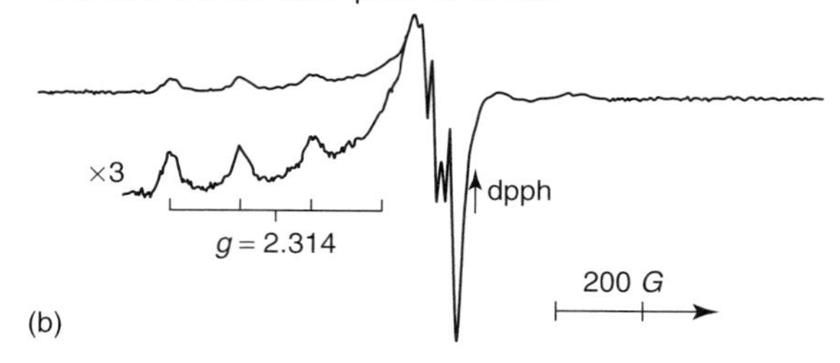


**Figure 8.7** EPR spectra of (a) dehydrated CuZSM-5, (b) after 50 torr NO adsorption for 10 minutes. (Copyright Elsevier 2000; Reproduced with permission from [47].)

Mathisen *et al.* compared the redox behavior of copper ions in CuZSM-5 in the SCR of $NO_x$ by propene [48]. In the case of CuZSM-5 prepared by ion exchange, $Cu^{II}$ can be reversibly reduced to $Cu^{I}$ and backward by propene and $NO_x$, respectively, as shown by XAS. No copper oxide species or metallic copper clusters were detected. On the basis of these findings, the authors postulate that the redox mechanism and not copper dimers are responsible for SCR activity. As to the CuAPO-5 material, acid sites and not framework copper ions are suggested to be catalytically active toward $NO_x$ reduction.

Ganemi *et al.* tested highly siliceous copper ion-exchanged ZSM-5 in the direct decomposition of NO [49]. The optimal Si/Al ratio was established to allow for one ion exchange site per channel intersection; 200% Cu exchange was shown to yield best performance. Both unidentate and bidentate $NO_3^-$ were detected in the IR spectrum. In the latter, the $NO_3$ species are bound to the bridged $Cu_2^+{-}O^{2-}{-}Cu_2^+$ sites, and the authors assume that these complexes block the sites active in NO decomposition. Another possible binuclear copper intermediate, that is, bis(μ-oxo)dicopper has been first predicted on the basis of theoretical considerations [50] and then identified in overexchanged CuZSM-5 materials under reaction conditions by UV–vis and EXAFS (extended X-ray adsorption fine structure) spectroscopies [51]. This complex takes a role of continuous $O_2$ production and releases and maintains a sustainable catalytic cycle. Beyond binuclear copper complexes, chain-like copper oxide structures were identified in CuZSM-5 with exchange rates of 75–100% based on the results of EPR and UV–vis spectroscopy [52]. As these species can be easily oxidized and reduced, their involvement in the reaction is likely.

Another zeolite-based catalyst active in catalytic reduction of $NO_x$ is FeZSM-5. High iron loadings up to Fe/Al = 1 can be achieved when sublimation of $FeCl_3$ is used to introduce the metal into the zeolite [53]. Oxygen-bridged binuclear iron complexes were proposed to be key intermediates in the reaction [53]. EPR spectroscopy indicated that reactive iron is present as $Fe^{3+}$ under oxidative catalytic conditions [54]. Different iron species exist in the catalysts under reaction conditions, that is, $Fe^{3+}$ in octahedral and tetrahedral coordination and iron oxide clusters, all with different ox-red behavior [55]. Formation of $NO_2/NO_3$ species coordinated to iron ions was demonstrated by IR as illustrated in Figure 8.8. These complexes are reduced by butane to yield the adsorbed cyanide and isocyanide groups, which decompose further to $N_2$ and $CO_2$ [56]. The importance of adsorbed $NO_2$ has been underlined in a combined IR/kinetic study focusing on the NO-assisted $N_2O$ decomposition [57]. A peroxide ion was proposed to be another active intermediate in the reaction based on the EPR and UV–Raman data [58]. A mechanism that involves the coordination of molecular oxygen as a peroxide ion with its further conversion into a bridged di-oxygen was proposed.

In the case of the overexchanged FeZSM-5 catalyst materials, the existence of binuclear iron complexes has been demonstrated by EXAFS spectroscopy [59]. Distorted octahedral $Fe^{III}$ sites in these complexes are highly reactive, capable of breaking the N–O bond [59], as was shown in the study of SCR of NO with isobutene [60]. Treatment with isobutene alone results in a slight reduction of

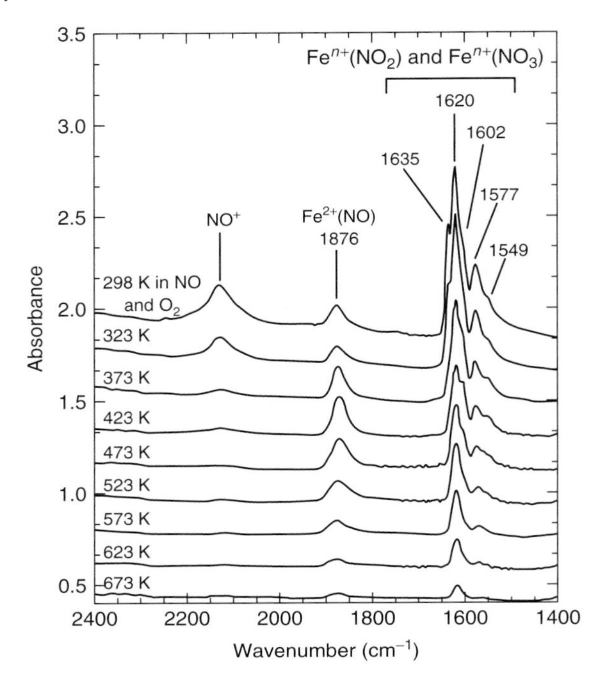


**Figure 8.8** Infrared spectra of FeZSM-5 taken during a temperature ramp exposed to 5000 ppm NO and 1% $O_2$ is passed over the catalyst after it had been exposed to this mixture for 20 minutes at room temperature. Nitrite and nitrate regions are indicated. (Copyright Springer 1999; Reproduced with permission from [56].)

iron. On the contrary, feeding the catalyst materials with NO and oxygen brings the average iron oxidation state up from 2.3 to 2.8. The number of Fe-neighboring atoms increases, indicating the partial reoxidation of binuclear sites or formation of N-coordinated species. In the SCR conditions, complete reoxidation of iron occurs, indicating that the reactivity originates from oxygen vacancies. Unfortunately, as EXAFS was unable to distinguish between light-scattering elements, the exact structure of the reactive complex could not be recovered. In another EXAFS study, the spectra of the catalyst before and after reaction with $N_2O$ were found to be identical, suggesting that only a very small concentration of sites, which is not detectable by EXAFS, is responsible for the activity [61].

*In situ* Mössbauer spectroscopy also gives evidence toward formation of binuclear iron complexes. The spectra of the active species were shown to be very similar to those of the enzymatic Fe–Fe complex. These species comprise more than 60% of the total metal content. Reduced $Fe^{2+}$ ions were shown to reversibly oxidize to $Fe^{3+}$ by nitrous oxide, generating active $\alpha$-oxygen species bringing about oxidation activity [62]. Interestingly, both iron atoms are capable of $\alpha$-oxygen generation, hence Dunbkov *et al.* argued that $\alpha$-sites are, in fact, monatomic entities in a paired arrangement spectroscopically registered as dinuclear complexes.

Unlike in the case of hydrocarbons, when ammonia is used as a reducing agent, ammonium nitrite was shown to be formed by IR spectroscopy [63]. At higher temperatures, nitrite decomposes to $N_2$ and $H_2O$. The effect of water on the reaction rate was rationalized. At low temperatures, water competes with ammonia and $NO_x$ for adsorption sites, decreasing the nitrite formation. Furthermore, Pérez-Ramírez *et al.* reported on the SCR of $N_2O$ by carbon monoxide [64]. According to the proposed mechanism, which is based on the combined data from UV–vis and EPR spectroscopy, CO removes oxygen species from the Fe–O–Fe complexes, thus, liberating then for $N_2O$ adsorption. A correlation between the concentration of isolated $Fe^{3+}$ and the $N_2O$ conversion was established.

Schwidder *et al.* assessed the role of Brønsted acidity in the NO SCR [65]. Several Fe-MFI catalysts with similar structure of Fe sites were found to exhibit a significantly different performance in SCR of NO with isobutane and with $NH_3$. The dramatic differences observed have been attributed to their strongly different acidity properties, which have been characterized by IR of adsorbed pyridine. The variations indicate an essential role of Brønsted sites in this reaction.

Zeolites exchanged with metals other than copper and iron, for example, nickel and cobalt were also shown to be active in the $DeNO_x$ process [66, 67]. Mihaylov *et al.* studied the process of NO adsorption on NiY and NiZSM-5 zeolites [68]. Only mononitrosyl species were shown to form in the case of NiY zeolite, whereas for NiZSM-5 dinitrosyls were also detected. Nitrites and nitrates formed on NiZSM-5 upon coadsorption of $NO/O_2$ are highly reactive toward SCR by methane: this finding has been rationalized in terms of electrophilicity and coordinative saturation of Ni ions in zeolite environments. The lower coordination number of $Ni^{2+}$ in ZSM-5 framework allows for simultaneous coordination of the nitrate and the reducing hydrocarbon molecule.

### 8.4.2
### Methanol-to-Olefin Conversion

Methanol is a valuable chemical, which can be made from the synthesis of gas and further converted into light alkenes and gasoline-range hydrocarbons. The zeolite-catalyzed conversion of methanol to hydrocarbons (MTH) is commonly referred to as *MTO* and *MTG* (methanol to gasoline) depending on the desired products. The MTH chemistry and its commercial potential have been known for decades. In 1986, a New Zeeland MTG facility was started up, but, due to a drop in crude oil prices, only the methanol synthesis step was left on stream. The Topsøe integrated gasoline synthesis (TIGAS), based on H-ZSM-5 as catalyst, was demonstrated on a pilot scale, but has never been scaled up. Later, more attention was paid to the MTO reaction, that is, polymer-grade ethene and propene from methanol using the UOP/Norsk hydro technology based on the silicoaluminophosphate H-SAPO-34.

In a simplified route of the MTO reaction, methanol is dehydrated to form dimethyl ether and water, followed by formation of alkenes. The long-standing question has been the specific mechanism behind this latter step, in which C–C bonds are formed from oxygenates (methanol/dimethyl ether). In the first *in situ* IR study of MTO chemistry, ZSM-5 catalysts were exposed to methanol and dimethyl ether at elevated temperatures [69]. Methoxy groups formed upon protonation of organic molecules by zeolite hydroxyl groups and subsequent elimination of water were identified for both reactants. These species were suggested to be essential for the formation of the first C–C bond. The reactivity of a methoxy species depends on the C–O bond strength, which is determined by the acid strength of the zeolite hydroxyl species. The authors noticed that the conclusions made are most probably relevant to the initial step of the reaction [69]. Comparison of various HZSM-5 zeolites containing different concentrations of framework and extra-framework aluminum has shown that several adsorbed species formed upon methanol adsorption, that is, methanol hydrogen-bonded to Brønsted acid sites, chemisorbed methanol as methoxy groups formed on Brønsted acid sites, silanol groups and extra-framework AlOH sites [70]. Dimethylether was found to form hydrogen bonds with silanol groups and Brønsted acid sites, and forms the same methoxy species generated from methanol at higher temperatures. Analysis of the IR spectra and the initial hydrocarbon products for different zeolite samples indicated that the Brønsted acid sites and the methoxy groups formed on them are the key to hydrocarbon formation. Extra-framework aluminum and the methoxy groups formed at silanol and AlOH were not found to play any direct role in the catalytic reaction.

Solid-state MAS NMR has also proved to be an excellent tool to study reactive intermediates participating in the MTO reaction pathways. A number of ex *situ* studies, in which the catalyst samples were exposed to methanol, thermally treated, and subsequently quenched, exist. In the pioneering work by Anderson and Klinowski [71], the speciation of the aromatics formed in the zeolite channels was carried out. Interestingly, the distribution of the polymethylated benzenes was found to significantly deviate from thermodynamic equilibrium. The 1,2,4,5-tetramethylbenzene is formed in larger quantities than the thermodynamically favored 1,2,3,5-tetramethylbenzene. The authors rationalized this finding in terms of the smaller size of the former (6.1 vs 6.7 Å), which fits into the ZSM-5 channel intersection dimensions. Other aromatic species formed included xylenes and smaller amounts of tri-, penta-, and hexamethylbenzenes.

Direct observation of the induction period (when an equilibrium between methanol, dimethyl ether, and water was established) and the onset of hydrocarbon synthesis with the primary product being ethylene have been observed with *in situ* NMR [72]. These findings suggested ethylene to be the "first" olefin or, in other words, the first C–C bond formed. The formation of significant amounts of methane was explained by abstraction of the hydride by the active surface methoxy species. When multinuclear solid-state NMR was applied to study the adsorption of methanol on the H-ZSM-5 zeolite, formation of hydrogen-bonded neutral methanol molecules as well as partially protonated clusters with three methanol molecules coordinated a bridged hydroxyl group [73]. These clusters with up to four methanol molecules could be observed also in zeolite Y [74]. However, in the case of H-$\beta$, silanol groups also participate in methanol absorption, forming weak H-bonds. Significantly large adsorbate complexes with up to seven molecules per SiOHAl or SiOH group were shown to form in the latter case.

It was established that the initial introduction of olefins into the feed reduces the induction period and enhances the catalytic activity through the formation of the aromatic species [75]. Haw *et al.* investigated the interaction of ethene with ZSM-5 with MAS NMR using a pulse-quench catalytic reactor [76]. During the induction period, signals of cyclopentenyl carbenium ion appeared. These species are suggested to be key intermediates in the working MTO catalysts, as they act as reservoirs of cyclic dienes which are easier to methylate.

Further NMR work, first published in the patent literature [77, 78], has given direct insight into the chemical nature of the catalytic scaffold formed within the zeolite channels, in which the formation and breakage of carbon–carbon bonds takes place. This so-called hydrocarbon pool was found to contain benzenes, naphthalenes, and methylated derivatives thereof. Direct evidence toward the formation of methylated aromatics was reported by Song *et al.* for SAPO-34 [79]. The average number of methyl groups per aromatic ring was shown to reach a maximum of about 4, which are consumed during the reaction (Figure 8.9) [80]. No induction period was observed when the catalyst was pretreated with methanol pulse, allowing the aromatic pool to be formed. In a subsequent study, the naphthalene moiety methylated with up to four groups was found to be the major participants in the process [81]. Wang *et al.* have demonstrated in a continuous flow experiment that surface methoxy groups are very active and capable of methylating toluene and cyclohexane, paving the way toward the hydrocarbon pool [82]. An experiment in which the feed of $^{13}$C-enriched methanol was switched to $^{12}$C-methanol has also substantiated the relevance of the alkylated aromatics [83]. Upon the switch, the $^{13}$C NMR signal of alkyl group decreased, indicating that those species participate in alkylation and splitting-off reactions.

A coupling of NMR to optical fiber-based UV–vis spectroscopy accomplished by Hunger *et al.* allowed the identification of the formation of cyclic compounds and carbenium ions characterized by absorption at 300–400 nm at temperatures as low as 413 K [84]. The same combination was also used to study the coking and regeneration of H-SAPO-34 [85]. At temperatures up to 623 K, polyalkylaromatics and enylic carbenium ions were registered from the NMR and UV–vis spectra, respectively. On average, about 0.4 aromatic rings per chabazite cage are formed in the reaction with one to four methyl groups per aromatic ring, depending on the temperature. At 673 K, the formation of polymethylated anthracenes has become evident, which are thought to be responsible for the deactivation of the catalyst occurring in these conditions. Treatment in air at 773 K was found necessary to nearly completely remove both substituted aromatic and polyaromatic compounds.

UV–vis and confocal fluorescence microspectroscopy is another tool to assess not only the nature of the carbonaceous species leading to deactivation of MTO

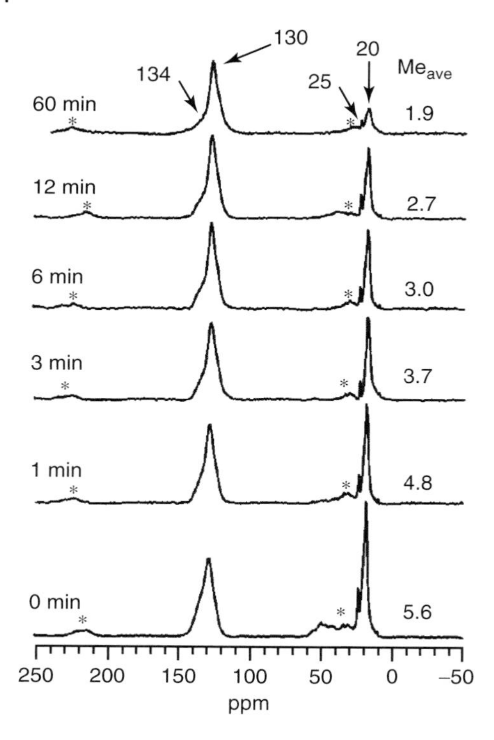


**Figure 8.9** $^{13}$C CP-MAS NMR spectra (75 MHz) showing the loss of methyl groups as a function of time from methylbenzenes trapped in the SAPO-34 cages at 673 K. For each case, a fresh catalyst bed was used to convert 0.1 ml. The average numbers of methyl groups per ring, $Me_{ave}$, are shown. (Copyright American Chemical Society 2001; Reproduced with permission from [81].)

catalysts but also their preferential location within a zeolite particle [86]. Mores *et al.* have demonstrated that in ZSM-5 coke initially forms in the near-surface area of the crystals and gradually diffuses into the particle (Figure 8.10). In the case of SAPO-34, the formation of aromatic coke precursors is limited to the near-surface region of SAPO-34 crystals, thereby creating diffusion limitations for the coke front moving toward the middle of the crystal during the MTO reaction. From the spectral and spatially resolved data, graphite-like coke deposited on the external crystal surface in the case of ZSM-5 and aromatic species formed inside the zeolite channels (for both catalysts) could be distinguished.

IR spectroscopy was applied to investigate the influence of the acid site density on performance in the MTO reaction [87]. It was shown that within the series of mordenite samples with Si/Al ratio varying in the range 5–105 rapid accumulation of polyaromatic compounds occurs in the Al-enriched catalyst materials. The formation of compounds with three or four fused aromatic rings leads to the loss of the active alkylbenzene intermediates as well as to the blockage of mordenite channels. On the other hand, the sparse distribution of acid sites in more siliceous

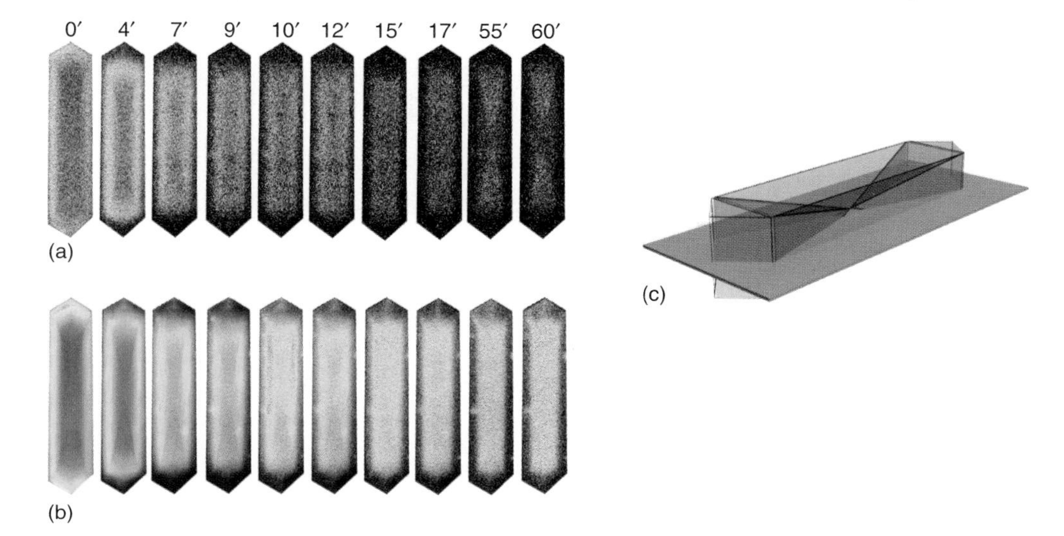


**Figure 8.10** Fluorescent carbonaceous species formed in H-ZSM-5 crystal during MTO reaction with time-on-stream. Confocal slices are taken at laser excitation (a) 488 nm and (b) 561 nm and the schematic representation of the slice where the confocal fluorescence measurement has been performed (c). (Copyright Wiley-VCH 2008; Reproduced with permission from [85].)

samples suppresses the condensation of alkylbenzenes in their pores and lengthens the catalyst lifetime.

In the work of Park *et al.*, several zeolites with different pore topologies and acidities were compared, namely, SAPO-34, ZSM-5, H-$\beta$, mordenite, and HY [88]. All the zeolites were shown to exhibit initial high conversion of methanol; however, their deactivation rates varied significantly, in more detail, activity of SAPO-34, ZSM-5, and H-$\beta$ were maintained even after several hours on stream, whereas mordenite showed rapid deactivation. IR spectroscopy was used to follow the formation of the aromatic intermediates. Intensity of the band at 1465 $cm^{-1}$ assigned to methylbenzenes was found to strongly correlate with the number of strong acid sites, leading to the conclusion that only these sites are relevant for the MTO process. The absorption band at around 1589 $cm^{-1}$ due to polyaromatic compounds dominated in the case of HY, as shown in Figure 8.11. It is suggested that deactivation is related to the condensation of alkylbenzenes to large molecules in large cavities of these materials, and subsequent pore blockage. On the contrary, smaller cages of SAPO-34, ZSM-5, and H-$\beta$ zeolites suppressed the formation of polyaromatics as evidenced from the IR spectra shown in Figure 8.11. As a result, the catalysts show a longer lifetime. On the basis of this spectroscopic and kinetic data, the authors deduced a relation between the pore and cavity geometry, product distribution, and deactivation behavior.

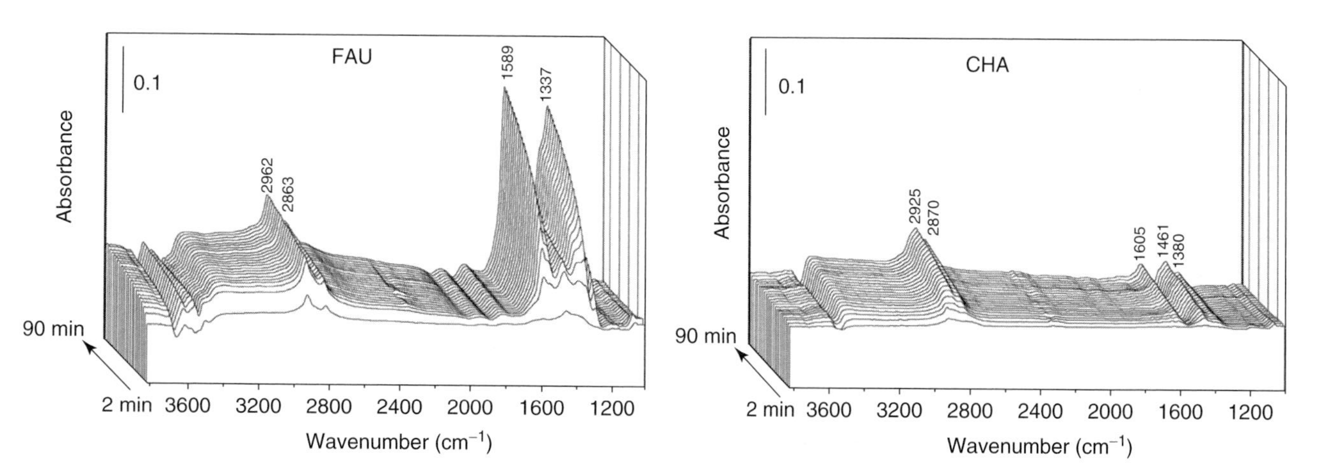


**Figure 8.11** *In situ* IR spectra of materials adsorbed and occluded on FAU (a) and SAPO-34 (b) zeolites during the MTO reaction: reaction temperature = 623 K. (Copyright Elsevier 2009; Reproduced with permission from [88].)

## 8.5 IR Microspectroscopy

One of the emerging investigation methods in the field of catalysis is microspectroscopy. Examples include UV–vis, Raman, X ray, fluorescence, and magnetic resonance imaging (MRI) techniques. One of these methods is IR microscopy, which allows obtaining vibrational spectra from sample regions as small as small as a few micrometers. Large (up to hundreds of micrometers) zeolite crystals are commonly used for IR microspectroscopic studies on molecular diffusion, template decomposition, and catalytic reactivity.

The usage of IR microspectroscopy in the field of zeolites has to the best of our knowledge been pioneered by van Bekkum *et al.* to analyze element distribution in boron-ZSM-5. Later on, in a first *in situ* study, the template decomposition in ZSM-5 crystals focused on the template decomposition process in ZSM-5 crystals [89]. Three steps were distinguished: (i) increasing the mobility of the tetrapropylammonium cations; (ii) partial oligomerization of the template fragments (both up to 613 K), and (iii) above 633 K, decomposition of the template via Hoffman elimination, with the formation of either dipropylamine or dipropylammonium ions, depending upon the aluminum concentration in the zeolite.

Schüth used polarized IR microscopy to reveal the molecular orientation of adsorbates in silicalite crystals [90]. It was shown that the IR spectrum of the adsorbed p-xylene strongly depends on the light polarization. Analysis of the IR bands and their polarization dependence indicated that the on-adsorbed *p*-xylene molecules align with the straight ZSM-5 channels. The method also elegantly revealed areas with different pore orientation. However, these findings were not interpreted, as the intergrowth architecture of ZSM-5 crystals has not been unambiguously known. Acidity of zeolites can also be probed in a spatially resolved manner using pyridine as a probe molecule [91]. Both OH and pyridine ring vibration spectral regions showed that acidic sites are more abundant in the center of the ZSM-5 crystal. Interestingly, no correlation of the acidic with Al concentration gradient measured with X-ray microprobe analysis has been observed (Figure 8.12).

IR microspectroscopy has proven to be an excellent tool to follow the diffusion of organic molecules in zeolites, especially when combined with interference microscopy. In a first study of this kind, an approach for the measurement of adsorbate concentrations in zeolites, based on IR [92, 93], was adopted. With an IR microscope, diffusion coefficients of toluene were determined in individual ZSM-5 crystals varying in their degree of intergrowth [94]. Severely intergrown crystals showed diffusion coefficients three orders of magnitude lower then perfect single crystals. It was proposed that this significant effect is due to different channel orientation in single and twinned crystals.

Chmelik *et al.* reported on the effect of surface modification of MFI crystals [95]. In an attempt to disentangle the effect of surface and bulk diffusion barriers, the crystals were treated with methylated silanes, which block the pores and effectively increase the surface barrier. The rates of diffusion into and out of the crystals were

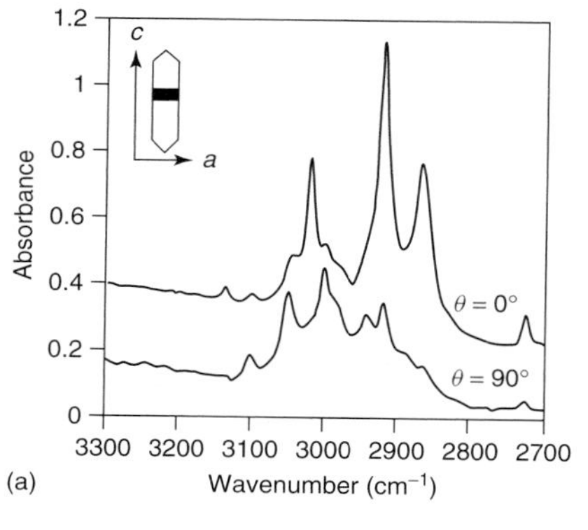


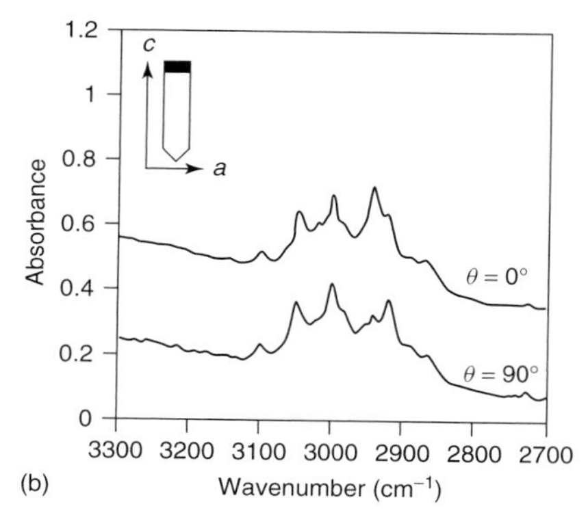


**Figure 8.12** Spatially resolved polarized IR spectra of a *p*-xylene loaded silicalite I crystal. $\theta = 0^\circ$ corresponds to polarization perpendicular to the crystal's long axis. Crystal orientated with (010) parallel to beam axis. (a) Middle of crystal. (b) End of crystal. (Copyright American Chemical Society 1992; Reproduced with permission from [89].)

shown to be correlated well with the strength of the surface resistance, with the untreated samples exhibiting the fastest uptake/release rates and crystals exposed to tripropylchlorosilane revealing very slow or no uptake, which indicates a nearly complete pore blockage. In a recent report, IR microspectroscopy was applied to study alkane diffusion and adsorption in metal-organic frameworks (MOFs) [96].

It should be noted that the most prohibiting limitation of the conventional IR microscope is its spatial resolution, which cannot be improved beyond ~20 μm with the conventional light source. One of the ways to overcome this problem is to use synchrotron light with brightness 100–1000 times higher than that of a globar light source. The improved characteristics of a synchrotron-based setup allow bringing the resolution to diffraction-limited values. Using this method, the first *in situ* study of the zeolite-catalyzed styrene oligomerization was carried out, where carbocationic reaction intermediates were identified on the basis of the spatially resolved IR spectra [97]. Some typical IR spectra and the related 2D IR map during the styrene oligomerization reaction are illustrated in Figure 8.13. Moreover, using inherently polarized synchrotron light, the orientation of the product molecules in the zeolite channels was determined. More specifically, it was found that the carbocationic fluorostyrene dimer was located in the straight channels of the ZSM-5 crystal.

## 8.6
## Concluding Remarks and Look into the Future

It is now commonly accepted that *in situ* characterization studies of a catalyst material at work are crucial if one is to comprehend the key catalytic steps in the reaction mechanism [96]. If these steps are well understood, critical improvements

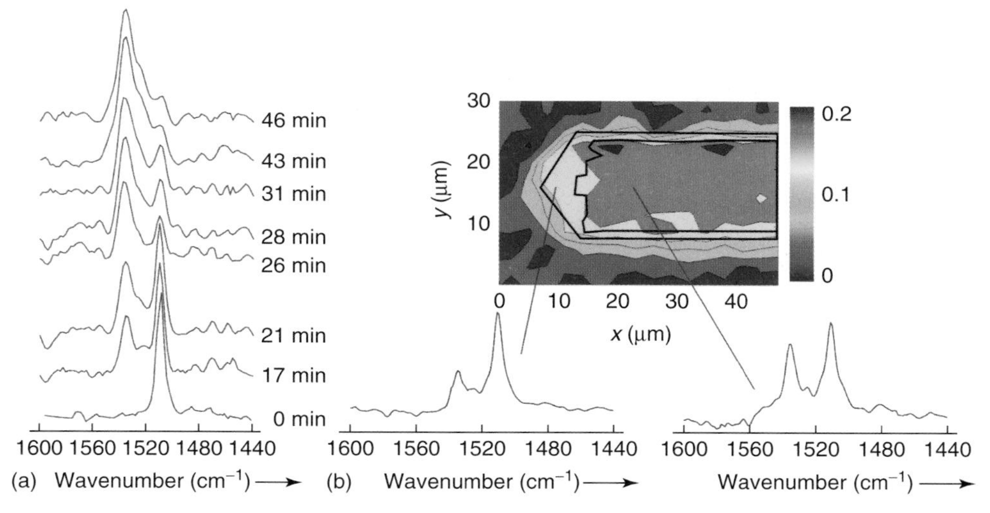


**Figure 8.13** (a) Spatially resolved IR spectra of an individual H-ZSM-5 crystal (5 μm × 5 μm area) taken *in situ* during the 4-fluorostyrene oligomerization reaction. A band at 1534 cm$^{-1}$ (marked by arrow) is due to the carbocationic fluorostyrene dimer. (b) Intensity of the IR band at 1534 cm$^{-1}$ mapped over the crystal after reaction and IR spectra taken from the edge and the body of the crystal, demonstrating differences in the intensity ratio of the bands. (Copyright Wiley-VCH 2008; Reproduced with permission from [96].)

can be made to the catalytic process in order to make it more efficient. Continued improvements in the spectroscopic instrumentation performance (i.e., better time/spatial resolution brought about by more powerful light sources and better detection systems), the embracing of new – often complementary – spectroscopic techniques, and the development of new spectroscopic-reaction setups means that there are more opportunities than ever to obtain fundamental insight into catalytic processes within molecular sieves, making these truly exciting times for the catalyst scientist. Vibrational spectroscopy takes a special as well as central position in the development of structure–activity relationships for heterogeneous catalysis as it is one of the few characterization methods, which provide detailed molecular insight in the adsorbed species on the catalyst surface, including reaction intermediates as well as deactivation products. As shown in this chapter, this has been especially fruitful in characterizing acid and redox active centers in molecular sieves. It would be highly welcomed if this vibrational information could be confined to the nanoscale of molecular sieves in order to complement the detailed knowledge available from electron microscopy methods. A way to do so is to combine vibrational spectroscopy measurements with near-field optical methods. This is the field of scanning near field optical microscopy (SNOM), including tip-enhanced Raman spectroscopy and scanning near field infrared microscopy (SNIM). Although these methods are still in their infancy and appropriate *in situ* cells are not yet developed, mainly due to the instability of the reflection tip, it would offer the possibility to relate spatial heterogeneities in catalytic solids to specific reactivity and deactivation patterns. With this knowledge in mind, advanced nanoscale structuring of catalytic solids could become within reach.

## Acknowledgment

E.S. and B.M.W. acknowledge the Netherlands Organization for Scientific Research (NWO-CW) for financial support (Veni, Vici and Top grants).

## References

1. Banares, M.A. and Wachs, I.E. (2002) *J. Raman Spectrosc.*, **33**, 359.
2. Weckhuysen, B.M. (2003) *Phys. Chem. Chem. Phys.*, **5**, 4351.
3. Eischens, R.P., Plisken, W.A., and Francis, S.A. (1954) *J. Chem. Phys.*, **24**, 1786.
4. Eischens, R.P. and Pliskin, W.A. (1958) in *Advances in Catalysis and Related Subjects*, Vol. 10 (eds D.D. Eley, W.G. Frankenburg, and V.I. Komarewsky), Academic Press, New York, p. 2.
5. Paukshtis, E.A. and Yurchenko, E.N. (1983) *Russ. Chem. Rev.*, **52**, 242.
6. Lercher, J.A., Gründling, C., and Eder-Mirth, G. (1996) *Catal. Today*, **27**, 353.
7. Knozinger, H. and Huber, S. (1998) *J. Chem. Soc., Faraday Trans.*, **94**, 2047.
8. Gil, B., Zones, S.I., Hwang, S.J., Bejblova, M., and Čejka, J. (2008) *J. Phys. Chem. C*, **112**, 2997.
9. Hoffmann, P. and Lobo, J.A. (2007) *Microporous Mesoporous Mater.*, **106**, 122.
10. Holm, M.S., Svelle, S., Joensen, F., Beato, P., Christensen, C.H., Bordiga, S., and Bjorgen, M. (2009) *Appl. Catal. A: Gen.*, **356**, 23.
11. Makarova, M.A. and Dwyer, J. (1993) *J. Phys. Chem.*, **97**, 6337.
12. Zecchina, A., Spoto, G., and Bordiga, S. (2005) *Phys. Chem. Chem. Phys.*, **7**, 1627.
13. Bevilacqua, M., Alejandre, A.G., Resini, C., Casagrande, M., Ramirez, J., and Busca, G. (2002) *Phys. Chem. Chem. Phys.*, **4**, 4575.
14. Zholobenko, V.L., Makarova, M.A., and Dwyer, J. (1993) *J. Phys. Chem.*, **97**, 5962.
15. Datka, J., Sulikowski, B., and Gil, B. (1996) *J. Phys. Chem.*, **100**, 11242.
16. Covarrubias, C., Quijada, R., and Rojas, R. (2009) *Microporous Mesoporous Mater.*, **117**, 118.
17. Zholobenko, V.L., Lukyanov, D.B., Dwyer, J., and Smith, W.J. (1998) *J. Phys. Chem. B*, **102**, 2715.
18. del Campo, A.E.S., Gayubo, A.G., Aguayo, A.T., Tarrio, A., and Bilbao, J. (1998) *Ind. Eng. Chem. Res.*, **37**, 2336.
19. Martins, G.A.V., Berlier, G., Coluccia, S., Pastore, H.O., Superti, G.B., Gatti, G., and Marchese, L. (2007) *J. Phys. Chem. C*, **111**, 330.
20. Bordiga, S., Regli, L., Cocina, D., Lamberti, C., Bjorgen, M., and Lillerud, K.P. (2005) *J. Phys. Chem. B*, **109**, 2779.
21. Parry, E.R., (1963) *J. Catal.*, **2**, 371.
22. Karge, H.G. and Geidel, E. (2004) in *Molecular Sieves: Characterization I* (eds H.G. Karge and J. Weitkamp), Springer, Berlin.
23. Lavalley, J.C. (1996) *Catal. Today*, **27**, 377.
24. Barzetti, T., Selli, E., Moscotti, D., and Forni, L. (1996) *J. Chem. Soc., Faraday Trans.*, **92**, 1401.
25. Maache, M., Janin, A., Lavalley, J.C., and Benazzi, E. (1995) *Zeolites*, **15**, 507.
26. Datka, J., Gil, B., and Kubacka, A. (1997) *Zeolites*, **18**, 245.
27. Buzzoni, R., Bordiga, S., Ricchiardi, G., Lamberti, C., Zecchina, A., and Bellussi, G. (1996) *Langmuir*, **12**, 930.

28. Oliviero, L., Vimont, A., Lavalley, J.C., Sarria, F.R., Gaillard, M., and Mauge, F. (2005) *Phys. Chem. Chem. Phys.*, **7**, 1861.
29. Nesterenko, N., Thibault-Starzyk, F., Montouillout, V., Yuschenko, V., Fernandez, C., Gilson, J., Fajula, F., and Ivanova, I. (2004) *Microporous Mesoporous Mater.*, **71**, 157.
30. Thibault-Starzyk, F., Stan, I., Abelló, S., Bonilla, A., Thomas, K., Fernandez, C., Gilson, J., and Pérez-Ramírez, J. (2009) *J. Catal.*, **264**, 11.
31. Ordomsky, V.V., Murzin, V.Y., Monakhova, Y.V., Zubavichus, Y.V., Knyazeva, E.E., Nesterenko, N.S., and Ivanova, I.I. (2007) *Microporous Mesoporous Mater.*, **105**, 101.
32. Oumi, Y., Takahashi, J., Takeshima, K., Jon, H., and Sano, T. (2007) *J. Porous Mater.*, **14**, 19.
33. Engelhardt, G., Fahlke, B., Magi, M., and Lippmaa, E. (1983) *Zeolites*, **3**, 292.
34. Engelhardt, G., Fahlke, B., Mägi, M., and Lippmaa, E. (1985) *Zeolites*, **5**, 49.
35. Sankar, G., Okubo, T., Fan, W., and Meneau, F. (2007) *Faraday Discuss.*, **136**, 157.
36. Vistad, O.B., Akporiaye, D.E., and Lillerud, K.P. (2001) *J. Phys. Chem. B*, **105**, 12437.
37. Vistad, O.B., Akporiaye, D.E., Taulelle, F., and Lillerud, K.P. (2003) *Chem. Mater.*, **15**, 1639.
38. Grandjean, D., Beale, A.M., Petukhov, A.V., and Weckhuysen, B.M. (2005) *J. Am. Chem. Soc.*, **127**, 14454.
39. O'Brien, M., Beale, A., Catlow, C., and Weckhuysen, B.M. (2006) *J. Am. Chem. Soc.*, **128**, 11744.
40. Fan, F., Feng, Z., Li, G., Sun, K., Ying, P., and Li, C. (2008) *Chem. Eur. J.*, **14**, 5125.
41. Tompsett, G., Panzarella, B., Conner, W., Yngvesson, K., Lu, F., Suib, S., Jones, K., and Bennett, S. (2006) *Rev. Sci. Instrum.*, **77**, 124101.
42. Szanyi, J. and Paffett, M.T. (1996) *J. Catal.*, **164**, 232.
43. Dedecek, J., Sobalik, Z., Tvaruzkova, Z., Kaucky, D., and Wichterlova, B. (1995) *J. Phys. Chem.*, **99**, 16327.
44. Wichterlova, B., Dedecek, J., and Vondrova, A. (1995) *J. Phys. Chem.*, **99**, 1065.
45. Pietrzyk, P., Gil, B., and Sojka, Z. (2007) *Catal. Today*, **126**, 103.
46. Kucherov, A.V., Gerlock, T.L., Jen, H.W., and Shelef, M. (1995) *Zeolites*, **15**, 9.
47. Park, S.K., Kurshev, V., Luan, Z.H., Lee, C.W., and Kevan, L. (2000) *Microporous Mesoporous Mater.*, **38**, 255.
48. Mathisen, K., Nicholson, D.G., Beale, A.M., Sanchez-Sanchez, M., Sankar, G., Bras, W., and Nikitenko, S. (2007) *J. Phys. Chem. C*, **111**, 3130.
49. Ganemi, B., Bjornbom, E., and Paul, J. (1998) *Appl. Catal. B: Environ.*, **17**, 293.
50. Goodman, B.R., Schneider, W.F., Hass, K.C., and Adams, J.B. (1998) *Catal. Lett.*, **56**, 183.
51. Groothaert, M.H., van Bokhoven, J.A., Battiston, A.A., Weckhuysen, B.M., and Schoonheydt, R.A. (2003) *J. Am. Chem. Soc.*, **125**, 7629.
52. Yashnik, S., Ismagilov, Z., and Anufrienko, V. (2005) *Catal. Today*, **110**, 310.
53. Chen, H. and Sachtler, W.M.H. (1998) *Catal. Today*, **42**, 73.
54. Kucherov, A., Montreuil, C., Kucherova, T., and Shelef, M. (1998) *Catal. Lett.*, **56**, 173.
55. Schwidder, M., Grunert, W., Bentrup, U., and Bruckner, A. (2006) *J. Catal.*, **239**, 173.
56. Lobree, L.J., Hwang, I.C., Reimer, J.A., and Bell, A.T. (1999) *Catal. Lett.*, **63**, 233.
57. Pirngruber, G.D. and Pieterse, J.A.Z. (2006) *J. Catal.*, **237**, 237.
58. Gao, Z.X., Kim, H.S., Sun, Q., Stair, P.C., and Sachtler, W.M.H. (2001) *J. Phys. Chem. B*, **105**, 6186.
59. Battiston, A.A., Bitter, J.H., and Koningsberger, D.C. (2000) *Catal. Lett.*, **66**, 75.
60. Battiston, A.A., Bitter, J.H., and Koningsberger, D.C. (2003) *J. Catal.*, **218**, 163.
61. Pirngruber, G.D., Roy, P.K., and Weiher, N. (2004) *J. Phys. Chem. B*, **108**, 13746.
62. Dubkov, K.A., Ovanesyan, N.S., Shteinman, A.A., Starokon, E.V., and Panov, G.I. (2002) *J. Catal.*, **207**, 341.

63. Sun, Q., Gao, Z.X., Wen, B., and Sachtler, W.M.H. (2002) *Catal. Lett.*, **78**, 1.
64. Perez-Ramirez, J., Kumar, M.S., and Bruckner, A. (2004) *J. Catal.*, **223**, 13.
65. Schwidder, M., Kumar, M.S., Bentrup, U., Perez-Ramirez, J., Brueckner, A., and Gruenert, W. (2008) *Microporous Mesoporous Mater.*, **111**, 124.
66. Li, Y.J. and Armor, J.N. (1993) *Appl. Catal. B: Environ.*, **2**, 239.
67. Brosius, R. and Martens, J.A. (2004) *Top. Catal*, **28**, 119.
68. Mihaylov, M., Hadjiivanov, K., and Panayotov, D. (2004) *Appl. Catal. B: Environ.*, **51**, 33.
69. Forester, T. and Howe, R. (1987) *J. Am. Chem. Soc.*, **109**, 5076.
70. Campbell, S.M., Jiang, X.Z., and Howe, R.F. (1999) *Microporous Mesoporous Mater.*, **29**, 91.
71. Anderson, M. and Klinowski, J. (1989) *Nature*, **339**, 200.
72. Munson, E.J., Kheir, A.A., Lazo, N.D., and Haw, J.F. (1992) *J. Phys. Chem.*, **96**, 7740.
73. Hunger, M. and Horvath, T. (1996) *J. Am. Chem. Soc.*, **118**, 12302.
74. Hunger, M. and Horvath, T. (1997) *Catal. Lett.*, **49**, 95.
75. Dahl, I.M. and Kolboe, S. (1994) *J. Catal.*, **149**, 458.
76. Haw, J.F., Nicholas, J.B., Song, W.G., Deng, F., Wang, Z.K., Xu, T., and Heneghan, C.S. (2000) *J. Am. Chem. Soc.*, **122**, 4763.
77. Song, W.G., Fu, H., and Haw, J.F. (2001) *J. Am. Chem. Soc.*, **123**, 4749.
78. Xu, T. and White, J.L. (2004) US Patent 6,734,330.
79. Xu, T. and White, J.L. (2004) US Patent 6,743,747.
80. Song, W.G., Haw, J.F., Nicholas, J.B., and Heneghan, C.S. (2000) *J. Am. Chem. Soc.*, **122**, 10726.
81. Song, W.G., Fu, H., and Haw, J.F. (2001) *J. Phys. Chem. B*, **105**, 12839.
82. Wang, W., Buchholz, A., Seiler, M., and Hunger, M. (2003) *J. Am. Chem. Soc.*, **125**, 15260.
83. Seiler, M., Wang, W., Buchholz, A., and Hunger, M. (2003) *Catal. Lett.*, **88**, 187.
84. Hunger, M. and Wang, W. (2004) *Chem. Commun.*, 584.
85. Jiang, Y., Huang, J., Marthala, V.R.R., Ooi, Y.S., Weitkamp, J., and Hunger, M. (2007) *Microporous Mesoporous Mater.*, **105**, 132.
86. Mores, D., Stavitski, E., Kox, M.H.F., Kornatowski, J., Olsbye, U., and Weckhuysen, B.M. (2008) *Chem. Eur. J.*, **14**, 11320.
87. Park, J.W., Kim, S.J., Seo, M., Kim, S.Y., Sugi, Y., and Seo, G. (2008) *Appl. Catal. A: General*, **349**, 76.
88. Park, J.W. and Seo, G. (2009) *Appl. Catal. A: Gen.*, **356**, 180.
89. Nowotny, M., Lercher, J., and Kessler, H. (1991) *Zeolites*, **11**, 454.
90. Schuth, F. (1992) *J. Phys. Chem.*, **96**, 7493.
91. Schuth, F. and Althoff, R. (1993) *J. Catal.*, **143**, 388.
92. Niessen, W. and Karge, H.G. (1993) *Microporous Mesoporous Mater.*, **1**, 1.
93. Karge, H.G. and Niessen, W. (1991) *Catal. Today*, **8**, 451.
94. Muller, G., Narbeshuber, T., Mirth, G., and Lercher, J.A. (1994) *J. Phys. Chem.*, **98**, 7436.
95. Chmelik, C., Varmla, A., Heinke, L., Shah, D.B., Karger, J., Kremer, F., Wilczok, U., and Schmidt, W. (2007) *Chem. Mater.*, **19**, 6012.
96. Chmelik, C., Kaerger, J., Wiebcke, M., Caro, J., van Baten, J.M., and Krishna, R. (2009) *Microporous Mesoporous Mater.*, **117**, 22.
97. Stavritski, E., Kox, M.H.F., Swart, I., de Groot, F.M.F., and Weckhuysen, B.M. (2008) *Angew. Chem. Int. Ed.*, **47**, 3543.

# 9
# Textural Characterization of Mesoporous Zeolites

*Lei Zhang, Adri N.C. van Laak, Petra E. de Jongh, and Krijn P. de Jong*

## 9.1
## Introduction

Zeolites and related zeolite-analog materials have found major industrial applications as catalysts and adsorbents, especially in refinery and petrochemical processes where zeolite catalysts exhibit unique shape selectivity endowed by their crystalline yet highly porous structures with well-defined channels of molecular dimensions [1–5]. However, as one side effect of their microporous structure, zeolite catalysts often suffer from restricted diffusion of guest species [6, 7]. Mass transport to and from the active sites inside the micropores (known as *configurational diffusion*) is much slower than that of molecular and Knudsen diffusion, which has led to lower catalyst utilization, and sometimes fast deactivation due to coke formation. Different approaches have been proposed to alleviate the diffusion limitation and enhance the accessibility of the internal sites [8–21]. One strategy is to synthesize novel zeolitic structures with a larger pore size. Various large-pore zeolites and zeolite analogs have been obtained, for example, VPI-5 [22], UTD-1 [23], SSZ-53 and SSZ-59 [24], ITQ-15 [25], ITQ-21 [26], ITQ-33 [27], and ITQ-37 [28]. However, despite the fact that a considerable amount of knowledge has been gained on the formation mechanisms of zeolites and numerous types of theoretical structures have been proposed for predicative synthesis, synthesis by design of novel zeolite structure is still a challenging topic, and most new structures are discovered by trial-and-error processes.

Another strategy is to decrease the effective intracrystalline diffusion path length, which has given rise to the so-called hierarchical zeolites, that is, zeolites featuring multiscale porosity other than microporosity [12, 15]. This can be achieved by generating intracrystalline mesopores or, alternatively, by downsizing the zeolite crystals, thereby increasing the intercrystalline porosity. Among the typical examples are the well-known ultrastable Y (USY) family materials featuring mesopores [29, 30], which can be obtained by dealumination via steam treatments and are being widely used in the petrochemical industry. Over the past decades, a wealth of

*Zeolites and Catalysis, Synthesis, Reactions and Applications. Vol. 1.*
Edited by Jiří Čejka, Avelino Corma, and Stacey Zones

ISBN: 978-3-527-32514-6

methods have been practised and proven to be effective for generating additional meso- and macropores in zeolite materials. Intercrystalline pores can be tuned by assembling nanosized zeolites or by depositing them onto a porous support, for example, mesoporous silicas [31–33]. Methods for obtaining intracrystalline pores can be generally categorized into two types, that is, templating methods [10] and postsynthesis treatments [11, 21]. For the templating method, various materials have been employed as space fillers during zeolite synthesis, including "soft templates," for example, surfactant [34–36], polymer [37, 38], and starch [39], and "hard templates," for example, carbon materials (carbon nanofiber [40], carbon black [41], carbon nanotube [42], carbon aerogel [43–45]), and nanosized $CaCO_3$ [46]. Postsynthesis treatment methods normally involve demetallization, that is, selective leaching of the framework-constituting species, for example, dealumination by steaming and/or acid leaching [21, 47–59], desilication by base leaching [11, 60–65], and detitanation with hydroperoxide [66–69], thereby leaving voids inside the zeolite particles. Depending on the framework structure and composition of the zeolite, different methods show different features in generating additional porous structure with regard to pore size, pore shape, pore volume, tortuosity, and connectivity. These methods may be complementary to one another. Recently, a synthesis approach combining two methods (template + base leaching) has also emerged [70]. Owing to the presence of the hierarchical porous structures, improved diffusion and catalysis properties have generally been observed [59, 71–76].

Despite the rapid enrichment of the toolkit for generating additional porosity, detailed and explicit characterization of these porous structures has received less attention. Although a wealth of techniques are now available for assessment of the porous structure [77], with regards to zeolites gas physisorption (nitrogen and argon) and transmission electron microscopy (TEM) are still the prevailing methods, which provide information on pore size, pore volume, and limited information on pore shape and connectivity. Hierarchy does not mean simple combination of micropores and meso- or macropores, but rather how the pores at different levels are organized and interconnected in three dimensions [12, 15], as the latter often determines the performance of the materials in practical uses [78]. Other potential techniques, which have been specially developed for evaluating porous materials, especially mesoporous materials, have been largely ignored for the characterization of mesoporous zeolite materials. It is the intention of the current chapter to highlight these techniques as well as some novel techniques and bring them to the front stage of characterization of hierarchical zeolites. As shown below, unprecedented information about the porous structure of hierarchical zeolites has been obtained by these characterization methods, which complement the results from the conventional gas physisorption and TEM analyses. We first review the various approaches for generating additional pores, then emphasize on the different techniques that have been used for characterization of these materials. The characterization methods are sectioned according to different techniques involved, although some of them actually rely on similar theories.

## 9.2 Methods for Generating Meso- and Macropores in Zeolites

Recent years have seen the rapidly growing diversity of the protocols for generating additional pores in zeolite materials, as well as the wide applicability of these methods to various zeolite structures. A number of excellent reviews concerning the synthesis approaches have been published [8–21]. In this section, a short summary of the synthesis methods of hierarchical zeolites featuring meso- and/or macropores is given.

### 9.2.1 Postsynthesis Modification

Postsynthesis modification generally involves selective leaching of the framework species by treating the zeolites with steam, acid, base, or complexing agents, thereby leaving additional voids. This can be achieved by dealumination, desilication, or selective leaching of other constituting elements of the zeolite framework.

#### 9.2.1.1 Dealumination

Among the various postsynthesis treatment methods, dealumination by steaming and/or acid leaching is probably the most renowned one. Steaming is usually performed at a temperature above 500 °C by contacting the ammonium or proton form of the zeolites with steam. During the treatment, Al–O–Si bonds undergo hydrolysis and aluminum is expelled from the framework, causing vacancy defects (silanol nest) and partial amorphization of the framework. Some less stable and mobile silicon species migrate and condense with silanols at other sites. Such a healing process results in the filling of some vacancies and large voids originating from expelled aluminum and mobile silicon species [79, 80], as depicted in Figure 9.1. In regions of high defect concentrations, spherical mesopores can coalescence into cylindrical pores. As amorphous debris is deposited on the mesopore surface or on the external surface of the treated zeolite crystals, which causes partial blockage of the micropores, subsequent mild acid washing is often necessary to remove these species. Diluted mineral acids (nitric acid and hydrochloric acid) or organic acid (for example, oxalic acid) can be used for this purpose. According to such a mechanism, the formation of mesopores is highly dependent on the Al concentration and the stability of Al sites against hydrolysis. Therefore, most work on steaming has been performed on zeolites with low pristine Si/Al ratios, for example, zeolite Y [29, 30, 48, 51, 80–88] and mordenite [82, 89–91]. Other examples include mazzite [92, 93], omega [50], ferrierite [94], and ZSM-5 [54, 95].

Dealumination can also be performed by only acid leaching with concentrated acid solutions. According to the same reasoning as indicated above, the effectiveness of this method also depends on the framework types and compositions of the zeolites used. In the case of mordenite, which has pseudo 1D channels, this method has been found to be especially effective and has been extensively investigated [59, 91, 96–99]. A correlation has been found between the extent of mesopore

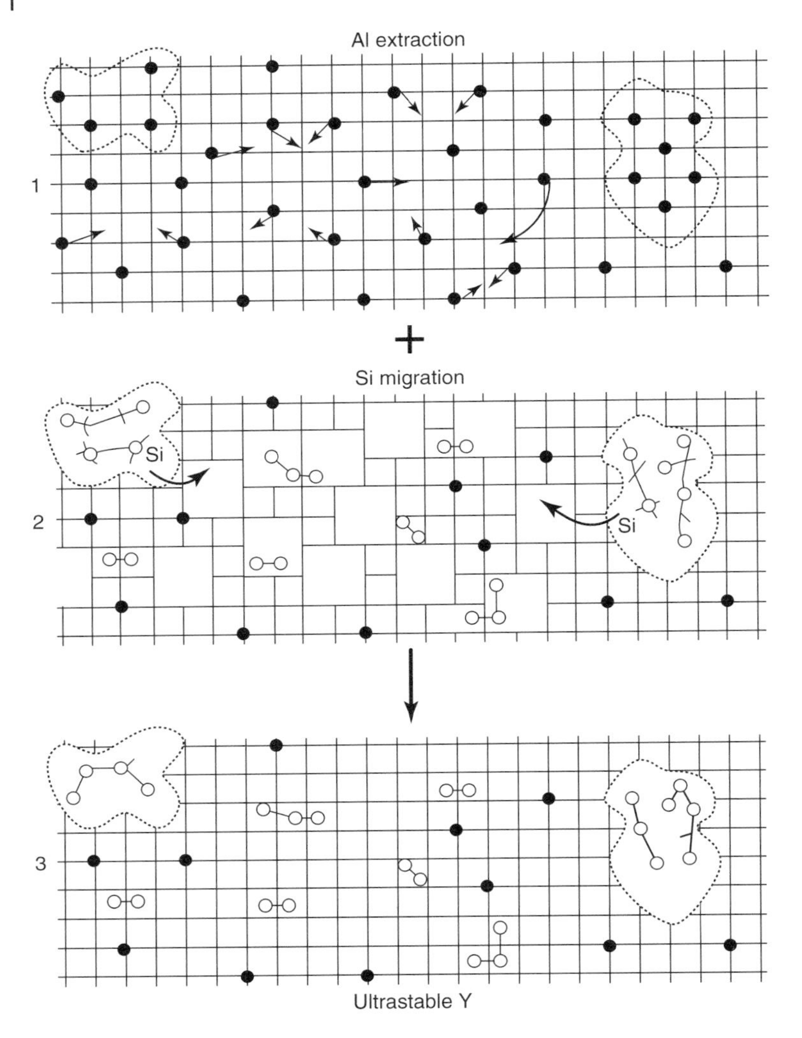


**Figure 9.1** Schematic presentation of the formation of mesopores. The grid represents the zeolite framework, the black dots are framework aluminum atoms, the open circles are aluminum atoms extracted from the framework, and the dotted lines indicate the mesopores. (Adapted from [79].)

generation and the so-called symmetry index, that is, the ratio of the XRD (X-ray diffraction) peak intensities ([111] + [241]/[350]) [100]. The symmetry index has been proposed as an indicator of the extent of stacking faults inside the framework. Dealumination is assumed to take place preferentially at these stacking faults. The mesoporous mordenites thus obtained exhibit a 3D structure and have been applied in industrial processes for the transalkylation of diisopropyl benzene and benzene to yield cumene and (hydro)isomerization of alkanes.

Dealumination leads, apart from mesopore generation, to an increase in the framework Si/Al ratios, thereby enhancing hydrophobicity and (hydro)thermal stability of the zeolites. One of the most well-known examples of mesoporous zeolites obtained by dealumination is the family of USY and related materials, which are invaluable industrial cracking catalysts. Meanwhile, the increased Si/Al ratios also lead to a decrease in the acid site density and consequently enhance the acidity of the acid sites, both exerting strong effects on the catalysis performance. Besides acids, other chemicals can also be used to withdraw aluminum from the framework; examples are $SiCl_4$ [83, 101], $(NH_4)_2SiF_6$ [83, 102], and EDTA (ethylenediaminetetraacetic acid) [57, 103]. However, caution should be taken to avoid severe framework collapse when the extraction of aluminum by EDTA or $(NH_4)_2SiF_6$ is faster than the migration of silicon species during the treatment [104].

#### 9.2.1.2 **Desilication**

As with aluminum, framework silicon can also be selectively extracted from the framework to generate mesopores. This has mostly been done by treating a zeolite with a base, for example, NaOH, KOH, LiOH, $NH_4OH$, and $Na_2CO_3$, or a specific acid like HF. While base treatment is usually used for the removal of amorphous gel impurities from zeolite crystals, its potential for generating mesopores has been long ignored. Dessau reported in 1992 that hollow crystals were obtained by refluxing large ZSM-5 crystals in $Na_2CO_3$ aqueous solution [60]. It was revealed that highly selective dissolution of the interior of the crystals had occurred, while the exterior surface remained relatively intact. This result hinted toward the inhibiting role of aluminum during the base treatment and gave direct evidence of aluminum zoning in large-crystal ZSM-5 synthesized in the presence of quaternary directing agents. Subotić *et al.* further investigated the role of aluminum during the base treatment of ZSM-5 [105, 106]. However, the evolution of the porous structure was not investigated in detail. Ogura *et al.* reported the first explicit evidence of mesopore formation in ZSM-5 crystals by NaOH treatment [62]. Groen *et al.* reported the detailed investigation of base treatment conditions for optimizing the mesopore formation in a series of papers [11, 107–122]. They found that for ZSM-5 crystals there appears to be an optimal window of Si/Al ratio in the parent zeolite, that is, $SiO_2/Al_2O_3 = 50–100$ (molar ratio) to achieve optimal mesoporosity with high mesopore surface areas up to 235 $m^2\ g^{-1}$, while still preserving the intrinsic crystalline and acidic properties (Figure 9.2) [119]. While ZSM-5 is still the most intensively investigated zeolite [39, 63–65, 123–128], desilication has recently also been applied to mordenite [74, 112], beta [113, 129], and ZSM-12 [75]. More recently, hollow nanoboxes of ZSM-5 and TS-1 were obtained by treating the zeolites with tetrapropylammonium hydroxide solution. A dissolution–recrystallization mechanism was proposed for the formation of such unique nanoporous structures [130, 131].

Similar to dealumination, desilication inevitably modulates the framework Si/Al ratios; however, in this case, it resulted in decreased Si/Al ratios. Moreover, some extraframework aluminum species are often observed after the base treatment [132]. Therefore, an additional acid treatment or ion-exchange step is needed to remove these species for opening the micropores and mesopores.

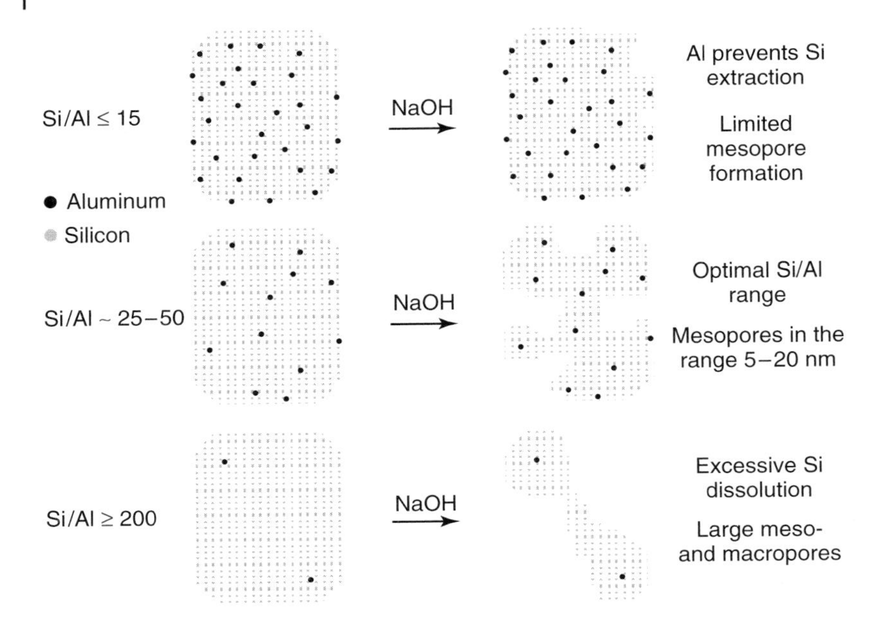


**Figure 9.2** Simplified schematic representation of the influence of the Al content on the mechanism of pore formation during the desilication treatment of MFI zeolites. (Reprinted with permission from [119].)

9.2.1.3 **Detitanation**

For zeolite containing other metals in the framework, mesopores can also be generated by similar selective leaching method. Schmidt *et al.* reported that mesoporous ETS-10 titanosilicate can be obtained by postsynthesis treatment with hydrogen peroxide [67–69]. The materials exhibited increased external surface areas and improved performances in the Beckmann rearrangement of cyclohexanone oxime to $\varepsilon$-caprolactam.

The various postsynthesis treatment methods mentioned above have proved effective for generating mesopores in zeolites. The intracrystalline diffusion path-length is shortened by virtue of the mesopores. Improved diffusion [59, 74, 75] and catalytic properties [39, 67, 112–114, 123–125, 133] have been generally observed. However, it is still difficult to generate the additional pores in a controllable way in terms of pore shape, pore volume, and connectivity. Moreover, it inevitably involves the change of framework composition, and partial amorphization of the crystalline structure. Therefore, in practical applications, it is sometimes difficult to disentangle the contribution of improved textural properties from that of varied framework composition. Another effect that has received less attention is the change of morphology and surface properties of the zeolite crystals by the postsynthesis treatment [62, 97], which can exert strong effects on the adsorption properties.

### 9.2.2
### Templating Method

A more straightforward method for generating mesopores is the templating method, which has been extensively used for the synthesis of mesoporous materials [16, 134, 135]. Different from the postsynthesis treatment method, templates are employed during the zeolite crystallization and selectively removed afterward. Therefore, the framework composition can be predetermined on the basis of the synthesis gel. Various types of template materials have been employed. On the basis of the structure and rigidity, they can be roughly categorized into hard template and soft template [10, 15].

#### 9.2.2.1 Hard Template

Hard templates here refer to materials with relatively rigid structure that are not supposed to deform during the zeolite synthesis. In this respect, carbon materials are ideal candidates for generating mesopores because of their inertness, rigidity, robustness, diversity, and easy removal by combustion [8, 12, 15, 16]. Carbon was initially used for making nanosized zeolites by the confined growth of zeolite crystals between the spaces of carbon particles [136–138]. Afterward, it was revealed that, by controlling the growth of the zeolite crystals, the carbon templates can be entrapped inside the zeolite crystals [41]. Mesoporous crystals with intracrystalline mesopores can be synthesized by the subsequent removal of carbon via combustion (Figure 9.3).

Carbon-templating routes have received wide attention and have developed rapidly during the last few years. Different types of carbon materials, for example, carbon black [40, 139–146], carbon nanofiber [40], carbon nanotube [42, 147], carbon

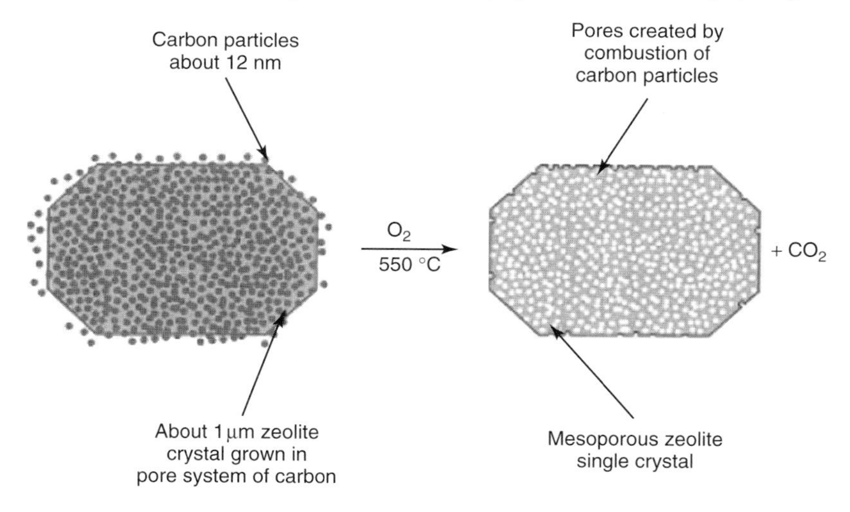


**Figure 9.3** Schematic presentation of the growth of zeolite crystals around carbon particles. (Reprinted with permission from [41].)

aerogel [43–45, 148–150], carbon by sugar decomposition [151, 152], ordered mesoporous carbon [153, 154], and colloid-imprinted carbon (CIC) [155–158] have been used for the synthesis of mesoporous zeolites with various framework types, for example, MFI [40–45, 136–140, 142–144, 148–159], FAU [44, 138, 150], MEL [72, 140, 145, 159, 160], BEA [138, 159], MTW [140, 146, 161], CHA [159], LTA [138], and AFI [159]. It is noted that using ordered nanoporous CIC as the templates, Fan *et al.* have synthesized MFI single crystals with ordered imprinted mesoporosity [158]. Recently, using the same concept, nanosized $CaCO_3$ [46] and resorcinol-formaldehyde aerogels [162, 163] have also been employed for synthesizing mesoporous zeolites. Similarly, zeolites featuring hierarchical structures can been synthesized via macrotemplating using polystyrene beads [164, 165], resin beads [166, 167], polyurethane foam [168], and even biological materials like bacteria [169], wood [170], and leaves [171, 172]. However, in most of the cases, the products are formed by coating of the templates with zeolite synthesis solution, and occur as polycrystalline ensembles of nanosized zeolites. Alternatively, such macrostructure can also be formed by assembling preformed colloidal zeolite precursor around the templates.

#### 9.2.2.2 **Soft Template**

The concept of soft template is adapted from the micelle templating synthesis of mesoporous materials [173, 174]. Ordered mesoporous materials have been synthesized by assembling zeolite "seed," with micelle templates. The "seed" can be synthesized either from the precursor solution [175–191], or by decomposition of zeolites [192–195], the so-called top-down approach. Such materials, albeit mostly do not exhibit a discernible zeolite phase, have shown enhanced thermal and hydrothermal stability and enhanced catalytic activity compared to conventional mesoporous materials with amorphous pore walls [175, 176, 180–187, 189, 192–196]. In some diffusion-limited reactions, their catalytic properties are comparable and even surpass those of the microporous zeolite counterparts [175, 185, 186, 189, 193, 194, 197]. Attempts to generate mesopores using supramolecular micelles during zeolite synthesis met little success and, in most of the cases, ended up with a phase-segregated composite of zeolite crystals and mesoporous materials with amorphous pore walls [198–200]. This stressed the importance of modulating the interplay between the formation of the mesoporous structures and the pore wall crystallization. Recently, this problem has been solved by an elegant approach using novel templates of alkoxysilane-containing surfactant or polymer [34, 35, 38]. The presence of the siloxy group increased the interaction between the templates and the pore wall, and helped in the retention of the mesoporous structure during the pore wall crystallization (Figure 9.4). The resulting zeolites showed a highly crystalline structure and uniform mesoporosity. Moreover, the mesopore size can be controlled in a similar manner as for mesoporous materials, for example, MCM-41, by using surfactant molecules with different chain lengths or polymers with different molecular weights. In this sense, the mesoporous zeolite obtained by this approach is a synergetic product, rather than a combination between zeolites and mesoporous materials. Other templates, for example, starch [39, 201] and

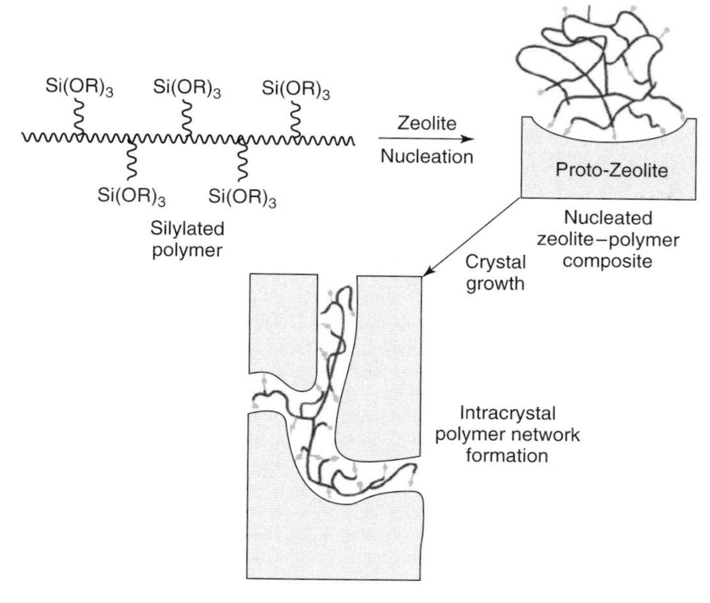


**Figure 9.4** Conceptional approach to the synthesis of a zeolite with intracrystal mesopores using a silylated polymer as the template. (Reprinted with permission from [38].)

polydiallyldimethylammonium chloride [37], have also been used for the synthesis of mesoporous zeolites.

Compared to the postsynthesis methods, the templating approach presents several advantages. First, mesopores can be generated without affecting the framework composition, thus making it possible to investigate separately the effects of textural and framework properties. Secondly, in principle, the pore volume, pore shape, and connectivity can be tuned in a more controllable way by choosing proper templates. However, the effects of these templates upon the zeolite composition and phase purity of the final product should be taken into account, both during the crystallization process as well as the template removal process.

### 9.2.3 Other Methods

Mesoporous zeolites can also be synthesized in the absence of templates. As mentioned before, in addition to the intracrystalline pores, mesopores can also arise from the stacking of nanosized zeolites as intercrystalline pores. However, conventional colloidal or nanosized zeolites suffer from the difficulties of separation. A solution to this is to synthesize assemblies of nanosized zeolites by controlled nucleation and growth of the zeolite crystals [202, 203].

Alternatively, small zeolite particles can also be deposited onto mesoporous support [32, 33], thereby generating mesoporous composite materials. Gagae *et al.* [204]

and Stevens *et al.* [205] have obtained zeolite nanocrystals embedded in a meso-matrix by acidifying zeolite seed or precursor solutions. Recently, a new synthesis approach, which combined the sol–gel transformation and the zeolite crystallization, was reported by Wang and coworkers [206–209]. By steam-assisted crystallization of a previously aged precursor gel, mesoporous composite TUD-C with integrated ZSM-5 nanocrystals was obtained [207]. Tetrapropylammonium hydroxide, which was used as the template for the formation of the microporous structure of zeolites, is believed to act also as a scaffolding agent for the mesopore formation. Later on, using a similar concept, another composite material TUD-M was synthesized by reversing the preparation steps, that is, a first hydrothermal treatment of the precursor gel followed by the sol–gel transformation [208].

Summarizing this section, we can see that numerous methods are available for generating meso- and macroporous zeolite materials. Each method features its own advantages over the others, and they are complimentary to each other. Another important issue concerning the different approaches for obtaining hierarchical zeolites, which is rarely addressed in the open literature, is the potential for large-scale production and the corresponding costs. For traditional processes, such as dealumination by steaming, engineers have created the means to scale-up from laboratory grams to the tons necessary in industrial processes. For desilication, the fast reaction time of 30 minutes could be problematic, resulting in inhomogeneously treated zeolites. The templating approaches have the disadvantage that they bring additional costs with them, not only in raw chemicals as, for instance, for the micelle route but also in the case of carbon tubes; an additional synthesis step for obtaining the template is also needed. Despite the fact that increasing the scale of a process generally leads to lower costs, upscaling and costs of zeolite production are aspects that should not be overlooked.

Nevertheless, in order to arrive at a deeper understanding of the efficiency and working mechanisms of these methods, as well as the impacts of these pores on the zeolite performance in practical use, it is vital to have a clear picture of the textural properties and structures of the hierarchical zeolites. In the following part of this chapter, we elaborate on the different characterization techniques for evaluating the textural porosity, with emphasis on some novel techniques (electron tomography (ET), optical microscopy), as well as some well-known techniques that have received less attention (thermoporometry, mercury porosimetry, etc.).

## 9.3
## Characterization of Textural Properties of Mesoporous Zeolites

### 9.3.1
### Gas Physisorption

Among the various techniques available for porous structure analysis, gas physisorption is still the standard and mostly used technique [210]. This has been due to the well-developed theory, as well as the easy operation and wide availability of the experimental equipment. This technique accurately determines the amount of gas adsorbed on a solid material at a certain temperature and pressure, and yields important information on the porous structure (pore volume, specific surface area, pore size distribution (PSD)) and the pore surface properties. Nitrogen and argon are the mostly used adsorbates. Usually isotherms are measured at liquid nitrogen temperature and pressures varying from vacuum to 1 bar. The presence of pores of different dimensions can be discerned from the shape of the isotherms. For microporous materials, like zeolites, Ar is advantageous over $N_2$ because the presence of a quadrupolar moment in $N_2$ can result in enhanced interaction with the heterogeneous surface of the zeolite framework, which results in difficulties in accessing the pore sizes and pore shapes [211]. As the interpretation of the recorded values usually relies on simplified models, the accuracy of the results is strongly dependent on the validity of the assumptions inherent to the model [210]. For example, in reports on zeolites, the BET (Brunauer–Emmett–Teller) specific surface area is often given [212]. However, since the micropore filling does not fulfill the conditions for multilayer adsorption, which is the basis of BET theory, the reported BET data do not represent a real physical surface area. In particular, for materials with multiscale pores, the interference of mesopores on the multilayer adsorption makes the interpretation more complicated. Nevertheless, for comparative studies, the BET surface area can be used as a value proportional to the volume adsorbed. For detailed description of gas physisorption analysis of zeolites, we refer to [213]. Here, we only discuss some specific phenomena related to zeolites featuring additional porosity.

Figure 9.5 shows typical nitrogen physisorption results of zeolite NaY and the samples obtained after different postsynthesis treatments [214]. The descriptions of the samples and the corresponding textural parameters derived from the isotherms are listed in Table 9.1. For NaY, a type I isotherm was observed, which is typical for chemisorption on nonporous materials, or physisorption on materials with only microporosity. After postsynthesis modification, type IV isotherms appeared, indicating the formation of mesopores. Moreover, different hysteresis loops were observed for samples obtained by different treatments. Hysteresis appearing in the multilayer range of physisorption isotherms is usually associated with capillary condensation in mesopores. Although the factors that affect the adsorption hysteresis loops are still not well understood, the shapes of hysteresis loops have often been identified with specific pore geometries. The steep hysteresis loop in sample HMVUSY (high-meso very ultra stable Y) indicates the presence of close-to-cylindrical mesopores with relatively uniform PSD, while the flat and wide hysteresis loop in XVUSY (eXtra very ultrastable Y) is more complex, and can be attributed to the presence of slit pores or ink-bottle-shaped pores with small openings, or due to the effect of a pore network, which is discussed in more detail below.

For materials with multiscale pores, especially for microporous materials like zeolites, the comparative analysis methods, like $t$ plot [215], $\alpha_s$ plot [213], and $\theta$ plot [216] methods, are generally used for the estimation of pore volume and specific surface area of pores of different dimensions by comparing the isotherms

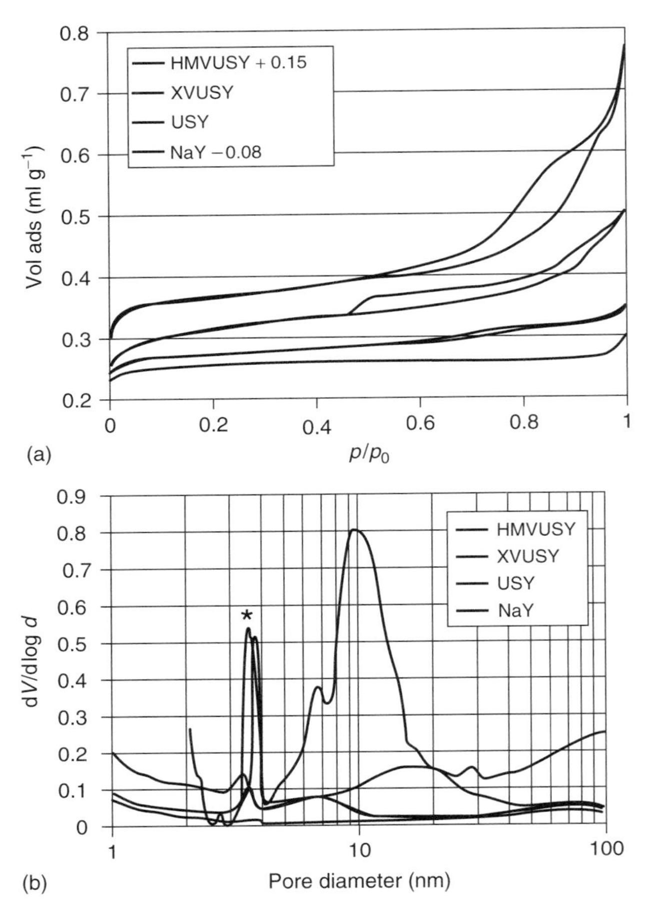


**Figure 9.5** Nitrogen physisorption isotherms and the BJH pore size distribution curves calculated by the desorption branches for several Y zeolites (from top to bottom: HMVUSY, XVUSY, USY, and NaY). For clarity, the isotherm of NaY has been shifted 0.08 $cm^3\ g^{-1}$ downward and the isotherm of the HMVUSY has been shifted upward by 0.15 $cm^3\ g^{-1}$. (Reprinted with permission from [214].)

of samples under investigation with those of reference samples. For the $t$-plot analysis, the multilayer film thickness of adsorbate ($t$-values) is determined on a reference nonporous solid with similar surface properties to the samples under investigation. Alternatively, standard reference plots can be used, for instance, the Harkins and Jura equation for silica and alumina substrates:

$$t\ (\mathrm{nm}) = 0.1\left[\frac{13.99}{0.034 - \log(p/p_0)}\right]^{1/2} \tag{9.1}$$

In a $t$-plot analysis, the volume adsorbed on the sample under investigation at different pressures is plotted against $t$ and the corresponding statistical average

**Table 9.1** Physical properties of NaY and the mesoporous Ys: USY, XVUSY, and HMVUSY[a].

| Sample | Si/Al bulk (at/at) | Si/Al XPS (at/at)[b] | $a_0$ (nm) | % Y[c] | $V_{micro}$ ($cm^3\ g^{-1}$)[d] | $V_{meso}$ ($cm^3\ g^{-1}$)[d] | $S_{ext}$ ($m^2\ g^{-1}$)[e] |
|---|---|---|---|---|---|---|---|
| NaY | 2.6 | 2.8 | 2.469 | 100 | 0.34 | 0.05 | 8 |
| USY | 2.6 | 1.1 | 2.450 | 87 | 0.26 | 0.11 | 63 |
| XVUSY | 39.3 | 71.3 | 2.423 | 72 | 0.28 | 0.25 | 120 |
| HMVUSY | 5.0 | 1.4 | 2.427 | 71 | 0.15 | 0.47 | 146 |

Results from [214].
[a] USY (ultrastable Y) was obtained by steaming treatment, XVUSY (eXtra very ultrastable Y) was obtained by steaming twice and acid leaching, HMVUSY (high-meso very ultrastable Y) was obtained by hydrothermal treatment.
[b] Si/Al ratios determined by XPS measurements.
[c] Relative crystallinity.
[d] $V_{micro}$ and $V_{meso}$: micropore volume and mesopore volume determined by *t*-plot analysis.
[e] Sum of external and mesopore surface area calculated from the *t* plot.

layer thickness is calculated from the standard isotherm obtained with a nonporous reference solid. For a nonporous sample, a straight line through the origin is expected. Deviation in shape of the *t* plot from linearity indicates the presence of pores of a certain dimension. By this means, values like the micropore volume, mesopore volume, macropore volume, and mesopore and external surface areas can be calculated. In a similar manner to that of the *t* plot, the textural properties can also be evaluated from the $\alpha_s$ plot, wherein $\alpha_s$ is defined by the ratios between the volume adsorbed in the sample under investigation and the volume adsorbed in the reference solid at a certain relative pressure [217]. For the assessment of microporosity, the thickness of the multilayer is irrelevant and it has been suggested to replace *t* by the "reduced" adsorption [210]. Nevertheless, both methods generally give consistent results for micropore assessment.

Using the *t*-plot method, the effects of the postsynthesis treatment on the porous structure was investigated for the zeolite Y samples shown in Figure 9.1. From Table 9.1, one can see that upon steaming and acid-leaching treatments, the mesopore volume increases significantly at the expense of their micropore volume. This has been attributed to the blocking of the micropores with extraframework species generated during the hydrothermal treatment and/or the partial amorphization of the framework.

From the isotherms, the PSD curves can be derived. For mesoporous materials, the BJH (Barret–Joyner–Halenda) method [218], which is based on the Kelvin equation for the hemispherical meniscus, is still the most frequently used method. The adsorption process in mesopores often associates with capillary condensation. Preferentially, the desorption branch of the isotherms is used for PSD analysis, as it is closer to the thermodynamic equilibrium [219]. Figure 9.5b shows the PSD curves calculated from the desorption branches of the isotherms using the

BJH method. For NaY, as expected, essentially no mesopores were observed. For HMVUSY, a centered peak was observed around 10 nm, which correlated well with the steep hysteresis observed in the isotherms. Peaks around 3–4 nm in the PSD curves appeared for all the treated samples, which corresponded with the sudden closing of the hysteresis loop in the isotherms at a partial pressure of 0.4–0.5. Such a coincidence does not imply similarity among the porous structures of the treated samples. Peaks in this range have often been observed with mesoporous materials, especially zeolites with mesopores, and have been erroneously attributed to the presence of true pores with sizes of 3–4 nm. It can actually be related to the nature of the adsorptive rather than solely the nature of the adsorbent. This phenomenon is often referred to as the *cavitation effect*, and has been discussed by Neimark and coworkers [220, 221].

As depicted in Figure 9.6, for spherical pores with narrow necks smaller than 4 nm, cavitation of the pores always occurs at a partial pressure corresponding to an apparent pore size of around 4 nm [222]. For nitrogen physisorption, pores with sizes smaller than 4 nm show no hysteresis and exhibit reversible adsorption and desorption isotherms. This is due to the instability of the hemispherical meniscus during desorption in pores with a critical size of ~4 nm, which is caused by the increased chemical potential of the pore walls provoking spontaneous nucleation of a bubble in the pore liquid. In ink-bottle pores as shown in Figure 9.6, the pore emptying during desorption is delayed because the meniscus in the necks is strongly curved and prevents evaporation. The tensile strength of the condensate in the cavities has a maximum limit, which corresponds to a partial pressure of

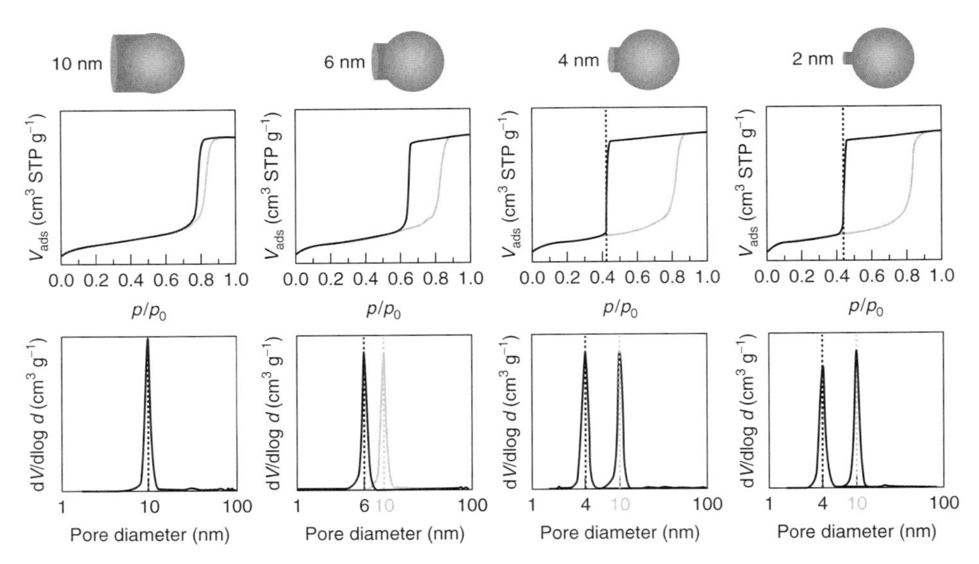


**Figure 9.6** $N_2$ adsorption and desorption isotherms at 77 K (middle) and corresponding PSD (bottom) as derived from BJH model of 10-nm cavities with entrance sizes between 2 and 10 nm. (Reprinted with permission from [222].)

0.4–0.5 for nitrogen at the boiling temperature [223, 224]. Below this limit, the condensed liquid no longer withstands the tension and evaporates, causing the so-called cavitation. Therefore, for hysteresis with steep desorption in the range of 0.4–0.5, it can be concluded that pores are present with cavity diameter larger than ~4 nm, and neck size smaller than ~4 nm. It is therefore necessary to have other evidences to better determine the size and shape of the mesopores. Similar effects can be expected for pore networks wherein larger pores are only accessible through smaller pores [225]. For discrimination between cavitation and capillary desorption, Ar adsorption can provide valuable evidence. In that case, cavitation occurs at a different partial pressure, for example, ~0.35 at 77 K. In summary, although the peaks in the BJH desorption curves may be considered as "artifacts," they actually, in many cases, point to the existence of mesoporous cavities in the zeolite crystals. In fact, this peak can be used to estimate the pore volume associated with these cavities in case independent information from ET or thermoporometry is available (see below).

Recent results have shown that, using the BJH method, pore size is often underestimated [226, 227] because the BJH method does not take into account the influence of solid–fluid interactions on capillary condensation [228]. For mesopores in zeolites, which are, in most of the cases, irregular, advanced calculation methods should be used. Recent developments of new methods, like BJH-BdB (Brockhoff–de Boer) [226], BdB-FHH (Frenkel–Halsey–Hill) [229], KJS (Kruk–Jaroniec–Sayari) [230], and NLDFT (Non-Local Density Functional Theory) [231], may provide more accurate estimation of the porous structure in mesoporous zeolites. Apart from the intracrystalline mesopores discussed above, mesopores can also arise from the intercrystalline spaces between nanosized zeolites. Intercrystalline pores often feature very similar hysteresis loops to those of intracrystalline pores. In that case, it is important to employ other characterization techniques for data interpretation.

### 9.3.2
### Thermoporometry

Another method for assessing mesoporosity is thermoporometry [232]. Similar to gas physisorption techniques, it is also based on the influence of the surface of the solid sample on the phase transitions of the adjoining medium. Thermoporometry relies on the depression of the triple point of a liquid filling a porous material. The triple-point temperature of these systems depends on the solid–liquid and the liquid–gas interfaces. The depression of the triple-point temperature is generally due to the strong curvature of the solid–liquid interface present within the small pores. Thus, the shift in the equilibrium temperature of the liquid–solid transformation allows the determination of the PSD (for example, 2–60 nm with water as the adsorbate) and pore volume [233–241]. Additionally, the shape of the mesopores can be estimated from this technique by comparing the corresponding crystallization and fusion thermograms. A full thermodynamic description of this phenomenon was given by Brun *et al.* [232]. In principle, any liquid adsorbate of known thermodynamic properties can be employed for thermoporometry, given that they exhibit suitable interaction with the walls of the porous materials. This hints at one important advantage of thermoporometry over other techniques: molecules relevant to the practical applications of the porous materials can be used as the adsorbates [234]. However, so far water and benzene are still the most investigated adsorbate molecules, as rather limited information is available for other adsorbates. Although thermoporometry has proven to be a powerful technique for the characterization of mesoporous materials, its application to mesoporous zeolites has been rarely reported. The only example, as far as we know, is reported by Janssen *et al.* on the characterization of mesoporous Y zeolites [242].

A series of zeolite Y samples were analyzed using differential scanning calorimeter (DSC) with water as the probe molecule. In a typical measurement, the sample is first strongly cooled, and then the heat flux during slow heating is recorded as a function of the temperature. The depression of the melting point of water, $\Delta T$(K), corresponds to the pore radius, $r$ (nm), according to the modified Gibbs–Thomson relation derived by Brun *et al.* [232]:

$$r = A - B/\Delta T \tag{9.2}$$

For heating scans, $A$ = 0.68 nm and $B$ = 32.33 nm K are used, while for cooling scans, $A$ = 0.57 nm and $B$ = 64.67 nm K are used. In the above equation, a correlation for a nonfreezing layer of 0.8 nm has already been included. Because the delayed nucleation, which is often encountered during cooling scans, can influence the pore size calculated from the freezing point depression, the melting point depression is more reproducible and most often used [235].

Figure 9.7 shows the DSC curves (heat flow vs temperature) during heating (a) and cooling (b) of water, MCM-41 and different Y zeolites: NaY, USY, XVUSY, and HMVUSY (see Table 9.1 for the description of the sample codes). The textural parameters shown in Table 9.2 are determined by thermoporometry. The amount of water used was about three times the total pore volume determined with nitrogen physisorption. Before the heat flows were measured during heating and cooling, the sample was first cooled to −60 °C in order to freeze all the water in the system. In the heating scan (−60 to −2 °C), no peaks were observed for pure (bulk) water. For the other samples, a peak arose up to about −2 °C due to the melting of the water in the intercrystalline spaces. Compared to NaY, additional peak appeared at about −10 °C for USY, XVUSY, and HMVUSY, and −40 °C for MCM-41, respectively, which indicated the presence of mesopores in these samples. For MCM-41, the pore size determined by the peak onset in the endotherm (−50 °C) according to the Gibbs–Thomson relation is 2.7 nm, and that determined by the peak maximum is 3.0 nm, both of which are in good agreement with the value calculated by nitrogen sorption (2.8 nm). For the mesoporous zeolites, the peaks are considerably wider than that of MCM-41, indicating the wide PSD, which is also consistent with the nitrogen physisorption results (Table 9.2). After subtraction of the background, the pore volumes can be calculated by integrating the peaks using the heating rate and the temperature dependence of both the ice density and the enthalpy of fusion [232]. These results are also compiled in Table 9.2. More importantly, in the exotherms of mesoporous Y samples, two peaks can be observed. The peaks

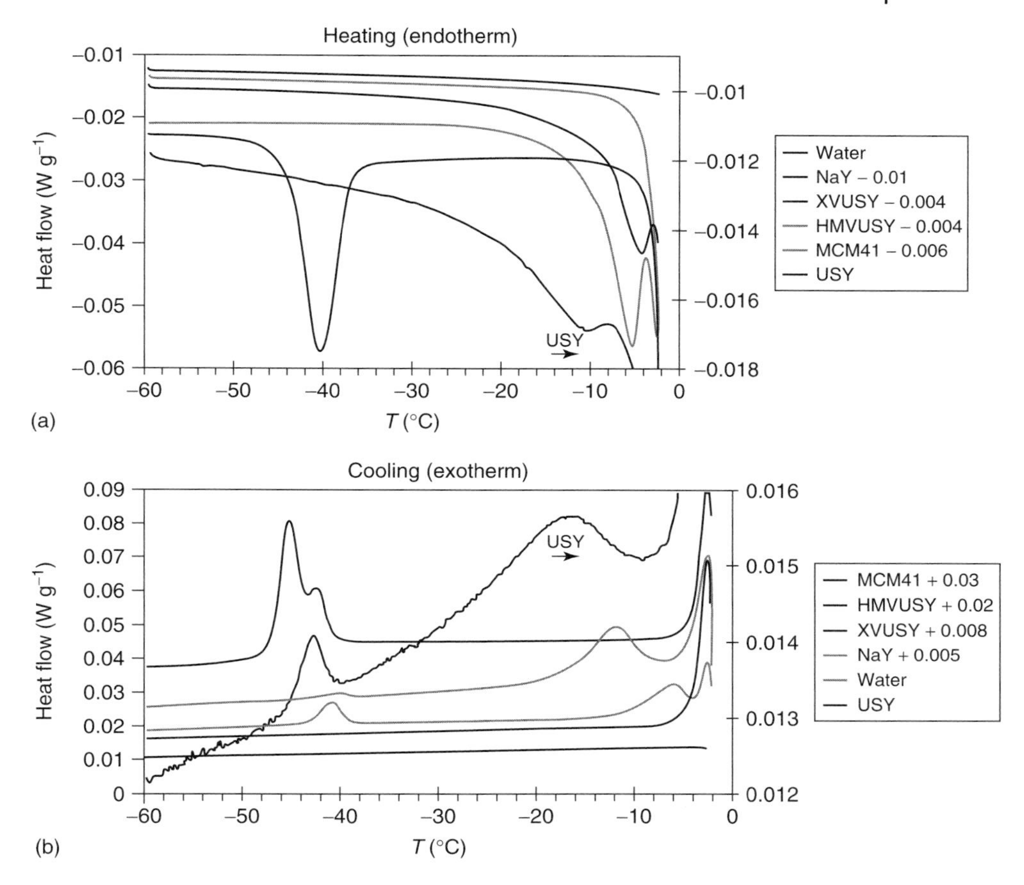


**Figure 9.7** DSC curves (heat flow vs temperature) during warming (a) and cooling (b) of water, NaY, XVUSY, HMVUSY, and MCM-41 (left axis) and USY (right axis). For clarity, some of the curves have been shifted upward or downward. The order of the samples in the legend (from top to bottom) is the order of the curves at the left side of the graphs. (Reprinted with permission from [242].)

at higher temperatures can be attributed to the heterogeneous nucleation in the cylindrical pores that extended to the external surface. The peaks around −40 °C, due to the absence of a counterpart in the endotherms, have been attributed to homogeneous nucleation of supercooled water trapped inside cavities, which are only accessible through micropores from the exterior. This water is not in contact with any ice crystals that could serve as nucleation sites. Such reasoning is supported by the observation that the pore volume calculated from the endotherm from −35 to −2 °C is smaller than that from −60 to −2 °C (0.028–0.033 ml g$^{-1}$). Therefore, in this way, the cylindrical pores and the cavities can be distinguished from thermoporometry. Thus, the relative contribution of the different pores to the total pore volumes, that is, cylindrical pores that extend to the external surface and

**Table 9.2** Peak onsets, peak maximums, and peak areas of the heating and cooling DSC curves of water-containing USY, XVUSY, HMVUSY, and MCM-41.

| Sample | $T_{onset}$ (°C) ±2 °C | Pore *D* (nm)[a] | $T_{max}$ (°C) | Pore D (nm)[a] | Peak area (J $g^{-1}$) | Pore volume ($cm^3$ $g^{-1b}$) | Mesopore diameter (nm)[c] | % Cavities[d] |
|---|---|---|---|---|---|---|---|---|
| | Heating (endotherm) | | | | | | $N_2$ physisorption | |
| USY[e] | −34 | 3 | −10.5 | 7.5 | 2.3 | 0.033 | 4–20 | 20 |
| USY[f] | −28 | 4 | −10.4 | 7.6 | 2.0 | 0.028 | | |
| XVUSY[e] | −32 | 3 | −4.1 | 17.1 | 14.7 | 0.18 | 4–40 | 29 |
| XVUSY[f] | −30 | 4 | −4.1 | 17.1 | 12.0 | 0.14 | | |
| HMVUSY[e] | −25 | 4 | −5.3 | 13.6 | 18.5 | 0.24 | 4–25 | 7 |
| HMVUSY[f] | −25 | 4 | −5.3 | 13.6 | 18.0 | 0.23 | | |
| MCM-41 | −50 | 2.7 | −40.3 | 3.0 | 16.3 | 0.62 | 2.8 | |
| | Cooling (exotherm) | | | | | | | |
| USY | −9.0[g] | >15.5 | −16.5 | 9.0 | n.d.[i] | | | |
| | −40.1 | n.p.[h] | −42.7 | n.p.[h] | 0.3 | | | |
| XVUSY | >−3.9[g] | >34.3 | −5.9 | 23.1 | n.d.[i] | | | |
| | −38.2 | n.p.[h] | −40.7 | n.p.[h] | 2.1 | | | |
| HMVUSY | >−7.0[g] | >19.6 | −11.8 | 12.1 | n.d.[i] | | | |
| | −38.2 | n.p.[h] | −40.2 | n.p.[h] | 0.6 | | | |
| MCM-41 | −38.5 | 4.5 | −42.5 | 4.2 | 15.9 | | | |

Results from [242].
[a]Calculated with Eq. (9.2).
[b]Calculated by correcting the peak area for the density of ice and the enthalpy of fusion.
[c]Mesopore diameter determined by nitrogen physisorption using the desorption branches of the isotherms.
[d]Proportion of the cavities volumes in the mesopore volumes.
[e]Heating from −60 to −2 °C.
[f]Heating from −35 to −2 °C after cooling from −2 to −35 °C.
[g]Not exactly determined due to overlap with the peak at −2 °C.
[h]n.p. = not possible to determine with Eq. (9.2) due to homogeneous nucleation.
[i]n.d.= not determined due to overlap with the peak at −2 °C.

the cavities, which are only accessible through micropores, can be calculated. For USY, the relative volume fraction of cavities according to thermoporometry is 15% (1 − 0.028/0.033, Table 9.2), which is in good agreement with that determined by nitrogen physisorption (20%, Table 9.2).

From this example, we can see that thermoporometry is a powerful technique for the characterization of mesoporous zeolites. Compared to gas physisorption, the sample preparation and measuring procedure in thermoporometry are much simpler and faster, and a fully automated apparatus including a sample changer for rapid product control is possible. However, thermoporometry is not yet a standardized method as is gas physisorption. For fragile samples, the possible deformation and even the collapse of the pore structure due to the expansion of liquid crystallization need to be taken into account. Some questions related to the data interpretation are still not well resolved, for example, the uncertainty about the thickness of the nonfreezable liquid layer [241], the enthalpy of fusion at low temperature [242], and the different contributions of pore shape factor and the pore size to the origin and width of the hysteresis between endotherms and exotherms [233]. Moreover, extensive data exist only for water and benzene, which limits the application of this method [237]. Nevertheless, as shown above, with thermoporometry it is possible to discriminate the cavities inside the mesoporous zeolites, wherein homogeneous nucleation takes place, and in the mesopores that extend to the external surface in which heterogeneous nucleation takes place. Moreover, the relative volume portion of different pores can be easily determined. The full potential of thermoporometry for characterizations of mesoporous zeolites is still to be explored, especially by using adsorbates relevant to the practical applications of the zeolites [234].

### 9.3.3 Mercury Porosimetry

Mercury porosimetry (MP) is a widely accepted method for the characterization of porous materials with macro- and mesopores [77]. It relies on the relationship between the pore radius $r$ and the hydrostatic pressure $P$ at which mercury can enter the pores. For a nonwetting liquid like mercury, a positive excess hydrostatic pressure $\Delta P$ is needed, and $\Delta P$ is varied inversely with $r$. With an increase in the pressure applied, mercury enters pores in decreasing order of size. According to this relation, the pore size can be calculated using the Washburn equation [243]:

$$r = -2\gamma \cos\theta / P \tag{9.3}$$

wherein $r$ is the pore radius of an equivalent cylindrical pore, $\gamma$ is the surface tension of mercury, $\theta$ is the contact angle between the mercury and the material, and $P$ is the applied pressure. Traditionally, a value of 484 mN m$^{-1}$ is used for $\gamma$ and 141° for $\theta$. Experimentally, the pressure can be varied between 0.1 and 2000 bars, corresponding to the pore radii in the range of 75 μm to 3.5 nm. Moreover, in combination with nitrogen physisorption, it is possible to differentiate between the mesopores that extend to the external surface and the cavities that are only accessible through micropores. While nitrogen physisorption probes both the micro- and mesopores, MP only accesses pores larger than 3.5 nm. Although MP is a well-developed method for meso- and macropore analysis, the reported examples of application of MP to mesoporous zeolites are rather limited. Lohse *et al.* [244] and Janssen *et al.* [214] used MP to characterize mesoporous Y zeolites obtained by dealumination. It has been shown that MP gave consistent results compared to those of nitrogen physisorption for pores larger than 4 nm in terms of pore volume as well as PSD. More recently, MP has also been applied to mesoporous ZSM-5 crystals obtained by desilication [245].

Figure 9.8 shows the PSD curves of the nontreated sample (Z-200-nt) and the base-treated sample (Z-200-at) obtained by MP and nitrogen physisorption. It is noted that the peaks around 2 nm in the PSD curve of the nitrogen physisorption are not related to a specific pores, but rather due to the so-called fluid-to-crystalline-like phase transition of the adsorbed phase in the MFI structure [246], which has

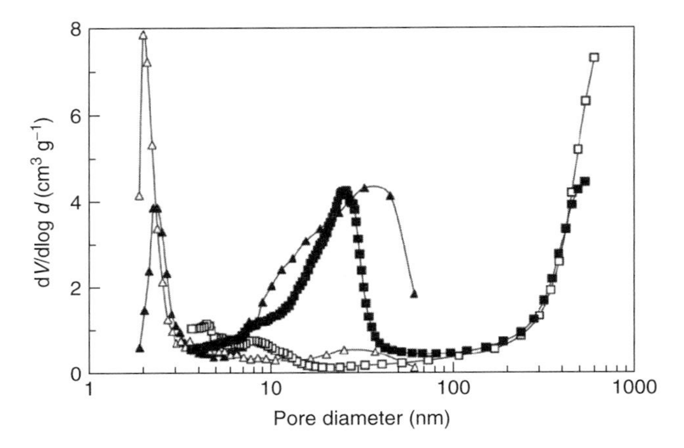


**Figure 9.8** Pore size distribution of the parent zeolite Z-200-nt (open symbols) and the base-treated zeolite Z-200-at (solid symbols) derived from (△, ▲) $N_2$ adsorption, and (□, ■) MP measurements. (Reprinted with permission from [245].)

been observed frequently. For pores larger than 100 nm, both nontreated and treated samples exhibited similar pore sizes, indicating that the morphology of the crystals did not change after the base treatment. In the mesopore region, both techniques gave comparable results. The pore volume in the range of 4–100 nm is 0.41 $cm^3\ g^{-1}$ by nitrogen physisorption and 0.31 $cm^3\ g^{-1}$ by MP. The lower pore volume from MP has been attributed to the elastic compression of the sample during the high-pressure measurements. MP gave a PSD centered at around 25 nm, while nitrogen physisorption gave a broader BJH PSD centered at 45 nm derived from the adsorption branch of the isotherm. The smaller size with a yet narrower PSD from MP is probably related to the pore-network effect. Because the pores are invaded in decreasing order of size in MP measurement, it implies that the sequential filling of the pores is dictated primarily by their mode of interconnection. Nevertheless, MP results provided evidence that the mesoporosity created by desilication is largely accessible from the external surface of the zeolite crystals, which is important for improved transport properties. In this context, MP provides a good estimation of the amount of the mesopores that extend to the external surface, which are expected to alleviate the diffusion limitation to a larger extent compared to the cavities inside the zeolite crystals. The above example also hints that caution should be taken when elastic or fragile samples are under investigation.

### 9.3.4 Electron Microscopy

#### 9.3.4.1 SEM and TEM

Electron microscopy (EM) is one of the most straightforward techniques for porous structure analysis, and has been used extensively, in many cases coupled with nitrogen physisorption, to give direct information on the textural properties [247]. Although a number of working modes are available, scanning electron microscopy (SEM) and TEM are the most frequently used techniques for porous materials. SEM works by rastering a beam of electrons over the sample surface and determining the intensity of the backscattered electrons or secondary electrons at different positions. Therefore, surface properties like morphology and composition can be obtained. With respect to zeolites with additional meso- and macropores, enhanced surface roughness is expected. In contrast to SEM, TEM relies on transmitted and diffracted electrons. As the electron wave passes through the sample, the interaction between electrons and the sample results in a change in the amplitude and phase of the electron wave. The overall change depends on the thickness and composition of the samples. Regions with higher porosity appear as a lighter area in the bright field TEM image due to the lower mass density. Therefore, a direct imaging of the pores can be achieved.

Ogura *et al.* reported the formation of mesoporous ZSM-5 crystals by base leaching with 0.2 M NaOH aqueous solution [62]. After base leaching, nitrogen physisorption measurements showed slightly lowered micropore volume (0.17 increased to 0.13 $cm^3\ g^{-1}$), yet considerably enhanced mesopore volume (0.07 increased to 0.28 $cm^3\ g^{-1}$). SEM images showed some grooves and voids on the surface of the treated sample (Figure 9.9), indicative of the etching of the materials by base solution. Such morphology with enhanced surface roughness has been frequently observed for mesoporous zeolites obtained by postsynthesis treatments [97], as well as by templating methods [143, 151, 159].

The nonuniform distribution of aluminum in ZSM-5 crystals has been a matter of debate. By treating the ZSM-5 crystals with Si/Al ratio of 35–115 with base solution (0.5 M $Na_2CO_3$, 16–20 hours at reflux), Dessau *et al.* obtained a series of hollow ZSM-5 crystals [60]. Figure 9.10 shows the SEM image of a typical treated example. After the alkaline treatment, the Si/Al ratio decreased from 62 in the original crystals to 8 in the treated ones. This result indicated the preferential dissolution of silicon species over aluminum species under basic conditions. The tetrahedral aluminum centers with negative charges are supposed to be inert to the attack of hydroxide. The formation of hollow crystals provided unambiguous evidence for aluminum zoning in ZSM-5 crystals, with higher aluminum concentration close to the exterior of the crystals. Similar phenomena were observed later by Groen *et al.* treating large ZSM-5 crystals (~20 μm, Si/Al ratio of 41) with 0.2 M NaOH solution [110]. With SEM–EDX (energy dispersive X-ray spectroscopy) analysis, spatially resolved elemental distribution was obtained in the crystals. As shown in Figure 9.11, a gradient of aluminum concentration can be observed in the parent crystals, while silicon is rather uniformly distributed. Point analysis over multiple crystals revealed that the aluminum concentration is 30 times higher at the edge than that in the center of the crystals. As a result of this, hollow crystals with inner macropores and relatively intact outer surface were obtained after the base treatment. Such a large void only moderately enhances the mesopore surface area, which was confirmed by nitrogen sorption measurements (5 vs 30 $m^2\ g^{-1}$, parent vs treated sample).

**Figure 9.9** SEM images of (a) the parent and (b) the alkaline-treated ZSM-5 crystals. (Reprinted with permission from [62].)

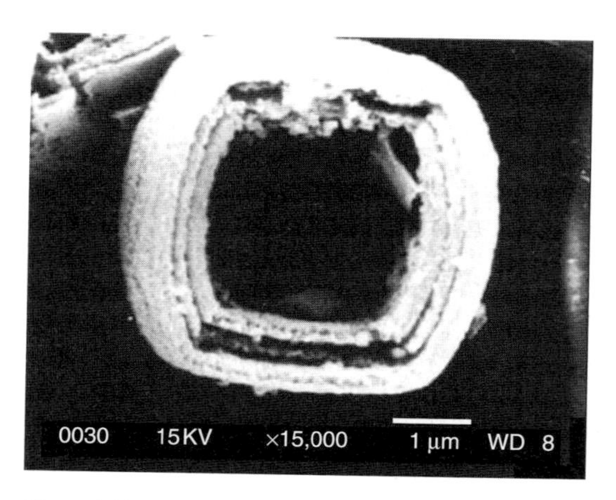


**Figure 9.10** SEM image of ZSM-5 crystal (Si/Al = 62) treated with $Na_2CO_3$ solution. (Reprinted with permission from [60].)

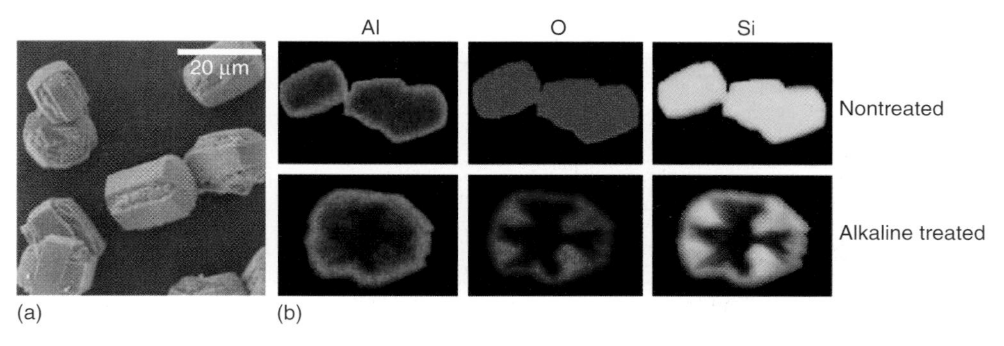


**Figure 9.11** (a) SEM micrograph of synthesized large ZSM-5 crystals and (b) SEM-EDX micrographs of the non-treated and alkaline-treated (0.2 M NaOH) ZSM-5 crystals (17 μm). Blue, red, and yellow colors represent aluminum, oxygen, and silicon concentrations, respectively. (Reprinted with permission from [11].)

Carbon templating has proven to be an effective method for generating well-defined mesopores in zeolites. Using carbon black as the template, Christensen *et al.* obtained mesoporous ZSM-5 crystals [41, 143]. Figure 9.12 shows the SEM images of the conventional zeolite (synthesized without carbon) and the mesoporous zeolite crystals. Although the typical coffin shape of MFI crystals was observed for both, the surface morphologies were very different with the mesoporous sample, showing a much rougher surface. The TEM image provided clear evidence of the highly mesoporous structure of the samples (Figure 9.12).

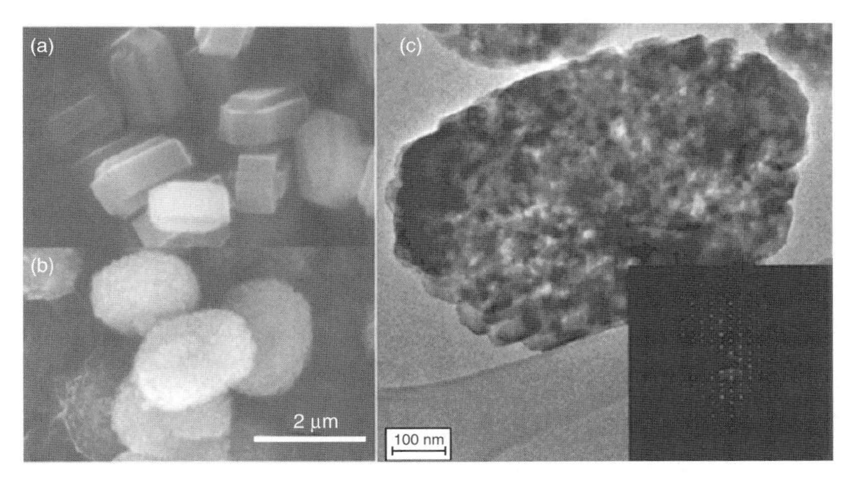


**Figure 9.12** (a, b) SEM images of the conventional (a) and mesoporous zeolite catalysts (b), (c) TEM of an isolated single ZSM-5 crystal and the diffraction pattern (inset) obtained from the same crystal. (Reprinted with permission from [41] and [143].)

From SEM and TEM analyses, the mesoporous crystals appeared to consist of agglomerated small crystals. However, electron diffraction measurement over the area covering the whole crystal indicated the single-crystal nature of the sample (inset of Figure 9.12c), which further supports the hypothesis that carbon particles were acting as the template and were entrapped during the synthesis. The advantages of the carbon-templating method are further illustrated by its wide applicability to other zeolites, as well as the diversity of available carbon sources. According to the synthesis mechanism using a hard template, under well-defined synthesis conditions, the size, shape, and tortuosity of the mesopores show fidelity with the morphology of the template materials. Consequently, the shape and size of the mesopores can be controlled by using proper template materials with a different morphology. This has been indeed observed for mesoporous silicalite-1 crystals with intracrystalline cylindrical pores synthesized using multiwall carbon nanotubes [42]. As shown in Figure 9.13, the zeolite crystals show single-crystal-like morphology with well-defined facets, however, with extensive cylindrical pores throughout the crystals. Moreover, the cylinder pores occurred as imprinted patterns of the carbon nanotube, further confirming the templating role of the carbon nanotubes.

For in-depth mesostructure analysis as well as the study of the formation mechanism of the mesopores, the detailed structure analysis with high resolution, probably down to atomic scale, is appreciated. In this regard, high-resolution transmission electron microscopy (HRTEM) is one of the suitable techniques. However, many zeolite specimens with suitable size and thickness for (HR)TEM analysis, especially those of mesoporous zeolites, are sensitive to electron beam irradiation. To overcome this limitation, Sasaki *et al.* equipped the electron microscope with a

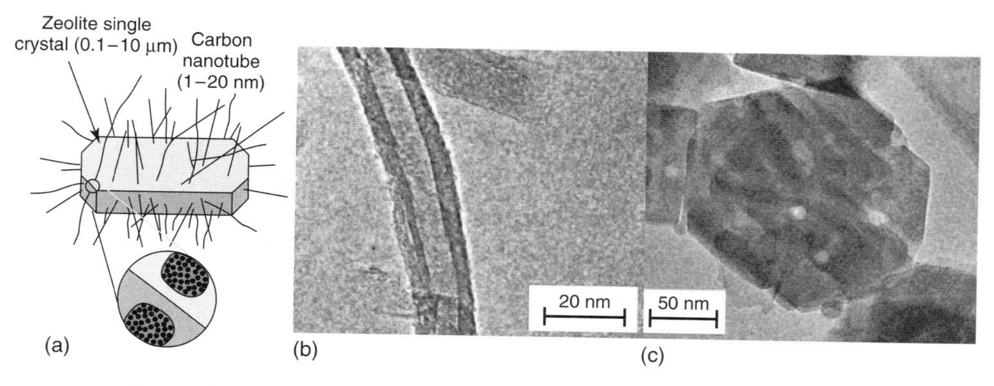


**Figure 9.13** (a) Schematic illustration of the synthesis principle for crystallization of mesoporous zeolite single crystals. (b) and (c) TEM images of the multiwall carbon nanotube (b) and the mesoporous silicalite-1 crystal obtained using carbon nanotubes as the template (c). (Reprinted with permission from [42].)

high-sensitivity SSC (slow scan CCD) camera, which allowed the HRTEM measurement to be performed with only 1/50 of the electron dose of the usual conditions [88]. Figure 9.14 displays the HRTEM image for the steam-dealuminated Y zeolite observed along the [110] direction, showing the mesopore channels as white contrast. By detailed analysis of the micrographs, they observed that microtwins existed along the mesopores. It was therefore suggested that the formation of mesopores start from the preexisting twin planes, which would present less stable regions in the crystals. In this context, the density of the mesopore can be controlled by adjusting the twin plane density in the parent crystal. With combined SEM and HRTEM measurements on EMT/FAU intergrowth zeolites, Terasaki *et al.* also observed that dealumination occurred preferentially in regions with a higher level of stacking disorder [248].

**3D TEM** TEM analysis has been one of the prevailing techniques and has provided a wealth of straightforward information on the structures of the mesopores in zeolites. The size and shape of the mesopores on individual particles can be

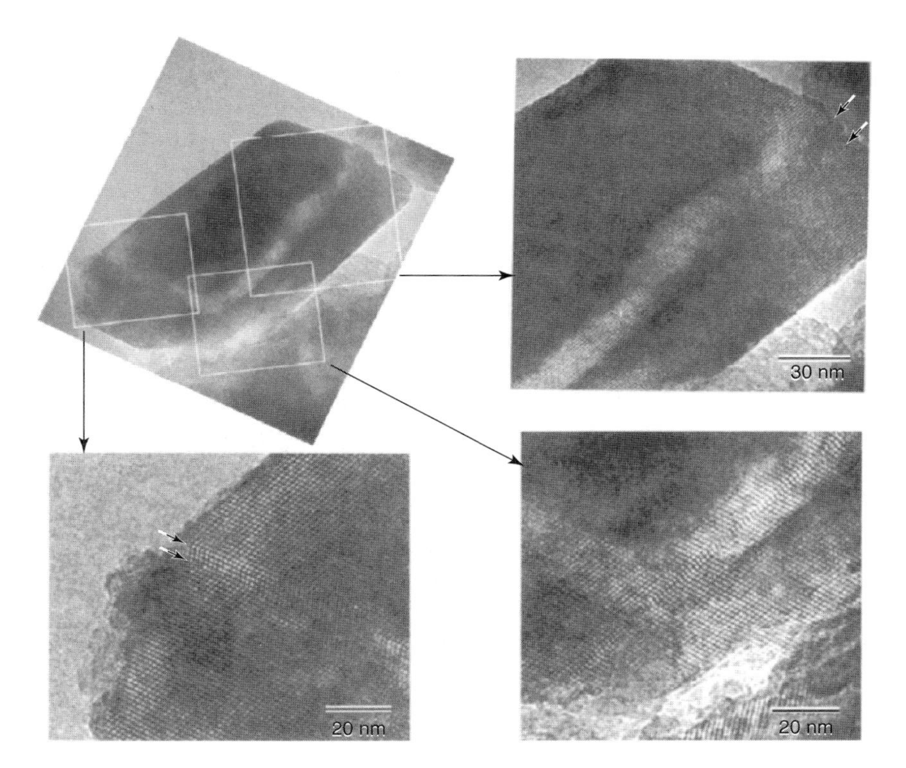


**Figure 9.14** HRTEM image for the dealuminated zeolite Y observed along the [110] zone axis. The white contrast with columnar shape shows the mesopore. (Reprinted with permission from [88].)

directly viewed with TEM, which complements the global information of ensembles obtained from other macroscopic techniques, like nitrogen physisorption. However, as conventional TEM image is a 2D projection of a 3D object, no unambiguous information on the shape, connectivity, location, and 3D orientation of mesopores can be derived. For example, in steamed Y zeolites, spherical and cylindrical pores have been frequently observed, while in 2D projection cylindrical pores may appear as spheres. Such discrepancy cannot be resolved by conventional TEM [249, 250]. Ultramicrotomy has been used for dealuminated Y zeolites to view the internal porous structure with ultrathin specimen (~20–50 nm) [85, 87]. However, fracture of the samples occurred during the sectioning process, which complicated the results interpretation. All these have necessitated a real 3D imaging of the intact sample. Toward this, the stereo-TEM has been applied for mesoporous zeolites [48]. By tilting the specimen of a steamed Y crystal from 0 to 50° with 10° intervals, Sasaki *et al.* observed that round pores observed in one orientation showed up as cylindrical pores when viewed after tilting over 50° [88]. Stereo-TEM has also been employed for imaging of mesoporous ZSM-5 crystals obtained with carbon nanotubes [147].

A step further toward 3D imaging is the development of ET [251]. With ET, a series of 2D TEM images is recorded by tilting the specimen in a large angular range. These images are subsequently aligned with respect to a common origin and tilt axis, and used for the computation of the 3D image of the object under investigation. With state-of-the-art equipment, a typical series of 141 images is collected in the range of −70 to 70° with angular increment per 1°. For a proper alignment of these images, markers like metal nanoparticles (for example, 5-nm gold particles) are often used. Alignment is achieved by least-square fitting of the positions of markers. According to the Projection–Slice theorem, the Fourier transform of a 2D projection is equal to a central slice through the 3D Fourier transform of the object [252]. By combining the series of the projections at different tilts and through an inverse Fourier transform process, the 3D morphology of the object can be obtained (Figure 9.15). Such a process, also known as *reconstruction*, provides the 2D images of the virtual slices of the object at different heights. For a detailed description of the basic principles and work flow of ET, we refer to the book by Deans *et al.* [252]. The applications of ET to catalysis materials have recently been exploited and highlighted in a recent review [253].

With respect to zeolites, ET has been applied to image the mesoporous structure of zeolites obtained by postsynthesis treatment [110, 214, 254–256] as well as by the carbon-templating method [40]. The first application of ET to zeolites, and also the first to heterogeneous catalysts, was reported by Koster *et al.* [254, 256]. Figure 9.16 shows the conventional TEM image and the digital thin slice (0.6 nm) obtained by ET of the acid-leached H-mordenite crystal, which is among one of the most important zeolites with industrial applications. Black dots correspond to the gold particles as markers for alignment. The ET analysis showed the existence of mesopores inside the crystal with unprecedented clarity, while in conventional TEM image such information is obscured by the variant sample thickness and/or by superimposition or different pores in the conventional TEM images.

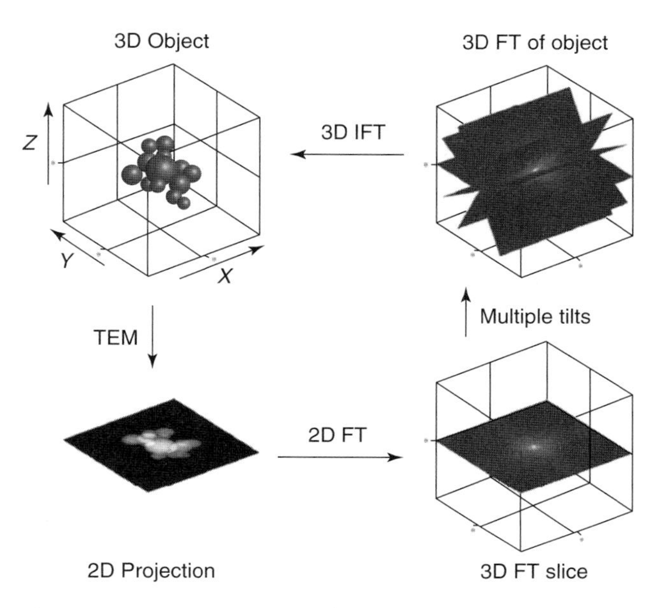


**Figure 9.15** ET reconstruction scheme using the Projection–Slice theorem. FT denotes the Fourier transform and IFT denotes the inverse Fourier transform. By combining projections recorded at multiple tilts, the 3D FT of the object is probed. An inverse Fourier transform then gives the reconstruction of the 3D object. (Reprinted with permission from [253].)

Another series of industrially important zeolites, that is, zeolite Y and the corresponding dealuminated counterparts have also been investigated with ET [255]. Figure 9.17 shows the 2D TEM and ET images of the NaY as well as the samples obtained after different treatments; that is, steaming (USY), seaming twice followed by acid leaching (XVUSY). As expected, no mesopores can be observed in the NaY sample. Compared to conventional TEM images, mesopores can be better resolved in the ET images. The diameters of the mesopores evaluated with ET are 3–20 nm for USY and 4–34 nm for XVUSY, which correlate well with the values estimated by nitrogen physisorption: 4–20 nm for USY and 4–40 nm for XVUSY, respectively. Moreover, ET images of a series of slices showed that, in USY and XVUSY, a significant fraction of the mesopores are cavities, although there were some cylindrical pores extending to the external surface. The presence of both types of mesopores was also indicated by the nitrogen physisorption measurements. More importantly, ET images of the steamed sample (USY) showed some dark cavities and dark bands surrounding the crystals, which could not be clearly distinguished in the 2D TEM image. Combining the nitrogen physisorption and XPS analysis (Si/Al atomic ratio), such areas were attributed to the amorphous materials enriched with aluminum (Si/Al = 1.1 at/at from XPS). On the basis of these observations, a mechanism for the mesopore formation during steaming and acid leaching was proposed as depicted in Figure 9.17g, which supported earlier studies [79], but added key aspects of mesopore formation. Thus, the mesoporous system

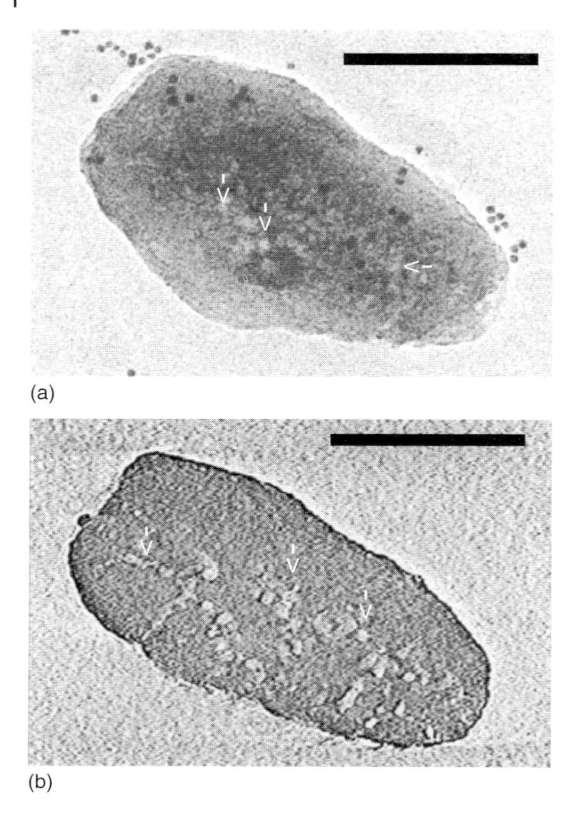


**Figure 9.16** (a) Conventional TEM image indicating the mesopores in the crystallite (white spots, arrows) and several gold beads (black dots, 5 nm in diameter) on the grid for the alignment. (b) Digital slice (0.6 nm thick) through the 3D reconstruction of the crystallite showing the mesopores inside the crystallite (arrows). Scale bar is 100 nm. (Reprinted with permission from [254].)

originated from cavities with amorphous materials deposited therein. During the second steaming and acid leaching, small cavities coalescenced to form larger cavities and cylindrical pores. The extraframework aluminum species, which were clearly visible in USY as dark areas, were removed in the subsequent treatment of XVUSY. By virtue of the detailed information provided by ET, a quantitative structural analysis on individual crystals is possible. As an example, PSD obtained by quantification results from ET is shown in Figure 9.18, which agreed well with that from nitrogen physisorption for bulk analysis [257]. Thus, results from both methods complement each other, whereas unlike gas physisorption ET does not assume a specific model for the pore shape and provides direct information on the shape, size, and connectivity of the mesopores.

ET has also been employed for the study of mesoporous zeolites formed by other methods. For small ZSM-5 crystals (400–700 nm in size), upon desilication by alkaline treatment, ET analysis showed rather uniform and interconnected

**Figure 9.17** 2D-TEM images of NaY (a), USY (c), and XVUSY (e), 3D ET slices of NaY (b, 1.7 nm), USY (d, 1.7 nm), and XVUSY (f, 1.25 nm), and the model for the generation of mesopores in zeolite Y (g). (Reprinted with permission from [255].)

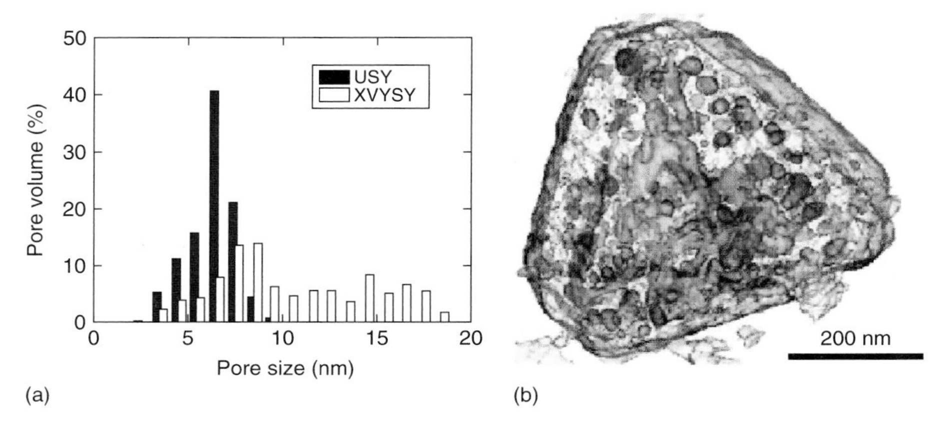


**Figure 9.18** (a) Mesopore size distributions of USY and XVUSY obtained by image analysis of electron tomograms. (b) 3D representation of XVUSY showing the mesopores through the transparent surface of the particle. (Adapted from [257].)

mesopores developed in the interior of the crystals, while the outer surface remained relatively unaffected [110]. These results provide additional evidence of the aluminum zoning in the small ZSM-5 crystals, which play a directing role for the mesopore formation during the base treatment.

As a last example, ET has also been applied to mesoporous zeolites obtained by the carbon-templating method [40]. For silicalite-1 samples synthesized with carbon nanofibers and carbon black aggregates, cylindrical mesopores were formed that start at the external surface of the zeolite crystals. However, the tortuosity of the mesopores templated by the carbon black aggregates was much higher than the cylindrical pores obtained by the carbon nanofibers. For larger zeolite crystals, a higher amount of carbon aggregates can be enclosed in the zeolites, leading to ink-bottle-type mesopores.

From the above examples, one can see that, compared to conventional TEM, ET can yield unprecedented and unambiguous information, both qualitatively and quantitatively, on the pore shape, pore size, and connectivity inside individual crystals, which complement the results from other bulk analysis and provide new insights into the formation mechanism of mesopores and opportunities for optimizing the textural properties of zeolites.

### 9.3.5
### NMR Techniques

#### 9.3.5.1 $^{129}$Xe NMR Spectroscopy

Nuclear magnetic resonance (NMR) spectroscopy is a powerful method enriched with a collection of techniques useful for structure characterization of zeolite materials. A wealth of information on the composition, structural defects, and pore surface properties of conventional zeolites are available with $^{27}Al$, $^{1}H$, and $^{29}Si$ NMR spectroscopies. With respect to the porous structure of zeolites, $^{129}Xe$ NMR has proven to be an important and valuable technique [258–260]. $^{129}Xe$ is an inert, nonpolar, spherical atom with a large electron cloud, which is sensitive to its environment. Depending on the various interactions with its surrounding, a wide range of chemical shift has been observed for $^{129}Xe$, based on which the so-called xenon NMR spectrometry has been developed. For xenon adsorbed in a porous solid, the observed $^{129}Xe$ chemical shift is the weighted average of different types of interactions on the NMR timescale [260]. For zeolites, the observed chemical shift of xenon is the sum of several terms corresponding to the various perturbations around it, including Xe–Xe collisions, electronic field exerted by cations, interactions with the pore surface, and the dimension and geometry of the cages and channels. Generally, the smaller the pore size, the higher is the chemical shift observed with respect to that of the gas-phase xenon due to the strongly restricted diffusion of xenon. While the chemical shift of xenon reflects the pore dimensions, the xenon isotherm can provide information on the pore volumes. Although considerable knowledge of $^{129}Xe$ NMR has been accumulated on zeolite structure, compositions, cations siting, and so on, it has been rarely applied to mesoporous zeolites. For dealuminated Y [261], mordenite [261, 262], and CSZ-[263] zeolites, a decreased $^{129}Xe$ shift has been observed, which has been attributed to the increased mean free path caused by the formation of meso- and macropores. More recently, laser-hyperpolarized (HP) $^{129}Xe$ NMR 2D exchange spectroscopy (EXSY) has been applied by Liu *et al.* to investigate the mesoporous ZSM-5 zeolites synthesized by the starch-templating method [201]. The optical pumping by laser for the production of HP xenon facilitates its working at very low concentration of xenon (~1%) under continuous flow [264]. Therefore, the observed $^{129}Xe$ chemical shift would mainly reflect the interaction between xenon atoms and the solid surface. The HP $^{129}Xe$ 2D-EXSY NMR technique is a powerful tool to study the dynamic process of the adsorbed xenon in porous solid, and has shown potential to reveal textural connectivity in such hierarchical zeolites.

Figure 9.19 shows the continuous-flow HP $^{129}Xe$ NMR spectra of mesoporous ZSM-5 with a Si/Al atomic ratio of 50 at variable temperatures. The peaks at 0 ppm are due to xenon from the gas phase. From 293 to 153 K, there is one downfield signal *a*, which was attributed to the xenon adsorbed in the micropores. The chemical shift of this signal increased with the decrease in the temperature due to the enhanced Xe-surface and Xe–Xe interactions at lower temperatures. At 153 K, a new upfield signal *b* emerged, which indicates that there is another type of pores with a wide size distribution in the mesoporous ZSM-5, and the fast exchange of xenon among these pores exists above 153 K. Figure 9.19b shows the 2D EXSY spectrum of the same sample at 143 K with mixing time $\tau_{mix} = 1$ ms. The 2D EXSY spectrum is generally obtained by monitoring frequencies before and after a so-called mixing time, $\tau_{mix}$, during which spin exchange and/or molecular reorientation motions can occur. Changes in the NMR frequencies can be observed as the off-diagonal intensities in the 2D spectrum. The cross-peaks indicate an exchange of xenon atoms between the corresponding environments on the

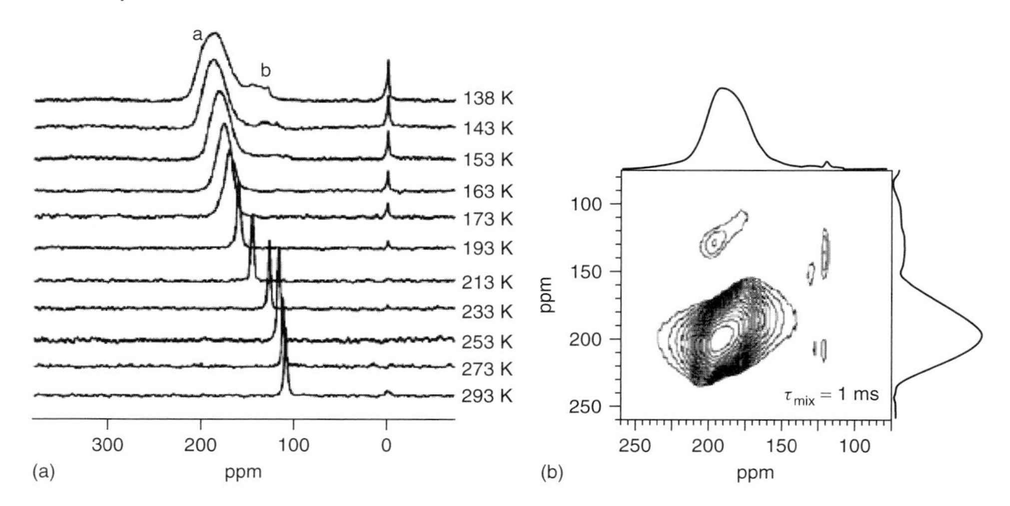


**Figure 9.19** (a) Laser-hyperpolarized $^{129}Xe$ NMR spectra of the Xe adsorbed in mesoporous ZSM-5 with an Si/Al ratio of 50 at a variant temperature from 293 to 138 K. (b) Laser-hyperpolarized $^{129}Xe$ 2D-EXSY NMR spectra of the same sample at 143 K with mixing times $\tau_{mix} = 1$ ms. (Reprinted with permission [201].)

diagonal within the period of $\tau_{mix}$. The cross-peaks observed in Figure 9.19b clearly demonstrated the exchange of xenon between the micropores and mesopores. No such cross-peaks can be observed for the mixing time shorter than 0.2 ms. The cross-peaks became stronger with longer $\tau_{mix}$ of 5 ms. Therefore, the exchange of xenon between these two types of porous domains takes place at a timescale of less than 1 ms. For a better understanding, the authors also prepared a sample by mixing conventional ZSM-5 with a commercial $SiO_2$ with similar mesopore size to that of the mesoporous ZSM-5 sample. The mixture sample also exhibited comparable textural properties with those of the mesoporous ZSM-5 sample, for example, micropore volumes of 0.10 $cm^3\ g^{-1}$ for both, mesopore volumes of 0.30 versus 0.34 $cm^3\ g^{-1}$, BET surface areas of 373 versus 350 $m^2\ g^{-1}$ for the mixture and the mesoporous zeolite respectively. In contrast, under the same experimental conditions, 2D EXSY spectra showed no exchange between the conventional ZSM and the silica at $\tau_{mix}$ of 1 ms. The exchange signals were only observable with $\tau_{mix} > 5$ ms. This indicates that the exchange of xenon in the mixture sample, which is actually interparticle diffusional exchange, occurred slower than that in the mesoporous ZSM-5 sample. Therefore, the microporous and mesoporous domains in the mesoporous zeolite were much closer and had a better connectivity than that in the physical mixture. This example showed that unprecedented information on the connectivity and communications between the micropores and mesopores can be obtained by the 2D EXSY $^{129}Xe$ NMR technique. The full potential of $^{129}Xe$ NMR for the characterization of mesoporous zeolites is still to be explored, especially in a quantitative way.

#### 9.3.5.2 PFG NMR

As the most important impact of generating textural mesopores in zeolites is to alleviate the diffusion limitation, a direct diffusion measurement of guest molecules, especially those molecules relevant to the practical applications of zeolites, is valuable. For various mesoporous zeolites, macroscopic diffusion measurements have indeed shown increased apparent diffusivities [71, 72, 74–76], which have been mostly attributed to the shortened diffusion path lengths. However, to gain a more insightful knowledge of the effects of the textual mesopores on the diffusion properties, a microscopic diffusion measurement is needed. In this context, pulsed field gradient (PFG) NMR is a versatile technique and has been widely used for studying intracrystalline diffusion in porous materials, especially zeolite materials.

The PFG NMR is based on the dependence of the Larmor frequency of spins on the amplitude of the applied field [6]. Superimposing a large constant magnetic field by a magnetic field gradient pulse, one can label the positions of the spins by the Larmor frequency and hence the accumulated phases due to the rotation with the specific Larmor frequency in the local field. If the molecules where the spins are located do not change their positions during the time of the diffusion measurements, the signal remains at the maximum level. However, if the molecular displacement occurs during the observation time interval, the measured signal decays to a certain extent. The attenuation of the NMR signal ($\Psi$) for the normal diffusion obtained with the 13-interval PFG NMR can be represented by the following equation [265]:

$$\Psi = \exp(-4\gamma^2\delta^2 g^2 Dt) \tag{9.4}$$

wherein $D$ is the diffusivity, $t$ is the effective diffusion time, $\delta$ is the duration of the applied gradient pulses with the amplitude g, and $\gamma$ is the gyromagnetic ratio. By measuring the attenuation of the signal as a function of the amplitude of the applied field gradient (g) with all the rest of the parameters constant, one can determine the diffusivity $D$.

Using PFG NMR, Kortunov *et al.* investigated the USY zeolite with *n*-octane and 1,3,5-triisopropylbenzene as the probe molecules [266]. USY was obtained by steaming treatment of ammonium-form Y zeolite (crystal size ~3 μm). Nitrogen physisorption showed the slightly decreased micropore volume (from 0.28 to 0.24 $cm^3\ g^{-1}$), yet increased mesopore volume (from 0.07 to 0.17 $cm^3\ g^{-1}$) after the steaming treatment. Figure 9.20 shows the measured dependencies of the effective *n*-octane diffusivities on the root mean square displacement $\langle r^2\rangle^{1/2}(\langle r^2(t)\rangle = 6Dt)$ in the ammonium-form Y as well as the USY. However, no clear difference was observed for $\langle r^2(t)\rangle^{1/2}$ smaller than ~1.5 μm (half of the crystal size), which revealed that the mesopores formed by steaming are essentially of no influence for the intracrystalline diffusion of *n*-octane. To further elucidate on this point, 1,3,5-triisopropylbenzene, whose size is larger than the micropore openings and hence not expected to enter the micropores, was used as the probe molecules. Again, similar diffusion behavior was observed between the parent and the steamed samples, with the intercrystalline diffusion dominating

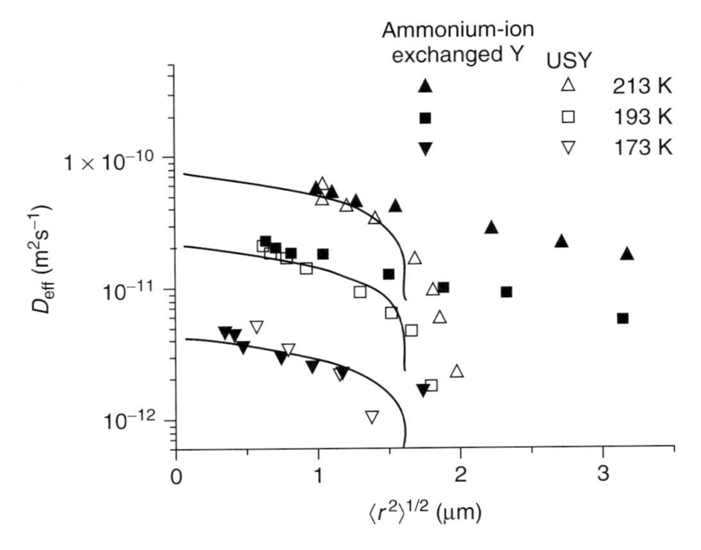


**Figure 9.20** Dependencies of the effective diffusivities on the mean square displacements measured for *n*-octane by PFG NMR (points) and those obtained by the dynamic Monte Carlo simulations (lines) for a cubic lattice with a size of $2.3 \times 2.3 \times 2.3\,\mu m^3$. The boundaries of the simulation lattice were assumed to be impenetrable for diffusing molecules. (Reprinted with permission from [266].)

in both cases. As for USY, it has been well established with 3D TEM that a significant part of the mesopores is actually cavities, which are only accessible via the micropores surrounding them. In line with this, the PFG NMR study showed that the intracrystalline diffusion is essentially unaffected by the mesopores in USY. The presence of the cavities induces only a moderate increase in the intracrystalline diffusivities, as the overall process is controlled by the slowest step, that is, the diffusion through the micropores. These results emphasize the importance of the shape and connectivity of the mesopores, in addition to the pore size and volume, when an alleviated diffusion limitation of zeolites is aimed at.

NMR techniques combined with a wide range of applicable probe molecules can provide important information on the mesopores in zeolites, especially the shape and connectivity of the mesopores. It is important to note that, other potential NMR techniques, which have been well investigated for mesoporous materials, have not been exploited for mesoporous zeolites. Among them, NMR cryoporosimetry is an especially informative one, which relies on the dependencies of the phase-transition behavior of confined guest species on the pore size of the solid matrix. For example, $^{1}H$ NMR spectra of water confined in mesopores have been used to characterize ordered mesoporous silica MCM-41 [267]. A simple relation has been found between the freezing point depression and the pore size.

### 9.3.6
### *In situ* Optical and Fluorescence Microscopy

*In situ* fluorescence microscopy has been widely applied to cellular biology, and has recently found potential as an advanced technique for studying heterogeneous catalysts under working conditions due to its high spatiotemporal resolution and sensitivity [268, 269]. Nonuniform catalytic behaviors over large, single ZSM-5 and SAPO-34 crystals have been unraveled using a combined technique of *in situ* optical and fluorescence microscopy [270–272]. The structures of the sub-building units of large CrAPO-5, SAPO-34, SAPO-5, and ZSM-5 crystals have also been suggested by imaging the zeolite crystals during the template decomposition process [273]. More recently, Kox *et al.* showed that the spatially resolved catalytic activity of mesoporous ZSM-5 crystals in the oligomerization of various styrene compounds can be accessed using such a combined measurement, which has provided spatially resolved direct evidence of the effect of mesopores during the catalytic events [133].

Figure 9.21 shows the SEM images of the parent ZSM-5 and the mesoporous ZSM-5 obtained by desilication. After desilication, the upper "lid" disappeared in

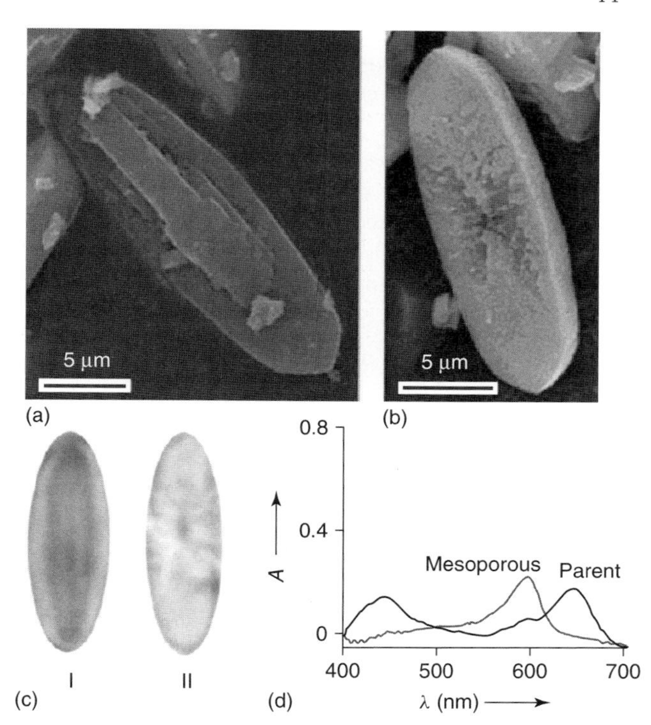


**Figure 9.21** (a,b) SEM images of the parent (a) and the mesoporous (b) H-ZSM-5 crystals. (c) Microphotographs of the parent (I) and mesoporous (II) crystals after oligomerization with 4-methoxystyrene. (d) The corresponding optical absorption spectra. (Reprinted with permission from [133].)

the mesoporous crystal. After the addition of 4-methoxystyrene and subsequent heat treatment at 373 K, as revealed by the optical micrographs, the parent sample was mostly colored in the upper lid, whereas the mesoporous sample exhibited more uniform coloration (Figure 9.21). It is therefore possible that the "lid" unit has a different pore orientation with the rest parts of the crystals, and restricts the diffusion of guest molecules into the interior parts of the crystals. After desilication, a higher accessibility of the inner parts of the crystals can be expected. Also shown in Figure 9.21 are the optical absorption spectra. The bands at 595 and 650 nm can be attributed to the dimeric and trimeric (or higher) oligomers, respectively. Upon desilication, the formation of trimeric or higher carbocations is restricted. This can be due to the shortened diffusion length in the mesoporous samples. Consequently, the chance of successive reactions over other Brønsted acid sites is decreased. The same type of reaction was conducted under confocal fluorescence microscopy for both crystals. In line with these observations, the fluorescence snapshots also showed more uniform coloration over the mesoporous crystals (Figure 9.22). Furthermore, the confocal fluorescence signal was more intense in the mesoporous sample, which can be rationalized by using the optical absorption spectra, that is, the optical absorption in the excitation wavelength (561 nm) directly corresponds to the fluorescence intensity of the crystals. Moreover, based on these results, a novel building scheme was proposed for ZSM-5 crystals (Figure 9.22C).

This example shows the impact of mesopores on the zeolite-catalyzed reactions directly. After the desilication treatment of ZSM-5 crystals, the microporous channels in large zeolite crystals are more accessible. Despite the fact that it is still difficult to get detailed information on the porous structure of mesoporous zeolites from optical and fluorescence microscopy due to the limited resolution (a few micrometers), *in situ* measurement under typical catalytic reaction conditions

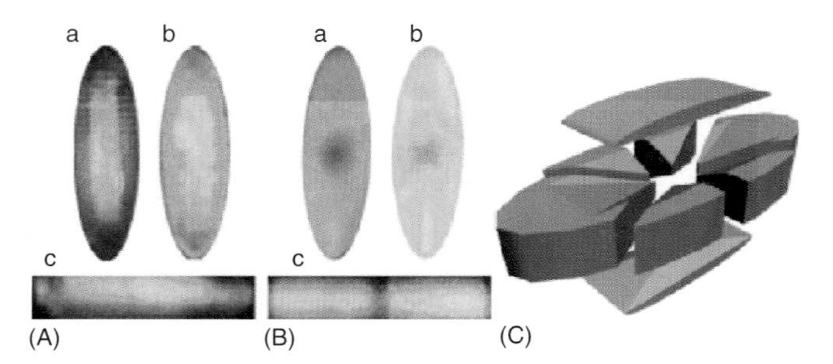


**Figure 9.22** (A,B) Confocal fluorescence images measured in the *in situ* spectroscopic cell after oligomerization of 4-methoxystyrene at 373 K on the parent (A) and mesoporous (B) crystals (excitation at 561 nm, detection at 580–640 nm,). Indices (a–c) correspond to the upper horizontal plane, middle horizontal plane parallel to the upper plane, and vertical intermediate plane, respectively. Measurements were performed ≈5 minutes after exposure to the styrene compound. (C) Schematic representation of the ZSM-5 crystal showing its individual building blocks. (Reprinted with permission from [133].)

using these techniques are expected to shed more insights into the practical applications of these mesoporous zeolites.

## 9.4 Summary and Outlook

The benefits of generating hierarchical pores in zeolite crystals have received general recognition in academic fields as well as industrial catalytic processes. A number of methods for generating additional pores in zeolites are now available and are being developed for a wider application to zeolites with different structure and compositions. Meanwhile, novel methods aiming for a higher extent of controllability and versatility still continue to come up. Different methods proceed according to different mechanisms, and therefore result in different end products in terms of textural properties and compositions. Nevertheless, whichever method is of choice in practice, a clear picture of the textural property and its evolution is always a prerequisite for a deeper understanding of the formation mechanisms of the mesopores and the impact of these pores on the practical applications of the zeolites, as well as for further optimizing the textural properties toward improved performance of the zeolites. It is clear that, not only the pore size and pore volume, which have been frequently reported as indicator values, but also the shape, location, and connectivity of these mesopores are important factors that determine the performances of zeolite.

We have summarized here the characterization methods that have been used for studying the textural properties of zeolites, more specifically mesoporous zeolites. While conventional methods like nitrogen physisorption and TEM have been well developed and are still prevailing, the limits of these methods are evident. Other characterization methods summarized in this review, which have shown potential in providing complementary information on shape and connectivity of the mesopores, should be further exploited for characterizing mesoporous zeolites. Among others, ET is a unique technique and it provides direct and detailed information on the textural properties in both a qualitative and a quantitative way. Other methods, like thermoporometry, mercury intrusion porosimetry, which may be easier to implement, however, have not been widely applied in mesoporous zeolites. One possible obstacle is the uncertainty in explaining the characterization results, which in many cases is caused by the complex nature of the mesopores with irregular shape and broad size distribution in most mesoporous zeolites, especially those obtained by postsynthesis treatments. Therefore, advances in the theories of these methods toward a more realistic model are highly anticipated. With respect to this, the knowledge gained from well-defined mesoporous materials, like MCM-41, is expected to facilitate their applications to mesoporous zeolites. One may also note that there are still a number of potential techniques for porous structure analysis that have not been applied to mesoporous zeolite characterization, for example, quasi-equilibrium thermodesorption, spectroscopic ellipsometry, light transmission, and radiation scattering of neutrons and X rays, to mention just a few. The progress of the characterization techniques, both experimentally and theoretically, will definitely benefit and accelerate the "designed synthesis" of mesoporous zeolites.

## Acknowledgments

We thank the financial support from the National Research School Combination Catalysis (NRSCC) and the ACTS/ASPECT program from The Netherlands Organization for Scientific Research (NWO-CW).

## References

1. Breck, D.W. (1974) *Zeolite Molecular Sieves, Structure, Chemistry and Uses*, Wiley-Intersciences, New York.
2. Barrer, R.M. (1978) *Zeolites and Clay Minerals as Sorbents and Molecular Sieves*, Academic, London.
3. van Bekkum, H., Flanigen, E.M., Jansen, J. C. (eds) (1991) *Introduction to Zeolite Science and Practice*, Elsevier, Amsterdam.
4. Corma, A. (1995) *Chem. Rev.*, **95**, 559–614.
5. van Steen, E., Claeys, M., and Callanan, L.H. (eds) (2005) *Recent Advances in the Science and Technology of Zeolites and Related Materials*, vol. 154B, Elsevier, Amsterdam.
6. Kärger, J. and Ruthven, D.M. (1992) *Diffusion in Zeolites and Other Microporous Solids*, John Wiley & Sons, Inc., New York.
7. Kärger, J. and Freude, D. (2002) *Chem. Eng. Technol.*, **25**, 769–778.
8. Čejka, J. and Mintova, S. (2007) *Catal. Rev.-Sci. Eng.*, **49**, 457–509.
9. Drews, T.O. and Tsapatsis, M. (2005) *Curr. Opin. Colloid Interface Sci.*, **10**, 233–238.
10. Egeblad, K., Christensen, C.H., Kustova, M., and Christensen, C.H. (2008) *Chem. Mater.*, **20**, 946–960.
11. Groen, J.C., Moulijn, J.A., and Pérez-Ramírez, J. (2006) *J. Mater. Chem.*, **16**, 2121–2131.
12. Hartmann, M. (2004) *Angew. Chem. Int. Ed.*, **43**, 5880–5882.
13. Kirschhock, C.E.A., Kremer, S.P.B., Vermant, J., Van Tendeloo, G., Jacobs, P.A., and Martens, J.A. (2005) *Chem. Eur. J.*, **11**, 4306–4313.
14. Meynen, V., Cool, P., and Vansant, E.F. (2007) *Microporous Mesoporous Mater.*, **104**, 26–38.
15. Pérez-Ramírez, J., Christensen, C.H., Egeblad, K., Christensen, C.H., and Groen, J.C. (2008) *Chem. Soc. Rev.*, **37**, 2530–2542.
16. Schüth, F. (2003) *Angew. Chem. Int. Ed.*, **42**, 3604–3622.
17. Schüth, F. (2005) *Annu. Rev. Mater. Res.*, **35**, 209–238.
18. Tao, Y.S., Kanoh, H., Abrams, L., and Kaneko, K. (2006) *Chem. Rev.*, **106**, 896–910.
19. Tosheva, L. and Valtchev, V.P. (2005) *Chem. Mater.*, **17**, 2494–2513.
20. Tosheva, L. and Valtchev, V.P. (2005) *C. R. Chim.*, **8**, 475–484.
21. van Donk, S., Janssen, A.H., Bitter, J.H., and de Jong, K.P. (2003) *Catal. Rev.-Sci. Eng.*, **45**, 297–319.
22. Davis, M.E., Saldarriaga, C., Montes, C., Garces, J., and Crowder, C. (1988) *Nature*, **331**, 698–699.
23. Freyhardt, C.C., Tsapatsis, M., Lobo, R.F., Balkus, K.J., and Davis, M.E. (1996) *Nature*, **381**, 295–298.
24. Burton, A., Elomari, S., Chen, C.Y., Medrud, R.C., Chan, I.Y., Bull, L.M., Kibby, C., Harris, T.V., Zones, S.I., and Vittoratos, E.S. (2003) *Chem. Eur. J.*, **9**, 5737–5748.
25. Corma, A., Díaz-Cabañas, M.J., Rey, F., Nicolooulas, S., and Boulahya, K. (2004) *Chem. Commun.*, 1356–1357.

26. Corma, A., Díaz-Cabañas, M., Martinez-Triguero, J., Rey, F., and Rius, J. (2002) *Nature*, **418**, 514–517.
27. Corma, A., Díaz-Cabañas, M.J., Jordá, J.L., Martínez, C., and Moliner, M. (2006) *Nature*, **443**, 842–845.
28. Sun, J.L., Bonneau, C., Cantín, A., Corma, A., Díaz-Cabañas, M.J., Moliner, M., Zhang, D.L., Li, M.R., and Zou, X.D. (2009) *Nature*, **458**, 1154–1157.
29. Choifeng, C., Hall, J.B., Huggins, B.J., and Beyerlein, R.A. (1993) *J. Catal.*, **140**, 395–405.
30. Beyerlein, R.A., ChoiFeng, C., Hall, J.B., Huggins, B.J., and Ray, G.J. (1997) *Top. Catal.*, **4**, 27–42.
31. Mavrodinova, V., Popova, M., Valchev, V., Nickolov, R., and Minchev, C. (2005) *J. Colloid Interface Sci.*, **286**, 268–273.
32. On, D.T. and Kaliaguine, S. (2002) *Angew. Chem. Int. Ed.*, **41**, 1036–1040.
33. On, D.T. and Kaliaguine, S. (2003) *J. Am. Chem. Soc.*, **125**, 618–619.
34. Choi, M., Cho, H.S., Srivastava, R., Venkatesan, C., Choi, D.H., and Ryoo, R. (2006) *Nat. Mater.*, **5**, 718–723.
35. Choi, M., Srivastava, R., and Ryoo, R. (2006) *Chem. Commun.*, 4380–4382.
36. Srivastava, R., Choi, M., and Ryoo, R. (2006) *Chem. Commun.*, 4489–4491.
37. Xiao, F.S., Wang, L.F., Yin, C.Y., Lin, K.F., Di, Y., Li, J.X., Xu, R.R., Su, D.S., Schlögl, R., Yokoi, T., and Tatsumi, T. (2006) *Angew. Chem. Int. Ed.*, **45**, 3090–3093.
38. Wang, H. and Pinnavaia, T.J. (2006) *Angew. Chem. Int. Ed.*, **45**, 7603–7606.
39. Mei, C.S., Wen, P.Y., Liu, Z.C., Liu, H.X., Wang, Y.D., Yang, W.M., Xie, Z.K., Hua, W.M., and Gao, Z. (2008) *J. Catal.*, **258**, 243–249.
40. Janssen, A.H., Schmidt, I., Jacobsen, C.J.H., Koster, A.J., and de Jong, K.P. (2003) *Microporous Mesoporous Mater.*, **65**, 59–75.
41. Jacobsen, C.J.H., Madsen, C., Houzvicka, J., Schmidt, I., and Carlsson, A. (2000) *J. Am. Chem. Soc.*, **122**, 7116–7117.
42. Schmidt, I., Boisen, A., Gustavsson, E., Ståhl, K., Pehrson, S., Dahl, S., Carlsson, A., and Jacobsen, C.J.H. (2001) *Chem. Mater.*, **13**, 4416–4418.
43. Fang, Y.M., Hu, H.Q., and Chen, G.H. (2008) *Microporous Mesoporous Mater.*, **113**, 481–489.
44. Tao, Y.S., Kanoh, H., and Kaneko, K. (2003) *J. Phys. Chem. B*, **107**, 10974–10976.
45. Tao, Y.S., Tanaka, H., Ohkubo, T., Kanoh, H., and Kaneko, K. (2003) *Adsorpt. Sci. Technol.*, **21**, 199–203.
46. Zhu, H., Liu, Z., Wang, Y., Kong, D., Yuan, X., and Xie, Z. (2008) *Chem. Mater.*, **20**, 1134–1139.
47. Meyers, B.L., Fleisch, T.H., Ray, G.J., Miller, J.T., and Hall, J.B. (1988) *J. Catal.*, **110**, 82–95.
48. Cartlidge, S., Nissen, H.U., and Wessicken, R. (1989) *Zeolites*, **9**, 346–349.
49. Horikoshi, H., Kasahara, S., Fukushima, T., Itabashi, K., Okada, T., Terasaki, O., and Watanabe, D. (1989) *Nippon Kagaku Kaishi*, 398–404.
50. Chauvin, B., Massiani, P., Dutartre, R., Figueras, F., Fajula, F., and Descourieres, T. (1990) *Zeolites*, **10**, 174–182.
51. Guisnet, M., Wang, Q.L., and Giannetto, G. (1990) *Catal. Lett.*, **4**, 299–302.
52. Wang, Q.L., Giannetto, G., and Guisnet, M. (1991) *J. Catal.*, **130**, 471–482.
53. Wang, Q.L., Giannetto, G., Torrealba, M., Perot, G., Kappenstein, C., and Guisnet, M. (1991) *J. Catal.*, **130**, 459–470.
54. Zholobenko, V.L., Kustov, L.M., Kazansky, V.B., Loeffler, E., Lohse, U., and Oehlmann, G. (1991) *Zeolites*, **11**, 132–134.
55. Beyerlein, R.A., Choifeng, C., Hall, J.B., Huggins, B.J., and Ray, G.J. (1994) in *Fluid Catalytic Cracking III – Materials and Processes*, vol. 571 (eds M.L.Occelli and P. Oconnor), ACS, Washington, D.C., pp. 81–97.
56. Nesterenko, N.S., Thibault-Starzyk, F., Montouillout, V., Yuschenko, V.V., Fernandez, C., Gilson, J.P., Fajula, F., and Ivanova, I. (2004) *Microporous Mesoporous Mater.*, **71**, 157–166.

57. Katada, N., Kageyama, Y., Takahara, K., Kanai, T., Begum, H.A., and Niwa, M. (2004) *J. Mol. Catal. A: Chem.*, **211**, 119–130.
58. van Bokhoven, J.A., Tromp, M., Koningsberger, D.C., Miller, J.T., Pieterse, J.A.Z., Lercher, J.A., Williams, B.A., and Kung, H.H. (2001) *J. Catal.*, **202**, 129–140.
59. van Donk, S., Broersma, A., Gijzeman, O.L.J., van Bokhoven, J.A., Bitter, J.H., and de Jong, K.P. (2001) *J. Catal.*, **204**, 272–280.
60. Dessau, R.M., Valyocsik, E.W., and Goeke, N.H. (1992) *Zeolites*, **12**, 776–779.
61. Lietz, G., Schnabel, K.H., Peuker, C., Gross, T., Storek, W., and Völter, J. (1994) *J. Catal.*, **148**, 562–568.
62. Ogura, M., Shinomiya, S.Y., Tateno, J., Nara, Y., Kikuchi, E., and Matsukata, H. (2000) *Chem. Lett.*, 882–883.
63. Ogura, M., Shinomiya, S.Y., Tateno, J., Nara, Y., Nomura, M., Kikuchi, E., and Matsukata, M. (2001) *Appl. Catal. A: Gen.*, **219**, 33–43.
64. Suzuki, T. and Okuhara, T. (2001) *Microporous Mesoporous Mater.*, **43**, 83–89.
65. Su, L.L., Liu, L., Zhuang, J.Q., Wang, H.X., Li, Y.G., Shen, W.J., Xu, Y.D., and Bao, X.H. (2003) *Catal. Lett.*, **91**, 155–167.
66. Goa, Y., Yoshitake, H., Wu, P., and Tatsumi, T. (2004) *Microporous Mesoporous Mater.*, **70**, 93–101.
67. Pavel, C.C., Palkovits, R., Schüth, F., and Schmidt, W. (2008) *J. Catal.*, **254**, 84–90.
68. Pavel, C.C., Park, S.H., Dreier, A., Tesche, B., and Schmidt, W. (2006) *Chem. Mater.*, **18**, 3813–3820.
69. Pavel, C.C. and Schmidt, W. (2006) *Chem. Commun.*, 882–884.
70. Holm, M.S., Egeblad, K., Vennestrom, P.N.R., Hartmann, C.G., Kustova, M., and Christensen, C.H. (2008) *Eur. J. Inorg. Chem.*, 5185–5189.
71. Cavalcante, C.L., Silva, N.M., Souza-Aguiar, E.F., and Sobrinho, E.V. (2003) *Adsorption*, **9**, 205–212.
72. Christensen, C.H., Johannsen, K., Törnqvist, E., Schmidt, I., Topsøe, H., and Christensen, C.H. (2007) *Catal. Today*, **128**, 117–122.
73. Hoang, V.T., Huang, Q.L., Malekian, A., Eic, M., Do, T.O., and Kaliaguine, S. (2005) *Adsorption*, **11**, 421–426.
74. Li, X.F., Prins, R., and van Bokhoven, J.A. (2009) *J. Catal.*, **262**, 257–265.
75. Wei, X.T. and Smirniotis, P.G. (2006) *Microporous Mesoporous Mater.*, **97**, 97–106.
76. Xu, B., Bordiga, S., Prins, R., and van Bokhoven, J.A. (2007) *Appl. Catal. A: Gen.*, **333**, 245–253.
77. Rouquerol, J., Avnir, D., Fairbridge, C.W., Everett, D.H., Haynes, J.H., Pernicone, N., Ramsay, J.D.F., Sing, K.S.W., and Unger, K.K. (1994) *Pure Appl. Chem.*, **66**, 1739–1758.
78. Gheorghiu, S. and Coppens, M.O. (2004) *AICHE J.*, **50**, 812–820.
79. Marcilly, C. (1986) *Pet. Technol.*, **328**, 12–18.
80. Kerr, G.T. (1967) *J. Phys. Chem.*, **71**, 4155–4156.
81. Morin, S., Gnep, N.S., and Guisnet, M. (1998) *Appl. Catal. A: Gen.*, **168**, 63–68.
82. Coster, D., Blumenfeld, A.L., and Fripiat, J.J. (1994) *J. Phys. Chem.*, **98**, 6201–6211.
83. Triantafillidis, C.S., Vlessidis, A.G., and Evmiridis, N.P. (2000) *Ind. Eng. Chem. Res.*, **39**, 307–319.
84. Boréave, A., Auroux, A., and Guimon, C. (1997) *Microporous Mater.*, **11**, 275–291.
85. Patzelová, V. and Jaeger, N.I. (1987) *Zeolites*, **7**, 240–242.
86. Zukal, A., Patzelová, V., and Lohse, U. (1986) *Zeolites*, **6**, 133–136.
87. Lynch, J., Raatz, F., and Dufresne, P. (1987) *Zeolites*, **7**, 333–340.
88. Sasaki, Y., Suzuki, T., Takamura, Y., Saji, A., and Saka, H. (1998) *J. Catal.*, **178**, 94–100.
89. Hong, Y., Gruver, V., and Fripiat, J.J. (1994) *J. Catal.*, **150**, 421–429.
90. Meyers, B.L., Fleisch, T.H., Ray, G.J., Miller, J.T., and Hall, J.B. (1988) *J. Catal.*, **110**, 82–95.
91. Lee, K.H. and Ha, B.H. (1998) *Microporous Mesoporous Mater.*, **23**, 211–219.
92. Dutartre, R., de Ménorval, L.C., Di Renzo, F., McQueen, D., Fajula,

F., and Schulz, P. (1996) *Mesoporous Mater.*, **6**, 311–320.
93. McQueen, D., Chiche, B.H., Fajula, F., Auroux, A., Guimon, C., Fitoussi, F., and Schulz, P. (1996) *J. Catal.*, **161**, 587–596.
94. Pellet, R.J., Casey, D.G., Huang, H.M., Kessler, R.V., Kuhlman, E.J., Oyoung, C.L., Sawicki, R.A., and Ugolini, J.R. (1995) *J. Catal.*, **157**, 423–435.
95. Rozwadowski, M., Komatowski, J., Wloch, J., Erdmann, K., and Golembiewski, R. (2002) *Appl. Surf. Sci.*, **191**, 352–361.
96. Tromp, M., van Bokhoven, J.A., Oostenbrink, M.T.G., Bitter, J.H., de Jong, K.P., and Koningsberger, D.C. (2000) *J. Catal.*, **190**, 209–214.
97. Giudici, R., Kouwenhoven, H.W., and Prins, R. (2000) *Appl. Catal. A: Gen.*, **203**, 101–110.
98. van Donk, S., Bitter, J.H., Verberckmoes, A., Versluijs-Helder, M., Broersma, A., and de Jong, K.P. (2005) *Angew. Chem. Int. Ed.*, **44**, 1360–1363.
99. Olken, M.M. and Garcés, J.M. (1992) in *Proceedings of 9th International Zeolite Conference*, vol. 2, Butterworth-Heinemann, Boston, p. 559.
100. Lee, G.-S.J., Garcés, J.M., Meima, G.R., and van der Aalst, M.J.M. (1989) US Patent 5243116.
101. Sulikowski, B., Borbély, G., Beyer, H.K., Karge, H.G., and Mishin, I.W. (1989) *J. Phys. Chem.*, **93**, 3240–3243.
102. Le van mao, R., Vo, N.T.C., Sjiariel, B., Lee, L., and Denes, G. (1992) *J. Mater. Chem.*, **2**, 595–599.
103. Datka, J., Kolidziejski, W., Klinowski, J., and Sulikowski, B. (1993) *Catal. Lett.*, **19**, 159–165.
104. Kerr, G.T., Chester, A.W., and Olson, D.H. (1994) *Catal. Lett.*, **25**, 401–402.
105. Čižmek, A., Subotić, B., Šmit, I., Tonejc, A., Aiello, R., Crea, F., and Nastro, A. (1997) *Microporous Mater.*, **8**, 159–169.
106. Čižmek, A., Subotić, B., Aiello, R., Crea, F., Nastro, A., and Tuoto, C. (1995) *Microporous Mater.*, **4**, 159–168.
107. Groen, J.C., Peffer, L.A.A., Moulijn, J.A., and Pérez-Ramírez, J. (2005) *Chem. Eur. J.*, **11**, 4983–4994.
108. Groen, J.C., Moulijn, J.A., and Pérez-Ramírez, J. (2007) *Ind. Eng. Chem. Res.*, **46**, 4193–4201.
109. Groen, J.C., Maldonado, L., Berrier, E., Bruckner, A., Moulijn, J.A., and Pérez-Ramírez, J. (2006) *J. Phys. Chem. B*, **110**, 20369–20378.
110. Groen, J.C., Bach, T., Ziese, U., Paulaime-Van Donk, A.M., de Jong, K.P., Moulijn, J.A., and Pérez-Ramírez, J. (2005) *J. Am. Chem. Soc.*, **127**, 10792–10793.
111. Groen, J.C., Moulijn, J.A., and Pérez-Ramírez, J. (2005) *Microporous Mesoporous Mater.*, **87**, 153–161.
112. Groen, J.C., Sano, T., Moulijn, J.A., and Pérez-Ramírez, J. (2007) *J. Catal.*, **251**, 21–27.
113. Groen, J.C., Abello, S., Villaescusa, L.A., and Pérez-Ramírez, J. (2008) *Microporous Mesoporous Mater.*, **114**, 93–102.
114. Groen, J.C., Bruckner, A., Berrier, E., Maldonado, L., Moulijn, J.A., and Pérez-Ramírez, J. (2006) *J. Catal.*, **243**, 212–216.
115. Groen, J.C., Pérez-Ramírez, J., and Peffer, L.A.A. (2002) *Chem. Lett.*, 94–95.
116. Groen, J.C., Zhu, W.D., Brouwer, S., Huynink, S.J., Kapteijn, F., Moulijn, J.A., and Pérez-Ramírez, J. (2007) *J. Am. Chem. Soc.*, **129**, 355–360.
117. Groen, J.C., Caicedo-Realpe, R., Abello, S., and Pérez-Ramírez, J. (2009) *Mater. Lett.*, **63**, 1037–1040.
118. Groen, J.C., Hamminga, G.M., Moulijn, J.A., and Pérez-Ramírez, J. (2007) *Phys. Chem. Chem. Phys.*, **9**, 4822–4830.
119. Groen, J.C., Jansen, J.C., Moulijn, J.A., and Pérez-Ramírez, J. (2004) *J. Phys. Chem. B*, **108**, 13062–13065.
120. Groen, J.C., Peffer, L.A.A., Moulijn, J.A., and Pérez-Ramírez, J. (2004) *Microporous Mesoporous Mater.*, **69**, 29–34.
121. Groen, J.C., Peffer, L.A.A., Moulijn, J.A., and Pérez-Ramírez, J. (2004) *Colloids Surf. A*, **241**, 53–58.
122. Groen, J.C., Peffer, L.A.A., Moulijn, J.A., and Pérez-Ramírez, J. (2005) in *Nanoporous Materials*, IV vol. 156 (eds

A.Sayari and M. Jaroniec), Elsevier, Amsterdam, pp. 401–408.

**123.** Gopalakrishnan, S., Zampieri, A., and Schwieger, W. (2008) *J. Catal.*, **260**, 193–197.

**124.** Bjorgen, M., Joensen, F., Holm, M.S., Olsbye, U., Lillerud, K.P., and Svelle, S. (2008) *Appl. Catal. A: Gen.*, **345**, 43–50.

**125.** Jin, L.J., Zhou, X.J., Hu, H.Q., and Ma, B. (2008) *Catal. Commun.*, **10**, 336–340.

**126.** Choi, D.H., Park, J.W., Kim, J.H., Sugi, Y., and Seo, G. (2006) *Polym. Degrad. Stab.*, **91**, 2860–2866.

**127.** Zhao, L., Shen, B.J., Gao, F.S., and Xu, C.M. (2008) *J. Catal.*, **258**, 228–234.

**128.** Song, Y.Q., Zhu, X.X., Song, Y., Wang, Q.X., and Xu, L.Y. (2006) *Appl. Catal. A: Gen.*, **302**, 69–77.

**129.** Pérez-Ramírez, J., Abello, S., Bonilla, A., and Groen, J.C. (2009) *Adv. Funct. Mater.*, **19**, 164–172.

**130.** Wang, Y.R., Lin, M., and Tuel, A. (2007) *Microporous Mesoporous Mater.*, **102**, 80–85.

**131.** Wang, Y.R. and Tuel, A. (2008) *Microporous Mesoporous Mater.*, **113**, 286–295.

**132.** Holm, M.S., Svelle, S., Joensen, F., Beato, P., Christensen, C.H., Bordiga, S., and Bjørgen, M. (2009) *Appl. Catal. A: Gen.*, **356**, 23–30.

**133.** Kox, M.H.F., Stavitski, E., Groen, J.C., Pérez-Ramírez, J., Kapteijn, F., and Weckhuysen, B.M. (2008) *Chem. Eur. J.*, **14**, 1718–1725.

**134.** Ciesla, U. and Schüth, F. (1999) *Microporous Mesoporous Mater.*, **27**, 131–149.

**135.** He, X. and Antonelli, D. (2002) *Angew. Chem. Int. Ed.*, **41**, 214–229.

**136.** Jacobsen, C.J.H., Madsen, C., Janssens, T.V.W., Jakobsen, H.J., and Skibsted, J. (2000) *Microporous Mesoporous Mater.*, **39**, 393–401.

**137.** Madsen, C. and Jacobsen, C.J.H. (1999) *Chem. Commun.*, 673–674.

**138.** Schmidt, I., Madsen, C., and Jacobsen, C.J.H. (2000) *Inorg. Chem.*, **39**, 2279–2283.

**139.** Chou, Y.H., Cundy, C.S., Garforth, A.A., and Zholobenko, V.L. (2006) *Microporous Mesoporous Mater.*, **89**, 78–87.

**140.** Kustova, M.Y., Rasmussen, S.B., Kustov, A.L., and Christensen, C.H. (2006) *Appl. Catal. B: Environ.*, **67**, 60–67.

**141.** Kustov, A.L., Hansen, T.W., Kustova, M., and Christensen, C.H. (2007) *Appl. Catal. B: Environ.*, **76**, 311–319.

**142.** Schmidt, I., Krogh, A., Wienberg, K., Carlsson, A., Brorson, M., and Jacobsen, C.J.H. (2000) *Chem. Commun.*, 2157–2158.

**143.** Christensen, C.H., Johannsen, K., Schmidt, I., and Christensen, C.H. (2003) *J. Am. Chem. Soc.*, **125**, 13370–13371.

**144.** Christensen, C.H., Schmidt, I., Carlsson, A., Johannsen, K., and Herbst, K. (2005) *J. Am. Chem. Soc.*, **127**, 8098–8102.

**145.** Kustova, M.Y., Hasselriis, P., and Christensen, C.H. (2004) *Catal. Lett.*, **96**, 205–211.

**146.** Wei, X.T. and Smirniotis, P.G. (2006) *Microporous Mesoporous Mater.*, **89**, 170–178.

**147.** Boisen, A., Schmidt, I., Carlsson, A., Dahl, S., Brorson, M., and Jacobsen, C.J.H. (2003) *Chem. Commun.*, 958–959.

**148.** Tao, Y.S., Kanoh, H., and Kaneko, K. (2003) *J. Am. Chem. Soc.*, **125**, 6044–6045.

**149.** Tao, Y., Hattori, Y., Matumoto, A., Kanoh, H., and Kaneko, K. (2005) *J. Phys. Chem. B*, **109**, 194–199.

**150.** Tao, Y.S., Kanoh, H., Hanzawa, Y., and Kaneko, K. (2004) *Colloids Surf. A*, **241**, 75–80.

**151.** Zhu, K., Egeblad, K., and Christensen, C.H. (2007) *Eur. J. Inorg. Chem.*, 3955–3960.

**152.** Kustova, M., Egeblad, K., Zhu, K., and Christensen, C.H. (2007) *Chem. Mater.*, **19**, 2915–2917.

**153.** Fang, Y.M. and Hu, H.Q. (2006) *J. Am. Chem. Soc.*, **128**, 10636–10637.

**154.** Fang, Y.M. and Hu, H.Q. (2007) *Catal. Commun.*, **8**, 817–820.

**155.** Li, H.C., Sakamoto, Y., Liu, Z., Ohsuna, T., Terasaki, O., Thommes, M., and Che, S.N. (2007) *Microporous Mesoporous Mater.*, **106**, 174–179.

**156.** Kim, S.S., Shah, J., and Pinnavaia, T.J. (2003) *Chem. Mater.*, **15**, 1664–1668.

**157.** Yoo, W.C., Kumar, S., Wang, Z.Y., Ergang, N.S., Fan, W., Karanikolos, G.N., McCormick, A.V., Penn, R.L., Tsapatsis, M., and Stein, A. (2008) *Angew. Chem. Int. Ed.*, **47**, 9096–9099.

**158.** Fan, W., Snyder, M.A., Kumar, S., Lee, P.S., Yoo, W.C., McCormick, A.V., Penn, R.L., Stein, A., and Tsapatsis, M. (2008) *Nat. Mater.*, **7**, 984–991.

**159.** Egeblad, K., Kustova, M., Klitgaard, S.K., Zhu, K.K., and Christensen, C.H. (2007) *Microporous Mesoporous Mater.*, **101**, 214–223.

**160.** Christensen, C.H., Schmidt, I., and Christensen, C.H. (2004) *Catal. Commun.*, **5**, 543–546.

**161.** Kustova, M.Y., Kustov, A., Christiansen, S.E., Leth, K.T., Rasmussen, S.B., and Christensen, C.H. (2006) *Catal. Commun.*, **7**, 705–708.

**162.** Li, W.C., Lu, A.H., Palkovits, R., Schmidt, W., Spliethoff, B., and Schüth, F. (2005) *J. Am. Chem. Soc.*, **127**, 12595–12600.

**163.** Tao, Y.S., Kanoh, H., and Kaneko, K. (2005) *Langmuir*, **21**, 504–507.

**164.** Holland, B.T., Abrams, L., and Stein, A. (1999) *J. Am. Chem. Soc.*, **121**, 4308–4309.

**165.** Rhodes, K.H., Davis, S.A., Caruso, F., Zhang, B.J., and Mann, S. (2000) *Chem. Mater.*, **12**, 2832–2834.

**166.** Naydenov, V., Tosheva, L., and Sterte, J. (2003) *Microporous Mesoporous Mater.*, **66**, 321–329.

**167.** Naydenov, V., Tosheva, L., and Sterte, J. (2002) *Chem. Mater.*, **14**, 4881–4885.

**168.** Lee, Y.J., Lee, J.S., Park, Y.S., and Yoon, K.B. (2001) *Adv. Mater.*, **13**, 1259–1263.

**169.** Zhang, B.J., Davis, S.A., Mendelson, N.H., and Mann, S. (2000) *Chem. Commun.*, 781–782.

**170.** Dong, A.G., Wang, Y.J., Tang, Y., Ren, N., Zhang, Y.H., Yue, J.H., and Gao, Z. (2002) *Adv. Mater.*, **14**, 926–929.

**171.** Valtchev, V., Smaihi, M., Faust, A.C., and Vidal, L. (2003) *Angew. Chem. Int. Ed.*, **42**, 2782–2785.

**172.** Valtchev, V.P., Smaihi, M., Faust, A.C., and Vidal, L. (2004) *Chem. Mater.*, **16**, 1350–1355.

**173.** Beck, J.S., Vartuli, J.C., Roth, W.J., Leonowicz, M.E., Kresge, C.T., Schmitt, K.D., Chu, C.T.W., Olson, D.H., and Sheppard, E.W. (1992) *J. Am. Chem. Soc.*, **114**, 10834–10843.

**174.** Zhao, D.Y., Huo, Q.S., Feng, J.L., Chmelka, B.F., and Stucky, G.D. (1998) *J. Am. Chem. Soc.*, **120**, 6024–6036.

**175.** Bagshaw, S.A., Baxter, N.I., Brew, D.R.M., Hosie, C.F., Nie, Y.T., Jaenicke, S., and Khuan, C.G. (2006) *J. Mater. Chem.*, **16**, 2235–2244.

**176.** Bagshaw, S.A., Jaenicke, S., and Khuan, C.G. (2003) *Catal. Commun.*, **4**, 140–146.

**177.** Liu, Y. and Pinnavaia, T.J. (2002) *Chem. Mater.*, **14**, 3–5.

**178.** Liu, Y. and Pinnavaia, T.J. (2002) *J. Mater. Chem.*, **12**, 3179–3190.

**179.** Liu, Y. and Pinnavaia, T.J. (2004) *J. Mater. Chem.*, **14**, 1099–1103.

**180.** Liu, Y., Zhang, W.Z., and Pinnavaia, T.J. (2000) *J. Am. Chem. Soc.*, **122**, 8791–8792.

**181.** Liu, Y., Zhang, W.Z., and Pinnavaia, T.J. (2001) *Angew. Chem. Int. Ed.*, **40**, 1255–1258.

**182.** Ooi, Y.S., Zakaria, R., Mohamed, A.R., and Bhatia, S. (2004) *Appl. Catal. A: Gen.*, **274**, 15–23.

**183.** Zhang, Z.T., Han, Y., Zhu, L., Wang, R.W., Yu, Y., Qiu, S.L., Zhao, D.Y., and Xiao, F.S. (2001) *Angew. Chem. Int. Ed.*, **40**, 1258–1262.

**184.** Han, Y., Xiao, F.S., Wu, S., Sun, Y.Y., Meng, X.J., Li, D.S., Lin, S., Deng, F., and Ai, X.J. (2001) *J. Phys. Chem. B*, **105**, 7963–7966.

**185.** Han, Y., Wu, S., Sun, Y.Y., Li, D.S., and Xiao, F.S. (2002) *Chem. Mater.*, **14**, 1144–1148.

**186.** Xiao, F.S., Han, Y., Yu, Y., Meng, X.J., Yang, M., and Wu, S. (2002) *J. Am. Chem. Soc.*, **124**, 888–889.

**187.** Sun, Y.Y., Han, Y., Yuan, L.N., Ma, S.Q., Jiang, D.Z., and Xiao, F.S. (2003) *J. Phys. Chem. B*, **107**, 1853–1857.

**188.** Di, Y., Yu, Y., Sun, Y.Y., Yang, X.Y., Lin, S., Zhang, M.Y., Li, S.G., and Xiao, F.S. (2003) *Microporous Mesoporous Mater.*, **62**, 221–228.

**189.** Lin, K.F., Sun, Z.H., Sen, L., Jiang, D.Z., and Xiao, F.S. (2004) *Microporous Mesoporous Mater.*, **72**, 193–201.

190. Xia, Y.D. and Mokaya, R. (2004) *J. Mater. Chem.*, **14**, 863–870.
191. Agúndez, J., Díaz, I., Márquez-Álvarez, C., Pérez-Pariente, J., and Sastre, E. (2003) *Chem. Commun.*, 150–151.
192. Wang, S., Dou, T., Li, Y.P., Zhang, Y., Li, X.F., and Yan, Z.C. (2005) *Catal. Commun.*, **6**, 87–91.
193. Ivanova, I.I., Kuznetsov, A.S., Yuschenko, V.V., and Knyazeva, E.E. (2004) *Pure Appl. Chem.*, **76**, 1647–1658.
194. Goto, Y., Fukushima, Y., Ratu, P., Imada, Y., Kubota, Y., Sugi, Y., Ogura, M., and Matsukata, M. (2002) *J. Porous Mater.*, **9**, 43–48.
195. Wang, H., Liu, Y., and Pinnavaia, T.J. (2006) *J. Phys. Chem. B*, **110**, 4524–4526.
196. Liu, Y. and Pinnavaia, T.J. (2004) *J. Mater. Chem.*, **14**, 3416–3420.
197. Han, Y., Li, N., Zhao, L., Li, D.F., Xu, X.Z., Wu, S., Di, Y., Li, C.J., Zou, Y.C., Yu, Y., and Xiao, F.S. (2003) *J. Phys. Chem. B*, **107**, 7551–7556.
198. Prokesova-Fojokova, P., Mintova, S., Čejka, J., Zilkova, N., and Zukal, A. (2006) *Microporous Mesoporous Mater.*, **92**, 154–160.
199. Karlsson, A., Stöcker, M., and Schmidt, R. (1999) *Microporous Mesoporous Mater.*, **27**, 181–192.
200. Huang, L.M., Guo, W.P., Deng, P., Xue, Z.Y., and Li, Q.Z. (2000) *J. Phys. Chem. B*, **104**, 2817–2823.
201. Liu, Y., Zhang, W.P., Liu, Z.C., Xu, S.T., Wang, Y.D., Xie, Z.K., Han, X.W., and Bao, X.H. (2008) *J. Phys. Chem. C*, **112**, 15375–15381.
202. Majano, G., Mintova, S., Ovsitser, O., Mihailova, B., and Bein, T. (2005) *Microporous Mesoporous Mater.*, **80**, 227–235.
203. Fang, Y.M., Hu, H.Q., and Chen, G.H. (2008) *Chem. Mater.*, **20**, 1670–1672.
204. Gagea, B.C., Liang, D., van Tendeloo, G., Martens, J.A., and Jacobs, P.A. (2006) *Stud. Surf. Sci. Catal.*, **162**, 259–266.
205. Stevens, W.J.J., Meynen, V., Bruijn, E., Lebedev, O.I., van Tendeloo, G., Cool, P., and Vansant, E.F. (2008) *Microporous Mesoporous Mater.*, **110**, 77–85.
206. Wang, J., Groen, J.C., and Coppens, M.O. (2008) *J. Phys. Chem. C*, **112**, 19336–19345.
207. Wang, J., Groen, J.C., Yue, W., Zhou, W., and Coppens, M.O. (2007) *Chem. Commun.*, 4653–4655.
208. Wang, J., Groen, J.C., Yue, W., Zhou, W., and Coppens, M.O. (2008) *J. Mater. Chem.*, **18**, 468–474.
209. Wang, J., Yue, W.B., Zhou, W.Z., and Coppens, M.O. (2009) *Microporous Mesoporous Mater.*, **120**, 19–28.
210. Sing, K.S.W., Everett, D.H., Haul, R.A.W., Moscou, L., Pierotti, R.A., Rouquerol, J., and Siemieniewska, T. (1985) *Pure Appl. Chem.*, **57**, 603–619.
211. Storck, S., Bretinger, H., and Maier, W.F. (1998) *Appl. Catal. A: Gen.*, **174**, 137–146.
212. Brunauer, S., Emmett, P.H., and Teller, E. (1938) *J. Am. Chem. Soc.*, **60**, 309–319.
213. Gregg, S.J. and Sing, K.S.W. (1982) *Adsorption, Surface Area and Porosity*, 2nd edn, Academic Press, London.
214. Janssen, A.H., Koster, A.J., and de Jong, K.P. (2002) *J. Phys. Chem. B*, **106**, 11905–11909.
215. Lippens, B.C., Linsen, B.G., and de Boer, J.H. (1964) *J. Catal.*, **3**, 32–37.
216. Jaroniec, M., Madey, R., Choma, J., McEnaney, B., and Mays, T.J. (1989) *Carbon*, **27**, 77–83.
217. Jaroniec, M., Kruk, M., and Olivier, J.P. (1999) *Langmuir*, **15**, 5410–5413.
218. Barret, E.P., Joyner, L.G., and Halenda, P.H. (1951) *J. Am. Chem. Soc.*, **73**, 373–380.
219. Ball, P.C. and Evan, R. (1984) *Langmuir*, **5**, 714–723.
220. Ravikovitch, P.I. and Neimark, A.V. (2002) *Langmuir*, **18**, 9830–9837.
221. Ravikovitch, P.I. and Neimark, A.V. (2002) *Langmuir*, **18**, 1550–1560.
222. Groen, J.C. (2007) *Mesoporus Zeolites Obtained by Desilication*, PhD thesis, Technische Universiteit Delft.
223. Kadlec, O. and Dubinin, M.M. (1969) *J. Colloid Interface Sci.*, **31**, 479–489.
224. Burgess, C.G.V. and Everett, D.H. (1970) *J. Colloid Interface Sci.*, **33**, 611–614.
225. Seaton, N.A. (1991) *Chem. Eng. Sci.*, **46**, 1895–1909.

**226.** Ojeda, M.L., Esparza, J.M., Campero, A., Cordero, S., Kornhauser, I., and Rojas, F. (2003) *Phys. Chem. Chem. Phys.*, **5**, 1859–1866.

**227.** Ravikovitch, P.I., Wei, D., Chueh, W.T., Haller, G.L., and Neimark, A.V. (1997) *J. Phys. Chem. B*, **101**, 3671–3679.

**228.** Ravikovitch, P.I. and Neimark, A.V. (2001) *J. Phys. Chem. B*, **105**, 6817–6823.

**229.** Lukens, W.W., Schmidt-Winkel, P., Zhao, D.Y., Feng, J.L., and Stucky, G.D. (1999) *Langmuir*, **15**, 5403–5409.

**230.** Kruk, M., Jaroniec, M., and Sayari, A. (1997) *Langmuir*, **13**, 6267–6273.

**231.** Neimark, A.V. (1995) *Langmuir*, **11**, 4183–4184.

**232.** Brun, M., Lallemand, A., Quinson, J.-F., and Eyraud, C. (1977) *Thermochim. Acta*, **21**, 59–88.

**233.** Denoyel, R. and Pellenq, R.J.M. (2002) *Langmuir*, **18**, 2710–2716.

**234.** Robens, E., Benzler, B., and Unger, K.K. (1999) *J. Therm. Anal. Calorim.*, **56**, 323–330.

**235.** Cuperus, F.P., Bargeman, D., and Smolders, C.A. (1992) *J. Membr. Sci.*, **66**, 45–53.

**236.** Iza, M., Woerly, S., Danumah, C., Kaliaguine, S., and Bousmina, M. (2000) *Polymer*, **41**, 5885–5893.

**237.** Ferguson, H.F., Frurip, D.J., Pastor, A.J., Peerey, L.M., and Whiting, L.F. (2000) *Thermochim. Acta*, **363**, 1–21.

**238.** Jallut, C., Lenoir, J., Bardot, C., and Eyraud, C. (1992) *J. Membr. Sci.*, **68**, 271–282.

**239.** Faivre, C., Bellet, D., and Dolino, G. (1999) *Eur. Phys. J. B*, **7**, 19–36.

**240.** Ishikiriyama, K., Todoki, M., and Motomura, K. (1995) *J. Colloid Interface Sci.*, **171**, 92–102.

**241.** Endo, A., Yamamoto, T., Inagi, Y., Iwakabe, K., and Ohmori, T. (2008) *J. Phys. Chem. C*, **112**, 9034–9039.

**242.** Janssen, A.H., Talsma, H., van Steenbergen, M.J., and de Jong, K.P. (2004) *Langmuir*, **20**, 41–45.

**243.** León y León, C.A. (1998) *Adv. Colloid Interface Sci.*, **76-77**, 341–372.

**244.** Lohse, U. and Mildebrath, M. (1981) *Z. Anorg. Allg. Chem.*, **476**, 126–135.

**245.** Groen, J.C., Brouwer, S., Peffer, L.A.A., and Pérez-Ramírez, J. (2006) *Part. Part. Syst. Char.*, **23**, 101–106.

**246.** Groen, J.C., Peffer, L.A.A., and Pérez-Ramírez, J. (2003) *Microporous Mesoporous Mater.*, **60**, 1–17.

**247.** Gai, P.L. and Buyes, E.D. (2003) *Electron Microscopy in Heterogeneous Catalysts*, Institute of Physics, Bristol.

**248.** Ohsuna, T., Terasaki, O., Watanabe, D., Anderson, M.W., and Carr, S.W. (1994) *Chem. Mater.*, **6**, 2201–2204.

**249.** Ersen, O., Hirlimann, C., Drillon, M., Werckmann, J., Tihay, F., Pham-Huu, C., Crucifix, C., and Schultz, P. (2007) *Solid State Sci.*, **9**, 1088–1098.

**250.** de Jong, K.P. and Koster, A.J. (2002) *ChemPhysChem*, **3**, 776–780.

**251.** Midgley, P.A. and Weyland, M. (2003) *Ultramicroscopy*, **96**, 413–431.

**252.** Deans, S.R. (1983) *The Radon Transform and Some of its Applications*, John Wiley & Sons, Inc., New York.

**253.** Friedrich, H., de Jongh, P.E., Verkleij, A.J., and de Jong, K.P. (2009) *Chem. Rev.*, **109**, 1613–1629.

**254.** Koster, A.J., Ziese, U., Verkleij, A.J., Janssen, A.H., and de Jong, K.P. (2000) *J. Phys. Chem. B*, **104**, 9368–9370.

**255.** Janssen, A.H., Koster, A.J., and de Jong, K.P. (2001) *Angew. Chem. Int. Ed.*, **40**, 1102–1104.

**256.** Koster, A.J., Ziese, U., Verkleij, A.J., de Graaf, J., Gues, J.W., and de Jong, K.P. (2000) *Stud. Surf. Sci. Catal.*, **130**, 329–334.

**257.** Ziese, U., Gommes, C.J., Blacher, S., Janssen, A.H., Koster, A.J., and de Jong, K.P. (2005) *Stud. Surf. Sci. Catal.*, **158**, 633–638.

**258.** Springuel-Huet, M.A., Bonardet, J.L., Gédéon, A., and Fraissard, J. (1999) *Magn. Reson. Chem.*, **37**, S1–S13.

**259.** Dybowski, C., Bansal, N., and Duncan, T.M. (1991) *Annu. Rev. Phys. Chem.*, **42**, 433–464.

**260.** Fraissard, J. and Ito, T. (1988) *Zeolites*, **8**, 350–361.

**261.** Kneller, J.M., Pietrass, T., Ott, K.C., and Labouriau, A. (2003) *Microporous Mesoporous Mater.*, **62**, 121–131.

**262.** Springuel-Huet, M.A. and Fraissard, J.P. (1992) *Zeolites*, **12**, 841–845.

263. Cotterman, R.L., Hickson, D.A., Cartlidge, S., Dybowski, C., Tsiao, C., and Venero, A.F. (1991) *Zeolites*, **11**, 27–34.
264. Raftery, D., MacNamara, E., Fisher, G., Rice, C.V., and Smith, J. (1997) *J. Am. Chem. Soc.*, **119**, 8746–8747.
265. Cotts, R.M., Hoch, M.J.R., Sun, T., and Markert, J.T. (1989) *J. Magn. Reson.*, **83**, 252–266.
266. Kortunov, P., Vasenkov, S., Kärger, J., Valiullin, R., Gottschalk, P., Elía, M.F., Perez, M., Stöcker, M., Drescher, B., McElhiney, G., Berger, C., Gläser, R., and Weitkamp, J. (2005) *J. Am. Chem. Soc.*, **127**, 13055–13059.
267. Schmidt, R., Hansen, E.W., Stöcker, M., Akporiaye, D., and Ellestad, O.H. (1995) *J. Am. Chem. Soc.*, **117**, 4049–4056.
268. Roeffaers, M.B.J., Hofkens, J., de Cremer, G., de Schryver, F.C., Jacobs, P.A., de Vos, D.E., and Sels, B.F. (2007) *Catal. Today*, **126**, 44–53.
269. Roeffaers, M.B.J., Sels, B.F., Uji-i, H., de Schryver, F.C., Jacobs, P.A., and de Vos, D.E. (2006) *Nature*, **439**, 572–575.
270. Mores, D., Stavitski, E., Kox, M.H.F., Kornatowski, J., Olsbye, U., and Weckhuysen, B.M. (2008) *Chem. Eur. J.*, **14**, 11320–11327.
271. Stavitski, E., Kox, M.H.F., and Weckhuysen, B.M. (2007) *Chem. Eur. J.*, **13**, 7057–7065.
272. Kox, M.H.F., Stavitski, E., and Weckhuysen, B.M. (2007) *Angew. Chem. Int. Ed.*, **46**, 3652–3655.
273. Karwacki, L., Stavitski, E., Kox, M.H.F., Kornatowski, J., and Weckhuysen, B.M. (2007) *Angew. Chem. Int. Ed.*, **46**, 7228–7231.

# 10
# Aluminum in Zeolites: Where is it and What is its Structure?

*Jeroen A. van Bokhoven and Nadiya Danilina*

## 10.1
## Introduction

Aluminum in a zeolite framework position introduces the negative charge in the framework that requires a cation to balance it. This gives zeolites much of their properties, such as cation exchange capacity and the ability to catalyze reactions. The number of active sites often scales with the number of aluminum atoms in the framework. The zeolite hydrophobicity and hydrophilicity also strongly depend on the aluminum content, which affect the adsorption of polar and apolar molecules into the zeolite [1, 2]. Thus, besides structure, the silicon to aluminum ratio determines the zeolite performance. The silicon to aluminum ratio can be accurately determined by using a combination of characterization methods, such as $^{29}$Si and $^{27}$Al magic-angle spinning nuclear magnetic resonance (MAS NMR) and elemental analysis. From the $^{27}$Al MAS NMR spectra the framework and extra-framework nature of aluminum can be also established. The aluminum coordination strongly varies with the pretreatment condition to which the zeolite was exposed. For example, exposing an acidic zeolite to moisture causes partial dealumination [3, 4]. Such structural changes are to a large extent reversible and strongly affect the catalytic performance of the zeolite. Steaming, a treatment that is used at a very large scale in industry to improve zeolite performance, causes dealumination, formation of new aluminum species, creation of Lewis acidity, and the formation of secondary mesopores. This treatment succeeds in improving the zeolite performance in terms of activity and stability in catalytic cracking, alkylation, acylation, and so on [5].

An additional factor that affects the zeolite properties is the distribution of aluminum. This can be regarded in two manners, which affect the zeolite performance in a different way. The first is the zoning of aluminum over the individual crystals and the second is the distribution over the crystallographic T sites. A T site is a crystallographic position, which is occupied by a tetrahedrally coordinated atom, usually silicon and aluminum. The number and degeneracy of T sites and the spatial aluminum distribution in a zeolite vary with zeolite structure type [6, 7].

*Zeolites and Catalysis, Synthesis, Reactions and Applications. Vol. 1.*
Edited by Jiří Čejka, Avelino Corma, and Stacey Zones

ISBN: 978-3-527-32514-6

For most of the zeolite structure types, the aluminum distribution remains unidentified.

Recent progress in the above-mentioned fields will be described using selected examples of the literature.

## 10.2 Structure of Aluminum Species in Zeolites

Framework aluminum is tetrahedrally coordinated and $^{27}$Al MAS NMR under hydrated conditions typically represents this species as a narrow resonance at a chemical shift around 60 ppm. The exact chemical shift depends on the local structure of the aluminum and varies with the averaged Al–O–Si angles according to

$$\delta_{\text{iso}} = -0.5\theta + 132 \tag{10.1}$$

This empirical formula has been successfully used to interpret $^{27}$Al MAS NMR data of various zeolites and in some cases, aluminums that occupy different crystallographic T sites were resolved, which is described further below.

The aluminum coordination strongly varies with the exact pretreatment condition. High temperature steam-activation or treatment of the zeolite in the presence of moisture results in partial collapse of the zeolite framework, which in addition to the tetrahedrally coordinated framework aluminum leads to various other aluminum species. These can be three- and five-coordinated and (distorted) tetrahedrally and octahedrally coordinated. Such species can be identified by different characterization tools, such as NMR [8–10], X-ray photoelectron spectroscopy (XPS) [11], and X-ray absorption spectroscopy (XAS) [12–14]. Because of its high resolution and widespread availability of magnets, NMR is the most used method. High field magnets and the use of advanced pulse schemes, such as multiple quantum-NMR, provide high-resolution data that yield the aluminum coordination unambiguously. Various schemes to provide a quantitative analysis have been described [15, 16].

Zeolite samples are generally hydrated before performing an NMR measurement, because NMR spectra of hydrated zeolites identify the framework aluminum as the narrow resonance around 60 ppm. Non-hydrated or dehydrated samples show very broad and often overlapping resonances, because aluminum is a quadrupolar nucleus, which is sensitive to its three-dimensional surrounding. Resonances can even become broadened beyond detection; the corresponding aluminum species are called *NMR-invisible*. However, recent technology and the correct use of experimental conditions should in general enable a quantitative detection of all aluminum in a sample [15]. Thus, the aluminum structure under dehydrated conditions has been successfully determined by $^{27}$Al MAS NMR. The framework and extra-framework aluminum of non-hydrated zeolites were determined of a steamed zeolite Y of silicon to aluminum ratio of 2.7 using a spin-echo pulse scheme [16]. Various aluminum species were detected: Tetrahedrally coordinated framework aluminums that were charge balanced by protons and/or sodium, respectively extra-framework aluminum, present in form of extra-framework aluminum cations and octahedrally coordinated aluminum in neutral extra-framework aluminum oxide clusters. The kind and amount of species significantly differed from those that are typically detected in steamed zeolites that were measured after exposure to moisture. In such sample, in addition to the typical tetrahedrally coordinated framework aluminum, a large amount of distorted tetrahedrally coordinated aluminum was detected, two types of octahedrally coordinated aluminum, which differed in their symmetry, and a small fraction of five-coordinated aluminum [17]. Moisture clearly plays a dominant role in affecting the aluminum coordination. *In situ* XAS at the Al K edge has identified a strong variation of the aluminum coordination as function of temperature [18]. The fraction of octahedrally coordinated aluminum, which is present in steamed zeolite Y (USY) when measured at room temperature in the presence of moisture, decreased upon increasing the temperature [3, 19]. About 50% of the octahedral aluminum lowered its coordination to tetrahedral aluminum upon heating to 400 °C. Heating a zeolite to a temperature in excess of 400 °C causes only small changes in the zeolite structure. An under-coordinated, most probably threefold coordinated aluminum species were detected even in the presence of small quantities of moisture [20]. Thus, when performing a catalytic reaction at high temperature, the structure that is typically measured by $^{27}$Al MAS NMR is not representative of the structure of the catalyst under catalytically relevant conditions. This might be one of the reasons, why the role of extra-framework aluminum species on catalytic performance has remained poorly understood for so long.

### 10.2.1
### Reversible versus Irreversible Structural Changes

A reversibility of the aluminum coordination in zeolites has been shown for many zeolites [3, 21–23]. The charge-compensating cation in zeolite beta was shown to strongly affect the aluminum coordination. Proton-exchanged beta showed a large fraction of up to 25% of octahedrally coordinated aluminum, which was quantitatively reverted to tetrahedral coordination after ion exchange with alkali cations and also after treatment with ammonia [21]. *In situ* Al K edge XAS [18] and $^{27}$Al MAS NMR [21–23] showed that part of the framework tetrahedrally coordinated aluminum in proton charge-balanced zeolites is converted into an octahedral coordination at room temperature. This octahedrally coordinated aluminum is unstable and reverts into the tetrahedral coordination by heating above 100 °C [18]. However, the presence of base-like ammonia is required to quantitatively restore the framework structure [21, 22]. Because of the ability of the aluminum atom to reversibly change its coordination, the octahedrally coordinated aluminum has been called *framework-associated aluminum*. The amount of framework-associated aluminum that forms in a zeolite depends on the silicon to aluminum ratio and the zeolite framework type [3]. The higher the aluminum content, the more octahedral aluminum forms. Among the different zeolites, the aluminum atoms in zeolite beta show a strong tendency of adopting the octahedral coordination.

### 10.2.2
### Cautionary Note

Various characterization methods require different pretreatments of the zeolite samples. Infrared spectroscopy is measured after heating a zeolite at high temperature to remove moisture from the sample, $^{27}$Al MAS NMR is measured after exposure to moisture to obtain high-resolution spectra, XPS is measured in vacuum, and so on. As a result, the structure that is probed differs.

### 10.2.3
### Development of Activity and Changing Aluminum Coordination

The formation of framework-associated octahedrally coordinated aluminum is related to a loss of catalytic performance [24, 25]. The development of the aluminum coordination was compared to the high-temperature propane cracking activity. Zeolite Y with silicon to aluminum ratio of 2.6 was treated according to Scheme 10.1.

Four samples were formed, which were characterized by NMR and infrared spectroscopy, X-ray diffraction, and nitrogen physisorption (Figures10.1–10.3 and Table 10.1). $NH_4Y$ contained only tetrahedrally coordinated aluminum as indicated by a narrow resonance around 60 pp m in the $^{27}$Al MAS NMR spectrum. However, exposure to moisture after the high-temperature treatment caused the formation of octahedrally coordinated aluminum indicated by the resonance at about 0 ppm (Figure 10.1). The transformation from framework tetrahedral to octahedral coordination occurs already at room temperature [18, 19, 23]. Removal of ammonia at high temperature and measuring an infrared spectrum shows the typical stretching hydroxyl groups in the supercage and sodalite cage at about 3640 and 3550 $cm^{-1}$, respectively (Figure 10.2). Also the XRD pattern showed loss of crystallinity indicated by the increased width of the diffraction peaks (Figure 10.3). Treatment with ammonia at 150 °C before measuring infrared, $^{27}$Al MAS NMR, and XRD yielded virtually identical spectra and patterns as the parent material. This

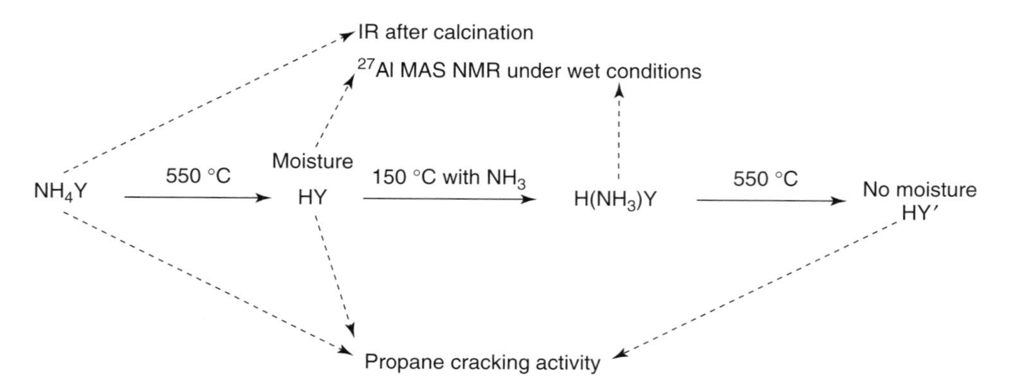


**Scheme 10.1** Treatment scheme of zeolite Y before kinetic measurement and characterization.

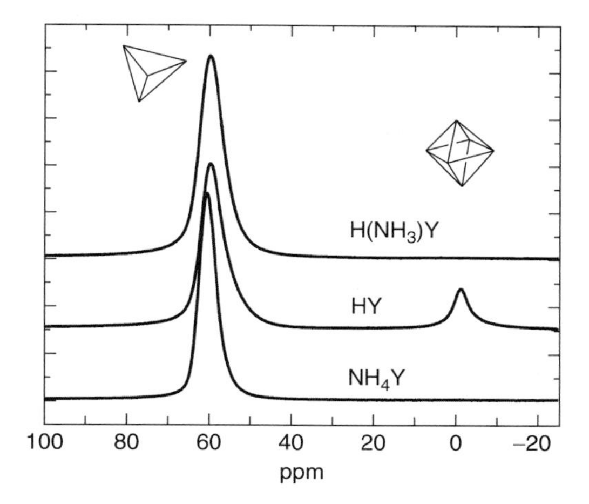


**Figure 10.1** $^{27}Al$ MAS NMR spectra of zeolite Y, treated according to Scheme 10.1.

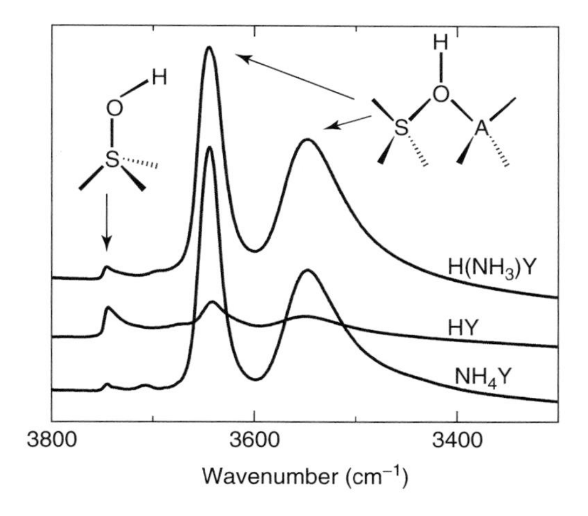


**Figure 10.2** FTIR spectra of zeolite Y, treated according to Scheme 10.1.

shows that the original structure can be completely recovered and that complete re-insertion of now-tetrahedral aluminum into the framework occurs. However, heating HY prior to treatment with ammonia causes additional structural damage that cannot be simply repaired by treatment under ammonia (Scheme 10.2). The catalytic activity of the parent $NH_4Y$ and ammonia-treated sample $H(NH_3)Y$ were virtually identical and much higher than that of the HY. That the catalytic activity in HY was much lower can be understood from the infrared spectrum, which shows a very low number of Brønsted acid sites (Figure 10.2). The heating of HY during the pretreatment of the infrared measurement, respectively, catalytic reaction causes a significant structural damage. A pretreatment of HY at high temperature (550 °C) before measuring nitrogen physisorption also showed a micropore volume and

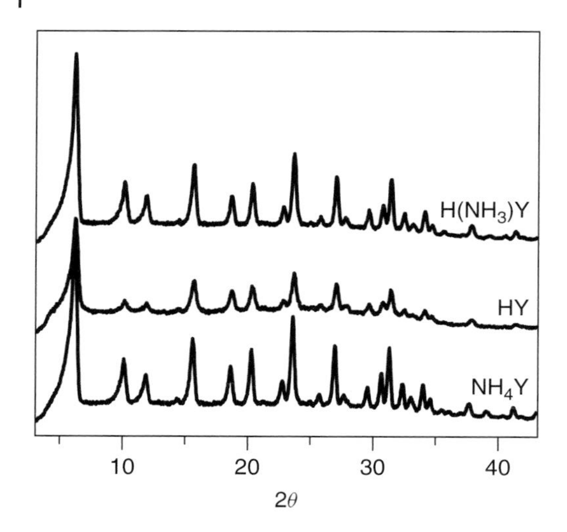


**Figure 10.3** XRD patterns of zeolite Y, treated according to Scheme 10.1.

**Table 10.1** Results of nitrogen physisorption for zeolite Y, treated according to Scheme 10.1.

| Sample | BET surface area ($m^{2g-1}$) | Micropore volume ($cm^{3g-1}$) |
|---|---|---|
| $NH_4Y$ | 810 | 0.32 |
| HY | 450 | 0.18 |
| $H(NH_3)Y$ | 770 | 0.31 |
| HY′ | 230 | 0.09 |

BET surface area of 0.09 $cm^3\ g^{-1}$ and 230 $m^2\ g^{-1}$ compared to 0.32 $cm^3\ g^{-1}$ and 810 $m^2\ g^{-1}$ for $NH_4Y$ (Table 10.1). Apparently, the defects that were formed after the exposure of the sample to moisture are starting points for the further structural collapse, which occurs upon heating. Unlike the initial formation of octahedrally coordinated aluminum, these structural changes are much more severe and irreversible. Scheme 10.2 (adapted from [26]) summarizes the structural changes that occur after the various treatments.

The role of steaming on generating highly active and selective catalysts has been debated in the literature for over decades [13, 27–29]. Because of the limitations in structure determination under catalytically relevant conditions in the past, structure–performance relations do not exist for many zeolite-catalyzed reaction systems. The complexity of chemical reactions over zeolites makes it likely that different reactions are affected differently by steaming. The extent of influence

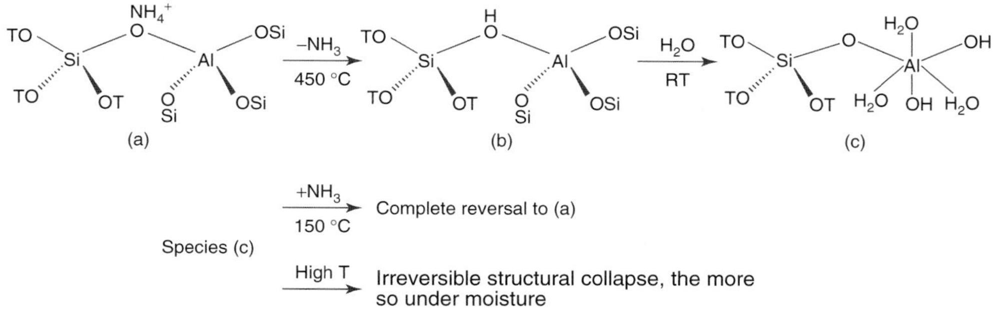

(a) $\xrightarrow[450\ °C]{-NH_3}$ (b) $\xrightarrow[RT]{H_2O}$ (c)

Species (c) $\xrightarrow[150\ °C]{+NH_3}$ Complete reversal to (a)

Species (c) $\xrightarrow{High\ T}$ Irreversible structural collapse, the more so under moisture

**Scheme 10.2** Development of zeolite structure under various conditions.

of extra-framework species to activate the reactants [30], of mesopores to alleviate diffusion limitation [31], and of isolation of the Brønsted acid sites [32] will depend on the kind of reaction that is performed as well as on the reaction conditions.

## 10.3 Where is the Aluminum in Zeolite Crystals?

Zeolite performance is affected by many parameters, mainly by zeolite structure type, silicon to aluminum ratio, and the distribution of aluminum. In a single crystal of zeolite of a particular structure type, the aluminum may not be homogeneously distributed throughout the crystal, a phenomenon that is called *zoning*. This spatial aluminum distribution should be differentiated from the aluminum and silicon distribution over the crystallographic T sites within the zeolite structure [33–35]. Although recent progress has been made in this field, the assignment of aluminum and silicon distributions to their occupancy of the crystallographic T sites remains one of the largest challenges in the structure determination of zeolites.

### 10.3.1 Aluminum Zoning

The physical properties and catalytic behavior of aluminosilicates are closely related to the aluminum content and distribution in their frameworks and over the crystal. It is well known that aluminum is often non-uniformly distributed in different zeolites [36, 37]. This aluminum zoning plays an important role in the design of catalytically active sites and their accessibility. The occurrence and kind of aluminum zoning in zeolites depend on many factors, such as zeolite framework [38, 39], silicon to aluminum ratio [40], crystal size [39], and synthesis parameters [35, 41–43].

The aluminum distribution on the surface and in the bulk of ZSM-5 has been most investigated, sometimes giving contradictory results. A heterogeneous aluminum distribution in large crystals of ZSM-5 was described by electron microprobe analysis [36, 39]. The crystals were synthesized in the presence of tetrapropylammonium ions and were about 40 μm in size and had silicon to aluminum ratio between 50 and 100. The analysis of a volume of about 2 $\mu m^3$ of the sample grinded with a diamond paste to remove the outer layers of the crystals showed a silicon-rich core and an aluminum-rich rim. The aluminum concentration in the rim was up to 10 times higher than in the core of the crystal. The zoning was symmetric and the surface aluminum concentration was fairly constant. The non-uniform aluminum distribution was explained by the crystallization mechanism of ZSM-5. Initially, an aluminum-rich gel is formed at the bottom of the reactor and the MFI phase nucleates from the silica-rich solution. At a later stage, aluminum, released from the gel, is progressively incorporated into the ZSM-5 framework. A similar behavior was reported for smaller ZSM-5 crystals of about 1 μm [37, 44]. The spatial aluminum distribution was found to be dependent on the synthesis route [41, 43]. By using different templates, the spatial distribution of aluminum in ZSM-5 crystals can thus be controlled. Tetrapropylammonium ions as template lead to strongly zoned profiles with the aluminum enriched in the outer shell of the crystals. In contrast, using 1,6-hexanediol as template or using a completely inorganic reaction mixture leads to homogeneous aluminum profiles. An explanation of the difference between the templates was given in terms of the interaction with the template molecule. Tetrapropylammonium ions in the presence of alkali ions interact preferentially with the silicate species, thus leading to incorporation of these species in the growing crystals. In a tetrapropylammonium ion-free system, the sodium ions, which interact strongly with aluminate species, are assumed to play the templating role. Nevertheless, there are studies confirming the homogeneous aluminum distribution across the tetrapropylammonium-ZSM-5 crystals of varying size [38, 42] or even a presence of a siliceous outer surface and an aluminum-rich interior [45]. In these studies surface-sensitive techniques, such as Auger electron spectroscopy and electron microprobe analysis, were used to obtain the depth profiles. It was reported that different Si/Na, Si/tetrapropylammonium ion, and Si/$H_2O$ ratios in the synthesis gel can lead to reverse silicon to aluminum ratio profiles across a crystal [41]. An aluminum-rich core was obtained, when low Si/Na and Si/tetrapropylammonium ion ratios and larger amount of water were used in the synthesis. The influence of the nature and the amount of alkali cations introduced into the synthesis gel on the aluminum distribution were also studied [46]. It was found that the counter-ion, such as lithium, sodium, potassium, rubidium, and cesium, does not influence the surface aluminum concentration and its distribution. In these studies, crystals of different size (0.4–100 μm) were compared. The surface analysis was performed using X-ray photoelectron spectroscopy and energy-dispersive X-ray spectroscopy, while the bulk composition was analyzed by proton-induced gamma-ray emission. The penetration depth for these techniques is about 10 nm, 1 μm, and 10 μm, respectively [41].

Only a few studies deal with aluminum zoning throughout the crystals of zeolites other than ZSM-5. Zeolite beta, synthesized using tetrapropylammonium hydroxide, showed an aluminum-rich core and silicon-rich shell, a behavior that is reverse compared to what is generally observed in ZSM-5 crystals [47]. The same, however, was observed for Na-X crystals [48]. Based on a theoretical study, aluminum-enrichment on the surface of the crystals was reported for zeolite Y [49]. On the other hand, applying Auger electron spectroscopy [38], secondary ion mass spectrometry [7], and fast atom bombardment mass spectrometry [50], it was shown that the aluminum distribution in zeolites Y, A, X, and ZSM-5 was homogeneous. It is obvious that the choice of the right technique is crucial to characterize the aluminum zoning. A reliable method may be scanning electron microscopy with energy-dispersive X-ray mapping [30, 32]. Because the penetration depth of the electron beam is about 1 μm, the crystal should be cut or the outer layers removed to insure the analysis is performed throughout the whole volume and not restricted only to the surface. Rubbing off surface layers or polishing the crystal result in a rough surface or it may break or crush crystals, which will affect the signal. A smooth cut of the crystal can be obtained by using a focused ion beam, Figure 10.4 shows a large crystal of ZSM-5, which was cut with a focused ion beam and its aluminum, silicon, and oxygen distribution using the energy-dispersive X-ray mapping (not published results). This crystal was synthesized in the presence of tetrapropylammonium ions as template. Most of the aluminum is concentrated in the 2–3 μm broad rim of the crystal at the expense of silicon.

Desilication of the crystals and subsequent scanning electron microscopy analysis were also used to visualize aluminum zoning [41, 51]. Base-treatment might change the elemental distribution and/or affect the defect concentration. Figure 10.5 shows the scanning electron micrograph of a base-treated ZSM-5 crystal, of which the original was shown in Figure 10.4. Their morphology, hollow coffins, confirmed that the crystals had a silicon-rich core, which was leached out, and an aluminum-rich rim.

**Figure 10.4** SE micrograph of the FIB cut crystal (a) and EDX (energy dispersive X-ray spectroscopy) maps, representing aluminum (b), silicon (c), and oxygen (d) distribution.

**Figure 10.5** SEM images of large crystals of NaOH-treated ZSM-5.

## 10.3.2
## Aluminum Distribution Over the Crystallographic T Sites

The chemical shift in $^{27}Al$ MAS NMR is determined by the local geometry of the aluminum atom. Equation (10.1) gave the empirical formula that relates the averaged T-O-T angle to chemical shift. In certain cases, the framework aluminum atoms are partially resolved in $^{27}Al$ MAS NMR [52]. Figure 10.6 shows an example of zeolite beta. Two partially overlapping resonances, referred to as *Td1* and *Td2*, represent the tetrahedrally coordinated framework aluminum. The different chemical shift of Td1 and Td2 originates from the different local structure of the aluminum atoms that occupy different T sites, notably the averaged T-O-T

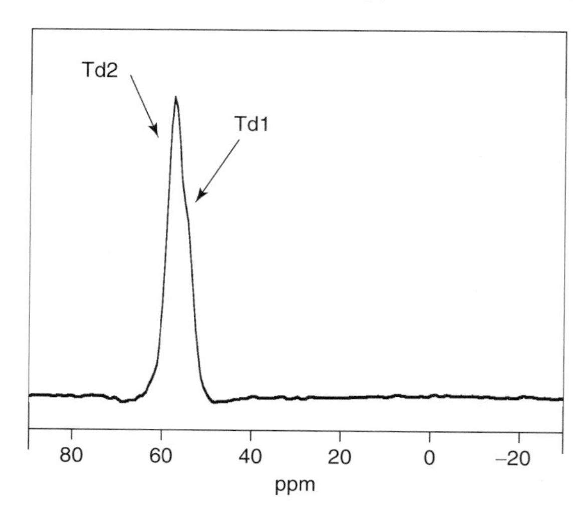


**Figure 10.6** $^{27}Al$ MAS NMR of zeolite beta.

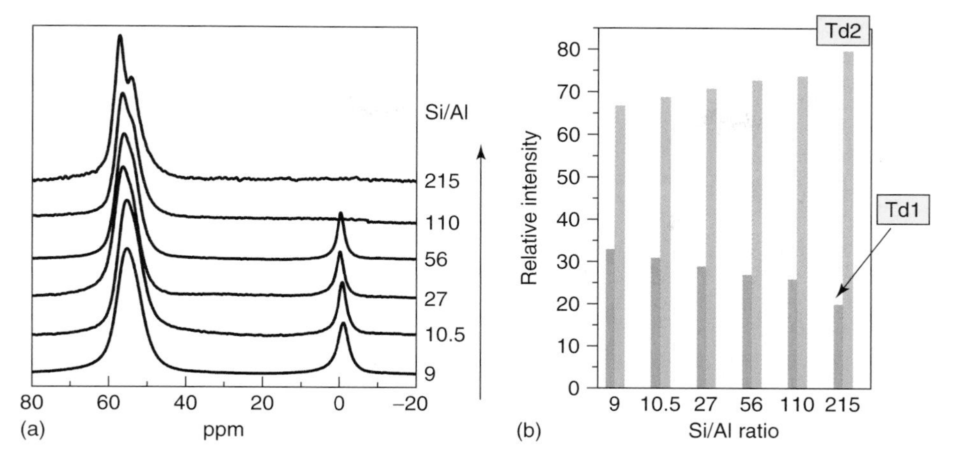


**Figure 10.7** (a) $^{27}$Al MAS NMR of zeolite beta of different silicon to aluminum ratios and (b) relative intensities of Td1 (blue) and Td2 (magenta).

angle (Eq. (10.1)) [22]. Zeolite beta can be synthesized with different silicon to aluminum ratio. Quantification of the $^{27}$Al MAS NMR spectra of these indicated a change in the relative ratio of intensity of the two contributions (Figure 10.7a,b). The higher the silicon to aluminum ratio was, the higher the relative intensity of Td2. This shows that depending on the aluminum content of the framework, the aluminum has the tendency to occupy different crystallographic sites. Similar results were obtained for ZSM-5 [53], mordenite [54], and $NH_4$-Y [55]. In a previous section, the instability of framework aluminum in proton-exchanged zeolites was described. After converting the zeolite beta to the proton form, a large fraction of octahedrally coordinated aluminum appeared (Figure 10.7a). Interestingly, only one of tetrahedral resonances decreased intensity, viz. Td2. A more harsh treatment even removed most of the framework aluminum that was associated with Td2, while the intensity of the resonance Td1 only slightly changed. This indicates that the stability of framework aluminum depends on the specific aluminum occupancy of T sites. Moreover, the higher the silicon to aluminum ratio is, the lower the fraction of octahedrally coordinated aluminum (Figure 10.7a), in all cases, originating mainly from Td2. The fewer aluminum atoms occupy the framework, the more stable they are to moisture as described above. Obviously, a zeolite with very high silicon to aluminum ratio is hydrophobic, which prevents water to adsorb in the pores, which thus protects the framework aluminum from being exposed to water.

More recently, $^{27}$Al MQ MAS NMR on ZSM-5 samples that were synthesized using different methods identified that the aluminum distribution depends on the synthesis parameters [53]. Using mixed quantum mechanics/molecular dynamics calculations a more precise correlation between chemical shift and occupancy of T sites was proposed [56]. Throughout the set of samples, 10 resonances were identified, the assignment of which was tentative for only 4 resonances. Thus, a much more precise assignment of aluminum onto crystallographic sites could be performed. This represents a significant improvement in interpreting $^{27}$Al MAS NMR spectra compared to the previous studies. The effect of using different organic molecules as structure-directing agents on the distribution of aluminum over the T sites was recently investigated for ferrierite [57, 58]. Different ferrierite samples were synthesized in the absence of sodium cations by using 1-benzyl-1-methylpyrrolidinium, tetramethylammonium cations, and pyrrolidine. The distribution of the acid sites of the zeolite was associated to the aluminum distribution. The acidity of the materials was investigated by FTIR spectroscopy and pyridine adsorption, and furthermore, correlated to their activity in m-xylene and n-butene isomerization. The distribution of bridging hydroxyl groups that were located in the ferrierite cage accessible through windows of eight-membered rings or in the 10-membered ring channel varied with the specific combination of templates used in the synthesis. The dependence of aluminum distribution over the crystallographic T sites in ZSM-5 on the silicon to aluminum ratio and the source of silicon and aluminum was shown by using cobalt(II) ions as probes for the so-called "Al pairs," [Al-O-(Si-O)$_{1,2}$-Al] [34, 35, 59]. $Co^{2+}$ ions in dehydrated zeolites are suggested to be coordinated only to framework oxygen atoms and balanced by two framework $AlO_2^-$ tetrahedrals. The concentration of $Co^{2+}$ in Co-exchanged zeolites corresponds to the concentration of the "Al pairs" in the zeolite framework. The distribution of $Co^{2+}$ ions at cationic sites can be estimated from UV–Vis spectra.

The non-random aluminum distribution was also found for mordenite [54, 60, 61], zeolite beta [60], zeolite omega [62], faujasite [55], ferrierite [57, 58], and merlinoite [63]. NMR spectroscopy was mainly used to determine the aluminum and silicon distribution. $CD_3CN$ and benzene adsorption measurements and DFT quantum chemical calculations showed that aluminum is preferentially sitting in the T site of mordenite, located at the bottom of the side pockets [60, 61]. By fitting the results of the numerical simulated annealing calculations to the experimental data obtained by $^{27}$Al and $^{29}$Si MAS NMR, it was suggested that aluminum preferentially occupies the tetrahedral site in the six-membered rings of zeolite omega [62]. Although the majority suggests the aluminum distribution in the zeolite framework to be non-random, a few groups concluded a random aluminum distribution based on the $^{27}$Al MQ, $^{29}$Si, and $^{1}$H MAS NMR experiments [64–67].

The above-mentioned studies were able to partially resolve the aluminum distribution and further progress is expected to be made. Recently, an alternative method to determine the silicon and aluminum distribution over the T sites has been proposed [33]. The method is based on the formation of an X-ray standing wave, which is generated in a single crystal of a zeolite. When an X-ray beam is diffracted, the incoming X-ray and the diffracted one interfere with each other. They thus generate a standing wave, which is parallel to the d-spacing. Thus, when looking perpendicular to the d-planes, there are certain positions in the unit cell, which experience X-ray intensity by positive interference and areas that have no or less intensity by negative interference. Figure 10.8 shows a schematic representation of a standing wave along a unit cell. The yellow planes represent the

**Figure 10.8** Schematic representation of an X-ray standing wave in the scolecite structure.

maximal intensity of the standing wave that is generated by interference between the initial and diffracted X-ray of the (040) reflection in zeolite scolecite, which is a naturally occurring zeolite that has two crystallographic T sites. Now, if an element is positioned where X-ray intensity is high, thus, at the position of one of the yellow planes in Figure 10.6, this atom will be excited and emits fluorescence radiation, whose energy is typical of its Z number. By detecting the fluorescence radiation in an energy-dispersive manner, aluminum and silicon atoms can be distinguished. Changing the energy of the incoming X-ray will vary the interference pattern and move the intensity perpendicularly to the d-planes. Alternatively, varying the incoming angle of the X-ray to the d-plane will also cause the standing wave to propagate through the unit cell. In such an experiment, the occupancy of the crystallographic sites by aluminum or silicon can be unambiguously determined by simulation of the intensity modulation using classical scattering theory. This synchrotron experiment has been explored on zeolite scolecite. It was unambiguously illustrated that aluminum preferentially occupies a single T site. Simulations showed that there are no fundamental reasons, why such measurement could not be performed on zeolite crystals with a structure as complex as that of ZSM-5 if it has silicon to aluminum ratio of about 25. The most important limitation is the required minimal crystal size, which depends on the state of the art equipment at the synchrotron beam line.

Important additional information about the local structure surrounding aluminum sites has come from simulations [68–72]. The simulated annealing, the procedure of standard Monte Carlo finite temperature techniques, based on the statistical mechanics and combinatorial optimization [73], was used to study the aluminum distribution in zeolite omega [62]. The aluminum partitioning ratio between the two crystallographically unequivalent sites in the framework of zeolite omega was obtained from $^{27}$Al MAS NMR spectra and fitted by trial and error method to the numerically calculated values. It has been concluded that aluminum is non-randomly distributed in the framework, which is in a good agreement with the experimental results described above. The possible aluminum and proton sitting in the zeolite ITQ-7 was proposed, combining force field atomistic simulations and FTIR experiments [74]. It was assumed that aluminum distribution is controlled by the energetics during the synthesis process and that the interaction between the structure-directing agent and the zeolite framework has to be involved in the model. Based on this assumption, variation of the structure-directing agent can be employed to control the distribution of framework negative charge in zeolites. Similar results were obtained for FAU and EMT zeolite structures using a combination of $^{29}$Si MAS NMR and model generation by computer algorithm [6]. These findings were recently fortified by the experimental results [57, 58]. Recently, an alternative theoretical approach to determine the aluminum distribution in the zeolite framework was presented [68]. Experimentally accessible properties, such as adsorption and diffusion, are crucially dependent on the aluminum distribution and associated cation distribution and can be directly correlated to those. Molecular simulations have been performed to obtain the adsorption and diffusion properties of alkanes in a variety of zeolitic structures by varying the position of the aluminum atoms. Henry coefficients of linear alkanes were computed for MOR, FER, and TON zeolite structures at a fixed silicon to aluminum ratio by varying the aluminum distribution. Direct comparison with the experimentally obtained coefficients allowed conclusions about preferential aluminum sitting. Nevertheless, theoretical approaches using *ab initio* calculations [75] or Monte Carlo simulations [76–79] face severe limitations. Small clusters used for *ab initio* computations cannot be used for description of the spatial distribution of aluminum. Monte Carlo simulations are based on the assumption that aluminum distribution in zeolites is random. Moreover, theoretical studies cannot reflect the potential dynamics effect occurring during zeolite synthesis.

The overall conclusion is that the aluminum distribution in the zeolite framework is neither random, nor controlled by a simple rule, but that it depends on the conditions of the zeolite synthesis. The aluminum distribution is not dominated by their relative stability in the various sites, but by the conditions of the synthesis. Such prospect opens interesting possibilities to tune the catalytic performance in terms of activity and selectivity.

## 10.4
## Summary and Outlook

Aluminum in zeolites is essential for their performance. This makes its analysis indispensable in terms of its structure, its position within a crystal, and its occupancy of crystallographic T sites. The structure of a zeolite is strongly dependent on the pretreatment as well as on the conditions of measurement. The zeolite structure is especially sensitive to moisture. At room temperature, part of the tetrahedrally coordinated aluminum in a proton-exchanged zeolite can be converted into octahedral aluminum, the amount of which depends on zeolite structure type and silicon to aluminum ratio. This change in coordination is completely reversible after treatment with a base, such as ammonia. Zeolite activation by steam causes large structural changes. The amount and kind of aluminum species depend on the extent of hydration and temperature. Hydration increases the amount of high-coordinated aluminum. Dehydration and/or increasing the temperature convert part of the octahedral coordination into a tetrahedral one. Moreover, at temperature higher than 400 °C, three-coordinated aluminum forms. These defect sites and the framework-associated octahedrally coordinated aluminum make the zeolite less stable and more sensitive to further heating. This sensitivity of aluminum species to different pretreatment conditions makes it difficult to understand their influence on catalytic performance. To determine structure–selectivity relationship for aluminum species in catalytic processes, *in situ* characterization techniques are required.

Aluminum zoning has been widely reported. Its extent mainly depends on the synthesis conditions. Using tetrapropylammonium ions in the synthesis of ZSM-5 produces a silicon-rich core and aluminum-rich outer rim. Other templates might provide different distributions, however. This enables control of the distribution in terms of zoning of the aluminum. For the characterization of aluminum zoning, the choice of the right technique is crucial.

Large, but incomplete, progress has been obtained in determining the aluminum and silicon occupancy of the crystallographic T sites. This distribution is not dominated by the thermodynamic stability of the T site occupying either aluminum or silicon, but it is determined by the synthesis conditions. NMR, X-ray scattering (X-ray standing waves), and computer simulation techniques are emerging that promises obtaining further insight.

The prospect of being able to control the aluminum zoning as well as its occupancy over the crystallographic T sites opens exciting possibilities to synthesize zeolite material with unique properties and unprecedented activity and high selectivity.

The structure of catalytically active sites, notably those of Lewis acid sites, and the location of aluminum in the crystalline framework remain two of the most pressing questions in structure-analysis in the field of zeolites. The physical properties and catalytic behavior of aluminosilicates depend on the aluminum content, the structure of aluminum species, and aluminum distribution in the zeolite framework and over the crystal. The aluminum content can be unambiguously obtained from the elemental analysis, but it is still challenging to identify and characterize the structure of the aluminum species, especially under catalytically relevant conditions. Moreover, to analyze the aluminum zoning and foremost to clearly assign the aluminums to the specific T sites remains challenging. In the field of NMR, there is a constant development of advanced techniques, such as two-dimensional multiple-quantum magic-angle spinning NMR, $^{27}$Al magic-angle spinning/multiple-quantum MAS and $^{27}$Al–$^{14}$N transfer of population in double-resonance NMR [80], and 2D $^{27}Al \rightarrow ^{29}Si$ rotor-assisted population transfer and Carr-Purcell-Meiboom-Gill heteronuclear correlation NMR [81], which promise to being able to further clarify the structure of aluminum. As the aluminum coordination strongly varies with the pretreatment conditions, it is very important to characterize it *in situ*. *In situ* Al K edge (XANES) X-ray absorption near edge structure [20] and *in situ* NMR [82] have opened up this possibility. It is likely that the structure of aluminum in zeolites under pretreatment and reaction conditions will be unambiguously determined in the near future. Thus, the structure of the catalytically active site and its interaction with reactants and intermediates under reaction conditions will be accessible. The structure of Lewis acid sites requires particular attention and a combination of methods that enable to determine the Lewis acid character in terms of amount and strength of the aluminum species, such as infrared and NMR spectroscopy, are particularly promising.

The knowledge of the aluminum distribution over the crystallographic T sites and thus the location of the catalytically active sites would allow exactly correlating the catalytic performance of aluminosilicates in terms of activity, selectivity, and stability. Numerous techniques have been applied or developed to characterize the aluminum distribution over the crystallographic T sites in a zeolite framework. The various methods based on NMR, X-ray standing waves, UV–Vis, theory, and so on, are likely to further develop and a combined approach is most promising.

## Acknowledgment

We acknowledge the Swiss National Science Foundation for financial support.

## References

1. Barthomeuf, D. (1980) *Stud. Surf. Sci. Catal.*, **5**, 55.
2. Llewellyn, P.L., Grillet, Y., and Rouquerol, J. (1994) *Langmuir*, **10**, 570.
3. Omegna, A., van Bokhoven, J.A., and Prins, R. (2003) *J. Phys. Chem. B*, **107**, 8854.
4. Xu, B., Rotunno, F., Bordiga, S., Prins, R., and van Bokhoven, J.A. (2006) *J. Catal.*, **241**, 66.
5. Szostak, R. (1991) *Stud. Surf. Sci. Catal.*, **58**, 153.
6. Feijen, E.J.P., Lievens, J.L., Martens, J.A., Grobet, P.J., and Jacobs, P.A. (1996) *J. Phys. Chem.*, **100**, 4970.
7. Dwyer, J., Fitch, F.R., Machado, F., Qin, G., Smyth, S.M., and Vickerman, J.C. (1981) *J. Chem. Soc., Chem. Commun.*, 422.
8. Pfeifer, H., Freude, D., and Hunger, M. (1985) *Zeolites*, **5**, 274.
9. Anderson, M.W. and Klinowski, J. (1989) *Nature*, **339**, 200.
10. Ramdas, S. and Klinowski, J. (1984) *Nature*, **308**, 521.
11. Collignon, F., Jacobs, P.A., Grobet, P., and Poncelet, G. (2001) *J. Phys. Chem. B*, **105**, 6812.
12. Joyner, R.W., Smith, A.D., Stockenhuber, M., and van den Berg, M.W.E. (2004) *Phys. Chem. Chem. Phys.*, **6**, 5435.
13. van Bokhoven, J.A., Kunkeler, P.J., van Bekkum, H., and Koningsberger, D.C. (2002) *J. Catal.*, **211**, 540.
14. van Bokhoven, J.A., Sambe, H., Ramaker, D.E., and Koningsberger, D.C. (1999) *J. Phys. Chem. B*, **103**, 7557.

15. Kraus, H., Müller, M., Prins, R., and Kentgens, A.P.M. (1998) *J. Phys. Chem. B*, **102**, 3862.
16. Jiao, J., Kanellopoulos, J., Wang, W., Ray, S.S., Foerster, H., Freude, D., and Hunger, M. (2005) *Phys. Chem. Chem. Phys.*, **7**, 3221.
17. van Bokhoven, J.A., Roest, A.L., Koningsberger, D.C., Miller, J.T., Nachtegaal, G.H., and Kentgens, A.P.M. (2000) *J. Phys. Chem. B*, **104**, 6743.
18. van Bokhoven, J.A., van der Eerden, A.M.J., and Koningsberger, D.C. (2002) *Stud. Surf. Sci. Catal.*, **142**, 1885.
19. Omegna, A., Prins, R., and van Bokhoven, J.A. (2005) *J. Phys.Chem. B*, **109**, 9280.
20. van Bokhoven, J.A., van der Eerden, A.M.J., and Koningsberger, D.C. (2003) *J. Am. Chem. Soc.*, **125**, 7435.
21. Bourgeatlami, E., Massiani, P., Direnzo, F., Espiau, P., Fajula, F., and Courieres, T.D. (1991) *Appl. Catal.*, **721**, 139.
22. van Bokhoven, J.A., Koningsberger, D.C., Kunkeler, P., van Bekkum, H., and Kentgens, A.P.M. (2000) *J. Am. Chem. Soc.*, **122**, 12842.
23. Wouters, B.H., Chen, T.H., and Grobet, P.J. (1998) *J. Am. Chem. Soc.*, **120**, 11419.
24. Katada, N., Kanai, T., and Niwa, M. (2004) *Microporous Mesoporous Matter.*, **75**, 61.
25. Xu, B., Sievers, C., Hong, S.B., Prins, R., and van Bokhoven, J.A. (2006) *J. Catal.*, **244**, 163.
26. Xu, B. (2007) Structure-performance relationships in solid-acid aluminosilicates, Doctoral thesis, ETH, 16922.
27. Kung, H.H., Williams, B.A., Babitz, S.M., Miller, J.T., Haag, W.O., and Snurr, R.Q. (2000) *Top. Catal*, **10**, 59.
28. Beyerlein, R.A., Choi-Feng, C., Hall, J.B., Huggins, B.J., and Ray, G.J. (1997) *Top. Catal*, **4**, 27.
29. Zholobenko, V.L., Kustov, L.M., Kazansky, V.B., Loeffler, E., Lohse, U., and Oehlmann, G. (1991) *Zeolites*, **11**, 132.
30. Lukyanov, D.B. (1991) *Zeolites*, **11**, 325.
31. Williams, B.A., Babitz, S.M., Miller, J.T., Snurr, R.Q., and Kung, H.H. (1999) *Appl. Catal. A*, **177**, 161.
32. Xu, B., Bordiga, S., Prins, R., and van Bokhoven, J.A. (2007) *Appl. Catal. A*, **333**, 245.
33. van Bokhoven, J.A., Lee, T.-L., Drakopoulos, M., Lamberti, C., Thiess, S., and Zegenhagen, J. (2008) *Nat. Mater.*, **7**, 551.
34. Dedecek, J., Kaucky, D., and Wichterlova, B. (2001) *Chem. Commun.*, 970.
35. Gabova, V., Dedecek, J., and Čejka, J. (2003) *Chem. Commun.*, 1196.
36. von Ballmoos, R. and Meier, W.M. (1981) *Nature*, **289**, 782.
37. Derouane, E.G., Gilson, J.P., Gabelica, Z., Mousty-Desbuquoit, C., and Verbist, J. (1981) *J. Catal.*, **71**, 447.
38. Suib, S.L. and Stucky, G.D. (1980) *J. Catal.*, **65**, 174.
39. Chao, K.-J. and Chern, J.-Y. (1988) *Zeolites*, **8**, 82.
40. Sklenak, S., Dedecek, J., Lo, C., Wichterlova, B., Gabova, V., Sierka, M., and Sauer, J. (2009) *Phys. Chem. Chem. Phys.*, **11**, 1237.
41. Debras, G., Gourgue, A., and Nagy, J.B. (1985) *Zeolites*, **5**, 369.
42. Lin, J.-C. and Chao, K.-J. (1986) *J. Chem. Soc., Faraday Trans. 1*, **82**, 2645.
43. Althoff, R., Schulz-Dobrick, B., Schüth, F., and Unger, K. (1993) *Microporous Mater.*, **1**, 207.
44. Dessau, R.M., Valyocsik, E.W., and Goeke, N.H. (1992) *Zeolites*, **12**, 776.
45. Hughes, A.E., Wilshier, K.G., Sexton, B.A., and Smart, P. (1983) *J. Catal.*, **80**, 221.
46. Nagy, J.B., Bodart, P., Collette, H., El Hage-Al Asswad, J., and Gabelica, Z. (1988) *Zeolites*, **8**, 209.
47. Perez-Pariente, J., Martens, J.A., and Jacobs, P.A. (1986) *Appl. Catal.*, **31**, 35.
48. Weeks, T.J. and Passoja, D.E. (1977) *Clays Clay Miner.*, **25**, 211.
49. Corma, A., Melo, F.V., and Rawlence, D.J. (1990) *Zeolites*, **10**, 690.
50. Dwyer, J., Fitch, F.R., Qin, G., and Vickerman, J.C. (1982) *J. Phys. Chem.*, **86**, 4574.
51. Groen, J.C., Bach, T., Ziese, U., Paulaime-van Donk, A.M., de Jong, K.P., Moulijn, J.A., and Perez-Ramirez, J. (2005) *J. Am. Chem. Soc.*, **127**, 10792.

52. Abraham, A., Lee, S.H., Shin, C.H., Hong, S.B., Prins, R., and van Bokhoven, J.A. (2004) *Phys. Chem. Chem. Phys.*, **6**, 3031.
53. Han, O.H., Kim, C.S., and Hong, S.B. (2002) *Angew. Chem. Int. Ed. Engl.*, **41**, 469.
54. Lu, B., Kanai, T., Oumi, Y., and Sano, T.J. (2007) *Porous Mater.*, **14**, 89.
55. Koranyj, T.I. and Nagy, J.B. (2007) *J. Phys. Chem. C*, **111**, 2520.
56. Sklenak, S., Dědeček, J., Li, C., Wichterlová, B., Gábová, V., Sierka, M., and Sauer, J. (2007) *Angew. Chem. Int. Ed. Engl.*, **46**, 7286.
57. Pinar, A.B., Marquez-Alvarez, C., Grande-Casas, M., and Perez-Pariente, J. (2009) *J. Catal.*, **263**, 258.
58. Marquez-Alvarez, C., Pinar, A.B., Garcia, R., Grande-Casas, M., and Perez-Pariente, J. (2009) *Top. Catal*, **52**, 1281.
59. Dedecek, J., Kaucky, D., Wichterlova, B., and Gonsiorova, O. (2002) *Phys. Chem. Chem. Phys.*, **4**, 5406.
60. Koranyj, T.I. and Nagy, J.B. (2005) *J. Phys. Chem. B*, **109**, 15791.
61. Bodart, P., Nagy, J.B., Debras, G., Gabelica, Z., and Jacobs, P.A. (1986) *J. Phys. Chem.*, **90**, 5183.
62. Li, B., Sun, P., Jin, Q., Wang, J., and Ding, D. (1999) *J. Mol. Catal. A*, **148**, 189.
63. Kennedy, G.J., Afeworki, M., and Hong, S.B. (2002) *Microporous Mesoporous Mater.*, **52**, 55.
64. Smith, J.V. (1971) in *Molecular Sieve Zeolites-1* (eds E.M.Flaningen and L.B. Sand), American Chemical Society, Washington, DC, p. 171.
65. Sarv, P., Fernandez, C., Amoureux, J.-P., and Keskinen, K. (1996) *J. Phys. Chem.*, **100**, 19223.
66. Jarman, R.H., Jacobson, A.J., and Melchior, M.T. (1984) *J. Phys. Chem.*, **88**, 5748.
67. Raatz, F., Roussel, J.C., Cantiani, R., Ferre, G., and Nagy, J.B. (1987) in *Innovation in Zeolite Materials Science* (eds P.J.Gorbet, W. Morties, W.J. Vansant and E.F. Schultz-Ekloff), Elsevier Science Publishers B.V., Amsterdam, p. 301.
68. Garcia-Perez, E., Dubbeldam, D., Liu, B., Smit, B., and Calero, S. (2007) *Angew. Chem. Int. Ed. Engl.*, **46**, 276.
69. (a) Ehresmann, J.O., Wang, W., Herreros, B., Luigi, D.P., Venkatraman, T.N., Song, W.G., Nicholas, J.B., and Haw, J.F. (2002) *J. Am. Chem. Soc.*, **124**, 10868; (b) Brandle, M., Sauer, J., Dovesi, R., and Harrison, N.M. (1998) *J. Chem. Phys.*, **109**, 10379.
70. Eichler, U., Brandle, M., and Sauer, J. (1997) *J. Phys. Chem. B*, **101**, 10035.
71. Beerdsen, E., Smit, B., and Calero, S. (2002) *J. Phys. Chem. B*, **106**, 10659.
72. Beerdsen, E., Dubbledam, D., Smit, B., Vlugt, T.J.H., and Calero, S. (2003) *J. Phys. Chem. B*, **107**, 12088.
73. Kirkpatrick, S., Gelatt, C.D., and Vecchi, M.P. Jr. (1983) *Science*, **220**, 671.
74. Sastre, G., Fornes, V., and Corma, A. (2002) *J. Phys. Chem. B*, **106**, 701.
75. Derouane, E.G. and Fripiat, J.G. (1985) *Zeolites*, **5**, 165.
76. Takaishi, T. and Kato, M. (1995) *Zeolites*, **15**, 689.
77. Takaishi, T., Kato, M., and Itabashi, K. (1995) *Zeolites*, **15**, 21.
78. Feng, X. and Hall, W.K. (1997) *Catal. Lett.*, **46**, 11.
79. Rice, M.J., Chakraborty, A.K., and Bell, A.T. (1999) *J. Catal.*, **186**, 222.
80. Abraham, A., Prins, R., van Bokhoven, J.A., van Eck, E.R.H., and Kentgens, A.P.M. (2009) *Solid State Nucl. Magn. Reson.*, **35**, 61.
81. Kennedy, G., Wiench, J.W., and Pruski, M. (2008) *Solid State Nucl. Magn. Reson.*, **33**, 76.
82. Hunger, M. (2004) *Catal. Today*, **97**, 3.

# 11
# Theoretical Chemistry of Zeolite Reactivity

*Evgeny A. Pidko and Rutger A. van Santen*

## 11.1
## Introduction

Nowadays, advances in chemical theory rely to a significant extent on computations. Novel theoretical concepts arising directly from experiments usually require further investigations using computational modeling. This is necessary to develop a conclusive molecular-level picture of the observed phenomenon. As a result, computational methods are nowadays widely and extensively applied in the chemical, physical, biomedical, and engineering sciences. They are used in assisting the interpretation of experimental data and increasingly in predicting the properties and behavior of matter at the atomic level.

Computational simulation techniques can be divided into two very broad categories. The first is based on the use of interatomic potentials (force fields). These methods are usually empirical and do not consider explicitly any electron in the system. The system of interest is described with functions (normally analytical), which expresses its energy as a function of nuclear coordinates. These are then used to calculate structures and energies by means of minimization methods, to calculate ensemble averages using Monte Carlo simulations, or to model dynamic processes via molecular dynamics simulations using classical Newton's law of motion. The current capabilities and challenges for the corresponding computational methods have been recently reviewed in excellent papers by Woodley and Catlow [1] and by Smit and Maesen [2, 3].

The second category includes quantum chemical methods based on the calculation of the electronic structure of the system. Such methods are particularly important for processes that depend on bond breaking or making, which include, of course, catalytic reactions. Hartree Fock (HF), Density Functional Theory (DFT), and post-HF *ab initio* approaches have been used in modeling zeolites, although DFT methods have predominated in recent applications.

This chapter focuses on illustrating the power and capabilities of modern quantum chemical techniques, and aims at highlighting the key areas and challenges

*Zeolites and Catalysis, Synthesis, Reactions and Applications. Vol. 1.*
Edited by Jiří Čejka, Avelino Corma, and Stacey Zones

ISBN: 978-3-527-32514-6

in theoretical chemistry of zeolites and microporous materials. The text is organized as follows. First, we briefly discuss the methodological aspects of quantum chemical modeling of zeolites (Section 11.2). This will involve a brief review of the capabilities and limitations of the traditional electronic structure methods and structural models used for studying microporous materials and their reactivity. Section 11.3 will illustrate the current advances in quantum chemical calculations of zeolite-catalyzed transformations of hydrocarbons. The approaches toward improvement of DFT with respect to a better description of weak van der Waals interactions will be discussed. In the next Section 11.4, the methanol-to-olefin (MTO), which is an industrially important process will be used to illustrate the power of computational methods in revealing the molecular mechanism of such complex catalytic reaction. Sections 11.5 and 11.6 will show how novel reactivity and structural concepts can be derived from theoretical calculations to rationalize the experimental observations. In these sections, we will discuss recent chemical concepts of the confinement-induced reactivity of microporous materials (Section 11.5) and nonlocalized charge compensation of extra-framework cations in zeolites (Section 11.6). The latter section also illustrates recent progress in revealing local structure of zeolites from *ab initio* calculations. The text is then concluded with an outlook on future challenges of computational chemistry of zeolites.

## 11.2
## Methodology

Quantum chemical electronic structure methods have a long history of applications to investigation of structural and reactivity properties of zeolites. In general, the goal of quantum chemical methods is to predict the structure, energy, and properties of a system containing many particles. The energy of the system is expressed as a direct function of the exact position of all atoms and forces that act upon the electrons and the nuclei of each atom. The exact quantum mechanical calculation of electronic structure is possible only for a very limited number of systems and thus, a number of simplifying approximations resulting in different quantum chemical methods are required to solve it for larger systems. Below, we present a simple overview of the important limitations of various quantum chemical methods as applied to the zeolite chemistry. More detail and in-depth discussion on the electronic structure calculations can be found in a number of very good references [4–9].

Electronic structure methods can be categorized as *ab initio* wavefunctions-based methods, *ab initio* density functional or semiempirical methods. All of them can be applied for the solution of different problems in zeolite chemistry. Wavefunction methods start with the HF solution and have well-prescribed methods that can be used to increase its accuracy. One of the deficiencies of the HF theory is that it does not treat dynamic electron correlation, which refers to the fact that the motion of electrons is correlated so as to avoid one another. The neglect of this effect can cause very serious errors in the calculated energies, geometries, vibrational, and other molecular properties.

### 11.2.1 *Ab initio* Methods

There are numerous so-called post-HF methods for treating correlated motion between electrons. One of the most widely used approaches is based on the definition of the correlation energy as a perturbation. In other words, the configurational interactions are treated as small perturbations to the Hamiltonian. Using this expansion the HF energy is equal to the sum of the zero and first order terms, whereas the correlation energy appears only as a second order term. The second order Møller–Plesset perturbation theory (MP2) typically recovers 80–90% of the correlation energy, while MP4 provides a reliably accurate solution to most system.

Another approach to treat the electron correlation is the configuration interactions method (CI). The general solution strategy for it is to construct a trial wavefunction that is comprised of a linear combination of the ground-state wavefunctions and excited-state wavefunctions. The trial wavefunction can include the exchange of one, two, or three electrons from the valence band into unoccupied orbitals; these are known as *configuration interaction singles* (CIS), *configuration interaction doubles* (CID), and *CI triples* and allow for single, single/double, and single/double/triple excitations.

Coupled cluster (CC) methods differ from perturbation theory in that they include specific corrections to the wavefunction for a particular type to an infinite order. CC theory therefore must be truncated. The lowest level of truncation is usually at double excitations (CCSD) since the single excitations do not extend the HF solution. CC theory can improve the accuracy for thermochemical calculations to within 1 kcal $mol^{-1}$.

Despite using these methods a very high accuracy can be achieved, almost all of the post-HF methods are prohibitive for calculation of reliable models of zeolites due to very high computational costs. Only the computationally cheapest post-HF methods can be currently used for studying zeolites. Thus, the application of the post-HF methods in zeolite chemistry is limited to the MP2 method, which still can be applied to calculations of only rather small zeolite models. One notes however that when the resolution of identity (RI) approximation is used [10], the resulting RI–MP2 method can in principle be used for calculations of systems containing more than hundred of atoms.

### 11.2.2 DFT Methods

A more attractive method is DFT. DFT is "*ab initio*" in the sense that it is derived from the first principles and does not usually require adjustable parameters. These methods formally scale with increase in the number of basis functions (electrons) as $N^3$, and thus, permit more realistic models compared to the higher-level post-HF methods, which usually scale as $N^5$ for MP2 and up to $N^7$ for MP4 and CCSD(T). On the other hand, the theoretical accuracy of DFT is not as high as the higher-level *ab initio* wavefunction methods.

The DFT is attributed to the work of Hohenberg and Kohn [11], who formally proved that the ground-state energy for a system is a unique functional of its electron density. Kohn and Sham [12] extended the theory to practice by showing how the energy can be portioned into kinetic energy of the motion of the electrons, potential energy of the nuclear–electron attraction, electron–electron repulsion, which involves with Coulomb as well as self interactions, and exchange correlation, which covers all other electron–electron interactions. The energy of an $N$-particle system can then be written as

$$E[\rho] = T[\rho] + U[\rho] + E_{xc}[\rho] \tag{11.1}$$

Kohn and Sham demonstrated that the $N$-particle system can be written as a set of $n$-electron problems (similar to the molecular orbitals in wavefunction methods) that could be solved self-consistently [12].

Although the DFT is in principle an exact approach, unfortunately, because the exact expression for the electron density functional as well as for the exchange correlation energy are not known, lots of assumptions and approximations are usually made. The most basic one is the local density approximation (LDA). It assumes that the exchange–correlation per electron is equivalent to that in a homogeneous electron gas, which has the same electron density at a specific point $r$. The LDA is obviously an oversimplification of the actual density distribution and usually leads to overestimation of calculated bond and binding energies.

The nonlocal gradient corrections to LDA functional improve the description of electron density. In this case, the correlation and exchange energies are functionals of both the density and its gradient. The gradient corrections take on various different functionals such as B88 [13], PW91 [14], PBE [15], for example. However, the accuracy of those is typically less than what can be expected from high-level *ab initio* methods.

One notes that the HF theory provides a more exact match of the exchange energy for single determinant systems. Thus, numerous hybrid functionals have been developed, where the exchange functional is a linear combination of the exact HF exchange and the correlation (and exchange) calculated from pure DFT. The geometry and energetics calculated within this approach (B3LYP and B3PW[16], MPW1PW91 [17], PBE0 functional [18], etc.) are usually in a good agreement with the experimental results and with those obtained using the post-HF methods. However, hybrid functionals still fail to describe chemical effects mainly based on electron–electron correlation such as dispersion and other weak interactions [19, 20].

### 11.2.3
### Basis Sets

As it was mentioned above, the energy in the DFT methods is formally a function of the electron density. However, in practice the density of the system $\rho(r)$ is written as a sum of squares of the Kohn–Sham orbitals:

$$\rho(r) = \sum |\psi_i(r)|^2 \tag{11.2}$$

This leads to another approximation usually done both in DFT and wavefunction-based methods. It consists in representation of each molecular orbital by a specific orthonormal basis set. The true electronic structure of a system can be in principle mathematically represented by an infinite number of basis functions. However, due to computational limitations, these functions are truncated in practice and described by a finite number of basis sets resulting in some potential loss in accuracy. A wide range of different basis sets currently exists and the choice of a certain one strongly depends on the solution method used, the type of the problem considered, and the accuracy required in each particular case. These functions can take on one of several forms. The most commonly used approaches use either a linear combination of local atomic orbitals, usually represented by Gaussian-type functions (GTO), or a linear combination of plane-waves (PWs) as basis sets. GTO basis sets are widely used in calculations of molecular systems, which also include cluster models of zeolites discussed in more detail in Section 11.2.4 of this chapter. GTO basis sets are implemented in various available quantum chemical softwares (Gamess-UK [21], Gaussian 03 [22], Tubomole [23], etc.). Since many years, they have been used in implementation of HF- and post-HF methods, as well as in DFT. PWs are more popular in simulations of solids (e.g., zeolite crystals). This is mainly due to the fact that their application to periodic systems is straightforward. The corresponding calculations are faster both for computing energies and gradients as compared to the approaches employing GTO basis sets. As a result, the PW approach is widely used in computer programs such as CASTEP [24], CPMD [25], and VASP [26], and so on, which have found a wide range of applications in studying various periodic systems usually by means of "pure" DFT (without exact HF exchange). In addition to the PW codes developed to model condensed phase systems, the CRYSTAL 06 [27] program that utilizes GTO basis sets can be used for studying both periodic and molecular systems within the same formalism irrespectively of the dimensionality of the system. It is important to note that in general, when a sufficiently large number of basis functions are employed, the computational results obtained using either GTO or PW basis sets are essentially the same [28].

When the PW basis set expansion of the electronic wavefunction is used, the number of PW components needed to correctly describe the behavior of the wavefunction near the nucleus is prohibitively large. To address this problem, within the PW approach the core electrons are described using the pseudopotential approximation. In this case, it is assumed that the core electrons do not significantly influence the electronic structure and properties of atoms and therefore, the ionic potential that arises from the nuclear charge and frozen core electron density is replaced by an effective pseudopotential. Although within the GTO approach, core electrons can be treated explicitly, the pseudopotential approximation can also be employed to reduce number of basis functions in calculations without dramatic loss of accuracy. This is in particular useful for the description of heavy atoms.

## 11.2.4
## Zeolite Models

The approximations done in order to compute energies by different quantum chemical methods as well as the use of finite basis set for the description of molecular orbitals are not the only factors leading to limited accuracy. When modeling zeolites, one can seldom take into account all of the atoms of the system (Figure 11.1a,b). Typically, a limited subset of the atoms of the zeolite is used to construct an atomistic model. The size of the model describing the reaction environment can be critical for obtaining the reliable results. Indeed, although the coordination of small molecules such as CO, $CH_4$, $H_2O$ to isolated cations can be studied very accurately using even CC theory, obviously the results thus obtained can hardly represent adsorption to the exchangeable cations in zeolites. This is an example of a competition of "model" versus "method" accuracy.

The current progress in computational chemistry also made it possible to use rather efficiently periodic boundary conditions in DFT calculations of solids. This

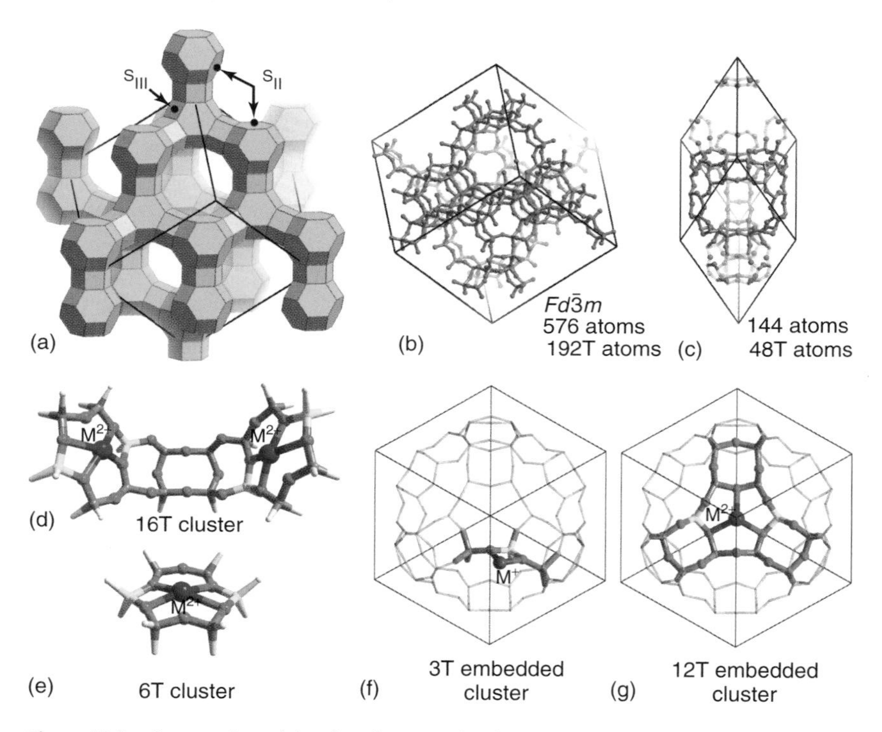


**Figure 11.1** Structural models of zeolite with faujasite topology (a) crystallographic unit cell with the *Fd3m* symmetry (b), a smaller rhombohedral lower-symmetry unit cell (c), cluster models containing 16 (d) and 6 (e) T-atoms representing, respectively, local structure of two and one $S_{II}$ faujasite sites with exchangeable cations stabilized at them, parts (f) and (g) show 3T and 12T clusters representing cation sites $S_{III}$ and $S_{II}$, respectively, embedded into the rhombohedral faujasite unit cell.

allows theoretical DFT studies of structure and properties of some zeolites usually with relatively small unit cells using a real crystal structure as a model (Figure 11.1c). Such periodic DFT calculations of zeolites are, however, mostly limited to the use of LDA and GGA density functionals.

Another method for the zeolite modeling is a so-called cluster approach. Here, only a part of the zeolite framework, containing finite number of atoms, is considered, whereas the influence of the rest atoms of the zeolite lattice is neglected (Figure 11.1d,e). A minimum requirement to the zeolite model is that it involves the reactive site or the adsorption site together with its environment. In this case, the model is built of several $TO_4$ units to mimic the local structure of a part of the zeolite. Although this approach results in some loss of "model" accuracy, it can be very useful for the analysis of different local properties of zeolites such as elementary reaction steps, adsorption, and so on. In addition, in the case of cluster modeling the higher-level *ab initio* methods as well as hybrid density functionals can be successfully used.

Hybrid quantum chemical embedding schemes form a wide range of popular methods for computational modeling of zeolites. They allow the combination of two or more computational techniques in one calculation and make it possible to investigate the chemistry of such systems like zeolites with high precision. The region of the system where the chemical process takes place (similar to that used for cluster modeling) is treated with an appropriately accurate method, while the remainder of the system is treated at a lower level of theory (Figure 11.1f,g). The main difficulty comes in linking two different regions together. The link region is usually defined in order to provide an adequate transfer of information between the inner and outer regions.

The energy for this system is then calculated as

$$E_{\text{hybrid}}(\text{System}) = E_{\text{high}}(\text{Model}) + E_{\text{low}}(\text{System}) - E_{\text{low}}(\text{Model}) \tag{11.3}$$

where $E_{\text{high}}$ (Model) refers to the energy calculated at higher level for the inner core region only. $E_{\text{low}}$ (System) – $E_{\text{low}}$ (Model) refers to the difference in energy between the full system and the core region both calculated at the low level of theory. Although usually quantum chemical methods (QM) are used for the description of the core region, while the rest atoms of the system are treated by molecular mechanics, it can be any lower level method which is faster than that used for the core. Therefore, using cluster embedding one can rather accurately investigate the local chemistry of zeolites and also take into account the longer range effects.

## 11.3
## Activation of Hydrocarbons in Zeolites: The Role of Dispersion Interactions

One of the main challenges in modeling of chemical reactions in zeolites is to predict accurately energies of adsorption and reaction profiles for hydrocarbon transformations in microporous materials. The current method of choice for modeling reactivity of such complex catalytic materials as zeolites is DFT. However, commonly used density functionals fail to describe correctly the long-range dispersion interactions [20]. The dominant interactions between the hydrocarbon species and the zeolite walls correspond to weak van der Waals interaction of dispersive nature, which therefore cannot be correctly computed within the conventional DFT. This may result not only in inaccurate computed energetics of chemical reactions but also in wrong prediction of stability or reactivity trends for systems, where the impact of dispersive interactions on the total stabilization energy of the reaction intermediates and transition states is not uniform along the reaction coordinates. Note that dispersion is an intermolecular correlation effect. As it has been mentioned above, the simplest electronic structure method that explicitly describes electron correlation is MP2 theory. However, MP2 calculations for periodic systems are presently feasible only for very small unit cells containing only few atoms in combination with small basis sets.

Recently, an embedding scheme to introduce local corrections at post-HF level to DFT calculations on a periodic zeolite model has been proposed [29, 30]. This approach allows an accurate modeling of structural and electrostatic properties of the zeolite reaction environment by using the periodic DFT calculations. The refinement for the self-interaction effects and van der Waals interactions between the adsorbed reactants and the zeolite walls is achieved by applying resolution of identity implementation of the MP2 method (RI–MP2) to a cluster model representing the essential part of the framework that is embedded into the periodic model of the zeolite. The thus designed MP2:DFT approach is suited for studying reactions between small or medium-sized substrate molecules and very large chemical systems as zeolite crystals and allows quantitative computing reaction energy profiles for transformations of hydrocarbons in microporous matrices with near chemical accuracy.

To illustrate this, let us consider interaction of isobutene with Brønsted acid site of a zeolite (Scheme 11.1) as a prototype of Brønsted acid-catalyzed transformations of hydrocarbons. This reaction is not only interesting from the practical point of view due to its relation to the skeletal isomerization of butenes [31]. It also attracts attention of many theoretical groups because of the fundamental question of whether it is possible to form and stabilize the *tert*-butyl carbenium ion in zeolite microporous matrix as a reaction intermediate [32].

There are several studies that report DFT calculations on protonation of isobutene employing rather small cluster models to mimic local surrounding Brønsted acid site in a zeolite (see e.g., [33, 34]). A very strong dependency of the relative stabilities of the protonated products on the level of computations and more importantly on the size of the cluster model was found. The only minima on the potential energy surface obtained within the cluster modeling approach corresponds to the covalently bound alkoxides, while the carbenium ions are present as very short-lived transition states.

On the other hand, when the long-range interactions with the zeolite framework and its structural details are explicitly included in the computation either within the embedded cluster approach with a very large part of the zeolite lattice as a low-level model [35] or by using periodic DFT [36, 37], a local minimum on the

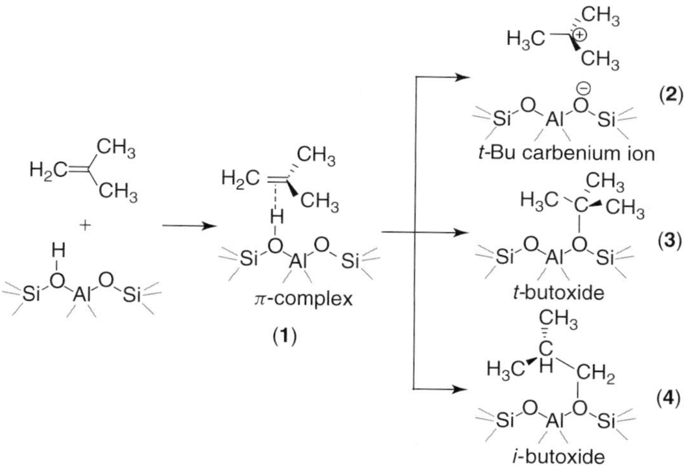

π-complex
(1)
t-Bu carbenium ion (2)
t-butoxide (3)
i-butoxide (4)

**Scheme 11.1** Protonation of isobutene on a zeolitic Brønsted acid site.

potential energy surface corresponding to the *tert*-butyl cation can be located for various zeolite topologies.

Indeed, applying DFT under periodic boundary conditions to the realistic system containing isobutene adsorbed in ferrierite a rather different picture was observed [37] as compared to the situation when a small cluster model was used to mimic the zeolite active site [34]. Only the π-complex of butane (**1**) with the Brønsted acid site of the zeolite was found to be more stable than the isolated alkene separated from the zeolite [37]. The stabilization however was rather minor (PBE/PW, $FER_{pbc}$, Table 11.1). The DFT-computed adsorption energies did not exceed a few kilojoules per mole that is much less than what would be expected for such a system. The existence of the local minimum on the potential energy surface corresponding to the *tert*-butyl carbocation (**2**) was reported. Its stability was shown to be at least comparable to that of covalently bound tertiary butoxide (**3**). Inclusion of zero-point vibrations and finite temperature effects further stabilized the carbenium ions relative to the covalently bound alkoxides. It was concluded that already at 120 K formation of *tert*-butyl cation in H-ferrierite becomes thermodynamically more favorable than formation of the covalently bound species [37]. However, this theoretical prediction lacks the experimental support, because simple carbenium ions have never been observed by either NMR or infrared spectroscopy upon olefin adsorption to hydrogen forms of zeolites [38]. This inconsistency may not be ascribed to any deficiency of the zeolite model used in the computational studies, and therefore, must be due to the inaccuracies of the computational method (DFT) used either in respect to description of the self-interaction effect or dispersive interactions.

Tuma and Sauer [30] computed the relative stabilities of the possible products of interaction of isobutene with H-ferrierite by means of the MP2:DFT hybrid method. A cluster model containing 16T atoms at the intersection of 8-membered

**Table 11.1** Calculated reaction energies ($\Delta$ E, kJ mol$^{-1}$) for the formation of the $\pi$ complex of isobutene, of the tert-butyl carbenium ion, and of the tert-butoxide and iso-butoxide in acidic zeolites.

| | B3LYP/ DZ, 3T [34] | PBE/CBS, 16T [39] | M06–L/ CBS, 16T [39] | MP2/CBS, 16T [39] | B3LYP: MM, $FER_{pbc}$ [40] | MP2//B3 LYP: MM $FER_{pbc}$ [40] | PBE/PW, $FER_{pbc}$ [30] | PBE+D, $FER_{pbc}$ [41] | MP2:DFT $FER_{pbc}$ [30] | Best estimate $FER_{pbc}$ [30] |
|---|---|---|---|---|---|---|---|---|---|---|
| (1) | −28 | −13 | −61 | −63 | −49 | −79 | −16 (−10)[a] | −92 | −77 (−44)[b] | −78 |
| (2) | – | 57 | 8 | 41 | – | – | 8 (36)[a] | −67 | −13 (−8)[b] | −21 |
| (3) | −35 | 10 | −67 | −67 | −62 | −67 | 19 (17)[a] | −78 | −66 (20)[b] | −48 |
| (4) | −54 | 10 | −59 | −67 | −145 | −94 | −3 (5)[a] | −94 | −80 (−27)[b] | −73 |

[a] Values in parenthesis are taken from [37].
[b] Values in parenthesis are corrected for BSSE.

and 10-membered channels of H-FER including the Brønsted acid site was defined for the MP2 level within the full periodic model that is in turn was described within DFT. The MP2 calculations on cluster models were performed using local basis set constructed from Gaussian functions. To avoid possible errors associated with the use of the limited-sized localized basis sets, the computational results were corrected for basis set superposition error (BSSE) and extrapolated to the complete basis set (CBS) limit. Furthermore, the results obtained within the embedded cluster approach were extrapolated to infinite cluster size (i.e., to the periodic limit). To access the reliability of the chosen theoretical methods for hydrocarbon reactions in zeolites, comparison with the results of CCSD(T) calculations was performed. It was concluded that the MP2 method allows chemically accurate description of the system considered.

Indeed, it was clearly shown that the post-HF corrections to the reaction energy profiles obtained by pure DFT (PBE exchange–correlation functional) are substantial (Table 11.1). More importantly, they are not uniform for different structures formed within the zeolite pores. When dispersion is included at the MP2 level, the adsorption energy of isobutene changes from $-16\,\mathrm{kJ\ mol^{-1}}$ to the realistic value of $-78\,\mathrm{kJ\ mol^{-1}}$. Stabilization of the covalently bound alkoxides due to van der Waals interaction with the zeolite walls is even larger (best estimate, Table 11.1). Surprisingly, it was found that the impact of dispersion interactions on the stabilities of the protonated species is the lowest for the tert-butyl carbenium ion. The corresponding reaction energy is lowered only by $30\,\mathrm{kJ\ mol^{-1}}$. As a result, the carbenium ion structure was shown to be the least stable species among the structures considered, whereas periodic PBE calculations predict this species to be only $15\,\mathrm{kJ\ mol^{-1}}$ less stable than the iso-butoxide species. When dispersion is included this energy gap becomes three times larger and reaches $52\,\mathrm{kJ\ mol^{-1}}$ [30]. Unfortunately, despite all the efforts made to reach high computational accuracy, the reported MP2:DFT results were not corrected for the finite temperature effects. Therefore, no definite conclusion can be made on the relative stabilities of the species formed upon isobutene protonation in ferrierite. Nevertheless, this large energy difference suggests that although there is a chance that at higher temperatures the equilibrium will shift toward the formation of tert-butyl carbenium ion, at ambient temperatures in line with the experimental observations the formation of covalently bound alkoxide species is preferred.

Very recently, the applications of the MP2:DFT approach to computational studies of reactivity of zeolites have been extended to the investigation of methylation of small alkenes with methanol over zeolite HZSM-5 [42]. Besides the highly sophisticated theoretical methods employed in this study, to the best of our knowledge this is the first *ab initio* periodic study of reactivity of microporous materials with such a complex structure as MFI. Similarly to the above-considered protonation of olefins, the zeolite framework has been represented by a periodically repeated MFI unit cell, while the interactions due to the confinement of the reacting species in the microporous space and the energetics of their transformations at the Brønsted acid site have been refined by applying the MP2 correction to a cluster model embedded in the periodic structure (Figure 11.2).

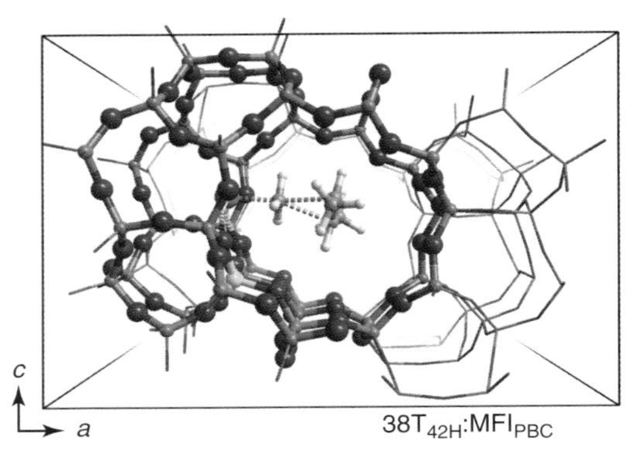


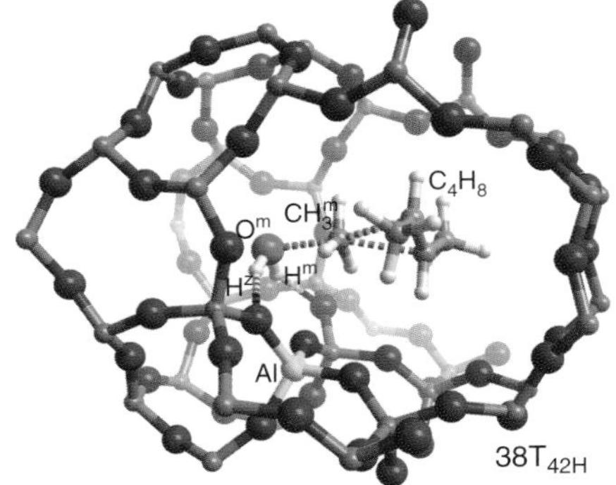


**Figure 11.2** Transition state structure for *t*-2-butene methylation with methanol over HZSM-5 zeolite. (a) Depicts the corresponding periodic MFI model shown along the straight channel. Highlighted atoms correspond to the largest 38T embedded cluster (enlarged in (b), boundary H atoms are omitted for clarity) treated at the MP2 level of theory in [42] (created with permission using the supplementary materials provided with the [42], *http://pubs.acs.org/doi/suppl/10.1021/ja807695p*).

The reaction chosen for the computational study is of high relevance to the industrially important MTO process. Reaction rates and activation barriers for the methylation of small alkenes over HZSM-5 are directly available from experimental studies [43, 44]. Thus, this reaction and the respective experimental data can be used to compare performance, accuracy, and predicting power of the currently widely used pure DFT methods and of the more advanced quantum chemical techniques (e.g., DFT+D and MP2 methods). A simplified schematic energy diagram [42] for this reaction is depicted in Figure 11.3. Several assumptions must be made at this step to compare the experimental and the computational results. The experimental kinetic studies indicate that olefin methylation is a first order reaction with respect to alkene concentration and zero order with respect to methanol concentration. The resulting experimental barriers [43, 44] represent apparent activation energies with respect to the state, in which methanol is adsorbed at the Brønsted acid site of a zeolite and alkene is in the gas-phase (Figure 11.3a). Secondly, the methylation reaction is assumed to take place via an associative one-step mechanism rather than a two-step consecutive process involving the formation of a methoxy group covalently bound to the zeolite walls.

Although previous theoretical studies performed using a small 4T cluster model [45] have contributed significantly to the molecular-level understanding of the mechanistic details of this catalytic process, the thus computed apparent activation barriers were significantly overestimated and could not even reproduce the experimentally observed trends in their dependency on the alkene chain length ($B3LYP_{4T}$ and $PBE_{4T}$ in Figure 11.3b). The former effect is mainly caused by the well-known drawback of the cluster modeling approach that is mainly due to the lack of the electrostatic stabilization of the polar transition states by the zeolite

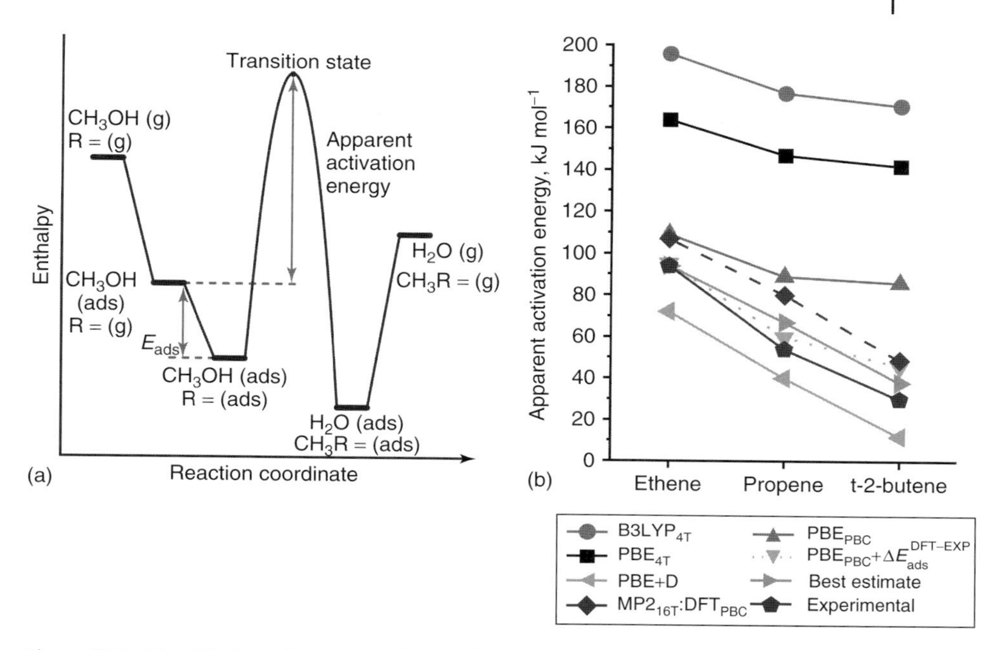


**Figure 11.3** Simplified reaction energy diagram for an alkene methylation with methanol over acidic zeolite (a) and the respective apparent activation barriers computed using various methods (b) [42].

lattice [46–48]. Indeed, the apparent activation barriers calculated using a periodically repeated MFI unit cell decrease substantially ($PBE_{PBC}$ in Figure 11.3b). For ethane the calculated barrier is only 15 kJ mol$^{-1}$ higher than that obtained from the experiment. Interestingly, this value corresponds almost exactly to the difference between the experimental and DFT-computed adsorption energies of ethylene on HZSM-5. Therefore, the absence of the hydrocarbon chain-length dependency of the apparent activation barriers calculated within DFT is primarily associated with its poor description of the dispersion effects. Indeed, the implication of the local post-HF correction within the MP2:DFT approach significantly improves the qualitative picture, although the thus obtained results still deviate by 8–20 kJ mol$^{-1}$ from the experimental values. This mismatch is further reduced after the extrapolation of the high-level correction results to the periodic and CBS limits ("best estimate" in Figure 11.3b), while the subsequent corrections for ZPE and finite temperature effects allow reproduction of the experimental apparent enthalpy barriers with nearly chemical accuracy (deviation between 0 and 13 kJ mol$^{-1}$). One notes that these deviations are contributed by uncertainties in both the computational and the experimental results. It has been convincingly shown [42] that the errors associated with various fitting and modeling procedures used in the theoretical study are of the same order as the uncertainties in the energetics derived from the experimental data. This means that the MP2:DFT method by Sauer and coworkers [29, 30, 42] allows to compute energy parameters for the reactions in zeolites that quantitatively agree with the experimental data.

Nevertheless, although the proposed DFT:MP2 scheme allows the very accurate calculations of adsorption and reaction energies in microporous space, the associated computations are still too demanding to be used for comprehensive studies and for an in-depth theoretical analysis of various factors that influence the selectivity and reactivity patterns of the zeolite catalyst. The authors state in [42] that "the hybrid DFT:MP2 method is computationally expensive and not suited for routine studies on many systems." Thus, there is still a strong desire for a robust computational tool aspiring to provide with reliable predictions for hydrocarbon transformations in zeolites that must combine efficiency and chemical accuracy of DFT methods along with the proper account for van der Waals dispersive interactions. This is reflected by the fact that the improvement of DFT toward a better description of nonbonding interactions is currently an active research area in theoretical chemistry.

The most pragmatic solution for this problem is to involve in the calculations force fields based on the empirically fitted interatomic potential. The state-of-the-art examples of those show extremely good results of quantitative quality for the prediction of structural properties of microporous materials and for the description of the processes that are mainly influenced by the nonbonding interactions [1–3]. The computational simplicity of the force field approach allows simulations of even dynamical properties of chemical systems composed of more than $10^6$ atoms at time scales up to nanoseconds. However, again due to the simplistic form of the interatomic potentials, they cannot be directly used to describe processes associated with bond breaking and making, that is, chemical reactivity. Thus, there are numerous approaches that in one way or another make use of the empirically derived nonbonding interatomic potentials combined with the electronic structure calculations to amend the results of DFT toward better description of van der Waals interactions.

One can see from the results presented in Figure 11.3 that the periodic DFT calculations ($DFT_{PBE}$) can be improved substantially by adding the contribution from van der Waals interactions at the initial state obtained as a difference between the underestimated DFT-predicted adsorption energies and those obtained from the experiment ($\Delta E_{ads}^{DFT-EXP}$). An associated computational procedure has been proposed by Demuth *et al.* [49] and Vos *et al.* [48]. It involves the correction of the periodic DFT results for van der Waals interactions using an add-on empirical 6–12 Lennard–Jones potential (Eq. (11.4)) acting between the atoms of the confined hydrocarbon molecule and of the microporous matrix. The correction in this case is applied for the fixed DFT optimized structures.

$$E_{vdW}(r_{in}) = \sum \left( \frac{A_{ij}}{r_{ij}^{12}} - \frac{B_{ij}}{r_{ij}^{6}} \right) \tag{11.4}$$

A similar m ethod that provides the possibility to optimize structures with inclusion of van der Waals interactions is the density functional theory plus damped dispersion (DFT+D) approach [50]. This scheme consists in adding a semiempirical term E(D) to the DFT energy E(DFT) resulting in the dispersion-corrected energy E(DFT+D). E(D) in this case is expressed as a sum over pairwise interatomic interactions computed using a force-field-like potential truncated after the first term (Eq. 11.5).

$$E(D) = -s_6 \sum \frac{c_{ij}}{r_{ij}^6} f_D(r_{ij}) \tag{11.5}$$

where $c_{ij}$ are the dispersion coefficients, the damping function $f_D(r_{ij})$ removes contributions for short-range interactions, while the global scaling parameter $s_6$ depends on the particular choice of the exchange–correlation functional. The DFT+D approach has been parameterized for many atoms and a wide variety of functionals and can be used in a combination with popular quantum chemical programs [41]. When applied to chemical processes in microporous materials, this approach has been shown to provide realistic adsorption energies for hydrocarbons in all-silica zeolites [41]. Although the DFT+D significantly improves the pure DFT results for the reaction energies (Table 11.1) and activation barriers (Figure 11.3b) for the conversion of hydrocarbons over acidic zeolites, these still significantly deviate from the higher-level *ab initio* MP2:DFT or experimental result. The DFT+D apparent activation energies are systematically underestimated by ~20–30 kJ mol$^{-1}$ (Figure 11.3b), while the qualitative trend in the hydrocarbon chain dependency is perfectly predicted. On the other hand, the thus computed relative stabilities of the products of protonation of isobutene differ from the best estimate values from MP2:DFT substantially (Table 11.1).

All of the above-considered computational techniques involve rather computationally demanding periodic DFT calculations of a large zeolite unit cells as the base for the geometry optimization of the structures of intermediates and transition states. Although the results thus obtained do not suffer from the artificial effects associated with the model accuracy, these methods may be unfeasible for such tasks as thorough computational screening of the catalytic performance of zeolite-based catalysts. In this case, the use of a hybrid QM:force field (QM:MM) approach may help to reduce the associated computational requirements. This method may be viewed as a "lower-level" analog of the MP2:DFT approach. In this case, the *ab initio* part (usually treated by DFT methods) describing the bond rearrangement at the zeolite active site is intentionally limited to a small part of the zeolite, while the van der Waals and electrostatic interactions with the remaining zeolite lattice are described using a computationally less demanding force field methods. This methodology allows fast and rather accurate calculation of the heats of adsorption and reaction energies of various hydrocarbons in zeolites [40, 51, 52]. However, when using the conventional DFT as the "high-level" method, the correct description of the dispersion contribution to the adsorption energy of longer-chain hydrocarbons requires the use of very small, usually containing only 3T atoms, cluster model [40]. The energetics can be substantially improved by correcting the DFT results by single-point MP2 calculations. The thus computed energetics (MP2//B3LYP : force field) of isobutene protonation in H-FER zeolite agree reasonably well with those obtained using the MP2:DFT scheme (Table 11.1).

The performance of DFT itself may also be substantially improved by parameterization of the exchange–correlation functionals. Zhao and Truhlar have recently reported a family of meta-GGA functionals (M05 [53], M06 [54], and related functionals) in which the performance in describing nonbonding interactions (Table 11.1) as well as in predicting reaction energies and activation barriers is significantly improved compared to the conventionally used GGA and hybrid functionals. The hybrid methods involving a combination of such density functionals and well-parameterized force fields are anticipated to be very efficient and accurate for the investigations of zeolite-catalyzed reactions [39].

Nevertheless, the simplifications involved in the above methods, such as the assumption of pairwise additivity of van der Waals interactions, the presence of empirically fitted parameters both in the force fields and in the parameterized density functionals can lead to unreliable results for systems dissimilar to the training set. The recently proposed nonlocal van der Waals density functional (vdW-DF) [55] is derived completely from first principles. It describes dispersion in a general and seamless fashion, and predicts correctly its asymptotic behavior. Self-consistent implementations of this method both with PW [56] and Gaussian basis sets have been reported [57]. Until now, the vdW-DF method has been successfully applied to weakly bound molecular complexes, polymer crystals, and molecules adsorbed on surfaces (see e.g., [56, 57] and references therein). However, to the best of our knowledge, its applicability to modeling chemical reactions within zeolite pores has not been investigated yet.

Summarizing, there is a strong desire for the efficient and accurate computational tool for studying chemical reactivity of zeolites that is able to correctly predict effects due to nonbonding interactions in the microporous matrix. Most of the currently available computational techniques involve numerous approximations and often contain empirically fitted parameters. In this respect, the hybrid MP2:DFT method by Tuma and Sauer [29, 30, 37] is useful to generate reliable datasets for various chemical processes in zeolite, on which parameterization of force fields, QM:MM methods, as well as assessment of the performance of various exchange–correlation functionals and various dispersion correction schemes can be based. The correct description of weak nonbonding interactions within the intermediates and transition states involved in catalytic conversions of hydrocarbons in zeolites is important not only for the fundamental understanding of these processes but also for the generating reliable microkinetic models able to predict the reactivity and selectivity patterns of microporous catalysts in various reactions.

## 11.4
## Molecular-Level Understanding of Complex Catalytic Reactions: MTO Process

Molecular-level determination of the reaction mechanism of complex catalytic transformations in zeolites based solely on experimental studies is usually an extremely challenging task. Theoretical methods based on quantum chemical calculations are in contrary ideally suited for revealing the molecular mechanism and for identifying the elementary reaction steps of such processes. This section illustrates the recent advances in understanding the molecular-level picture of the industrially important MTO process from quantum chemical calculations.

The MTO process catalyzed by acidic zeolites has been subject of extensive experimental studies driven by the possibility of converting almost any carbon-containing feedstock (i.e., natural gas, coal, biomass) into a crucial petrochemical feedstock like ethylene and propylene. The actual reaction mechanism of this process has been a topic of intense debates for the last 30 years [58, 59]. Initially, the research was focused on the formation of the first C—C bond via the combination of two or more methanol molecules to produce alkene and water [58, 59]. Such a "direct" mechanism involved only methanol and $C_1$ derivatives. An alternative mechanism has been suggested by Dahl and Kolboe [60] that assumed the formation of some "hydrocarbon pool" species that is continually adding and splitting reactants and products. Recently, both the experimental results by Haw *et al.* [61, 62] and quantum chemical studies by Lesthaeghe *et al.* [63, 64] provide evidence for the preference of the latter mechanism.

Indeed, Lesthage *et al.* [63, 64] screened practically all of the possible direct C—C coupling reaction routes over a zeolitic Brønsted acid site modeled using a small 5T cluster at the B3LYP/6-31G(d) level of theory. The combination of the thus obtained results in complete pathways and calculation of barrier heights as well as the rate coefficients clearly showed that there is no successful pathway leading to the formation of ethylene or any intermediate containing a C—C bond from only methanol. These results are in line with the experimental observations of the very low activity of methanol and DME (dimethyl ether) over HZSM-5 zeolite in the absence of organic impurities acting as a hydrocarbon pool species [61, 62]. It was concluded that the failure of the direct C—C coupling mechanism is mainly due to the low stability of the ylide intermediates and the highly activated nature of the concerted C—C bond formation and C—H bond breaking involved in these mechanisms. Both of these effects were attributed to the low basicity of the framework zeolite oxygens that cannot efficiently stabilize the respective species.

The more likely pathway involves organic reaction centers trapped in the zeolite pores which act as cocatalysts. In particular, experimental studies have proved formation of various cyclic resonance-stabilized carbenium ions (Scheme 11.2) upon the MTO process in the microporous space. Stable dimethylcyclopentenyl (**5**) and pentamethylbenzenium (**6**) cations in HZSM-5 [59, 65] and hexamethylbenzenium (**7**) and heptamethylbenzenium (**8**) cations in zeolite HBEA [66, 67] have been detected by various spectroscopic methods. Obviously, the catalytic activity is therefore influenced both by the nature of the hydrocarbon pool species and by the

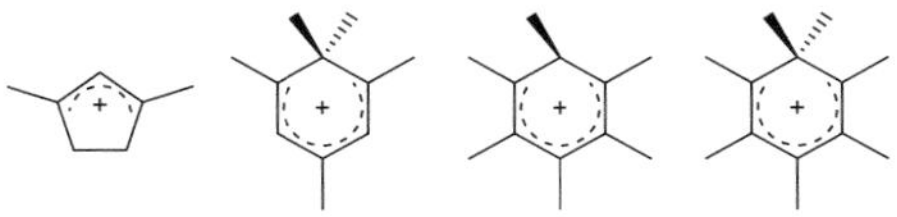

(**5**) (**6**) (**7**) (**8**)

**Scheme 11.2** Stable cyclic carbocations experimentally detected in zeolites.

topology of the zeolite framework that determines the preferred pathway for the transformation of these bulky species.

An attempt to separate these effects was done by using quantum chemical calculations [68]. The geminal methylation of various methylbenzenes with methanol over a zeolitic Brønsted acid site (Figure 11.4a) was modeled with a 5T cluster model. Note that such a small cluster may be viewed as a general representation of any aluminosilicate. It does not mimic structural features of any particular zeolite and therefore completely neglects all of the effects due to the sterics and electrostatics of the microporous space. Although the activation energies computed were substantially overestimated, the obtained results indicate the increasing reactivity of larger polymethylbenzenes (Figure 11.4b).

The effect of the zeolite topology has been included by extending the calculations to larger clusters containing 44T and 46T atoms, where the catalytically active site and the reactants were still modeled at DFT level using an embedded 5T cluster, while the remaining part of the cluster model was treated at HF level [68]. It has been concluded that the structural features of the zeolite framework play a major role in the reaction kinetics. The reaction rates for the geminal methylation of hexamethylbenzene follow the order: CHA >> MFI > BEA (Figure 11.4c). These striking differences in reactivity can be attributed to the molecular recognition features of the zeolite cages (Figure 11.4d–f). Indeed, the size and the shape of the chabazite cage were shown to be ideal for this reaction step. The larger pores of

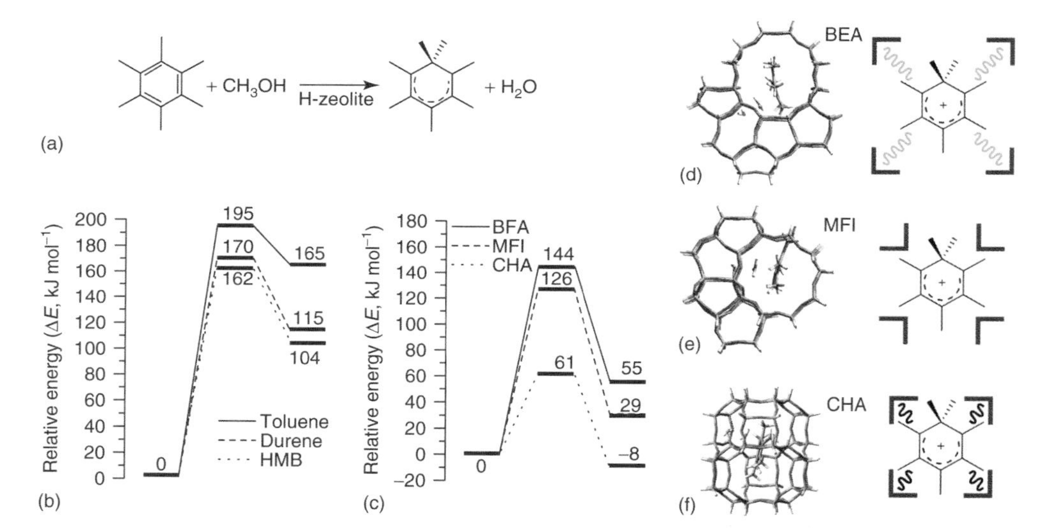


**Figure 11.4** Methylation of polymethylbenzenes in acidic zeolites (a) and the computed activation barriers and reaction enthalpies depending on the nature of the organic molecule (b) and on the zeolite topology (c). The molecular recognition effects are illustrated with the schematic representation of the resulting carbenium ions in the confined space and with the structures of the transition states for methylation of HMB in zeolites BEA (d), MFI (e), and CHA (f). (Adapted from [68, 69] with the help of the authors.)

beta zeolite (BEA) are not able to provide an effective electrostatic stabilization of the organic species. On the other hand, the small pores of HZSM-5 zeolite (MFI) restrict the formation of large carbocations, suggesting thus that the hydrocarbon pool in this case involves less bulky species. Combining the effects of such a transition state shape selectivity of MFI framework and those caused by the nature of the hydrocarbon pool species, the highest reactivity was predicted for the less sterically demanding 1,2,4-trimethylbenzene and pentamethylbenzene. This is in line with the experimental observations of Svelle *et al.* [70] on the higher reactivity of the smaller methylbenzenes in HZSM-5 zeolite.

Finally, combining the theoretical and experimental results a complete catalytic cycle for the conversion of methanol to olefins (MTO) in HZSM-5 zeolite (Figure 11.5) was presented by McCann *et al.* [71]. For each elementary reaction step, activation barriers and reaction rates were computed. This complete route convincingly explains the formation of the experimentally detected cations (**5**) and (**6**) all the way from toluene, multiple scrambling of labeled carbons from the methanol feed into the hydrocarbon pool species, as well as the substantial selectivity to isobutene during the MTO process in HZSM-5.

Very recently, the mechanism of MTO reaction over the hexamethylbenzene species encapsulated in the chabazite cages of the aluminophosphate HSAPO-34 (CHA topology) has been thoroughly investigated by means of periodic DFT calculations [72]. The results obtained indicate that the catalytic reactivity of this system is to a significant extent controlled by the intrinsic acidity of the zeolite catalysts. A plausible role of water is suggested, which can act as the promoter for the $H^+$ shift between the confined organic species and the zeolite framework to facilitate the elimination of alkenes. The computed apparent activation barrier of the rate-determining steps for the production of ethene was 230 kJ $mol^{-1}$ and that for the production of propene was 206 kJ $mol^{-1}$. The authors thus concluded that the MTO process over the hexamethylbenzene/HSAPO-34 catalysts is selective toward propene. The reported activation energies seem to be significantly overestimated. This can be due to either the actual choice of the hydrocarbon pool species or the deficiencies of the computational methodology used in this study.

Note that the hydrocarbon pool mechanism of the MTO process in zeolites involves a tight confinement of the hydrocarbon molecules in the microporous space. Their transformations during the catalytic process substantially alter their conformation, size, and length of the side-chains, and so on. Thus, the impact of intermolecular van der Waals interactions between the encapsulated organic molecules and the zeolite walls in different reaction intermediates and transition states on their stability may significantly vary along the reaction coordinate. This in turn may have a great influence not only on the exact computed values of enthalpies and activation barriers of the elementary reaction steps but also on the qualitative trends in the energetics of the MTO catalytic process. These effects have been completely neglected in the computational studies discussed above and therefore the influence of dispersion on various elementary reaction steps of the MTO process in zeolites still has to be investigated.

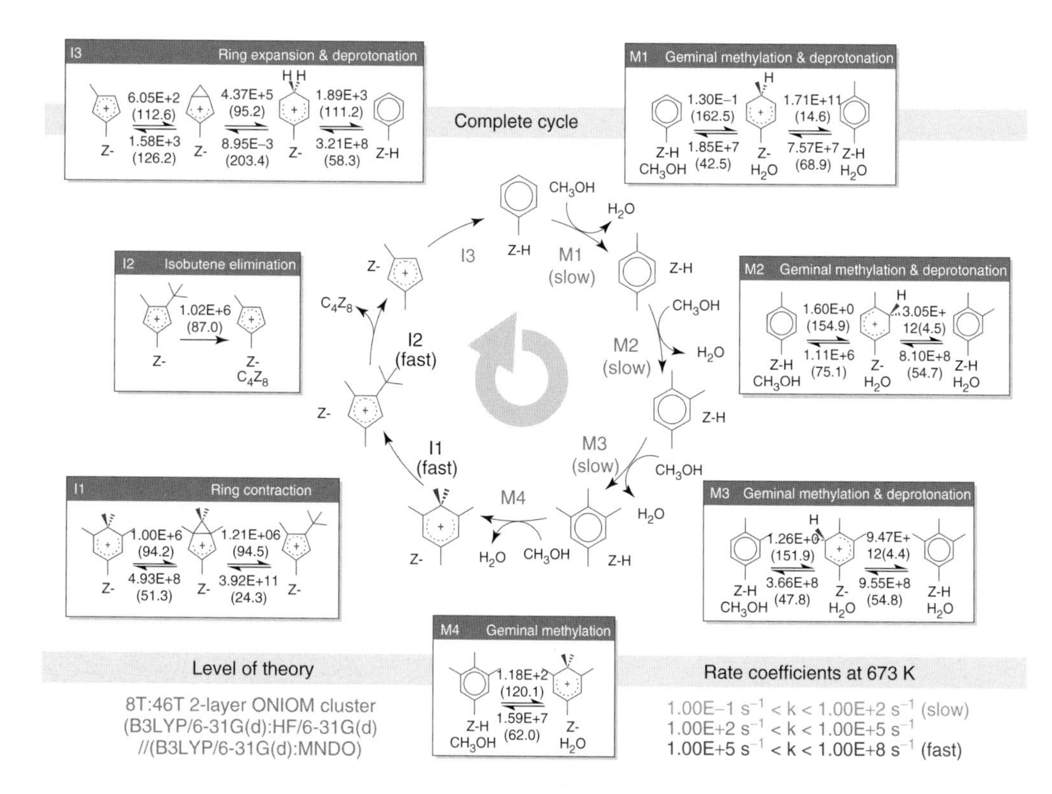


**Figure 11.5** Full catalytic cycle for carbon-atom scrambling and isobutene formation from methanol through a combined methylbenzene/cyclopentenyl cation pool mechanism in HZSM-5. Calculated rate constants at 673 K are given in per second and reaction barriers at 0 K (in brackets) are given in kilojoules per mole. (Adapted from [71] with the help of the authors. Copyright Wiley-VCH Verlag GmbH & Co. KGaA. Reproduced with permission.)

Summarizing, the computational work performed until now has greatly improved our fundamental understanding of the elementary reaction steps involved to produce light alkenes from aromatic reaction centers during the MTO process over acidic zeolites. However, the thorough investigation of the stability and reactivity of other possible hydrocarbon pool species confined within the pores of zeolite is needed to rationalize the experimentally observed selectivity patterns. The extension of the theoretical studies to other zeolite catalysts with different topologies and framework composition is also very important, because the resulting different geometrical constraints and different acidic properties will probably lead to different major catalytic cycles. Furthermore, the mechanism of the initial formation of the hydrocarbon pool cocatalysts in the zeolite pores, and related the question of how the first C–C bond is formed have not been addressed yet. Rationalization of these issues is necessary for the creation of a complete molecular-level picture of the MTO process over acidic zeolites. The resulting fundamental understanding of various factors underlying the reactivity of zeolites in the MTO process will ultimately create a possibility to rationally design a catalyst with specific local spatial environment and framework composition toward the optimal catalytic performance and desired product selectivity.

## 11.5 Molecular Recognition and Confinement-Driven Reactivity

The chemical reactivity of zeolites mainly arises from the presence of various "defect" sites in the ideal full-silica framework. Such defects can be either framework sites formed upon the isomorphous substitution of silicon with other ions ($Sn^{4+}$, $Ti^{4+}$, etc.) in the zeolite lattice or extra-framework cationic species that compensate for the negative charge on the framework due to the presence of lower-valent ions in the framework ($Al^{3+}$, $Ga^{3+}$, etc.). Depending on the type and the properties of such sites the chemistry of the resulting catalyst can vary dramatically. Besides the intrinsic chemical properties, steric effects in the form of shape selectivity are well known to be important for the reactivity of the zeolite-based materials [3, 73, 74]. Depending on the size of the zeolite cavities and channels only molecules below a certain size or of a specific shape are allowed to reach the active site or to leave the zeolite matrix. In addition, steric constraints imposed by the zeolite structure on a particular reaction intermediate or a transition state can define formation of certain product molecules. The effects of the zeolite topology on the relative stabilities of different reaction intermediates formed in the microporous space have been illustrated in the previous Sections 11.3 and 11.4. It should be noted that in all these cases, it was assumed that the chemical reaction is controlled by the zeolite topology and the intrinsic properties of the active species, whereas their intrazeolite arrangement is usually neglected.

For the case of low-silica zeolites the latter factor can become crucial. The high density of the exchangeable ions in these materials can allow formation of multiple noncovalent interactions with the adsorbed molecules resulting in their specific orientation and activation in the zeolite cage. This resembles the way enzymes activate substrates [75]. They use molecular recognition features to orient confined molecules at the active site resulting in their specific chemical activation. Similar effects have been also reported for numerous examples in coordination chemistry and homogeneous catalysis [76]. This section illustrates the importance of molecular recognition features in zeolites for the adsorption properties and reactivity.

The term *molecular recognition* is usually applied when selective binding and formation of specific activated complexes is a result of multiple attractive noncovalent interactions. Therefore, one expects the associated effects to have a significant impact on the adsorption properties of microporous materials.

Recently, adsorption of various probe molecules such as CO [77–79], $CH_3CN$ [80], $CO_2$ [81], and $N_2O_4$ [82, 83] in zeolites has been investigated by periodic DFT calculations. The obtained results indicate that the vast majority of adsorption complexes in cation-exchanged Al-rich zeolites cannot be described in terms of the interactions with a specific type of single cation sites. Instead, they usually involve numerous intermolecular interactions between the confined molecules and the exchangeable cations in the zeolite.

Even for such a small molecule as CO, formation of the linearly bridged CO adsorption complexes were detected in Na- and K-containing ferrierites with Si/Al ratio of 8 using a combined computational and spectroscopic approach [78]. Interaction with dual adsorption sites was found to be slightly more favorable ($\sim$5–10 kJ mol$^{-1}$) as compared to coordination to a single exchangeable cation. Furthermore, very recently evidence for the formation of similar complexes in high-silica Na-ZSM-5 and K-ZSM-5 has been reported [79]. Systematically, lower CO adsorption energies for the potassium-containing zeolites have been reported. This is in line with the expected lower Lewis acidity of the larger alkaline cations. The probability of the specific interaction of CO with multiple cation sites was shown to depend on the Al content and hence on the density of the exchangeable cations in the microporous matrix, on their ionic radius and on the zeolite topology.

Similarly, a preference for the formation of bridging adsorption complexes of $CO_2$ in Na-FER zeolite with high concentration of the exchangeable cations was demonstrated by the combination of variable-temperature IR spectroscopy and periodic DFT calculations [81]. Nevertheless, it was concluded that the maximum loading of $CO_2$ apparently does not depend on the concentration of $Na^+$ in the ferrierite zeolite.

Molecular adsorption of $N_2O_4$ on Na-, K-, and Rb-containing zeolites Y and X was also investigated by means of periodic DFT calculations [82, 83]. Bonding within the adsorption complex formed upon adsorption of a nonpolar $N_2O_4$ on alkali-exchanged faujasite is very weak and corresponds to the induced polarization of the adsorbed species by the extra-framework cations. Strength of such interactions correlates with the size of the alkali ion. The smaller the cation is, the stronger its polarizing ability and Lewis acidity are. As a result, similar to the cases of CO and $H_2$ adsorption on alkali-exchanged zeolites [79, 84], one expects a decrease of the $N_2O_4$ adsorption energies simultaneously with the increase of the ionic radius

**Table 11.2** Adsorption energies (kJ mol$^{-1}$) of $N_2O_4$ molecules at $S_{II}$ and $S_{III}$ cations[a] in the cage of alkaline-exchanged zeolites X [82] and Y [83].

| | $S_{II}$/X | $S_{III}$/X | $S_{II}{}^a$/Y | $S_{II}{}^b$/Y | $S_{III}$/Y |
|---|---|---|---|---|---|
| $Na^+$ | −10 | −11 | −11 | −11 | −33 |
| $K^+$ | −10 | −15 | −11 | −10 | −16 |
| $Rb^+$ | −21 | −26 | −18 | −16 | −29 |

[a]The $S_{II}$ and $S_{III}$ sites of the faujasite lattice are illustrated in Figure 11.1a.

of the zeolitic cations. The computational results show, however, the opposite trend (Table 11.2).

This phenomenon can be best rationalized with the example of lower-silica modification of faujasite that is zeolite X [82]. A high density of the extra-framework cations in the cage of zeolite X allows formation of multiple interactions between the adsorbed $N_2O_4$ molecule and the exchangeable ions (Figure 11.6a–c). The size of the intrazeolite cations strongly influences the probability of such multicentered binding. Indeed, in the case of NaX zeolite, the smaller sodium ions neighboring the primary adsorption site are located too far to form interatomic contacts of a notable strength with the adsorbed molecule and are able to only slightly polarize it. On the other hand, when the ionic radius of the cations increases numerous additional adsorbate–adsorbent interactions can be formed. Whereas, in the cases of NaX and KX (Figure 11.6a,b, respectively), one can distinguish the primary and the secondary interactions in the corresponding adsorption complexes, the interatomic distances for the respective contacts between the $N_2O_4$ confined in the cage of RbX with the accessible extra-framework $Rb^+$ cations become very similar (Figure 11.6c). In the latter case, not only the exchangeable cations nearest to the main adsorption site but almost all $Rb^+$ in the faujasite supercage interact with the adsorbed $N_2O_4$. As a result, despite the expected lower polarizing ability of the softer cations and hence the weaker individual interatomic contacts formed with them upon adsorption, the overall interaction energy increases along with the increase of the ionic radius of the exchangeable cations.

Besides the higher probability for the formation of the multiple-centered adsorption complexes, the unexpected enhancement of the interaction energy with the larger exchangeable cations can be also contributed by the differences in the effective shielding of those when stabilized at cation sites of zeolites. Indeed, larger alkaline or alkali-earth cations ($K^+$, $Rb^+$, $Ca^{2+}$, etc.) at the $S_{II}$ sites (six-membered rings, Figure 11.1a) of faujasite are only slightly shielded by the surrounding framework oxygens. As a result, the effective polarization of the adsorbed molecules in their field increases as compared to the cases of smaller exchangeable cations ($Li^+$, $Mg^{2+}$, etc.), which are strongly shielded at the $S_{II}$ positions of faujasite. This has

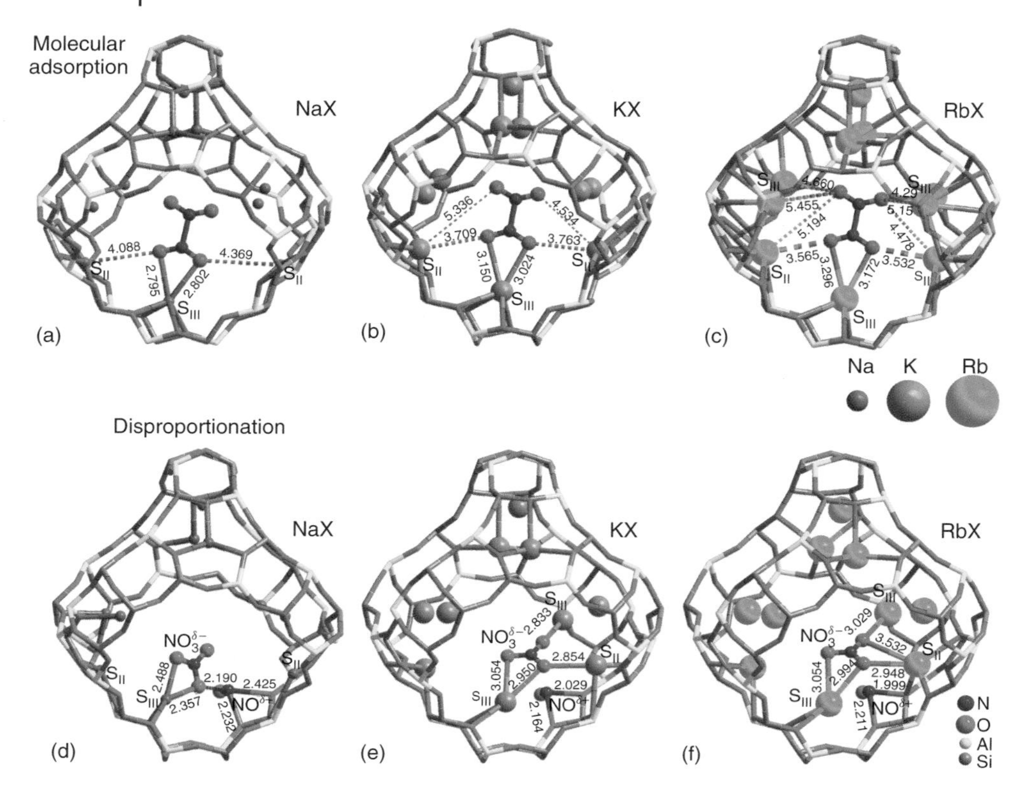


**Figure 11.6** Molecular adsorption of $N_2O_4$ (a–c) and formation of $NO_3^{\delta-} \cdots NO^{\delta+}$ ion pair (d–f) due to $N_2O_4$ disproportionation in alkali-exchanged zeolites X. (Adapted from [85].)

been recently demonstrated by a combination of IR spectroscopy and cluster *ab initio* calculations of adsorption of light alkanes on alkali-earth exchanged zeolite Y [86, 87]. Furthermore, it has been shown that the differences in the effective shielding of the exchangeable cations in zeolites can indirectly affect the preferred conformation in the resulting adsorption complexes via favoring additional secondary van der Waals interactions between the adsorbed molecules and the framework oxygens of the cation site. In the case of $N_2O_4$ molecular adsorption, the combination of both effects due to the effective shielding and due to the formation of multiple interaction with the exchangeable cations results in the partial restoration of the expected trends in the adsorption energies when the primary adsorbate–adsorbent interaction takes place with the weakly shielded alkaline cation stabilized at the $S_{III}$ site (four-membered ring, Figure 11.1a) of zeolite Y (Table 11.2).

These effects are much more pronounced when polar species are formed in the cation-exchanged zeolite matrices. Disproportionation of $N_2O_4$ in alkali-exchanged faujasites results in the formation of rather polar $NO_3^{\delta-} \cdots NO^{\delta+}$ species. Interaction of the involved charged fragments with the corresponding counterion of the zeolite matrix facilitates the cleavage of the contact ion pair. Periodic DFT calculations [82, 83] show the dominant role of the exchangeable cations for the reactivity of alkaline zeolite in the $N_2O_4$ disproportionation reaction. Stability of the reaction products was found to correlate very well with the size, intrazeolite arrangement, and mobility of the exchangeable cations. It was shown that in the case of Na-containing faujasites, the smaller size of the sodium cations and their limited mobility do not allow formation of the appropriate configuration of the active site to stabilize the negatively charged $NO_3^{\delta-}$(Figure 11.6d). In contrast, the larger and mobile $Rb^+$ cations shape a perfect stabilizing environment for the nitro group (Figure 11.6f). Thus, it has been concluded that the molecular recognition features of the microporous matrix facilitate the charge separation upon the $N_2O_4$ disproportionation in a fashion of a polar solvent. The cooperative effect of the extra-framework cations, their intrazeolite arrangement and mobility induced by adsorption, as well as the steric properties of the zeolite cage are crucial for the reactivity of the cation-exchanged zeolites in this reaction.

Recently a concept of confinement-induced reactivity, which is interrelated with the above-discussed molecular recognition features of the intrazeolite void space, has been put forward to rationalize the photo-catalytic activity of rather inert alkali-earth containing faujasites in oxidation of alkenes [85, 88]. The chemical reaction in this case is initiated by a visible-light induced electron transfer between the adsorbed olefin and dioxigen. Experimental studies indicated a significant decrease of the energy of such an electron excitation when the reagents are loaded into the zeolite matrix containing alkaline- or alkali-earth cations [89–91]. This effect has been initially attributed to the stabilization of the hydrocarbon·$O_2$ charge-transfer state by the interaction of the strong electrostatic field in the zeolite cage with the large dipole moment generated upon the excitation. In contrary, a theoretical analysis of the initial step of photochemical activation of 2,3-dimethyl-2-butene (DMB) and $O_2$ coadsorbed in a 16T cluster model (Figure 11.1d) representing a part of Ca-, Mg-, and Sr-exchanged supercage of zeolite Y indicate that the effect of the electrostatic field of the zeolite cavity is only minor for the reactivity of these microporous materials [88]. It has been proposed that the role of the zeolite in the photo-oxidation of alkenes with molecular oxygen is the complexation of the reagents to the extra-framework cations in a pre-transition state configuration. This results in confinement of these molecules with a specific orientation. The resulting formation of a $\pi - \pi$ intermolecular complex (Figure 11.7) substantially increases the probability of the photo-initiation of the oxidation reaction. The relative orientation and the distance between the adsorbed reagents, which depend strongly on the size of the exchangeable cations in the adsorption site, are crucial for the chemical reactivity of these systems. It has been concluded that a high density, specific location, and size of the exchangeable cations in the zeolite result in molecular recognition of the adsorbed species and their chemical activation.

Thus, molecular recognition features of the intrazeolite space are crucial both for the adsorption properties and for the chemical reactivity of the microporous

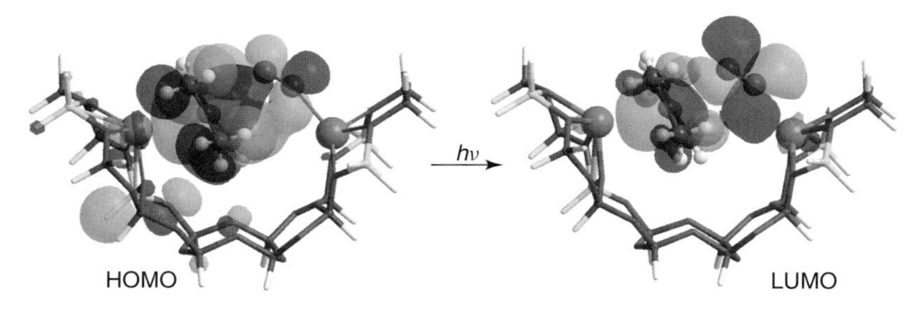


**Figure 11.7** Frontier molecular orbitals of the DMB · $O_2$ adsorption complex in CaY involved in the intermolecular electron excitation. (Adapted from [85].)

materials. The presence of multiple interaction sites in the zeolite cage can control the preferred conformations of the adsorbed molecules and their chemical transformations. Further theoretical studies are needed to understand better the impact and the role of the multiple interaction sites in the conventional zeolite-catalyzed reactions. Indeed, the computational studies discussed so far suggest that there is a possibility of interaction between the distant active species and bulkier reaction intermediates formed in the zeolite channels. We anticipate that the reactivity and selectivity trends predicted by computational studies may alter substantially when a realistic chemical composition of the zeolite catalysts is considered.

## 11.6
## Structural Properties of Zeolites: Framework Al Distribution and Structure and Charge Compensation of Extra-framework Cations

Catalytically active species, that is, protons, metal cations or more complex cationic aggregates, compensate for the negative charge of the microporous aluminosilicate framework. The crystallographic position of the corresponding negatively charged aluminum-occupied oxygen tertrahedra ($[AlO_4]^-$) governs to a large extent the location of the active sites that in turn may have a significant influence on the catalytic activity and selectivity [92, 93]. Thus, understanding the exact Al sitting and the factors which govern its distribution in zeolites is strongly desired. The related question is how the Al distribution influences structural and chemical properties of confined cationic species as well as their location in the zeolite matrix. An in-depth discussion of the problem of Al sitting in zeolites is provided in Chapter 14 of this book. Therefore, in this section we limit ourselves to only a brief illustration of the recent progress made in revealing the location of framework aluminum in high-silica zeolites by combined spectroscopic and quantum chemical studies, while the main focus will be on the recent theoretical concept of the structures of extra-framework cations.

Recently, a tool based on the complementary use of the high resolution of $^{27}$Al 3Q MAS (magic-angle spinning) NMR spectroscopy and DFT calculations has been proposed for studying the local geometry of framework $[AlO_4]^-$ units and for the identification of the Al sitting in high-silica zeolites [94–96]. The respective experimental technique allows identification and quantification of the $^{27}Al$ resonances corresponding to individual T-sites in the zeolite framework. High-silica zeolite models with different framework Al distribution were optimized using a hybrid DFT:force field method where the higher-level method is applied to cluster models containing the Al atom surrounded by at least five coordination shells. The remaining part of the zeolite was then described by a force field method. A bare charged framework of ZSM-5 zeolite (MFI topology) that includes neither cations nor water molecules was used as a relevant model. Each model contained only one type of Al substitution. After the structure determination, NMR shielding tensors were calculated for the atoms of the optimized clusters using the gauge-independent atomic orbital (GIAO) methods [97]. The resulting calculated isotropic $^{27}Al$ NMR shifts were then used to assign and rationalize the experimental observations.

The results thus obtained allowed to assign the observed $^{27}Al$ resonances to the particular T-sites in the MFI framework [94–96]. Although a trend has been detected for smaller $^{27}Al$ isotropic chemical shifts with increasing average T–O–T angle [95] that had been previously proposed as a tool for the identification of the Al sitting in zeolites [98], the corresponding correlation was shown to be not suitable for the assignment purposes. A very important conclusion was made that the local geometry of framework $[AlO_4]^-$ tetrahedra cannot be deduced directly from the experimental NMR data, whereas such information can only be obtained from the theoretical calculations [95].

Concerning the relative location of the anionic $[AlO_4]^-$ sites in the framework, the very recent study by Dědeček *et al.* [96] has shown that the influence of the second Al site in the next nearest or in the next-next-nearest framework position on the $^{27}Al$ isotropic shift is not uniform. Thus, it has been concluded that this combined NMR and DFT approach is only suitable for determining the Al sitting in zeolites with only very low local density of aluminum. In other words, the concentration of the Al–O-$(SiO)_n$–Al ($n$ = 1 or 2) sequences in the framework must be negligible.

Revealing the Al distribution and the rules, which govern Al sitting, in high-silica zeolites is very important to understand the molecular structure and chemical reactivity of cationic species confined in the microporous matrix. For univalent cations, a well-accepted model is localization of the positively charged ions in the vicinity of the negatively charged $[AlO_4]^-$ framework tetrahedral units [99]. For cations with a higher charge, this model requires close proximity of the aluminum substitutions in the zeolite framework. This requirement may not always be met, especially for high-silica zeolites. Indeed, even closely located $[AlO_4]^-$ tetrahedra may not necessarily face the same zeolitic ring or even channel preventing thus a direct charge compensation of multivalent cations (see e.g., [96]).

An alternative model involves indirect compensation of the multiple-charge cations by distantly placed negative charges of the zeolite lattice. This concept has been initially put forward to account for the high reactivity of Zn-modified ZSM-5 in alkane activation [100, 101]. In this case, part of the exchangeable $Zn^{2+}$ cations is located in the vicinity of one framework anionic $[AlO_4]^-$ site, while the other negative site required for the overall charge neutrality is located at a longer distance, where it does not directly interact with the extra-framework positive charge. The existence of such species has been supported by spectroscopic [100–103] and theoretical studies [104, 105]. However, a firm theoretical evidence for a structural model, in which the positions of multivalent cations in zeolite are not dominated by the direct interaction between the mononuclear $M^{2+}$ cation and the framework charge, has not yet been presented.

A useful structural model includes then the presence of extra-framework oxygen-containing anions that coordinate to the metal (M), resulting in the formation of multinuclear cationic complexes with a formal charge of +1 such as $[M^{3+} = O^{2-}]^+$, $[M^{2+}{-}OH^-]^+$, and so on For example, formation of the isolated gallyl $GaO^+$ ions was proposed to be responsible for the experimentally observed enhancement of the dehydrogenation catalytic activity of ZSM-5 modified with $Ga^+$ upon the stoichiometric treatment with $N_2O$ [106]. However, a comprehensive computational analysis of possible reaction paths for the light alkane dehydrogenation indicated that the isolated $GaO^+$ ions cannot be responsible for the catalytic activity [107]. An alternative interpretation is the formation of multiple-charged extra-framework oligomeric $(GaO)_n{}^{n+}$ cations. Stability and reactivity of such species in high-silica mordenite have been studied by periodic DFT calculations [108, 109]. It has been shown that isolated gallyl ions tend to oligomerize resulting in formation of oxygen-bridged $Ga^{3+}$ pairs. The stability of the resulting cationic complexes does not require proximate Al substitution in the framework (Figure 11.8). The theoretical calculations indicate that the oligomers with a higher degree of aggregation can be in principle formed in oxidized Ga-exchanged zeolites [109].

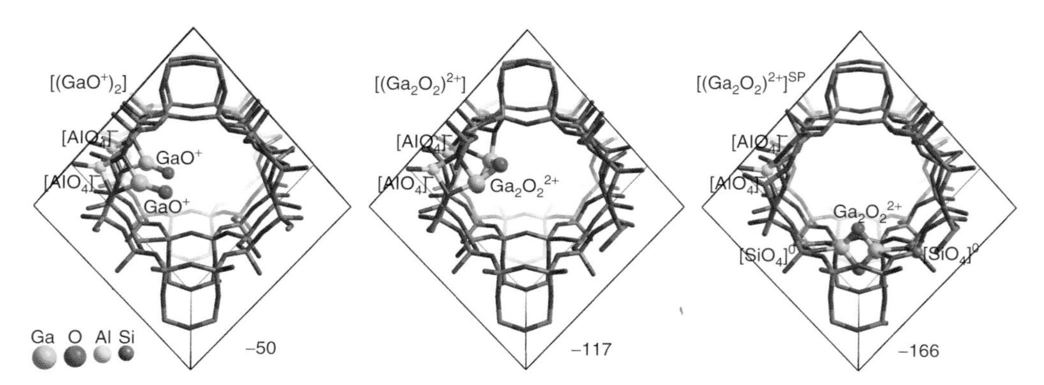


**Figure 11.8** Structures of $(GaO)_2{}^{2+}$ isomers in a high-silica (Si/Al = 23) mordenite model. The numbers under the structures correspond to the DFT-computed reaction energies ($\Delta E$ in kilojoules per mole) for the stoichiometric oxidation of two exchangeable $Ga^+$ sites with $N_2O$ toward the respective cations [109].

It has been concluded that the formation of the favorable coordination environment of the metal centers via interaction with basic oxygen anions dominates over direct charge compensation and leads to clustering of the extra-framework species. The presence of multiply charged bi- or oligonuclear metal oxide species in zeolites does not require the immediate proximity of an equivalent number of negative framework charges. Nonlocalized charge compensation is expected to be a common feature of high-silica zeolites modified with metals ions. The corresponding theoretical concept is argued to be useful to develop new structural models for the intrazeolitic active sites involving multiple metal centers.

Finally, implication of the concept of nonlocalized charge compensation allowed considering formation of other possible multinuclear Ga-containing intrazeolite species in high-silica zeolite matrix. The performance of different cationic oxygen- and sulfur-containing Ga clusters in light alkane dehydrogenation has been analyzed by means of periodic DFT calculations [110]. From the results thus obtained a structure–reactivity relationship of a remarkable predictive power has been derived for their catalytic performance. It has been shown that the computed activation free energies as well as the free energy changes of the important elementary steps of the catalytic ethane dehydrogenation cycle scale linearly with the values of free energy change of the active site regeneration (Figure 11.9). The later parameter has been chosen as the reactivity descriptor because it reflects strength of the associated active Lewis acid–base pairs. The relationship presented points to the optimum composition and structure of the intrazeolite Ga cluster – $[H\text{-}Ga(O)(OH)Ga]^{2+}$ – for the catalytic dehydrogenation of light alkanes [110]. Similar active species have been previously proposed to account for the enhanced dehydrogenation activity of Ga-modified ZSM-5 zeolite upon water cofeeding [109, 111].

Nevertheless, so far the preference for the nonlocalized charge compensation in zeolites has been convincingly shown only for Ga-containing species stabilized

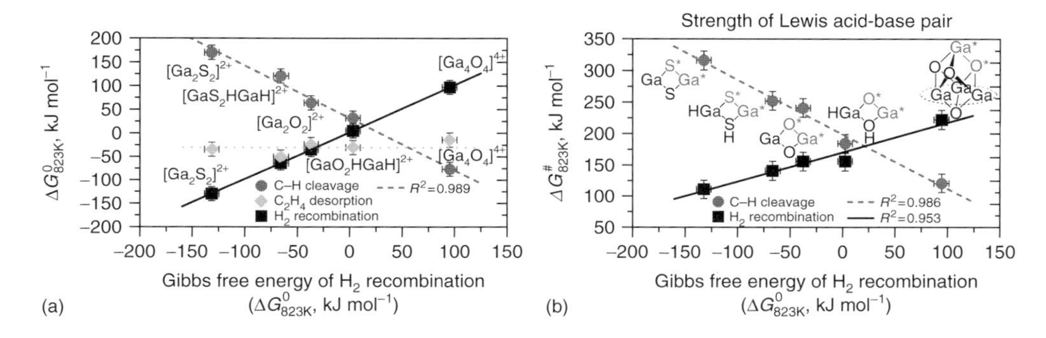


**Figure 11.9** (a) Gibbs free energies ($\Delta G^{\circ}_{823\ K}$) of elementary reaction steps of $C_2H_6$ dehydrogenation and (b) activation Gibbs free energies ($\Delta G^{\#}_{823\ K}$) of the C–H activation and of the $H_2$ recombination step plotted against the respective values of $\Delta G^{\circ}_{823\ K}$ of $H_2$ recombination [110].

in a single zeolite topology. Although, these results have been anticipated to be valid also for other metal ions in different microporous matrices, the expansion of theoretical studies to other metal-containing zeolites with different topologies and framework composition is needed to generalize this theoretical structural concept. Further theoretical efforts in understanding the fundamental factors that govern Al sitting and their relative distribution in zeolite framework are strongly desired to create a resolved molecular-level picture of the structural properties of these microporous materials.

## 11.7 Summary and Outlook

Computational modeling is becoming one of the key contributors to the zeolite science. Theoretical methods play a pivotal role in assisting the interpretation of the experimental data, revealing the structural and chemical properties of microporous materials, and in developing the molecular-level understanding of mechanistic aspects of catalytic reactions in confined space. Obviously, it was impossible to review all of the computational methods and areas of their application in zeolite sciences on pages available here. In this chapter we attempted to illustrate the current capabilities and limitations of promising quantum chemical methods as applied to zeolite sciences. The power of quantum chemical techniques in rationalizing the complex chemical processes in microporous matrices and in developing novel useful chemical and structural concepts is highlighted.

There are two major future challenges remaining in the computational chemistry of zeolites. Thanks to the great development of theoretical methodologies and the rapid growth in hardware performance we are now able to model rather accurately various aspects of the chemical processes taking place in the zeolite void space. This allows us to unravel molecular details of many known processes and to understand the fundamental factors that determine and control the chemical reactivity of microporous catalysts. The next step is the development of *ab initio*-based computational approaches that will serve as a tool for the prediction of chemical reactivity of zeolite catalysts. This however is not a trivial task taking into account the high complexity of the associated chemical processes, many aspects of which are yet not well understood.

The second challenge is to develop novel hierarchical approaches integrating various levels of theory into one multiscale simulation to cover the disparate length and time scales and allow a comprehensive theoretical kinetic description of a working catalyst. The *ab initio* electronic structure calculations discussed in this chapter provide important molecular-level information about the details of the elementary reactions involved in the catalytic processes. Combining the results of such calculations with the statistical simulations (e.g., kinetic Monte Carlo) to account for the interplay between all elementary processes involved in the catalytic cycle at the mesoscopic scale, and with the macroscopic theories, which describe the effects of heat and mass transfer, would ultimately result in the complete description of the zeolite catalysis.

## References


1. Woodley, S.M. and Catlow, R. (2008) *Nat. Mater.*, **7**, 937.
2. Smit, B. and Maesen, T.L.M. (2008) *Chem. Rev.*, **108**, 4125.
3. Smit, B. and Maesen, T.L.M. (2008) *Nature*, **451**, 671.
4. Jensen, F. (1999) *Introduction to Computational Chemistry*, Wiley-Interscience, New York.
5. Levine, I.N. (1983) *Quantum Chemistry*, Allyn and Bacon, Boston.
6. Leach, A.R. (1996) *Molecular Modeling: Principles and Applications*, Pearson Education, Harlow.
7. Foresman, J.B. and Frish, A. (1996) *Exploring Chemistry with Electronic Structure*, 2nd edn, Gaussian, Pittsburg.
8. Parr, R.G. and Yang, W. (1989) *Density Functional Theory of Atoms in Molecules*, Oxford University Press, New York.
9. Young, D.C. (2001) *Computational Chemistry: A Practical Guide for Applying Techniques to Real-World Problems*, Wiley-Interscience, New York.
10. Feyereisen, M., Fitzgerald, G., and Komornicki, A. (1993) *Chem. Phys. Lett.*, **208**, 359.
11. Hohenberg, P. and Kohn, W. (1964) *Phys. Rev.*, **136**, B864.
12. Kohn, W. and Sham, L. (1965) *Phys. Rev.*, **140**, A1133.
13. Becke, A.D. (1988) *Phys. Rev. A*, **38**, 3098.
14. Perdew, J.P., Chevary, J.A., Vosko, S.H., Jackson, K.A., Pederson, M.R., Singh, D.J., and Fiolhais, C. (1992) *Phys. Rev. B*, **46**, 6671.
15. Perdew, J.P., Burke, K., and Ernzerhof, M. (1996) *Phys. Rev. Lett.*, **77**, 3865.
16. Becke, A.D. (1993) *J. Chem. Phys.*, **98**, 5648.
17. Adamo, C. and Barone, V. (1998) *J. Chem. Phys.*, **108**, 664.
18. Adamo, C. and Barone, V. (1999) *J. Chem. Phys.*, **110**, 6158.
19. Johnson, E.R. and DiLabio, G.A. (2006) *Chem. Phys. Lett.*, **419**, 333.
20. Zhao, Y. and Truhlar, D.G. (2005) *J. Chem. Theory Comput.*, **1**, 415.
21. Guest, M.F., Bush, I.J., van Dam, H.J.J., Sherwood, P., Thomas, J.M.H., van Lenthe, J.H., Havenith, R.W.A., and Kendrick, J. (2005) *Mol. Phys.*, **103**, 719.
22. Frisch, M.J., Trucks, G.W., Schlegel, H.B., Scuseria, G.E., Robb, M.A., Cheeseman, J.R., Montgomery, J.A. Jr., Vreven, T., Kudin, K.N., Burant, J.C., Millam, J.M., Iyengar, S.S., Tomasi, J., Barone, V., Mennucci, B., Cossi, M., Scalmani, G., Rega, N., Petersson, G.A., Nakatsuji, H., Hada, M., Ehara, M., Toyota, K., Fukuda, R., Hasegawa, J., Ishida, M., Nakajima, T., Honda, Y., Kitao, O., Nakai, H., Klene, M., Li, X., Knox, J.E., Hratchian, H.P., Cross, J.B., Bakken, V., Adamo, C., Jaramillo, J., Gomperts, R., Stratmann, R.E., Yazyev, O., Austin, A.J., Cammi, R., Pomelli, C., Ochterski, J.W., Ayala, P.Y., Morokuma, K., Voth, G.A., Salvador, P., Dannenberg, J.J., Zakrzewski, V.G., Dapprich, S., Daniels, A.D., Strain, M.C., Farkas, O., Malick, D.K., Rabuck, A.D., Raghavachari, K., Foresman, J.B., Ortiz, J.V., Cui, Q., Baboul, A.G., Clifford, S., Cioslowski, J., Stefanov, B.B., Liu, G., Liashenko, A., Piskorz, P., Komaromi, I., Martin, R.L., Fox, D.J., Keith, T., Al-Laham, M.A., Peng, C.Y., Nanayakkara, A., Challacombe, M., Gill, P.M.W., Johnson, B., Chen, W., Wong, M.W., Gonzalez, C., and Pople, J.A. (2003) *Gaussian 03, Revision B.05*, Gaussian, Inc., Pittsburgh.
23. Ahlrichs, R. (1989) *Chem. Phys. Lett.*, **162**, 165.
24. Segall, M.D., Lindan, P.J.D., Probert, M.J., Pickard, C.J., Hasnip, P.J., Clark, S.J., and Payne, M.C. (2002) *J. Phys. Condens. Matter.*, **14**, 2717.

25. (a) Marx, D. and Hutter, J. (2000) *Modern Methods and Algorithms of Quantum Chemistry*, NIC, FZ Jülich, p. 301; (b) Andreoni, W. and Curioni, A. (2000) *Parallel Comput.*, **26**, 819.
26. (a) Kresse, G. and Hafner, J. (1994) *Phys. Rev. B*, **49**, 14251; (b) Kresse, G., and Furthmüller, J. (1996) *Comput. Mater. Sci.*, **6**, 15; (c) Kresse, G. and Furthmüller, J. (1996) *Phys. Rev. B*, **54**, 11169.
27. Dovesi, R., Saunders, V.R., Roetti, C., Orlando, R., Zicovich-Wilson, C.M., Pascale, F., Civalleri, B., Doll, K., Harrison, N.M., Bush, I.J., D'Arco, Ph., and Llunell, M. (2006) *CRYSTAL2006 User's Manual*, Universita di Torino, Torino, (2008) *http://www.crystal.unito.it.*
28. Tosoni, S., Tuma, C., Sauer, J., Civalleri, B., and Ugliengo, P. (2007) *J. Chem. Phys.*, **127**, 154102.
29. Tuma, C. and Sauer, J. (2004) *Chem. Phys. Lett.*, **387**, 388.
30. Tuma, C. and Sauer, J. (2006) *Phys. Chem. Chem. Phys.*, **8**, 3955.
31. de Ménorval, B., Ayrault, P., Gnep, N.S., and Guisnet, M. (2005) *J. Catal.*, **230**, 38.
32. Kato, T. and Reed, C.A. (2004) *Angew. Chem. Int. Ed. Engl.*, **43**, 2907.
33. Boronat, M. and Corma, A. (2008) *Appl. Catal. A Gen.*, **336**, 2.
34. Correa, R.J. and Mota, C.J.A. (2002) *Phys. Chem. Chem. Phys.*, **4**, 375.
35. Boronat, M., Viruela, P.M., and Corma, A. (2004) *J. Am. Chem. Soc.*, **126**, 3300.
36. Rozanska, X., van Santen, R.A., Demuth, T., Jutschka, F., and Hafner, J. (2003) *J. Phys. Chem. B*, **107**, 1309.
37. Tuma, C. and Sauer, J. (2005) *Angew. Chem. Int. Ed. Engl.*, **44**, 4769.
38. Haw, J.F., Nicholas, J.B., Xu, T., Beck, L.W., and Ferguson, D.B. (1996) *Acc. Chem. Res.*, **29**, 259.
39. Zhao, Y. and Truhlar, D.G. (2008) *J. Phys. Chem. C*, **112**, 6860.
40. de Moor, B.A., Reyniers, M.-F., Sierka, M., Sauer, J., and Marin, G.B. (2008) *J. Phys. Chem. C*, **112**, 11796.
41. Kerber, T., Sierka, M., and Sauer, J. (2008) *J. Comp. Chem.*, **29**, 2088.
42. Svelle, S., Tuma, C., Rozanska, X., Kerber, T., and Sauer, J. (2009) *J. Am. Chem. Soc.*, **131**, 816.
43. Svelle, S., Rønning, P.O., and Kolboe, S. (2004) *J. Catal.*, **224**, 115.
44. Svelle, S., Rønning, P.O., Olsbye, U., and Kolboe, S. (2005) *J. Catal.*, **234**, 385.
45. Svelle, S., Arstad, B., Kolboe, S., and Swang, O. (2003) *J. Phys. Chem. B*, **107**, 9281.
46. Rozanska, X., Saintigny, X., van Santen, R.A., and Hutschka, F. (2001) *J. Catal.*, **202**, 141.
47. Rozanska, X., van Santen, R.A., Hutschka, F., and Hafner, J. (2001) *J. Am. Chem. Soc.*, **123**, 7655.
48. Vos, A.M., Rozanska, X., Schoonheydt, R.A., van Santen, R.A., Hutschka, F., and Hafner, J. (2001) *J. Am. Chem. Soc.*, **123**, 2799.
49. Demuth, T., Benco, L., Hafner, J., Toulhoat, H., and Hutschka, F. (2001) *J. Chem. Phys.*, **114**, 3703.
50. (a) Grimme, S. (2004) *J. Comput. Chem.*, **25**, 1463; (b) Grimme, S. (2006) *J. Comput. Chem.*, **27**, 1787.
51. Joshi, Y.V. and Thomson, K.T. (2008) *J. Phys. Chem. C*, **112**, 12825.
52. de Moor, B.A., Reyniers, M.-F., Sierka, M., Sauer, J., and Marin, G.B. (2009) *Phys. Chem. Chem. Phys.*, **11**, 2939.
53. Zhao, Y., Schultz, N.E., and Truhlar, D.G. (2005) *J. Chem. Phys.*, **123**, 161103.
54. Zhao, Y. and Truhlar, D.G. (2008) *Theor. Chem. Acc.*, **120**, 215.
55. (a) Dion, M., Rydberg, H., Schröder, E., Langreth, D.C., and Lundqvist, B.I. (2004) *Phys. Rev. Lett.*, **92**, 246401; (b) Dion, M., Rydberg, H., Schröder, E., Langreth, D.C., and Lundqvist, B.I. (2005) *Phys. Rev. Lett.*, **95**, 109902(E).
56. Thonhauser, T., Cooper, V.R., Li, S., Puzder, A., Hyldgaard, P., and Langreth, D.C. (2007) *Phys. Rev. B*, **76**, 125112.
57. Vydrov, Q.A., Wu, Q., and van Voorhis, T. (2008) *J. Chem. Phys.*, **129**, 014106.
58. Stocker, M. (1999) *Microporous Mesoporous Mater.*, **29**, 3.
59. Haw, J.F., Song, W.G., Marcos, D.M., and Micholas, J.B. (2003) *Acc. Chem. Res.*, **36**, 314.

60. (a) Dahl, I.M. and Kolboe, S. (1993) *Catal. Lett.*, **20**, 329; (b) Dahl, I.M. and Kolboe, S. (1994) *J. Catal.*, **149**, 458.
61. Song, W.G., Marcus, D.M., Fu, H., Ehresmann, J.O., and Haw, J.F. (2002) *J. Am. Chem. Soc.*, **124**, 3844.
62. Marcus, D.M., McLachlan, K.A., Wildman, M.A., Ohresmann, J.O., Kletnieks, P.W., and Haw, J.F. (2006) *Angew. Chem. Int. Ed. Engl.*, **45**, 3133.
63. Lesthaeghe, D., van Speybroeck, V., Marin, G.B., and Waroquier, M. (2006) *Angew. Chem. Int. Ed. Engl.*, **45**, 1714.
64. Lesthaeghe, D., van Speybroeck, V., Marin, G.B., and Waroquier, M. (2007) *Ind. Eng. Chem. Res.*, **46**, 8832.
65. Sassi, A., Wildman, M.A., Ahn, H.J., Prasad, P., Nicholas, J.B., and Haw, J.F. (2002) *J. Phys. Chem. B*, **106**, 2294.
66. Bjorgen, M., Bonino, F., Kolboe, S., Lillerud, K.-P., Zecchina, A., and Bordiga, S. (2003) *J. Am. Chem. Soc.*, **125**, 15863.
67. Song, W.G., Nicholas, J.B., Sassi, A., and Haw, J.F. (2002) *Catal. Lett.*, **81**, 49.
68. Lesthaeghe, D., van Speybroeck, V., Marin, G.B., and Waroquier, M. (2007) *Angew. Chem. Int. Ed. Engl.*, **46**, 1311.
69. Lesthaeghe, D., van Speybroeck, V., Marin, G.B., and Waroquier, M. (2007) *Stud. Surf. Sci. Catal.*, **170**, 1668.
70. Svelle, S., Joensen, F., Nerlov, J., Olsbye, U., Lillerud, K.-P., Kolboe, S., and Bjorgen, M. (2006) *J. Am. Chem. Soc.*, **128**, 14770.
71. McCann, D.M., Lesthaeghe, D., Kletnieks, P.W., Guenther, D.R., Hayman, M.J., Van Speybroeck, V., Waroquier, M., and Haw, J.F. (2008) *Angew. Chem. Int. Ed. Engl.*, **47**, 5179.
72. Wang, C.-M., Wang, Y.-D., Xie, Z.-K., and Liu, Z.-P. (2009) *J. Phys. Chem. C*, **113**, 4584.
73. Corma, A. (2003) *J. Catal.*, **216**, 298.
74. Degnan, T.F. Jr. (2003) *J. Catal.*, **216**, 32.
75. Ringe, D. and Petsko, G.A. (2008) *Science*, **320**, 1428.
76. Das, S., Brudvig, G.W., and Crabtree, R.H. (2008) *Chem. Commun.*, 413.
77. OteroAreán, C., Rodrígue Delgado, M., López Bauçà, C., Vrbka, L., and Nachtigall, P. (2007) *Phys. Chem. Chem. Phys.*, **9**, 4657.
78. Garrone, E., Bulánek, R., Frolich, K., Otero Areán, C., Rodrígues Delgado, M., Turnes Palomino, G., Nachtigallová, D., and Nachtigall, P. (2006) *J. Phys. Chem. B*, **110**, 22542.
79. Otero Areán, C., Rodrígues Delgado, M., Frolich, K., Bulánek, R., Pulido, A., Fiol Bibiloni, G., and Nachtigall, P. (2008) *J. Phys. Chem. C*, **112**, 4658.
80. Nachtigallová, D., Virbka, L., Budský, O., and Nachtigall, P. (2008) *Phys. Chem. Chem. Phys.*, **10**, 4189.
81. Pulido, A., Nachtigall, P., Zukal, A., Domínguez, I., and Čejka, J. (2009) *J. Phys. Chem. C*, **113**, 2928.
82. Pidko, E.A., Mignon, P., Geerlings, P., Schoonheydt, R.A., and van Santen, R.A. (2008) *J. Phys. Chem. C*, **112**, 5510.
83. Mignon, P., Pidko, E.A., van Santen, R.A., Geerlings, P., and Schoonheydt, R.A. (2008) *Chem. Eur. J.*, **14**, 5168.
84. Otero Areán, C., Nachtigallová, D., Nachtigall, P., Garrone, E., and Rodrígues Delgado, M. (2007) *Phys. Chem. Chem. Phys.*, **9**, 1421.
85. Pidko, E.A. and van Santen, R.A. (2010) *Int. J. Quantum Chem.*, **110**, 210.
86. Pidko, E.A. and van Santen, R.A. (2006) *ChemPhysChem*, **7**, 1657.
87. Pidko, E.A., Xu, J., Mojet, B.L., Lefferts, L., Subbotina, I.R., Kazansky, V.B., and van Santen, R.A. (2006) *J. Phys. Chem. B*, **110**, 22618.
88. Pidko, E.A. and van Santen, R.A. (2006) *J. Phys. Chem. B*, **110**, 2963.
89. Frei, H. (2006) *Science*, **313**, 209.
90. Vasenkov, S. and Frei, H. (1997) *J. Phys. Chem. B*, **101**, 4539.
91. Blatter, F., Sun, H., Vasenkov, S., and Frei, H. (1998) *Catal. Today*, **41**, 297.
92. Bhan, A., Allian, A.D., Sunley, G.J., Law, D.J., and Iglesia, E. (2007) *J. Am. Chem. Soc.*, **129**, 4919.
93. Bhan, A. and Iglesia, E. (2008) *Acc. Chem. Res.*, **41**, 559.
94. Sklenak, S., Dědeček, J., Li, C., Wichterlová, B., Gábová, V., Sierka,

M., and Sauer, J. (2007) *Angew. Chem. Int. Ed. Engl.*, **46**, 7286.

95. Sklenak, S., Dědeček, J., Li, C., Wichterlová, B., Gábová, V., Sierka, M., and Sauer, J. (2009) *Phys. Chem. Chem. Phys.*, **11**, 1237.
96. Dědeček, J., Sklenak, S., Li, C., Wichterlová, B., Gábová, V., Brus, J., Sierka, M., and Sauer, J. (2009) *J. Phys. Chem.*, **113**, 1447.
97. Wolinski, K., Hinton, J.H., and Pulay, P. (1990) *J. Am. Chem. Soc.*, **112**, 8251.
98. Lippmaa, E., Samson, A., and Magi, M. (1986) *J. Am. Chem. Soc.*, **108**, 1730.
99. Centi, G., Wichterlova, B., and Bell, A.T. (eds) (2001) *Catalysis by Unique Metal Ion Structures in Solid Matrices*, NATO Science Series, Kluwer Acadamic, Dordrecht.
100. Kazansky, V.B. and Serykh, A.I. (2004) *Phys. Chem. Chem. Phys.*, **6**, 3760.
101. Kazansky, V. and Serykh, A. (2004) *Microporous Mesoporous Mater.*, **70**, 151.
102. Kazansky, V.B., Serykh, A.I., and Pidko, E.A. (2004) *J. Catal.*, **225**, 369.
103. Kazansky, V.B. and Pidko, E.A. (2005) *J. Phys. Chem. B*, **109**, 2103.
104. Zhidomirov, G.M., Shubin, A.A., Kazansky, V.B., and van Santen, R.A. (2005) *Theor. Chem. Acc.*, **114**, 90.
105. Pidko, E.A. and van Santen, R.A. (2007) *J. Phys. Chem. C*, **111**, 2643.
106. Rane, N., Overweg, A.R., Kazansky, V.B., van Santen, R.A., and Hensen, E.J.M. (2006) *J. Catal.*, **239**, 478.
107. Pidko, E.A., Hensen, E.J.M., and van Santen, R.A. (2007) *J. Phys. Chem. B*, **111**, 13068.
108. Pidko, E.A., Hensen, E.J.M., Zhidomirov, G.M., and van Santen, R.A. (2008) *J. Catal.*, **255**, 139.
109. Pidko, E.A., van Santen, R.A., and Hensen, E.J.M. (2009) *Phys. Chem. Chem. Phys.*, **11**, 2893.
110. Pidko, E.A. and van Santen, R.A. (2009) *J. Phys.Chem. B*, **113**, 4246.
111. Hensen, E.J.M., Pidko, E.A., Rane, N., and van Santen, R.A. (2007) *Angew. Chem. Int. Ed. Engl.*, **46**, 7273.

# 12
# Modeling of Transport and Accessibility in Zeolites

*Sofía Calero Diaz*

## 12.1
## Introduction

Modeling plays an important role in the field of zeolites and related porous materials. The use of molecular simulations allows the prediction of adsorption and diffusion coefficients in these materials and also provides important information about the processes taking place inside the porous structures at the molecular level [1–10]. Hence, molecular modeling is a very good complement to experimental work in the understanding of the molecular behavior inside the pores.

Despite the great interest and the applicability of zeolites, there are still many facets of the molecular mechanisms for a given reaction inside their pores that are poorly understood. These mechanisms are important for (i) the way the zeolite adsorbs, diffuses, and concentrates the adsorbates near the specific active sites; (ii) the interactions between the zeolite and the adsorbates and the effect on the electronic properties of the system; (iii) the chemical conversion at the active site; and (iv) the way the zeolite disperses the final product. Detailed knowledge on the molecular mechanisms involved will eventually lead to an increase in the reaction efficiency.

This chapter focuses on molecular modeling of transport and accessibility in zeolites. It describes how simulations have contributed to a better understanding of these materials and provides a summary of the state of the art as well as of current challenges. The chapter is organized as follows. First, common models and potentials are briefly described. This is followed by a general overview on current simulation methods to compute adsorption, diffusion, free energies, surface areas, and pore volumes. The chapter continues with some examples on applications of molecular modeling to processes of interest from the industrial and environmental point of view. Finally, the chapter closes with a summary and some remarks on future challenges.

*Zeolites and Catalysis, Synthesis, Reactions and Applications. Vol. 1.*
Edited by Jiří Čejka, Avelino Corma, and Stacey Zones

ISBN: 978-3-527-32514-6

## 12.2
## Molecular Models

To perform molecular simulations in zeolites, adequate models for all the atoms and molecules involved in the system are needed together with inter- and intramolecular potentials to describe the interactions between them. This section deals with a discussion on the most common models and potentials used in the literature for the zeolite framework, the nonframework cations, and the guest molecules.

### 12.2.1
### Modeling Zeolites and Nonframework Cations

The zeolite framework is usually built from silicon, aluminum, and oxygen with the crystallographic positions of these atoms taken from the dehydrated structures [11]. Zeolites with a Si/Al ratio higher than 1 can be obtained from random substitution of aluminum by silicon, either ignoring [12, 13] or taking into account distribution rules [14–18]. The aluminum atoms can also be assigned to energetic and entropic preferential positions [19–24], or using theoretical approaches that identifies experimentally accessible properties dependent on the aluminum distribution and associated cation distribution [25, 26]. Substitution of silicon for aluminum generates a negative net charge in the zeolite framework that needs to be compensated by either nonframework protons or cations in order to make the zeolite charge neutral. Some models explicitly distinguishes silicon from aluminum assigning different charges not only to those atoms but also to the oxygen atoms bridging two silicon atoms, and the oxygen atoms bridging one silicon and one aluminum atom [16, 27]. The charge distribution on the oxygen framework is often considered static in such a way that the polarization of oxygen by nearby extra framework cations is neglected. The extra framework cations can either remain fixed [28, 29] or move freely and adjust their position depending on their interactions with the system [16, 27, 30]. The latter requires potentials to predict the distribution of cations in the bare or loaded zeolite. The cation motions have to be sampled using displacements and random insertions that bypass energy barriers.

A force field is described as a set of functions needed to define the interactions in a molecular system. A wide variety of force fields that can be applied to zeolites exist. Among them, we find universal force field (UFF) [31], Discover (CFF) [32], MM2 [33], MM3 [34, 35], MM4 [36, 37], Dreiding [38], SHARP [39], VALBON [40], AMBER [41], CHARMM [42], OPLS [43], Tripos [44], ECEPP/2 [45], GROMOS [46], MMFF [47], Burchart [48], BKS [48], and specialty force fields for morphology predictions [49] or for computing adsorption [50]. In one approach, force fields are designed to be generic, providing a broad coverage of the periodic table, including inorganic compounds, metals, and transition metals. The diagonal terms in the force-constant matrix or these force fields are usually defined using simple functional forms. Owing to the generality of parameterization, these force fields are normally expected to yield reasonable predictions of molecular structures. However, emphasis was given to improving the accuracy in predicting molecular properties while maintaining a fair broad coverage of the periodic table. To achieve this goal, the force fields require complicated functional forms [32, 34–37, 47], and the parameters are derived by fitting to experimental data or to *ab initio* data.

Surprisingly, these general force fields give very poor results for specialized systems as adsorption and diffusion in zeolites. It is for this reason that new force fields have been optimized for pure silica zeolites [51–53] and also for those with nonframework sodium, calcium, and protons [16, 54–56]. The new force field parameters provide quantitative good predictions for adsorption and molecular transport in these systems [16, 51–58]. Most molecular simulation studies in zeolites are performed using the Kiselev-type potentials, where the zeolite atoms are held rigid at the crystallographic positions [59]. However, some authors have also investigated the effect of flexibility, using a variety of potentials for the framework atoms [60–62] and testing the accuracy and viability by comparing the computed adsorption [63, 64], diffusion [62, 65, 66], IR spectra [60, 61], or structural parameters [67, 68] with experimental data.

### 12.2.2
### Modeling Guest Molecules

To model guest molecules, rigid or flexible models can be used. For simple molecules such as carbon dioxide, nitrogen, hydrogen, oxygen, or even water, rigid models with multipoles or polarization seem a good representation [30, 56, 57, 69–78]. Complex molecules such as hydrocarbons normally require flexible models. A variety of flexible models have been addressed in literature spanning from the simplicity and the efficiency of the united-atom models to the complexity and the accuracy of the full-atom models [38, 79–86]. These models typically include partial charges for all atoms, expressions for describing bond-bending, bond-stretching, and torsion motions, and Lennard-Jones or Buckingham potential parameters that are often obtained from the fitting to *ab initio* [47, 87, 88] or to experimental vapor–liquid equilibrium data [53, 80, 89]. This is illustrated in Figure 12.1 with a comparison of the experimental vapor–liquid equilibrium curve (liquid branch) for ethylene [90], propylene [90], carbon dioxide [91], and argon [92], with computed data using available models of literature [53, 56, 64, 69, 93, 94].

The interactions between the guest molecules and the zeolite and nonframework cations must be reproduced with efficient and accurate potentials. Although some authors have opted for more sophisticated models [4], a simple and computational efficient option is the Kiselev-type model [59]. This model is based on Lennard-Jones potentials for the van der Waals interactions and on partial charges on all atoms of the system for the coulombic interactions that can be neglected for nonpolar guest molecules. The interactions of the guest molecules with the zeolite are dominated by the dispersive forces between the guest and the oxygen atoms of the structure [59], so the van der Waals interactions of guest molecules with the Si or Al atoms are often neglected.

The development of transferable potentials provides accurate representation of the interactions of the experimental system that is being simulated remains

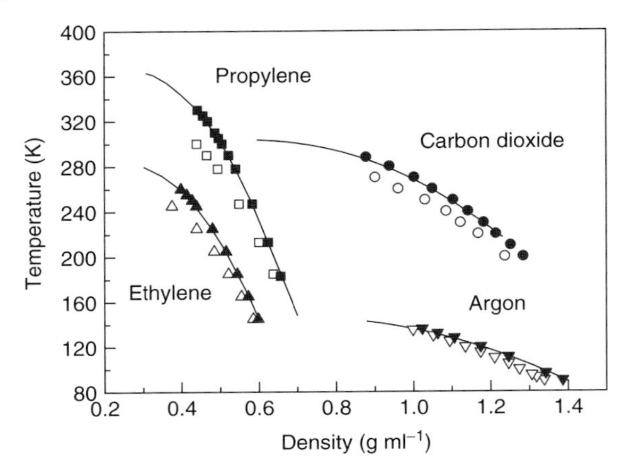


**Figure 12.1** Liquid branch of the vapor–liquid equilibrium curves for ethylene, propylene, carbon dioxide, and argon. Experimental data are shown as solid [90–92]. Simulated data using available models of literature are shown as open triangles [93], filled triangles [53], open squares [93], filled squares [53], open circles [69], filled circles [56], down open triangles [94], and down filled triangles [64].

a challenge. Accurate parameters can predict topology-specific adsorption and transport properties, and this can be an effective tool to resolve discrepancies among experimental datasets. Most efforts to improve the models are focused on the optimization of the Lennard-Jones parameters [16, 52–55], and Dubbeldam *et al.* reported a method to obtain accurate sets of parameters for united atom models, based on the fitting to experimental isotherms with inflection points [51]. Recent work has also highlighted that the partial charges assigned to the zeolite atoms are essential to reproduce experimental values of systems containing polar molecules [56, 57, 70, 95]. As an example, Desbiens *et al.* computed water adsorption in MFI using a host–guest Kiselev-type potential and a TIP4P model for water [96]. They found that this zeolite exhibits hydrophobic character only if the partial charge of the silicon atoms is kept below 1.7 a.u. The agreement with the experimental adsorption data drastically improved when the charge on the silicon atom was varied from 1.4 to 1.2. a.u. (Figure 12.2).

## 12.3 Simulation Methods

This section summarizes the specific methods to compute adsorption, free energies, surface areas, pore volumes, and diffusion in zeolites, and gives a short introduction of the most common simulation techniques. A more detailed description of these methods can be found in [97, 98]. In contrast to catalysis, modeling of adsorption and diffusion in zeolites is generally based on classical mechanics because quantum chemical calculation is still too expensive to compute statistical mechanical properties. In addition, adsorption in zeolites is dominated by dispersive

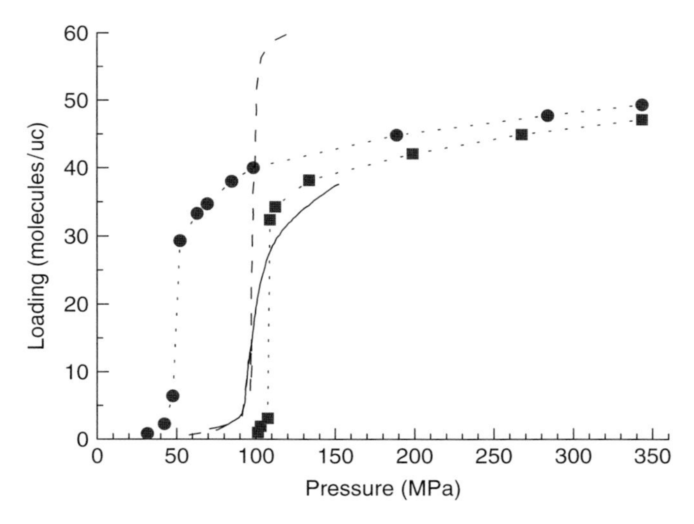


**Figure 12.2** Experimental and simulated liquid-phase adsorption isotherms of water in MFI at 300 K. Experimental data are shown in solid and dashed lines. Simulations were performed using the TIP4P water model with a silicon partial charge of 1.4 a.u. (circles) and 1.2 a.u. (squares). The dotted lines are a guide to the eye. Copyright Wiley-VCH Verlag GmbH & Co. KGaA. Reproduced with permission from Desbiens *et al.* [96].

interactions that are notoriously difficult to treat quantum mechanically. Dispersion using MP2 limits the amount of atoms so severely that in most cases one cannot afford to use it to compute binding energies of a molecule with the entire periodic unit cell of a zeolite. These properties are accessible with plane-wave codes (e.g., VASP) using GGA- and LDA-type basis sets, but one can argue about the validity of these results for the computation of binding energies of small adsorbates. This chapter focuses on classical force field–based methods for adsorption and diffusion, leaving catalysis and quantum approaches [88, 99, 100] beyond the scope of this work.

## 12.3.1
## Computing Adsorption

Adsorption isotherms, Henry coefficients, and isosteric heats of adsorption are important parameters for gas separation, catalysis, capture, and storage applications in zeolites that can be computed using molecular simulations. The simulations yield absolute values, which have to be converted to excess properties before comparing with experimental data [101, 102].

Adsorption isotherms are usually obtained using Monte Carlo simulations in the grand-canonical ensemble (GCMC). In this ensemble, the temperature, the volume, and the chemical potential of the adsorbed molecules are imposed. During the simulation, the molecules are exchanged with a reservoir at the same temperature and chemical potential. Hence, the number of molecules fluctuates and during the simulation, the average number of adsorbate molecules is computed. The chemical potential is directly related to fugacity, which is computed from the pressure using an equation of state [23, 97]. Recently, transition matrix Monte Carlo has been proposed as an efficient alternative for GCMC simulations [103, 104].

Henry coefficients are directly related to the excess free energy (or excess chemical potential) of the adsorbed molecules. The free energy of a molecule cannot be directly computed using Monte Carlo or molecular dynamics (MDs) simulations techniques. At low-to-intermediate loading, special simulation techniques such as the Widom test particle method can be applied. Details on this and other available techniques used to compute free energies can be found in the literature [97].

In the limit of zero coverage, the heat of adsorption $Q_{st}$ can be obtained from the derivative of the Henry constant from the Clausius–Clapeyron equation (see, e.g., [105]), but in practice, it is more efficient to obtain it directly from the energies of the system (see, e.g., [106]). Hence, both the Henry coefficient and the isosteric heat of adsorption are usually computed from the total energy of the system in the canonical ensemble with a fixed number of molecules ($N$), a given volume ($V$), and temperature ($T$). This requires two independent simulations to provide: (i) the energy of the single molecule inside the zeolite and (ii) the energy of the ideal gas situation [81, 106, 107]. This method of energy differences is very efficient for pure silica structures, but it is unsuited to compute the isosteric heats of adsorption in structures with nonframework cations. When cations move around the energies of cation–cation, cation-framework energies are very large and the final result consists of the subtraction of two very large energy values. The standard error is difficult to reduce in such a system. In 2008, Vlugt *et al.* resolved that impediment by proposing a new method making use of as much cancellation in energy terms as possible. The method provides significantly better and reliable results [108]. Figure 12.3 shows the isosteric heat of adsorption computed at 493 K for several *n*-alkanes in a LTA5A zeolite containing 32 sodium and 32 calcium cations per unit cell using the three methods. All force fields and models used for this comparison were taken from the literature [16, 52, 55].

The Monte Carlo simulations consist of random attempted moves that sample a given ensemble. Those moves are molecular translations and rotations for the NVT ensemble and additional molecular insertion and deletion for GCMC, and they are accepted or rejected with adequate criteria for the sampling of the corresponding ensemble probability function [97]. The NVT ensemble keeps fixed the number of particles (N), the volume (V), and the temperature (T) of the system. The conventional Monte Carlo method is very efficient for small molecules, but inefficient for the insertion of big molecules such as long-chain hydrocarbons in the system. The configurational-bias Monte Carlo (CBMC) technique was developed to make possible the insertion for molecules in moderately dense liquids, initially for lattice models [109] and subsequently extended to continuous models [110–112]. It improves the conformational sampling of the molecules and increases the efficiency of the chain insertions, required for the calculations of the adsorption isotherms,

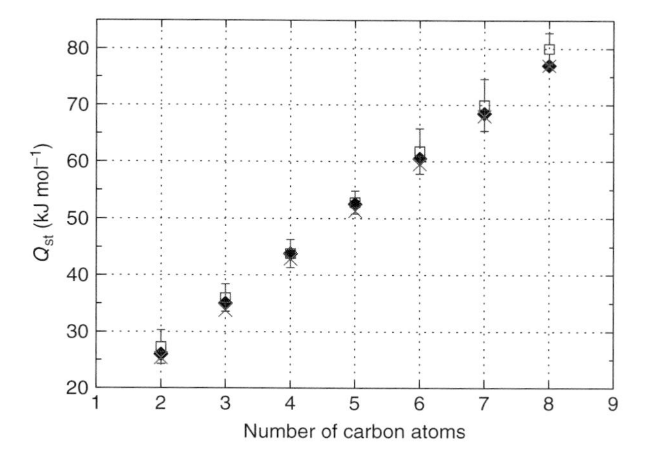


**Figure 12.3** Isosteric heat of adsorption at zero coverage for *n*-alkanes in zeolite LTA5A. The heats of adsorption were computed at 493 K using the method based on the Henry constant from the Clausius-Clapeyron equation [105] (♦), the method of energy differences [81, 107] (□), and the new method reported by Vlugt *et al.* [108] (×). The error bars are smaller than the symbol size, except for the method based on energy differences, for which the error bars are included.

the free energy, and the Henry coefficients by many orders of magnitude. In a CBMC simulation, instead of inserting a molecule randomly in the host system, the molecule is gradually inserted atom by atom, avoiding overlap with the zeolite. The rules that accept or reject the grown molecule are chosen in such a way that they exactly remove the bias caused by the growing scheme [97]. Over the last few years, several improvements of the CBMC scheme have been proposed [113–115].

In the last years, there have been many simulations studies of adsorption in zeolites. Detailed overviews on adsorption studies using molecular simulations were compiled in 2001 by Fuchs and Cheettham [4], and more recently by Smit and Maesen [10]. A few examples on simulation of adsorption in zeolites are also highlighted in Section 12.3 of this chapter.

### 12.3.2
### Computing Free Energy Barriers

The free energy methods used to obtain the Henry coefficients can also be applied to compute the free energy as a function of the position in the zeolite and hence to obtain information about the free energy barrier that the molecules have to cross if they hop from one position to another. To compute the free energy as a function of the position in the zeolite, it is necessary to relate a position in the zeolite channel or cage to a reaction coordinate $q$. The computed free energy at a given position $q$ contains both the potential energy and the entropy contribution that is directly related to the probability of the molecule to be found at $q$. A good example of the amount of information that can be provided using these methods

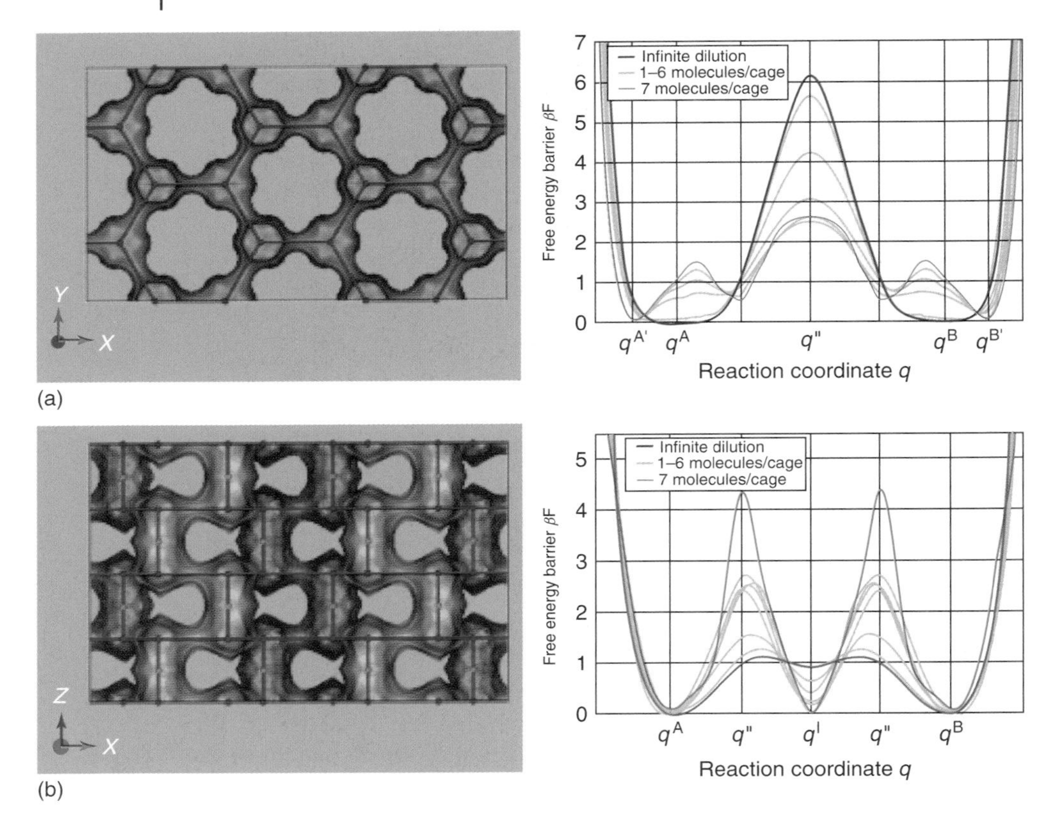


**Figure 12.4** Topology of the ERI-type zeolite showing the *xy*-direction (a) and the *z*-direction (b). Free energy profiles at 600 K for ethane in ERI at infinite dilution and 1, 2, 3, 4, 5, 6, and 7 molecules per unit cell. The free energy profiles were obtained in the *xy*-plane with $q^A$ the center of a cage and $q^B$ the center of a neighboring cage (c), and in the *z*-direction across a cage with $q^A$ the top of the cage, $q^I$ the middle of the cage, and $q^B$ the bottom of the cage (d). Reprinted with permission from Dubbeldam *et al.* [119]. Copyright © by the National Academy of Sciences.

can be taken from the studies of Dubbeldam *et al.* in ERI-type zeolites [116–119]. Figure 12.4 (taken from [119]) shows the computed free energies for ethane at several loadings in ERI-type zeolite. The maximum of the free energy is at $q = 0$ position, which corresponds to the dividing window $q^*$, and the minimum values, which correspond to the values deep inside the cages (indicated as cages a and b in the figure). The free barriers obtained indicate that in the *xy* direction the hopping takes place on the hexagonal lattice, and each hop in the *z* direction is preceded by a hop in the *xy*-direction [119]. The analysis was performed as a function of loading. A reversal of the main direction of diffusion was discovered that by examining the free energy, profiles could be related to the hopping interactions and preferences of individual molecules.

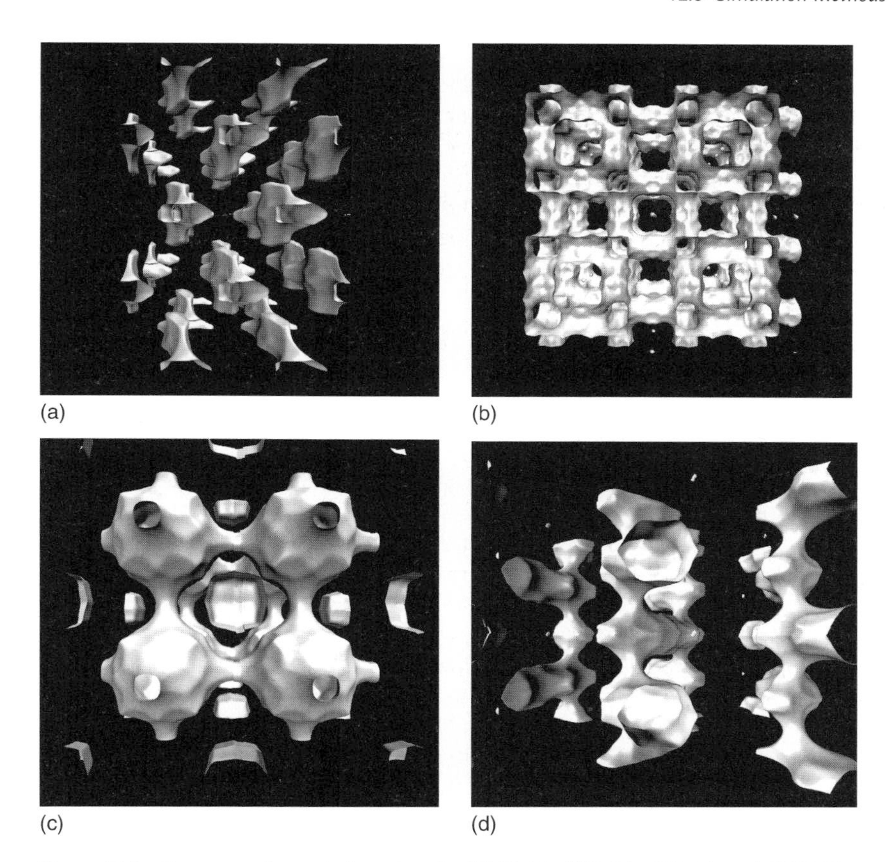


**Figure 12.5** Structure of a periodic unit cell of ITW (a), IWS (b), LTA (c), and IHW (d).

### 12.3.3
### Computing Volume-Rendered Pictures, Zeolite Surface Areas, and Zeolite Pore Volumes

Volume-rendered (as in Figure 12.4) and isocontour pictures, surface areas, and pore volumes are extremely helpful to characterize zeolites. Figure 12.5 shows the isocontour pictures of the ITW, IWS, LTA, and IHW zeolites. To obtain these pictures, the zeolite unit cell is divided into a grid. Second, the free energy of a test particle is computed for all of the grid points. To obtain the energy landscape, the chosen energy value of the three-dimensional dataset is visualized and the zeolite framework is made transparent to avoid overlap. Usually, a high energy value is used to generate the pores and the framework walls.

The surface area and the pore volume can be obtained indirectly from experimental isotherms using the BET theory and the saturation loading, respectively. Those parameters can also be computed from Monte Carlo simulations [120]. The surface area can be easily computed by rolling spherical test particles over the surface of the zeolite framework. If the spherical test particles are randomly shot into the zeolite, one can keep track of the fractions that do not overlap with the structure and compute the geometric the pore volume. These methods can be applied to both rigid and flexible frameworks and also to pure silica zeolites and for those containing extra framework cations. Since these methods depend on the test particle used, to properly compute the pore volume and the surface area, the procedure must be repeated for a variety of test particles of arbitrary size, including point atoms. Table 12.1 compiles the computed pore volumes obtained for several zeolites using the described method.

### 12.3.4 Computing Diffusion

The way the molecules diffuse in the zeolite pores strongly influences the behavior of the structure during adsorption, separation, and catalytic processes. In recent years, simulations of diffusion of adsorbed molecules in zeolites have spurred important advances that can be attributed to both the development of efficient algorithms and the massive increase of computer power. From a simulation point of view, computing diffusion coefficients is challenging, and several reviews on this topic have been published recently [7–9, 121, 122].

The first simulation studies for diffusion in confined systems focused on self-diffusivity [123] calculations for a single component using equilibrium MD simulations. With growing computer power, the simulation studies have gradually shifted toward computing self-diffusivities for mixtures [124–126] as well as transport diffusivities [127–134] that are the relevant diffusion coefficients for use in technological applications. Transport diffusivities can be obtained from three equivalent approaches that correspond to the Fick, the Onsager, and the Maxwell Stefan formulations [129, 135, 136]. The first attempts to compute transport diffusivities in zeolites were achieved for methane in MFI [130] and LTA [131] using nonequilibrium MD methods. Most recently, the increase of computer power made it possible to compute transport diffusivities from equilibrium MD simulations [132–134, 137, 138], and also to extend those calculations for obtaining diffusivities over loading, providing insights into the mechanisms that control the molecular traffic along the zeolite pores [139, 140].

Molecular simulations also play an important role in the understanding of multicomponent diffusion in zeolites by the development of theoretical models to predict multicomponent diffusion from single-component data [136, 141, 142]. This knowledge is essential for industrial applications, and experimental measurements for multicomponent adsorption and diffusion are difficult with little available data in literature. A good example of the usefulness of molecular simulations in this context was reported by Skoulidas and Sholl [143]. Using MD simulations, they analyzed the effect of molecular loading and pore topology of the self- and transport diffusivities for a variety of molecules in four zeolites (MFI, MTW, ISV, and ITE). On the basis of this, Beerdsen *et al.* were able to classify the pore topologies by matching the free energy of the molecules with the zeolite structure [144].

**Table 12.1** Zeolite pore volumes obtained from molecular simulations.

| Zeolite | Pore volume ($cm^3\,g^{-1}$) | Zeolite | Pore volume ($cm^3\,g^{-1}$) | Zeolite | Pore volume ($cm^3\,g^{-1}$) |
|---|---|---|---|---|---|
| ABW | 0.0211 | DOH | 0.1665 | MER | 0.1133 |
| AEL | 0.0902 | DON | 0.1661 | MFI | 0.1642 |
| AET | 0.1424 | EAB | 0.1945 | MFS | 0.1321 |
| AFG | 0.1287 | ECR | 0.2275 | MON | 0.0326 |
| AFI | 0.1620 | EDI | 0.0535 | MOR | 0.1501 |
| AFO | 0.0875 | EMT | 0.3423 | MTN | 0.1704 |
| AFR | 0.2386 | ERI | 0.2227 | MTT | 0.0733 |
| AFS | 0.2897 | EUO | 0.1458 | MWW | 0.2033 |
| AFT | 0.1823 | FAU | 0.3680 | MTW | 0.1109 |
| AFX | 0.2199 | FAU (NaX) | 0.3268 | NAT | 0.0025 |
| AFY | 0.3300 | FAU (NaY) | 0.3050 | OFF | 0.2238 |
| AHT | 0.0459 | FER | 0.1469 | PAU | 0.1620 |
| ANA | 0.0150 | GME | 0.2384 | RHO | 0.2517 |
| APC | 0.0331 | IHW | 0.1310 | RON | 0.0930 |
| APD | 0.0031 | ISV | 0.2863 | SAS | 0.2575 |
| AST | 0.2274 | ITE | 0.2271 | SAT | 0.1779 |
| ATN | 0.0938 | ITW | 0.0957 | SBE | 0.3408 |
| ATO | 0.0810 | IWS | 0.2443 | SFF | 0.2036 |
| ATT | 0.1170 | JBW | 0.0338 | SFG | 0.1390 |
| ATV | 0.0279 | KFI | 0.2327 | SGT | 0.1782 |
| AWW | 0.1801 | LEV | 0.2191 | SOD | 0.1314 |
| BEA (pol. A) | 0.2763 | LIO | 0.1811 | SOF | 0.1451 |
| BEA (pol. B) | 0.2721 | LOS | 0.2060 | STF | 0.2017 |
| BEC | 0.3289 | LOV | 0.0323 | STT | 0.1916 |
| BOG | 0.2407 | LTA (ITQ-29) | 0.2854 | STW | 0.2045 |
| BPH | 0.3235 | LTA4A | 0.2568 | THO | 0.0185 |
| CAN | 0.1278 | LTA5A | 0.2605 | TON | 0.0913 |
| CAS | 0.0155 | LTL | 0.1685 | TSC | 0.3702 |
| CHA | 0.2425 | LTN | 0.1894 | VFI | 0.2967 |
| CHI | 0.0069 | MAZ | 0.1718 | VSV | 0.0544 |
| CLO | 0.3279 | MEI | 0.2912 | WEN | 0.0735 |
| DDR | 0.1400 | MEL | 0.1546 | – | – |
| DFO | 0.2915 | MEP | 0.1357 | – | – |

Sodalite cages and nonaccessible pockets were blocked during the simulation.

The facility to analyze the molecular motion in the zeolite at a molecular level made molecular simulations a powerful tool to shed light on difficult problems such as single-file [145–147] and resonant diffusion [148], window [116, 117] and levitation [9] effects, or molecular traffic control [139, 149–151] and molecular path control [119, 152]. The development of special simulation techniques to compute diffusion in those processes that take place in a timescale inaccessible to MD is a general problem that has attracted great attention in the last years. Hence, approaches based on a sequence of rare events such as Bennet-Chandler [153], Ruiz-Montero [154], or transition path sampling [155] and those more recent such as transition interface sampling [156], temperature-accelerated dynamics [157], or long timescale kinetics Monte Carlo [157] can all be considered as rare event sampling (RES) methods. Earlier RES studies in zeolites [2] date back to the 1970s and they have been increasing in complexity over the years. In the 1990s, RES methods were applied to compute molecular diffusion at infinite dilution in MFI [158–160] and faujasite [161–163]. Those studies were later extended to other complex topologies such as LTA, LTL, ERI, and CHA not only for infinite dilution [119, 152, 164–166] but also for low, medium, and high loadings [119, 152, 166–171].

## 12.4 Molecular Modeling Applied to Processes Involving Zeolites

As shown in previous sections, molecular simulation is becoming a powerful tool for gaining an insight into the adsorption and diffusion processes taking place in the zeolite pores at a molecular level. Since zeolites are in widespread use in industrial and environmental applications, this section is devoted to emphasize some examples of the active role that molecular modeling is playing in technological processes as well as in green chemistry.

### 12.4.1 Applications in Technological Processes

The adsorption, diffusion, separation, and catalytic properties of a given zeolite depend on multiple factors such as the topology and the surface properties of the structure (hydrophobicity, hydrophilicity), the shape and size of its pores, the Si/Al ratio, the number, location, and type of nonframework cations it contains, and its level of hydration. Molecular simulations are being applied to the study of these properties and factors aiming to predict the zeolite materials features and their performance in industrial processes.

#### 12.4.1.1 Molecular Modeling of Confined Water in Zeolites

A current challenge is to provide information at a molecular level of the complex behavior of water when it is confined into the zeolites pores. This is the first step toward understanding the role of water in technological processes such as those involving ion exchange to reduce water hardness, for water/alcohol separations, and for water removal. Molecular modeling can be used to provide knowledge on the behavior of water confined in a given zeolite and also on the way water influences the properties of the structure. The effect of confined water on the properties of the zeolites depends not only on the level of hydration but also on the topology and chemical composition of the structure. Pure silica zeolites are highly hydrophobic, whereas the isomorphic substitution of silicon with aluminum and the subsequent neutralization of the framework with cations make the structure hydrophilic [172].

Recent simulation studies on hydrophilic zeolites have been applied to analyze the effect that the hydration level exerts on the framework stability and on the positions and mobility of the framework cations, as well as to understand the underlying mechanisms involved in the coordination of the water to the nonframework cations and to the zeolite frame. Simulation studies on hydrophobic structures are mostly focused on the structural and dynamical characteristics of water confined in the zeolite and on the search of optimal molecular sieves for gas separation in the presence of water [57, 70, 76, 95, 173–176]. Numerous studies of water in hydrophobic zeolites (mainly MFI) have been performed over the years concluding that water–water interactions prevail over water–zeolite interactions. Therefore, the topology of the zeolite greatly influences the water properties as it shapes the molecular aggregates inside the pores. This is clearly illustrated in Figure 12.6 that shows the snapshots that Puibasset and Pellenq [177] reported

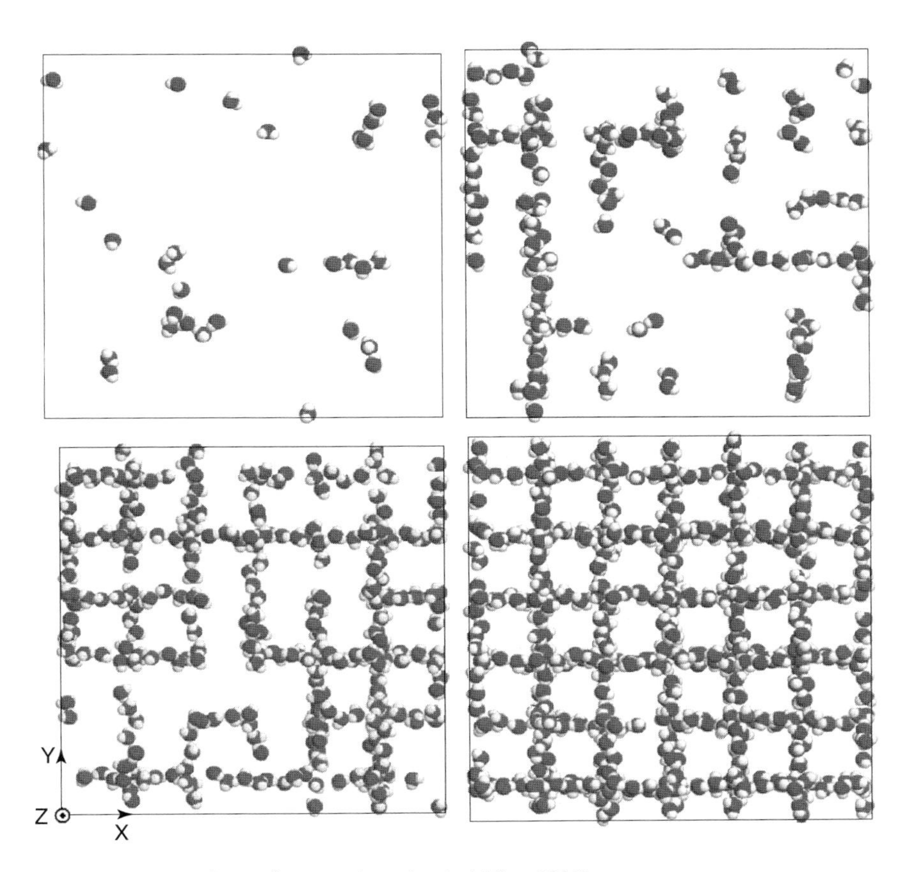


**Figure 12.6** Snapshots of water adsorption in MFI at 300 K for the relative pressures $P/P_0 = 0.0886$, 0.1477, 0.2045, and 0.3409. For shake of clarity, the zeolite framework was removed by the authors. Figure from [177].

for water configurations in MFI at four relative pressures. For the lower pressure, the adsorbed molecules of water are isolated in small clusters containing a few molecules. An increase of pressure leads isolated molecules to get together forming clusters that merge into larger ones until it creates a single (infinite) cluster at the higher pressure. Detailed information on simulation studies of water confined in hydrophobic and hydrophilic zeolites can be found in the review of Bougeard and Smirnov [178] as well as in very recent publications [57, 177, 179, 180].

#### 12.4.1.2 Molecular Modeling of Hydrocarbons in Zeolites

Zeolites are used for a variety of industrial applications. Among them, they are widely used in the petrochemical industry for isomerization and cracking of hydrocarbons [181–183]. Zeolites become catalytically active by substitution of aluminum for silicon into the framework. Each substitution creates a negative charge in the framework that is compensated by a counterion or a proton. The location of these counterions influences the adsorption and the catalytic properties for these materials [25, 26]. In addition, zeolites act as shape-selective catalysts with a high selectivity and reaction yield. The shape and size of their pores control both the way reactants enter the zeolite to the catalytically active sites and the way reaction products leave the zeolite. Since the catalytic and structural properties of zeolites depend on the diffusivity of the adsorbed molecules, many simulation studies are focused on adsorption and diffusion of hydrocarbons [10]. As described in Section 12.2, these studies are being performed using advanced simulation techniques that from day to day increase in both speed and efficiency. A large number of groups have computed adsorption and diffusion of alkanes in zeolites. These works as well as current developments on molecular modeling applied to shape-selective catalysis are compiled in a recent review of Smit and Maesen [10].

Many examples of hydrocarbon molecules adsorbed and converted by the zeolites can be found in the literature [10]. Among them, the so-called window effect is an interesting example where molecular modeling is helpful to shed light on the microscopic mechanisms involved in heterogeneous catalysis. Most hydrocracking processes result in a product distribution with a single maximum. However, in 1968, Chen *et al.* found that ERI-type zeolites yield a bimodal product distribution [184]. This window effect was attributed to the relative diffusion rates of the alkanes in the zeolite, though most recent measurements failed to reproduce these diffusion values [185, 186]. Molecular simulation calculations corroborated, in 2003, the existence of the controversial window effect, showing that diffusion rates can increase by orders of magnitude when the alkane chain length increases, so that the shape of the alkane is no longer commensurate with that of a zeolite cage [116, 118]. The first simulation studies were performed in structures consisting of channels (OFF) and in structures consisting of cages separated by small windows but differing in the size and shape of the cages as well as in the orientation of the windows with respect to the cages (ERI, CHA, and LTA). In contrast to the channel-like structures, the latter showed a nonmonotonic periodic dependence of the Henry coefficients and heats of adsorption. In addition, when a molecule is incommensurate with the cage structure, the diffusion rate increases by orders of magnitude. The simulation studies showed that the maximum length for a given hydrocarbon to fit in a single cage was of 13, 11, and 23 carbon atoms for ERI, CHA, and LTA, respectively. These values are directly related to the local minima in the Henry coefficients and heats of adsorption and to the local maxima in the diffusion coefficients [118]. The unusually low adsorption obtained for chains of alkanes similar or larger than the zeolite cage provided an alternative cracking mechanism that affords prediction of selectivity as a function of cage size and opens the possibility of length-selective cracking, where the obtained distribution is controlled by choosing zeolites with cages of suitable size [116–118].

#### 12.4.1.3 **Molecular Modeling of Separation of Mixtures in Zeolites**

Most industrial applications of adsorption involve mixtures. However, compared with the amount of literature about experimental studies on single-component adsorption in zeolites, fewer experimental studies have been published on mixtures [187–189]. This is due to the experimental difficulties to accurately determine the molecular composition of the adsorbed phase [190]. When experimental data are unavailable, molecular modeling becomes a very useful tool to predict the adsorption of mixtures in the zeolite as well as to find out the separation mechanisms that take place inside the pores. Molecular separations of mixtures in zeolites are based not only on the adsorption selectivity but also on the difference between the diffusivities of the components of the mixture. The selectivity for a given zeolite can be further optimized by changing the type and the amount of nonframework cations [15, 23, 55, 58, 191]. The first simulation studies on this topic were reported by Beerdsen *et al.* showing that an increase in the number of sodium cations in MFI- and MOR-type structures leads to an increase in the selectivity for adsorbing linear (MFI) and branched alkanes (MOR), respectively [23]. Further simulation studies in MFI showed that the adsorption of hydrocarbons increases with decreasing atomic weight of the nonframework cation [15].

Zeolite-based separation processes involve both mixture adsorption and diffusion that are strongly interrelated. To illustrate the idea, we have computed free energy profiles of propane and propylene in ITW at 500 K (Figure 12.7). Differences in the free energy barrier indicate that propane is virtually excluded from the zeolite, whereas propylene is adsorbed, supporting experimental findings. Free energies were computed using available well-tested models, force fields, and methods [52, 53, 116].

Simulation studies on zeolite-based separations are mostly applied to analyze the zeolite efficiency on the separation of mixtures of hydrocarbons, alcohol–water solutions, and natural gas purification. This section sums up current work performed on hydrocarbons and alcohol–water mixtures, whereas studies on natural gas purification are addressed in Section 12.4.2.2.

The separation of mixtures of hydrocarbons is an important activity in several technological and petrochemical processes. For a given separation task, CBMC simulations allow the efficient screening of zeolite topologies based on selective adsorption and foster the development of novel separation processes exploiting entropy and enthalpy effects [192–196]. Hence, for mixtures of linear alkanes,

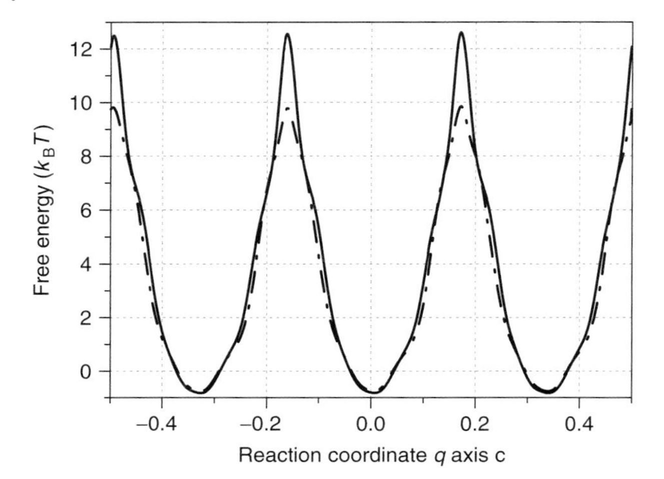


**Figure 12.7** Free energy profiles of propane (solid line) and propylene (dashed line) in ITW-type zeolite at 500 K.

smaller molecules are preferentially adsorbed at high loadings, because the few empty voids are easily filled by them (*size* entropy effect). For mixtures of isomers formed by linear and branched alkanes, selectivity is in favor of those that pack more efficiently within the zeolite channel (*configurational* and *length* entropy effects). For a detailed overview on molecular simulation studies of mixtures of hydrocarbons in zeolites, the reader is referred to the recent review of Smit and Maesen [10].

An example where a mixture separation can be mainly attributed to enthalpic effects is reported by Ban *et al.* [197]. The authors computed adsorption isotherms for mixtures of benzene and propylene in several zeolites finding very high selectivities for benzene in structures formed by channels such as MOR, BEA, and MWW. The selectivity for benzene decreases in zeolites with large cavities such as NaY and it is even lower in MFI, formed by intersecting channels, with a reversal of the selectivity in favor of propylene for 373 K. The selectivity behavior can be explained based on enthalpic effects; in MOR, both propylene and benzene fit in the main channel. However, benzene is preferentially adsorbed as its heat of adsorption is much larger. Similar features are found for BEA and MWW that almost exclusively adsorb benzene, independent of temperature. NaY shows larger cavities where the heat of adsorption of benzene decreases. This allows the adsorption of some molecules of propylene decreasing the adsorption selectivity for benzene. In MFI, benzene shows similar behaviour than in NaY. It is preferentially adsorbed at the intersections whereas propylene – that in absence of benzene can adsorb at both the intersections and the channels – is adsorbed in the channels (Figure 12.8).

The location of benzene at the intersections is a key factor in the catalysts performance because of "traffic junction" effects [198]. Hansen *et al.* presented a study using a combination of quantum chemistry, MC and MD simulations to

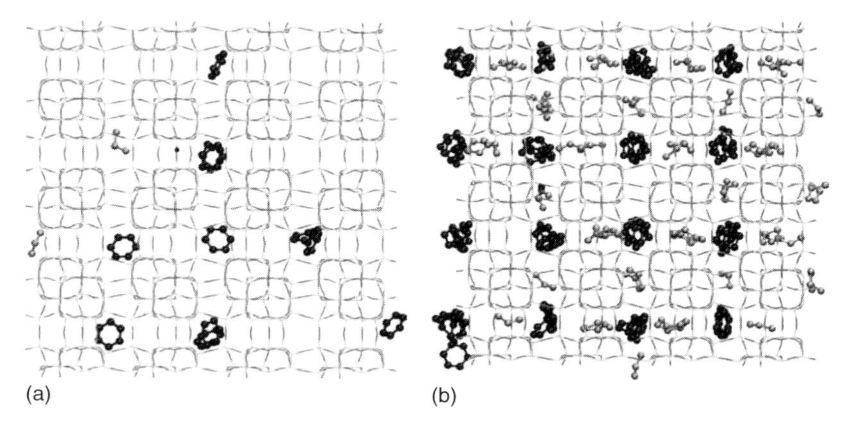


**Figure 12.8** Snapshots of an equimolar mixture benzene/propylene adsorbed in MFI at low (a) and high (b) loading. Reprinted with permission from Ban *et al.* [197]. Copyright 2009 American Chemical Society.

predict performance for reaction of benzene with propylene using MFI catalysts [199]. These simulations are helpful in interpreting published experimental data.

Zeolites are also widely used for solvent dehydration, since their polar nature make them optimal structures to separate water from organic compounds. The technological interest in finding zeolite membranes with high permselectivity of water to alcohols is leading researchers to study the processes involved on the water–alcohols separations using molecular simulations [71, 200, 201]. Hence, the water–alcohol adsorption and diffusion behavior in zeolites has been recently studied. Lu *et al.* [201] used GCMC simulations for the water–alcohol mixtures in MFI, MOR, CFI, and DON, and Yang *et al.* [71] analyzed the diffusion behavior of the mixture in MFI with equilibrium MD techniques. In spite of the improvements achieved on this topic, more efficient and realistic models still need to be developed as the overall agreement with equilibrium experimental data is only qualitative. Detailed information on the most recent models as well as on previous available models can be found in [202].

### 12.4.2 Applications in Green Chemistry

*Green chemistry* can be defined as the use of chemistry to reduce or eliminate hazardous substances. Among a number of examples, zeolites may play a useful role carbon dioxide capture and natural gas purification. This section explains the role that molecular simulations are currently playing in understanding the mechanisms involved in these processes at a molecular level.

#### 12.4.2.1 Carbon Dioxide Capture

The prediction of carbon dioxide adsorption on porous materials is becoming an imperative task not only for the need to develop cheap carbon dioxide capture

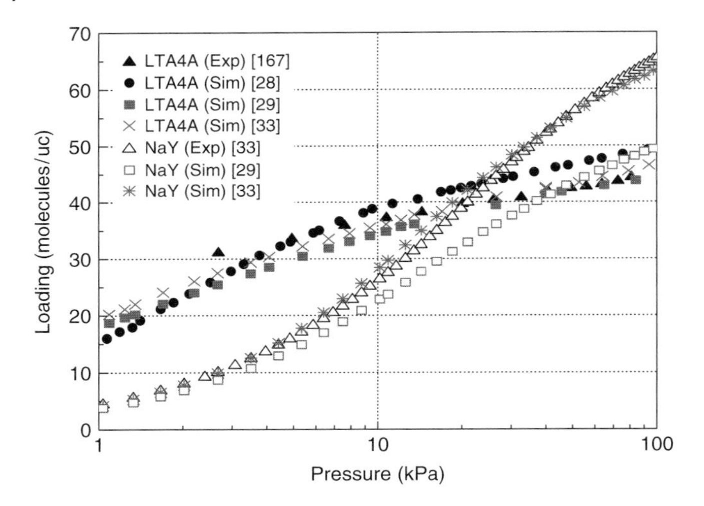


**Figure 12.9** Experimental [56, 207] and computed (using available force fields [28, 29, 56]) adsorption isotherms of carbon dioxide in LTA4A (with 96 aluminum atoms and sodium cations per unit cell) and in NaY (with 54 aluminum atoms and sodium cations per unit cell) at 298 K.

technologies but also for the improvement of gas separation processes such as natural gas purification. Recent studies have found that carbon dioxide is strongly adsorbed by a variety of zeolites and that the adsorption efficiency strongly depends on the zeolite type and composition [56, 203–205]. For instance, the zeolite basicity and electric field are essential factors on carbon dioxide adsorption that can be either induced or controlled by the nature and density of the cations within the zeolite pores [206]. At this point, molecular simulation can be extremely useful to systematically analyze the performance of the structures as a function of its size and shape, the type of the pore, the aluminum composition, and the nature of the nonframework cations. However, this task is seriously restricted by the lack of good and transferable force fields. Most of the works reporting molecular studies on carbon dioxide in zeolites are for all-silica structures and the few works reporting simulations in zeolites containing aluminum atoms compensate the net negative charge of the framework with sodium nonframework cations [28–30, 56]. Although force field development for nonframework cations other than sodium is still an open task, Garcia-Sanchez *et al.* [56] developed a general force field that outperforms previous force fields as it is more accurate, transferable between zeolite structures, and applicable to all Si/Al ratios. The adsorption isotherms computed with available force fields [28, 29, 56] and compared with experimental data [56, 207] are shown in Figure 12.9.

#### 12.4.2.2 **Natural Gas Purification**

The natural gas composition is around 95% methane, traces of heavier gaseous hydrocarbons such as ethane and propane, and other light gasses such as carbon dioxide and nitrogen. The presence of carbon dioxide in natural gas reduces the combustion power efficiency and contributes to greenhouse gas emissions. Therefore, the production of cheap and clean fuel from natural gas is based on efficient purification processes. Zeolites are suitable materials for storage, separation, and purification of natural gas for their thermal and mechanical stability, high carbon dioxide adsorption capacities, high rates of transport, and high selectivities. However, the mechanisms to identity the zeolite properties and performances on natural gas purification are not well defined yet and molecular simulations are currently being applied to make up for this deficiency.

Several simulation studies have been reported on the adsorption and diffusion of the natural gas components. They have investigated the idea of analyzing the influence of several factors such as temperature, pressure, chemical composition, zeolite topology, and pore size on the adsorption [204, 208–214] and transport [124, 203, 205, 215–218] of these mixtures. For example, Krishna *et al.* reported molecular simulations for the self-diffusion of pure and equimolar mixture of carbon dioxide and methane in DDR, CHA, and MFI structures [205], and in a follow-up study, by using MC and MD simulations, screened 12 zeolite topologies looking for the optimal structure to separate the mixture, thereby finding the highest permeation selectivities for DDR and CHA zeolites [214]. Furthermore, they observed that the separation selectivity on these two structures is particularly enhanced by what they called the *segregated nature of the mixture during adsorption*. The segregation effect results in a preferential adsorption of carbon dioxide at the windows regions, hindering the diffusion of methane between cages and improving in this way, the selectivity [218]. These studies may lead to future strategies to improve natural gas separation processes.

## 12.5 Summary and Outlook

Molecular modeling is already a powerful tool to accurately predict transport and accessibility in zeolites. However, efficient methods and good force fields capable of reproducing ideal experimental conditions for all zeolites are essential for this purpose. A number of simulation studies have been reported to exploit practical issues in zeolites. MC simulations predict adsorption isotherm, Henry coefficients, isosteric heats of adsorption, cation distribution, preferable adsorption sites, and siting of molecules within the pores that are in good agreement with experiment. For diffusion, there is also good agreement with experimental data available, though most of the simulation work in diffusion focuses on self-diffusivities and methods to compute the corrected diffusivities for slow-diffusion systems are still scarce.

Much progress has been made on the development of fast and efficient methods and accurate and transferable force fields. However, significant challenges remain in these areas. Future work is also expected to focus on the development of accurate potential models, particularly for different type of nonframework cations and for complex adsorbates. Finally, another on-going challenge is to combine methods based on both quantum and classical mechanics to the study of chemisorption and catalysis. The development of methods to adequately integrate diffusion, adsorption, and reaction kinetics will lead to the prediction of properties of new high-performance materials with increasing efficiency and speed. The microscopic information obtained from simulations provides the underlying knowledge from a molecular point of view that may guide to the development of more efficient processes, to fine-tune zeolites for a particular application, and also to steer the experimental effort in successful directions.

## Acknowledgments

This work is supported by the Spanish "Ministerio de Educación y Ciencia (MEC)" (CTQ2007-63229) and Junta de Andalucía (P07-FQM-02595). The author thanks J. M. Castillo, E. Garcia-Perez, A. Garcia-Sanchez, J. J. Gutierrez-Sevillano, and A. Martin-Calvo for their help with figures and especially for their joint work to provide the computed values listed in Table 12.1. The author is also thankful to T. J. H. Vlugt and S. Ban for Figure 12.8 and to D. Dubbeldam, T. J. H. Vlugt, J. A. Anta, and R. Krishna for a critical reading of the manuscript.

## References

1. Karger, J. and Ruthven, D.M. (1992) *Diffusion in Zeolites*, John Wiley & Sons, Ltd, New York.
2. Theodorou, D.N., Snurr, R.Q., and Bell, A.T. (1996)Molecular dynamics and diffusion in microporous materials,in *Comprehensive Supramolecular Chemistry: Solid-state Supramolecular Chemistry–Two- and Three-dimensional Inorganic Networks*, vol. 7, Pergamon, Oxford, p. 507–548.
3. Keil, F.J., Krishna, R., and Coppens, M.O. (2000) *Rev. Chem. Eng.*, **16**, 71.
4. Fuchs, A.H.and Cheetham, A.K. (2001) *J. Phys. Chem. B*, **105**, 7375.
5. Ramanan, H.and Auerbach, S.M. (2006)in *Conference of the NATO-Advanced-Study-Institute on Fluid Transport in Nanoporous Materials* (eds W.C.Conner and J. Fraissard), La Colle sur Loup, France, p. 93.
6. Auerbach, S.M. (2006)in *Conference of the NATO-Advanced-Study-Institute on Fluid Transport in Nanoporous Materials* (eds W.C.Conner and J. Fraissard), La Colle sur Loup, France, p. 535.
7. Dubbeldam, D.and Snurr, R.Q. (2006)3rd International Conference on Foundations of Molecular Modeling and Simulation (FOMMS), Blaine, p. 305.
8. Jobic, H.and Theodorou, D.N. (2007) *Microporous Mesoporous Mater.*, **102**, 21.
9. Yashonath, S.and Ghorai, P.K. (2008) *J. Phys. Chem. B*, **112**, 665.
10. Smit, B.and Maesen, T.L.M. (2008) *Chem. Rev.*, **108**, 4125.
11. Baerlocher, C., Meier, W.M., and Olson, D.H. (2007) *Atlas of Zeolite Structure Types*, 6th revised edn, Elsevier, London.
12. Auerbach, S.M., Bull, L.M., Henson, N.J., Metiu, H.I., and Cheetham, A.K. (1996) *J. Phys. Chem.*, **100**, 5923.
13. Buttefey, S., Boutin, A., and Fuchs, A.H. (2002) *Mol. Simul.*, **28**, 1049.
14. Dempsey, E., Kuehl, G.H., and Olson, D.J. (1969) *Phys. Chem.*, **73**, 387.
15. Beerdsen, E., Dubbeldam, D., Smit, B., Vlugt, T.J.H., and Calero, S. (2003) *J. Phys. Chem. B*, **107**, 12088.

16. Calero, S., Dubbeldam, D., Krishna, R., Smit, B., Vlugt, T.J.H., Denayer, J.F.M., Martens, J.A., and Maesen, T.L.M. (2004) *J. Am. Chem. Soc.*, **126**, 11377.
17. Löwenstein, W. (1954) *Am. Mineral.*, **39**, 92.
18. Garcia-Perez, E., Torrens, I.M., Lago, S., Dubbeldam, D., Vlugt, T.J.H., Maesen, T.L.M., Smit, B., Krishna, R., and Calero, S. (2005) *Appl. Surf. Sci.*, **252**, 716.
19. Sastre, G., Fornes, V., and Corma, A. (2002) *J. Phys. Chem. B*, **106**, 701.
20. Sastre, G., Fornes, V., and Corma, A. (2002) *J. Phys. Chem. B*, **106**, 701.
21. Sklenak, S., Dedecek, J., Li, C., Wichterlova, B., Gabova, V., Sierka, M., and Sauer, J. (2009) *Phys. Chem. Chem. Phys.*, **11**, 1237.
22. Macedonia, M.D., Moore, D.D., and Maginn, E.J. (2000) *Langmuir*, **16**, 3823.
23. Beerdsen, E., Smit, B., and Calero, S. (2002) *J. Phys. Chem. B*, **106**, 10659.
24. Mellot-Draznieks, C., Buttefey, S., Boutin, A., and Fuchs, A.H. (2001) *Chem. Commun.*, 2200.
25. Garcia-Perez, E., Dubbeldam, D., Liu, B., Smit, B., and Calero, S. (2007) *Angew. Chem. Int. Ed.*, **46**, 276.
26. Liu, B., Garcia-Perez, E., Dubbeldam, D., Smit, B., and Calero, S. (2007) *J. Phys. Chem. C*, **111**, 10419.
27. Jaramillo, E.and Auerbach, S.M. (1999) *J. Phys. Chem. B*, **103**, 9589.
28. Jaramillo, E.and Chandross, M. (2004) *J. Phys. Chem. B*, **108**, 20155.
29. Maurin, L.P.and Bell, R.G. (2005) *J. Phys. Chem. B*, **109**, 16084.
30. Akten, E.D., Siriwardane, R., and Sholl, D.S. (2003) *Energy Fuels*, **17**, 977.
31. Rappe, A.K., Casewit, C.J., Colwell, K.S., Goddard, W.A., and Skiff, W.M. (1992) *J. Am. Chem. Soc.*, **114**, 10024.
32. Hagler, A.T.and Ewig, C.S. (1994) *Comput. Phys. Commun.*, **84**, 131.
33. Allinger, N.L. (1977) *J. Am. Chem. Soc.*, **99**, 8127.
34. Allinger, N.L., Yuh, Y.H., and Lii, J.H. (1989) *J. Am. Chem. Soc.*, **111**, 8551.
35. Allinger, N.L., Li, F.B., Yan, L.Q., and Tai, J.C. (1990) *J. Comput. Chem.*, **11**, 868.
36. Allinger, N.L., Chen, K.S., and Lii, J.H. (1996) *J. Comput. Chem.*, **17**, 642.
37. Allinger, N.L. (2002) *Abstr. Pap. Am. Chem. Soc.*, **223**, 97.
38. Mayo, S.L., Olafson, B.D., and Goddard, W.A. (1990) *J. Phys. Chem.*, **94**, 8897.
39. Allured, V.S., Kelly, C.M., and Landis, C.R. (1991) *J. Am. Chem. Soc.*, **113**, 1.
40. Root, D.M., Landis, C.R., and Cleveland, T. (1993) *J. Am. Chem. Soc.*, **115**, 4201.
41. Weiner, S.J., Kollman, P.A., Case, D.A., Singh, U.C., Ghio, C., Alagona, G., Profeta, S., and Weiner, P. (1984) *J. Am. Chem. Soc.*, **106**, 765.
42. Brooks, B.R., Bruccoleri, R.E., Olafson, B.D., States, D.J., Swaminathan, S., and Karplus, M. (1983) *J. Comput. Chem.*, **4**, 187.
43. Jorgensen, W.L.and Tiradorives, J. (1988) *J. Am. Chem. Soc.*, **110**, 1657.
44. Clark, M., Cramer, R.D., and Vanopdenbosch, N. (1989) *J. Comput. Chem.*, **10**, 982.
45. Momany, F.A., McGuire, R.F., Burgess, A.W., and Scheraga, H.A. (1975) *J. Phys. Chem.*, **79**, 2361.
46. Hermans, J., Berendsen, H.J.C., Vangunsteren, W.F., and Postma, J.P.M. (1984) *Biopolymers*, **23**, 1513.
47. Halgren, T.A. (1992) *J. Am. Chem. Soc.*, **114**, 7827.
48. Burchart, E.D., Jansen, J.C., and Vanbekkum, H. (1989) *Zeolites*, **9**, 432.
49. Lifson, S., Hagler, A.T., and Dauber, P. (1979) *J. Am. Chem. Soc.*, **101**, 5111.
50. Momany, F.A., Carruthe., L.M., McGuire, R.F., and Scheraga, H.A. (1974) *J. Phys. Chem.*, **78**, 1595.
51. Dubbeldam, D., Calero, S., Vlugt, T.J.H., Krishna, R., Maesen, T.L.M., Beerdsen, E., and Smit, B. (2004) *Phys. Rev. Lett.*, **93**, 8.
52. Dubbeldam, D., Calero, S., Vlugt, T.J.H., Krishna, R., Maesen, T.L.M., and Smit, B. (2004) *J. Phys. Chem. B*, **108**, 12301.
53. Liu, B., Smit, B., Rey, F., Valencia, S., and Calero, S. (2008) *J. Phys. Chem. C*, **112**, 2492.
54. Calero, S., Lobato, M.D., Garcia-Perez, E., Mejias, J.A., Lago, S., Vlugt,

T.J.H., Maesen, T.L.M., Smit, B., and Dubbeldam, D. (2006) *J. Phys. Chem. B*, **110**, 5838.
55. Garcia-Perez, E., Dubbeldam, D., Maesen, T.L.M., and Calero, S. (2006) *J. Phys. Chem. B*, **110**, 23968.
56. Garcia-Sanchez, A., Ania, C.O., Parra, J.B., Dubbeldam, D., Vlugt, T.J.H., Krishna, R., and Calero, S. (2009) *J. Phys. Chem. C*, **113**, 8814–8820.
57. Castillo, J.M., Dubbeldam, D., Vlugt, T.J.H., Smit, B., and Calero, S. (2009) *Mol. Simul.*, **35**, 1067–1076.
58. Garcia-Sanchez, A., Garcia-Perez, E., Dubbeldam, D., Krishna, R., and Calero, S. (2007) *Adsorption Sci. Technol.*, **25**, 417.
59. Bezus, A.G., Kiselev, A.V., Lopatkin, A.A., and Du, P.Q. (1978) *J. Chem. Soc. Faraday Trans. 2*, **74**, 367.
60. Demontis, P., Suffritti, G.B., Quartieri, S., Fois, E.S., and Gamba, A. (1988) *J. Phys. Chem.*, **92**, 867.
61. Nicholas, J.B., Hopfinger, A.J., Trouw, F.R., and Iton, L.E. (1991) *J. Am. Chem. Soc.*, **113**, 4792.
62. Leroy, F., Rousseau, B., and Fuchs, A.H. (2004) *Phys. Chem. Chem. Phys.*, **6**, 775.
63. Vlugt, T.J.H.and Schenk, M. (2002) *J. Phys. Chem. B*, **106**, 12757.
64. Garcia-Perez, E., Parra, J.B., Ania, C.O., Dubbeldam, D., Vlugt, T.J.H., Castillo, J.M., Merkling, P.J., and Calero, S. (2008) *J. Phys. Chem. C*, **112**, 9976.
65. Zimmermann, N.E.R., Jakobtorweihen, S., Beerdsen, E., Smit, B., and Keil, F.J. (2007) *J. Phys. Chem. C*, **111**, 17370.
66. Bouyermaouen, A.and Bellemans, A. (1998) *J. Chem. Phys.*, **108**, 2170.
67. Hill, J.R.and Sauer, J. (1994) *J. Phys. Chem.*, **98**, 1238.
68. Hill, J.R.and Sauer, J. (1995) *J. Phys. Chem.*, **99**, 9536.
69. Harris, J.G.and Yung, K.H. (1995) *J. Phys. Chem.*, **99** (31), 12021.
70. Di Lella, A., Desbiens, N., Boutin, A., Demachy, I., Ungerer, P., Bellat, J.P., and Fuchs, A.H. (2006) *Phys. Chem. Chem. Phys.*, **8**, 5396.
71. Yang, J.Z., Chen, Y., Zhu, A.M., Liu, Q.L., and Wu, J.Y. (2008) *J. Memb. Sci.*, **318**, 327.
72. Guillot, B.and Guissani, Y. (2001) *J. Chem. Phys.*, **114**, 6720.
73. Pellenq, R.J.M., Roussel, T., and Puibasset, J. (2008) *J. Int. Adsorpt. Soc.*, **14**, 733–742.
74. Darkrim, F.and Levesque, D. (1998) *J. Chem. Phys.*, **109**, 4981.
75. Garberoglio, G., Skoulidas, A.I., and Johnson, J.K. (2005) *J. Phys. Chem. B*, **109**, 13094.
76. Trzpit, M., Soulard, M., Patarin, J., Desbiens, N., Cailliez, F., Boutin, A., Demachy, I., and Fuchs, A.H. (2007) *Langmuir*, **23**, 10131.
77. Cailliez, F., Stirnemann, G., Boutin, A., Demachy, I., and Fuchs, A.H. (2008) *J. Phys. Chem. C*, **112**, 10435.
78. Wender, A., Barreau, A., Lefebvre, C., Di Lella, A., Boutin, A., Ungerer, P., and Fuchs, A.H. (2006) *Adsorpt. Sci. Technol.*, **24**, 713.
79. Nath, S.K., Escobedo, F.A., and de Pablo, J.J. (1998) *J. Chem. Phys.*, **108**, 9905.
80. Martin, M.G.and Siepmann, J.I. (1998) *J. Phys. Chem. B*, **102**, 2569.
81. Smit, B.and Siepmann, J.I. (1994) *J. Phys. Chem.*, **98**, 8442.
82. Jorgensen, W.L.and Swenson, C.J. (1985) *J. Am. Chem. Soc.*, **107**, 1489.
83. Vlugt, T.J.H., Zhu, W., Kapteijn, F., Moulijn, J.A., Smit, B., and Krishna, R. (1998) *J. Am. Chem. Soc.*, **120**, 5599.
84. Lubna, N., Kamath, G., Potoff, J.J., Rai, N., and Siepmann, J.I. (2005) *J. Phys. Chem. B*, **109**, 24100.
85. Zeng, Y.P., Ju, S.G., Xing, W.H., and Chen, C.L. (2007) *Sep. Purif. Technol.*, **55**, 82.
86. Zeng, Y.P., Ju, S.G., Xing, W.H., and Chen, C.L. (2007) *Ind. Eng. Chem. Res.*, **46**, 242.
87. Maple, J.R., Hwang, M.J., Stockfisch, T.P., and Hagler, A.T. (1994) *Isr. J. Chem.*, **34**, 195.
88. Zhao, Y.and Truhlar, D.G. (2008) *J. Phys. Chem. C*, **112**, 6860.
89. Martin, M.G.and Siepmann, J.I. (1999) *J. Phys. Chem. B*, **103**, 4508.
90. Smith, B.D.and Srivastava, R. (1986) *Thermodynamic Data for Pure Compounds: Part A, Hydrocarbons and Ketones*, Elsevier, Amsterdam.

91. Stoll, J., Vrabec, J., and Hasse, H. (2003)GVC/DECHEMA Annual Meeting, Mannhein, Germany, p. 891.
92. Vrabec, J., Stoll, J., and Hasse, H. (2001) *J. Phys. Chem. B*, **105**, 12126.
93. Wick, C., Martin, M., and Siepmann, J.I. (2000) *J. Phys. Chem. B*, **104**, 8008.
94. Skoulidas, A.I.and Sholl, D.S. (2002) *J. Phys. Chem. B*, **106**, 5058.
95. Desbiens, N., Boutin, A., and Demachy, I. (2005) *J. Phys. Chem. B*, **109**, 24071.
96. Desbiens, N., Demachy, I., Fuchs, A.H., Kirsch-Rodeschini, H., Soulard, M., and Patarin, J. (2005) *Angew. Chem. Int. Ed.*, **44**, 5310.
97. Frenkel, D.and Smit, B. (2002) *Understanding Molecular Simulations: From Algorithms to Applications*, 2nd edn, Academic Press, San Diego.
98. Allen, M.P.and Tildesley, D.J. (1987) *Computer Simulations of Liquids*, Clarendon Press, Oxford.
99. Bussai, C., Fritzsche, S., Haberlandt, R., and Hannongbua, S. (2004) *J. Phys. Chem. B*, **108**, 13347.
100. Schroder, K.P.and Sauer, J. (1996) *J. Phys. Chem.*, **100**, 11043.
101. Talu, O.and Myers, A.L. (2001) *AIChE J.*, **47**, 1160.
102. Myers, A.L.and Monson, P.A. (2002) *Langmuir*, **18**, 10261.
103. Chen, H.B.and Sholl, D.S. (2006) *Langmuir*, **22**, 709.
104. Shen, V.K.and Errington, J.R. (2005) *J. Chem. Phys.*, **122**, 8.
105. Karavias, F.and Myers, A.L. (1991) *Langmuir*, **7**, 3118.
106. Vlugt, T.J.H., Krishna, R., and Smit, B. (1999) *J. Phys. Chem. B*, **103**, 1102.
107. Woods, G.B., Panagiotopoulos, A.Z., and Rowlinson, J.S. (1988) *Mol. Phys.*, **63**, 49.
108. Vlugt, T.J.H., Garcia-Perez, E., Dubbeldam, D., Ban, S., and Calero, S. (2008) *J. Chem. Theory Comput.*, **4**, 1107.
109. Harris, J.and Rice, S.A. (1988) *J. Chem. Phys.*, **89** (9), 5898.
110. Depablo, J.J., Laso, M., and Suter, U.W. (1992) *J. Chem. Phys.*, **96**, 6157.
111. Siepmann, J.I.and Frenkel, D. (1992) *Mol. Phys.*, **75**, 59.
112. Frenkel, D., Mooij, G., and Smit, B. (1992) *J. Phys. Condens. Matter*, **4**, 3053.
113. Consta, S., Vlugt, T.J.H., Hoeth, J.W., Smit, B., and Frenkel, D. (1999) *Mol. Phys.*, **97**, 1243.
114. Consta, S., Wilding, N.B., Frenkel, D., and Alexandrowicz, Z. (1999) *J. Chem. Phys.*, **110**, 3220.
115. Houdayer, J. (2002) *J. Chem. Phys.*, **116**, 1783.
116. Dubbeldam, D., Calero, S., Maesen, T.L.M., and Smit, B. (2003) *Phys. Rev. Lett.*, **90**, 245901.
117. Dubbeldam, D., Calero, S., Maesen, T.L.M., and Smit, B. (2003) *Angew. Chem. Int. Ed.*, **42**, 3624.
118. Dubbeldam, D.and Smit, B. (2003) *J. Phys. Chem. B*, **107**, 12138.
119. Dubbeldam, D., Beerdsen, E., Calero, S., and Smit, B. (2005) *Proc. Natl. Acad. Sci. U.S.A.*, **102**, 12317.
120. Walton, K.S.and Snurr, R.Q. (2007) *J. Am. Chem. Soc.*, **129**, 8552.
121. Auerbach, S.M. (2000) *Int. Rev. Phys. Chem.*, **19**, 155.
122. Sholl, D.S. (2006) *Acc. Chem. Res.*, **39**, 403.
123. Demontis, P.and Suffritti, G.B. (1997) *Chem. Rev.*, **97**, 2845.
124. Snurr, R.Q.and Karger, J. (1997) *J. Phys. Chem. B*, **101**, 6469.
125. Gergidis, L.N.and Theodorou, D.N. (1999) *J. Phys. Chem. B*, **103**, 3380.
126. Jost, S., Bar, N.K., Fritzsche, S., Haberlandt, R., and Karger, J. (1998) *J. Phys. Chem. B*, **102**, 6375.
127. Krishna, R.and van Baten, J.M. (2008) *Chem. Eng. Sci.*, **63**, 3120.
128. Krishna, R.and van Baten, J.M. (2008) *Microporous Mesoporous Mater.*, **109**, 91.
129. Krishna, R.and van Baten, J.M. (2009) *Chem. Eng. Sci.*, **64**, 870.
130. Maginn, E.J., Bell, A.T., and Theodorou, D.N. (1993) *J. Phys. Chem.*, **97**, 4173.
131. Fritzsche, S., Haberlandt, R., and Karger, J. (1995) *Z. Phys. Chem.: Int. J. Res. Phys. Chem. Chem. Phys.*, **189**, 211.
132. Arya, G., Chang, H.C., and Maginn, E.J. (2001) *J. Chem. Phys.*, **115**, 8112.

133. Sholl, D.S. (2000) *Ind. Eng. Chem. Res.*, **39**, 3737.
134. Hoogenboom, J.P., Tepper, H.L., van der Vegt, N.F.A., and Briels, W.J. (2000) *J. Chem. Phys.*, **113**, 6875.
135. Krishna, R.and Baur, R. (2003) *Sep. Purif. Technol.*, **33**, 213.
136. Krishna, R.and van Baten, J.M. (2005) *J. Phys. Chem. B*, **109**, 6386.
137. Sastre, G., Catlow, C.R.A., Chica, A., and Corma, A. (2000) *J. Phys. Chem. B*, **104**, 416.
138. Sastre, G., Catlow, C.R.A., and Corma, A. (1999) *J. Phys. Chem. B*, **103**, 5187.
139. Derouane, E.G.and Gabelica, Z. (1980) *J. Catal.*, **65**, 486.
140. Krishna, R., van Baten, J.M., and Dubbeldam, D. (2004) *J. Phys. Chem. B*, **108**, 14820.
141. Krishna, R.and van Baten, J.M. (2006) *Chem. Phys. Lett.*, **420**, 545.
142. Skoulidas, A.I., Sholl, D.S., and Krishna, R. (2003) *Langmuir*, **19**, 7977.
143. Skoulidas, A.I.and Sholl, D.S. (2003) *J. Phys. Chem. A*, **107**, 10132.
144. Beerdsen, E., Dubbeldam, D., and Smit, B. (2006) *Phys. Rev. Lett.*, **96**, 4.
145. Hahn, K., Karger, J., and Kukla, V. (1996) *Phys. Rev. Lett.*, **76**, 2762.
146. Kukla, V., Kornatowski, J., Demuth, D., Gimus, I., Pfeifer, H., Rees, L.V.C., Schunk, S., Unger, K.K., and Karger, J. (1996) *Science*, **272**, 702.
147. Gupta, V., Nivarthi, S.S., Keffer, D., McCormick, A.V., and Davis, H.T. (1996) *Science*, **274**, 164.
148. Tsekov, R.and Evstatieva, E. (2005) *Adv. Colloid Interface Sci.*, **114**, 159.
149. Brauer, P., Brzank, A., and Karger, J. (2003) *J. Phys. Chem. B*, **107**, 1821.
150. Clark, L.A., Ye, G.T., and Snurr, R.Q. (2000) *Phys. Rev. Lett.*, **84**, 2893.
151. Harish, R., Karevski, D., and Schutz, G.M. (2008) *J. Catal.*, **253**, 191.
152. Dubbeldam, D., Beerdsen, E., Calero, S., and Smit, B. (2006) *J. Phys. Chem. B*, **110**, 3164.
153. Chandler, D. (1978) *J. Chem. Phys.*, **68**, 2959.
154. RuizMontero, M.J., Frenkel, D., and Brey, J.J. (1997) *Mol. Phys.*, **90**, 925.
155. Dellago, C., Bolhuis, P.G., and Geissler, P.L. (2002) *Adv. Chem. Phys.*, **123**, 1.
156. van Erp, T.S., Moroni, D., and Bolhuis, P.G. (2003) *J. Chem. Phys.*, **118**, 7762.
157. Sorensen, M.R.and Voter, A.F. (2000) *J. Chem. Phys.*, **112**, 9599.
158. June, R.L., Bell, A.T., and Theodorou, D.N. (1991) *J. Phys. Chem.*, **95**, 8866.
159. Forester, T.R.and Smith, W. (1997) *J. Chem. Soc., Faraday Trans.*, **93**, 3249.
160. Maginn, E.J., Bell, A.T., and Theodorou, D.N. (1996) *J. Phys. Chem.*, **100**, 7155.
161. Jousse, F.and Auerbach, S.M. (1997) *J. Chem. Phys.*, **107**, 9629.
162. Mosell, T., Schrimpf, G., and Brickmann, J. (1997) *J. Phys. Chem. B*, **101**, 9476.
163. Mosell, T., Schrimpf, G., and Brickmann, J. (1997) *J. Phys. Chem. B*, **101**, 9485.
164. Ghorai, P.K.R., Yashonath, S., and Lynden-Bell, R.M. (2002) *Mol. Phys.*, **100**, 641.
165. Schuring, A., Auerbach, S.M., Fritzsche, S., and Haberlandt, R. (2002) *J. Chem. Phys.*, **116**, 10890.
166. Dubbeldam, D., Beerdsen, E., Vlugt, T.J.H., and Smit, B. (2005) *J. Chem. Phys.*, **122**, 22.
167. Nagumo, R., Takaba, H., and Nakao, S.I. (2008) *J. Phys. Chem. C*, **112**, 2805.
168. Gupta, A.and Snurr, R.Q. (2005) *J. Phys. Chem. B*, **109**, 1822.
169. Tunca, C.and Ford, D.M. (1999) *J. Chem. Phys.*, **111**, 2751.
170. Tunca, C.and Ford, D.M. (2003) *Chem. Eng. Sci.*, **58**, 3373.
171. Beerdsen, E., Smit, B., and Dubbeldam, D. (2004) *Phys. Rev. Lett.*, **93**, 24.
172. Bowen, T.C., Noble, R.D., and Falconer, J.L. (2004) *J. Memb. Sci.*, **245**, 1.
173. Channon, Y.M., Catlow, C.R.A., Gorman, A.M., and Jackson, R.A. (1998) *J. Phys. Chem. B*, **102**, 4045.
174. Demontis, P., Stara, G., and Suffritti, G.B. (2003) *J. Phys. Chem. B*, **107**, 4426.
175. Fleys, M.and Thompson, R.W. (2005) *J. Chem. Theory Comput.*, **1**, 453.
176. Fleys, M., Thompson, R.W., and MacDonald, J.C. (2004) *J. Phys. Chem. B*, **108**, 12197.

177. Puibasset, J.and Pellenq, R.J.M. (2008) *J. Phys. Chem. B*, **112**, 6390.
178. Bougeard, D.and Smirnov, K.S. (2007) *Phys. Chem. Chem. Phys.*, **9**, 226.
179. Ockwig, N.W., Cygan, R.T., Criscenti, L.J., and Nenoff, T.M. (2008) *Phys. Chem. Chem. Phys.*, **10**, 800.
180. Demontis, P., Gulin-Gonzalez, J., Jobic, H., Masia, M., Sale, R., and Suffritti, G.B. (2008) *Acs Nano*, **2**, 1603.
181. Maesen, T.L.M., Krishna, R., van Baten, J.M., Smit, B., Calero, S., and Sanchez, J.M.C. (2008) *J. Catal.*, **256**, 95.
182. Maesen, T.L.M., Beerdsen, E., Calero, S., Dubbeldam, D., and Smit, B. (2006) *J. Catal.*, **237**, 278.
183. Maesen, T.L.M., Calero, S., Schenk, M., and Smit, B. (2004) *J. Catal.*, **221**, 241.
184. Chen, N.Y., Lucki, S.J., and Mower, E.B. (1969) *J. Catal.*, **13**, 329.
185. Magalhaes, F.D., Laurence, R.L., and Conner, W.C. (1996) *AIChE J.*, **42**, 68.
186. Cavalcante, C.L., Eic, M., Ruthven, D.M., and Occelli, M.L. (1995) *Zeolites*, **15**, 293.
187. Talu, O., Li, J.M., and Myers, A.L. (1995) *Adsorpt. J. Int. Adsorpt. Soc.*, **1**, 103.
188. Denayer, J.F., Ocakoglu, A.R., De Jonckheere, B.A., Martens, J.A., Thybaut, J.W., Marin, G.B., and Baron, G.V. (2003) *Int. J. Chem. Reactor Eng.*, **1**, A36.
189. Denayer, J.F.M., Ocakoglu, R.A., Huybrechts, W., Dejonckheere, B., Jacobs, P., Calero, S., Krishna, R., Smit, B., Baron, G.V., and Martens, J.A. (2003) *J. Catal.*, **220**, 66.
190. Ruthven, D.M. (1984) *Principles of Adsorption and Adsorption Processes*, Wiley-Interscience, New York.
191. Lachet, V., Boutin, A., Tavitian, B., and Fuchs, A.H. (1999) *Langmuir*, **15**, 8678.
192. Krishna, R., Smit, B., and Vlugt, T.J.H. (1998) *J. Phys. Chem. A*, **102**, 7727.
193. Calero, S., Smit, B., and Krishna, R. (2001) *Phys. Chem. Chem. Phys.*, **3**, 4390.
194. Krishna, R., Calero, S., and Smit, B. (2002) *Chem. Eng. J.*, **88**, 81.
195. Krishna, R., Smit, B., and Calero, S. (2002) *Chem. Soc. Rev.*, **31**, 185.
196. Calero, S., Smit, B., and Krishna, R. (2001) *J. Catal.*, **202**, 395.
197. Ban, S., Van Laak, A., De Jongh, P.E., Van der Eerden, J., and Vlugt, T.J.H. (2007) *J. Phys. Chem. C*, **111**, 17241.
198. Krishna, R.and van Baten, J.M. (2008) *Chem. Eng. J.*, **140**, 614.
199. Hansen, N., Krishna, R., van Baten, J.M., Bell, A.T., and Kew, F.J. (2009) *J. Phys. Chem. C*, **113**, 235.
200. Kuhn, J., Castillo, J.M., Gascon, J., Calero, S., Dubbeldam, D., Vlugt, T.J.H., Kapteijn, F., and Gross, J. (2009) *J. Phys. Chem. C*, **113**, 14290–14301.
201. Lu, L.H., Shao, Q., Huang, L.L., and Lu, X.H. (2007)11th International Conference on Properties and Phase Equilibria for Product and Process Design, Crete, Greece, p. 191.
202. Rutkai, G., Csanyi, E., and Kristof, T. (2008) *Microporous Mesoporous Mater.*, **114**, 455.
203. Krishna, R., van Baten, J.M., Garcia-Perez, E., and Calero, S. (2007) *Ind. Eng. Chem. Res.*, **46**, 2974.
204. Garcia-Perez, E., Parra, J.B., Ania, C.O., Garcia-Sanchez, A., Van Baten, J.M., Krishna, R., Dubbeldam, D., and Calero, S. (2007) *Adsorpt. J. Int. Adsorpt. Soc.*, **13**, 469.
205. Krishna, R., van Baten, J.M., Garcia-Perez, E., and Calero, S. (2006) *Chem. Phys. Lett.*, **429**, 219.
206. Bonenfant, D., Kharoune, M., Niquette, P., Mimeault, M., and Hausler, R. (2008) *Sci. Technol. Adv. Mater.*, **9**.
207. Ahn, H., Moon, J.H., Hyun, S.H., and Lee, C.H. (2004) *Adsorpt. J. Int. Adsorpt. Soc.*, **10**, 111.
208. Goj, A., Sholl, D.S., Akten, E.D., and Kohen, D. (2002) *J. Phys. Chem. B*, **106**, 8367.
209. Yue, X.P.and Yang, X.N. (2006) *Langmuir*, **22**, 3138.
210. Heuchel, M., Snurr, R.Q., and Buss, E. (1997) *Langmuir*, **13**, 6795.
211. Jia, Y.X., Wang, M., Wu, L.Y., and Gao, C.J. (2007) *Sep. Sci. Technol.*, **42**, 3681.
212. Babarao, R., Hu, Z.Q., Jiang, J.W., Chempath, S., and Sandler, S.I. (2007) *Langmuir*, **23**, 659.
213. Ghoufi, A., Gaberova, L., Rouquerol, J., Vincent, D., Llewellyn, P.L., and

Maurin, G. (2009) *Microporous Mesoporous Mater.*, **119**, 117.

214. Krishna, R.and van Baten, J.M. (2007) *Chem. Eng. J.*, **133**, 121.

215. Sanborn, M.J.and Snurr, R.Q. (2000) *Sep. Purif. Technol.*, **20**, 1.

216. Papadopoulos, G.K., Jobic, H., and Theodorou, D.N. (2004) *J. Phys. Chem. B*, **108**, 12748.

217. Babarao, R.and Jiang, J.W. (2008) *Langmuir*, **24**, 5474.

218. Krishna, R.and van Baten, J.M. (2008) *Sep. Purif. Technol.*, **61**, 414.

# 13
# Diffusion in Zeolites – Impact on Catalysis

*Johan van den Bergh, Jorge Gascon, and Freek Kapteijn*

## 13.1
## Introduction

Zeolite catalysts and adsorbents are widely accepted in industry. In 1948, commercial adsorbents based on synthetic aluminosilicates zeolite A and X were available [1]. Zeolite Y as FCC catalyst was commercially available in [2].

Aluminum-containing zeolites are inherently catalytically active in several ways. The isomorphic substituted aluminum atom within the zeolite framework has a negative charge that is compensated by a counterion. When the counterion is a proton, a Brønsted acid site is created. Moreover, framework oxygen atoms can give rise to weak Lewis base activity. Noble- metal ions can be introduced by an ion exchange with the cations after synthesis. Incorporation of metals like Ti, V, Fe, and Cr in the framework can provide the zeolite with activity for redox reactions. A well-known example of the latter type is titanium silicalite-1 (TS-1): a redox molecular sieve catalyst [3].

Not only the catalytic activity makes zeolites particularly interesting but also the location of the active sites within the well-defined geometry of a zeolite. Because of the geometrical constraints of the zeolite, the selectivity of a chemical reaction can be increased by three mechanisms: reactant selectivity, product selectivity, and transition state selectivity. In the case of reactant selectivity, bulky components in the feed do not enter the zeolite and will have no opportunity to react. When several products are formed within the zeolite, but only some are able to leave the zeolite, or some leave the zeolite more rapidly, the process is named product selectivity. When the geometrical constraints of the active site within the zeolite prohibit the formation of products or transition states leading to certain products, transition state selectivity is obtained [4, 5].

The importance of diffusion in (zeolite) catalysts arises from the fact that the catalytically active site needs to be reached by the reactants, and products need to move away from this site. As depicted in Figure 13.1 for a packed bed configuration, reactants need to move from the bulk to the active site. As commonly applied in catalysis, the macroporous ($d_{pore} > 50\,nm$) particles at the bed level consist of pelleted smaller particles (crystals in the case of zeolites) with

*Zeolites and Catalysis, Synthesis, Reactions and Applications. Vol. 1.*
Edited by Jiří Čejka, Avelino Corma, and Stacey Zones

ISBN: 978-3-527-32514-6

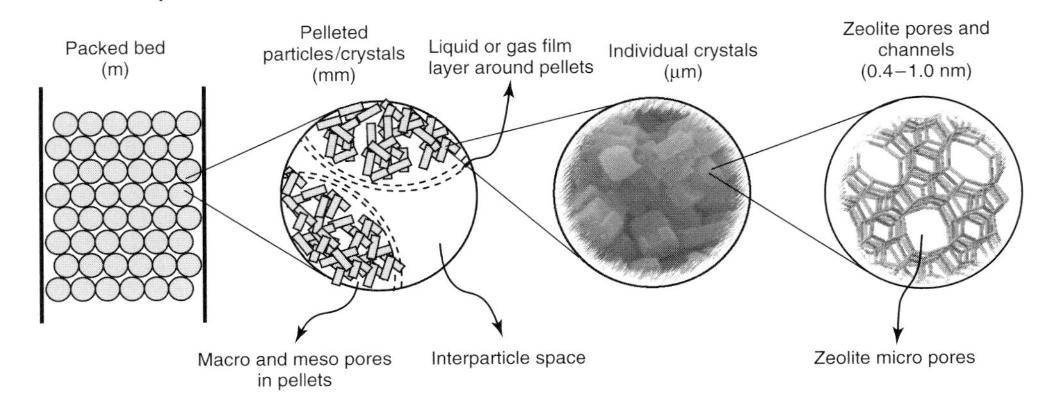


**Figure 13.1** The reactant pathway from bulk to active site encompasses diffusion processes on a broad length scale in common catalytic processes. (After Ruthven [6].)

meso- ($2\,\text{nm} < d_{\text{pore}} < 50\,\text{nm}$) or even micropores ($d_{\text{pore}} < 2\,\text{nm}$). External resistances, which can stand in the way of optimal catalysis, are film layers around the particles that can introduce external heat and mass transfer limitations. Additionally, macropores in the pelleted particles and meso- and micropores in the small particles or crystals can induce internal (diffusional) mass transport limitations. Furthermore, internal heat transfer limitations can occur due to poor heat removal from the (pelleted) particles.

Most transport limitations and associated catalyst effectiveness are based on solving the reaction–diffusion problem in catalyst particles. Transformation of the governing differential equation yields a dimensionless parameter, the Thiele modulus, expressing the ratio of the kinetics and diffusion rates for the particle. This reaction–diffusion problem and the Thiele concept are textbook material [7] and are not discussed in this chapter. Because of the small pore size of zeolite catalysts, diffusion limitations may quickly arise. This chapter focuses on internal diffusion (limitations) in zeolite catalyst particles that considerably deviates from diffusion in meso- and macroporous materials. The impact of these deviations with respect to the applicability of the Thiele concept to estimate the catalyst effectiveness is evaluated. Furthermore, results of recently introduced spatially resolved measurement techniques are discussed in relation to the performance of a zeolite catalyst at the subcrystal level. Finally, the state-of-the-art approaches to overcome, prevent, or utilize diffusion limitations are discussed.

## 13.2
## Diffusion and Reaction in Zeolites: Basic Concepts

Diffusion in the gas phase or in relative large pores (>100 nm [8]) is dominated by intermolecular collisions, and the flux of component $i$ can be described by the Maxwell–Stefan (MS) approach [9], in which forces acting on molecules (in diffusional processes, the gradient in thermodynamic potential) are balanced by the friction between molecules and, in case of porous materials, with a solid. In the latter case, this was coined the "dusty gas model." This model was extended to zeolites by Krishna [10]. The often used Fick's law is a simplification of the generalized MS equations for thermodynamically ideal systems [9].

In the case of a porous material, a correction needs to be made to account for the porosity ($\varepsilon$) and tortuosity ($\tau$) of the material, leading to an "effective" diffusivity.

$$N_i = -\frac{\varepsilon}{\tau} D_i \nabla C_i = -D_i^{\mathrm{eff}} \nabla C_i \tag{13.1}$$

In porous materials, when the mean free path of a molecule is in the order of or larger than the pore diameter (~10–100 nm [8]), molecule-wall collisions start to dominate and the diffusivity can be described by the Knudsen diffusion mechanism. A flux in such small pores can be presented as

$$N_i = -\frac{\varepsilon}{\tau} D_{kn,i} \nabla C_i, \quad D_{kn,i} = \frac{d_0}{3}\sqrt{\left(\frac{8RT}{\pi M_i}\right)} \tag{13.2}$$

In the case of zeolites, the pores approach molecular dimensions (~0.3–0.74 nm) and, consequently, mass transport through zeolite pores is determined by the interaction of the molecules with the zeolite pore wall (see also Chapter 24 on modeling transport in zeolites). Now molecules are adsorbed on the zeolite, lose their gaseous nature, and transport is often referred to as *surface* or *zeolitic diffusion* [8]. The flux can now also be represented in a Fickian way; the concentration ($q_i$) represents the adsorbed amount on the zeolite or loading. A common unit for the loading is moles per kilogram; therefore, the zeolite density ($\rho$) is added to arrive at consistent dimensions:

$$N_i = \rho D_i \nabla q_i \tag{13.3}$$

The tortuosity and porosity presented in Eq. (13.1) are not specified in Eq. (13.3); these are an inherent property of the diffusivity. Each zeolite has its own specific pore network with its own tortuosity and porosity: some zeolites are characterized by channels and intersections (e.g., MFI); others resemble systems of cages connected by windows (e.g., faujasite (FAU) and Linde type A (LTA)). Moreover, the pore network can be one-, two-, or three-dimensional with different pore sizes or connectivity in different directions, leading to diffusion anisotropy. Since zeolites are crystalline materials, this may imply that some crystal facets are better accessible than others, significantly influencing local transport and potentially giving rise to concentration profiles inside the zeolite.

### 13.2.1 Importance of Adsorption

The adsorbed phase ($q_i$) in Eq. (13.3) is related to the gas phase fugacity through an adsorption isotherm, of which the classical example is the Langmuir isotherm:

$$q_i = \frac{q_i^{\text{sat}} K_i f_i}{1 + K_i f_i} \tag{13.4}$$

An important difference between gas phase and adsorbed phase diffusion is the concentration level, being much higher in the case of adsorbed phase diffusion. When the gradient in chemical potential is taken as the fundamental driving force for diffusion [8, 9], a correction needs to be made to Eq. (13.3). Now, a so-called thermodynamic correction factor ($\Gamma_i$) is introduced; the diffusivity is referred to as *corrected* or *Maxwell–Stefan (MS)* diffusivity:

$$N_i = -\rho Đ_i \nabla \ln f_i = -\rho Đ_i \Gamma_i \nabla q_i, \quad \Gamma_i = \frac{\text{d} \ln f_i}{\text{d} \ln q_i} \tag{13.5}$$

For a single-site Langmuir isotherm, the thermodynamic correction factor is given by

$$\Gamma_i = \frac{\text{d} \ln f_i}{\text{d} \ln q_i} = \frac{1}{1 - \theta_i}, \quad \theta_i = \frac{q_i}{q_i^{\text{sat}}} \tag{13.6}$$

In the limit of low loading, the thermodynamic correction factor approaches 1 and the MS and Fickian diffusivity are equal. Although the MS diffusivity appears to be physically more correct, the Fickian diffusivity remains very important since this diffusivity can be directly assessed in diffusion measurements.

### 13.2.2 Self-Diffusivity

In the previous text, only transport diffusivities are considered: diffusion induced by a concentration gradient. However, in the absence of concentration gradients, molecules also move, this feature is called *Brownian motion* or *self-diffusion*. This self- or tracer diffusivity ($D_{\text{self}}$) is commonly defined for a 3D space by the following equation, in which the mean square displacement ($\langle r^2(t) \rangle$) is proportional with time ($t$):

$$\langle r^2(t) \rangle = 6 D_{\text{self}} t \tag{13.7}$$

Depending on the measurement technique, a self- or transport diffusivity (or both) can be obtained, whereas the transport diffusivity is required for design. Therefore, relating the two is often desired, and a direct relation is recently proposed and is discussed in more detail in the next section.

### 13.2.3
### Mixture Diffusion

Diffusion in mixtures appears to be best treated by the MS approach to mass transport [8, 9, 11]. Because of the relatively high concentrations adsorbed in the zeolite, interactions between molecules can play a significant role in terms of "speeding up" or "slowing down" other components. In the MS approach besides the interaction (or "friction") of the individual molecule with the zeolite, also the interaction between the different diffusing molecules is accounted for and balanced with the driving force for mass transport:

$$\rho\theta_i \nabla \ln f = \sum_{j=1}^{n} \frac{q_j N_i - q_i N_j}{q_j^{\text{sat}} q_i^{\text{sat}} Ð_{ij}} + \frac{N_i}{q_i^{\text{sat}} Ð_i}; \quad i = 1, 2 \ldots n \tag{13.8}$$

Within this approach, the estimation of $Ð_{ij}$ can be difficult; however, a reasonable estimation can be made through a logarithmic ("Vignes") interpolation [9, 12] based on the single-component exchange diffusivities and a correction factor $F$ for the confinement of the molecules in the narrow zeolite pores [13].

For a single-component system of tagged and untagged species, the saturation capacities are equal, and one can show [13] that the single-component exchange coefficient is related to the self-diffusivity and MS diffusivity as

$$\frac{1}{D_{\text{Self},i}} = \frac{1}{Ð_i} + \frac{1}{Ð_{ii}} \tag{13.9}$$

$$Ð_{ij} = F \cdot Ð_{ii}^{\frac{\theta_i}{\theta_i + \theta_j}} Ð_{jj}^{\frac{\theta_i}{\theta_i + \theta_j}} \tag{13.10}$$

For mesoporous systems, the factor $F$ equals 1, whereas for the microporous zeolites, $<1$ holds. This factor is fairly constant for a zeolite and depends on the pore size (see [13]).

It is evident that in the case of mixture diffusion, an accurate estimation of the individual component loading in the zeolite and the driving force is required to satisfactory model in such a system. For zeolitic systems, the ideal adsorbed solution theory (IAST) [14] provides an acceptable mixture prediction based on the single-component isotherms [8, 15], but when adsorption heterogeneity becomes manifest, IAST tends to fail [16–18].

At significant loading, the molecular interaction can play an important role, strongly influencing the reactant and product concentration profiles. When the loading is relatively low, the cross-correlation effects can often be ignored [19].

### 13.2.4
### Diffusion Measurement Techniques

For molecular gaseous and Knudsen diffusion, a reasonable estimation of the diffusivity can be made: in the case of Knudsen diffusion based on Eq. (13.2) and in the case of gaseous diffusion by relations like that of Fuller *et al.* [20]. Although some generalizations on the diffusivities in zeolites can be made based on their class (e.g., cage-like) and pore size [21], accurate predictions from simple methods are not feasible. Therefore, significant effort has been devoted to the development of new experimental techniques [22] and computer simulations (e.g., grand canonical Monte Carlo (GCMC) and molecular dynamics (MDs) [5, 23] to quantify adsorption and diffusion properties in zeolites. In the last decades, with improving experimental techniques and methodologies, many phenomena characteristic for zeolites have been clarified, but also many new peculiarities of diffusion in zeolites have been discovered. The latter underlines the difficulty to capture diffusion in zeolites in a "simple" and general engineering approach, which is desired for process design.

Excellent reviews on diffusion measurement techniques are available (e.g., [22, 24, 25]) and it is not our objective to repeat this in this chapter. Here, the focus is on the new, state-of-the-art time- and space-resolved techniques [26–28] that have provided exciting new insights and direct evidence for several phenomena in zeolite mass transport, for instance, surface barriers, diffusion anisotropy, and catalyst activity at the subcrystal level.

Although these state-of-the-art techniques have clear added value, this means that in no way other techniques have lost their value [29]. A remark needs to be made on the controversy of the difference in diffusivities obtained by macroscopic techniques (e.g., transient adsorption uptake, membrane experiments, or zero length chromatography) and microscopic techniques (e.g., quasi elastic neutron scattering (QENS) and pulsed field gradient nuclear magnetic resonance (PFG-NMR)). The measured diffusivities for the same system may differ in orders of magnitude [22, 30]; diffusivities measured by macroscopic techniques are sometimes found to be significantly lower. These differences can be explained by the length scale of the measurements and the fact that, consequently, in macrotechniques, also surface and internal barriers are part of the investigated domain, whereas in microtechniques, the measured domain can be much smaller and probes only the internal zeolite pore structure.

The consequence of these findings is twofold: diffusivities measured by macroscopic techniques are generally not representative for the true intracrystalline diffusivity and diffusivities measured by microscopic techniques are a priori not representative for a real system that may comprise internal and surface barriers. Therefore, interpretation and application of diffusivity data should be done with caution.

### 13.2.5
### Relating Diffusion and Catalysis

In the classical reaction engineering approach, to account for internal diffusion limitations, the concept of catalyst effectiveness factor is introduced, representing the ratio between the observed reaction rate versus the expected (intrinsic) reaction rate at bulk concentration and temperature:

$$\eta = \frac{r_{\text{observed}}}{r_{\text{intrinsic}}} \tag{13.11}$$

On solving the mass balance of a first-order irreversible reaction in a slab-like and spherical geometry, respectively, two simple relations to describe the catalyst effectiveness are obtained:

$$\eta = \frac{\tan\text{h}(\phi)}{\phi} \quad \text{and} \quad \eta = \frac{3}{\phi}\left[\frac{1}{\tan\text{h}(\phi)} - \frac{1}{\phi}\right] \tag{13.12}$$

which becomes a function only of the Thiele modulus [31] that can be calculated from the characteristic length for diffusion ($L$, ratio between volume and external surface area) of the particle, the intrinsic reaction rate constant based on the volume of the particle ($k$), and the effective diffusivity ($D_{\text{eff}}$). The Thiele modulus can be considered as the square root of the ratio of the characteristic time for diffusion and of reaction:

$$\phi = L\sqrt{\frac{k}{D_{\text{eff}}}} = \sqrt{\frac{kL^2}{D_{\text{eff}}}} = \sqrt{\frac{\tau_{\text{diff}}}{\tau_{\text{reaction}}}} \text{ with } L = \frac{V}{A} \tag{13.13}$$

For cylindrical and spherical geometry, similar relations are obtained as Eq. (13.12) and, thus, a powerful, simple approach is found to describe the catalyst effectiveness in an accurate way. A textbook example of a successful application of the effectiveness factor and the Thiele modulus to model diffusion limitations is given by Post *et al.* [32]. They were able to model the catalyst effectiveness of cobalt Fischer Tropsch catalysts with different pore and particles sizes accurately within this concept.

In the case of a zeolite catalyst, diffusion is sometimes described based on the adsorbed phase concentration, whereas the reaction rate is based on the bulk gas phase concentration or pressure. In this case, an adsorption constant ($K$) is added to the Thiele modulus to maintain its dimensionless character [33]:

$$\phi = L\sqrt{\frac{k}{KD_{\text{eff}}}} \tag{13.14}$$

This explains in part the large differences in diffusivity values for zeolites and meso/macroporous materials. Post *et al.* have also provided an experimental verification of the Thiele concept for the conversion of 2,2-dimethyl butane over a H-ZSM5 catalyst. At reaction temperatures ranging from 673 to 803 K, the measured effectiveness factor closely follows the predicted effectiveness calculated from the Thiele modulus [25, 33].

Figure 13.2 shows how the concentration profile across a zeolite crystal at different values of the Thiele modulus and the dependence of the effectiveness factor on the Thiele modulus for the specific case of a slab-like geometry (Eq. (13.12)). Full utilization of the catalyst particle ($\eta \to 1$) only takes place at very low values of the Thiele modulus ($\phi \to 0$). Contrarily, $\phi = 10$ renders $\eta = 0.1$, meaning that only 10% of the catalyst volume is effectively used in the reaction. Transport limitations not only affect activity, but may also impact selectivity and stability

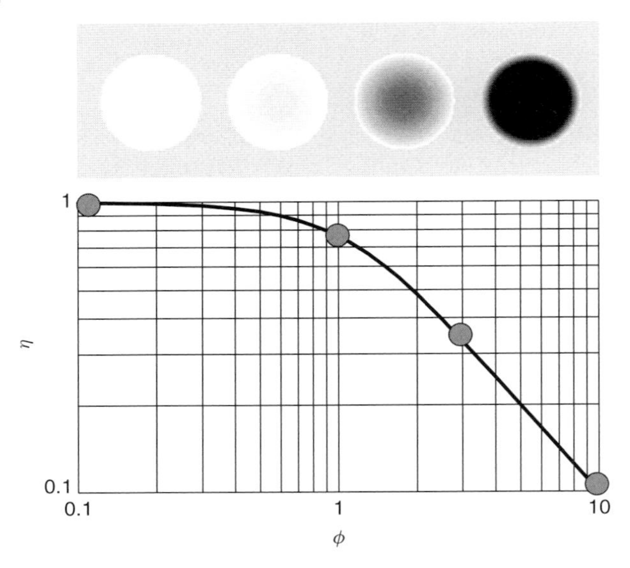


**Figure 13.2** Concentration profiles across a zeolite crystal (sphere geometry) at different values of the Thiele modulus $\phi$ (from left to right: $\phi = 0.1$, 1, 3, and 10) and the dependence of the effectiveness factor on the Thiele modulus. Low Thiele moduli lead to full catalyst utilization ($\phi \rightarrow 0, \eta \rightarrow 1$), whereas high Thiele moduli render a poorly utilized catalyst ($\phi \rightarrow \infty, \eta \rightarrow 1/\phi$). The reactant concentration across a zeolite crystal is exhausted ($c/c_s = 0$) close to the surface at $\phi = 10$, while being practically uniform and very similar to the surface concentration ($c/c_s = 1$) at $\phi = 0.1$. (After [34].)

[7, 34]. Especially in case of consecutive reactions, diffusion limitations should be avoided if the intermediate product is desired. To what extent the Thiele concept is applicable to zeolites is discussed in more detail in Section 13.3.4.

## 13.3 Diffusion in Zeolites: Potential Issues

Once the scenario has been set, now some special issues will be treated that can typically occur in zeolites, what the consequences are with respect to catalysis and, when possible, how to account for these effects in modeling studies.

### 13.3.1 Concentration Dependence of Diffusion

In contrast to gas phase reactions in meso- and macroporous materials, in the case of zeolites, relative high concentrations are common practice and the diffusivity is found to be (strongly) dependent on the occupancy of the zeolite. Kärger *et al.* [35] have identified five types of loading dependencies for self-diffusivities. Figure 13.3 shows some typical transport diffusivity behaviors as a function of the loading. This figure captures in principle the dependencies originally postulated by Kärger *et al.*,

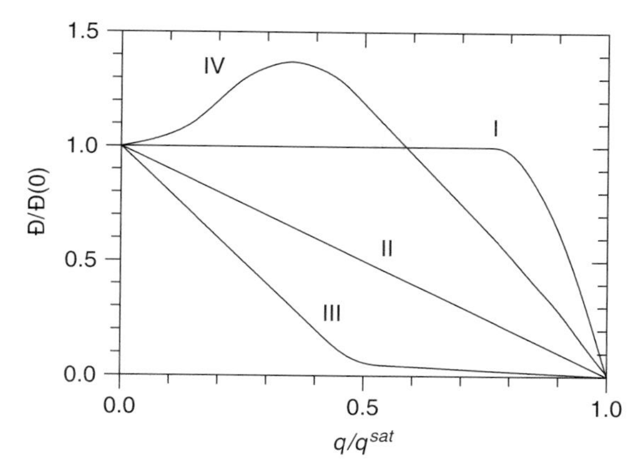


**Figure 13.3** Common dependencies of the Maxwell–Stefan diffusivity on the loading on the zeolite, normalized to that at zero loading.

and accounting for the observation that when approaching saturation loading, the diffusivity appears to become very low in general. It goes without saying that not every diffusivity – concentration dependency fits one of the four profiles. There is a broad spectrum of variation.

Type I can be described by Eq. (13.5), which represents a loading-independent diffusivity, except at very high occupancies. However, it is frequently observed that the diffusivity decreases with loading and approaches very small values when the zeolite becomes saturated. This can be interpreted as a reduction of free volume available for diffusion and can be incorporated in the diffusivity modeling with the addition of a term that corrects this. The diffusivity follows in this case the so-called "strong confinement" scenario, which is represented by type II in Figure 13.3 and can be described by

$$Đ_i = Đ_i(0)(1 - \theta_i) \tag{13.15}$$

It is noteworthy to mention that in the case the diffusivity obeys the strong confinement scenario (Eq. (13.15)) and the isotherm can be described by a single-site Langmuir isotherm, the thermodynamic correction factor (Eq. (13.6)) and confinement term cancel and Fickian-type diffusion is found (Eq. (13.3)). Most zeolite–host systems can be treated well with Eq. (13.5) either with or without the strong confinement correction. The diffusivities of many light gases in cage-like zeolites first show a strong increase in diffusivity with increasing loading before eventually going to very small diffusivity values (type III in Figure 13.3) when approaching saturation loading. Although much evidence of this phenomena comes from computer simulations (e.g., MD) [21, 36], clear experimental evidence of strongly increasing diffusivities with loading is also present [22, 37, 38]. Two possible explanations are given in literature. The first explanation relates this phenomenon to intermolecular repulsion effects [39, 40]: when the loading increases, the repulsion effects increase and, consequently, the activation barrier for diffusion is reduced leading to an increased diffusivity [41]. A second, alternative explanation is found in the heterogeneity of adsorption on the zeolite [42, 43]. In the case of diffusion of a gas phase in porous medium, the gas is present as a continuum and can be treated as such. But, in the case of a zeolite, the adsorbed phase can be very well segregated. A well-known example is the adsorption of alkanes in zeolite MFI. Adsorption of butane and hexane yields a clear two-step isotherm, where the two steps can be related to adsorption in the channels and intersections of MFI, respectively [44, 45]. In a similar manner, segregated adsorption is found between adsorption sites in the cage and in the connecting window for strongly confined molecules in small-pore zeolites consisting of cages connected by windows [17]. Under the assumption that molecules at the window determine mass transport in the zeolite pores, the observed loading dependency could be explained [42].

The last type of diffusivity behavior, type IV from Figure 13.3, is, for example, found for linear hydrocarbons in zeolite MFI [46] and seems to be related to the presence of two distinct adsorption sites in this zeolite system, that is, when the first of the two sites becomes fully occupied, the diffusivity is strongly reduced.

### 13.3.2 Single-File Diffusion

Another very interesting phenomenon is single-file diffusion. Recently, this topic has been thoroughly reviewed by Kärger [47]. This type of diffusion is characterized by the fact that molecules are not able to pass each other and maintain their initial sequence in time, which can be easily envisaged in 1D zeolite pores. Such a system is strongly correlated and the diffusion process is strikingly different from ordinary diffusion.

The mean square displacement ($\langle r^2(t) \rangle$) is proportional to the mobility factor ($F$) and the square root in time at sufficiently long observation times and infinite length [48]:

$$\langle r^2(t) \rangle = 2F\sqrt{t} \tag{13.16}$$

It is already intuitively clear that, in this type of diffusion, molecules can severely hinder each other and, therefore, the diffusivity will be low compared with ordinary self-diffusion. This is quantified in Eq. (13.16) where the mean square displacement increases with the square root in time and not with time, as found for "normal" self-diffusion (Eq. (13.7)).

In the simplest case, with equally spaced adsorption sites, the mobility factor is given by [48]

$$F = \lambda^2 \frac{1-\theta}{\theta} \frac{1}{\sqrt{2\pi\tau}} \tag{13.17}$$

which is a function of the mean residence time at an adsorption site ($\tau$) and the jump distance ($\lambda$). However, a real zeolite catalyst has a finite size and open ends and it has been found that the mean square displacement of such a system eventually becomes proportional with time, related by an effective diffusion coefficient dependent on the characteristic length of the channel (Eq. (13.18)) [49]. This type of diffusion is called *center of mass diffusion* and can be seen as a type of ordinary diffusion, although much slower.

$$\langle r^2(t) \rangle = 2D\frac{1-\theta}{\theta}\frac{\lambda}{L}t = 2D_{\text{eff}}t \tag{13.18}$$

Which model applies best is dependent on the length of the channel.

Experimental evidence of single-file diffusion was measured by microscopic methods. One of the first direct experimental evidence was provided by Gupta *et al.* [50] who measured the diffusivity of ethane on large $AlPO_4$-5 (AFI topology, 1D pore system) crystals by PFG-NMR. The crystals were large enough to measure the dependency of the mean square displacement with the square root of time. Also with QENS, direct evidence has been claimed. Jobic *et al.* [51] could describe their QENS spectra of cyclopropane in $AlPO_4$-5 and methane in ZSM-48 (a disordered 1D pore system) at higher loadings.

Single-file diffusion affects the concentration profiles inside the catalyst and, consequently, the catalyst effectiveness. Kärger *et al.* [48] introduced a characteristic diffusion time to extend its use to single-file systems:

$$\phi = \sqrt{3k\tau_{\text{intra}}} \tag{13.19}$$

where the intracrystalline diffusion time ($\tau_{\text{intra}}$) for single-file diffusion is given by [48]:

$$\tau_{\text{intra}} = \frac{L^2}{12D} \tag{13.20}$$

With insertion of an appropriate description of the diffusivity (Eq. (13.18) in the Thiele modulus, a qualitative description of a model system of a first-order irreversible reaction in a single-file diffusion regime was obtained. It must be noted that compared with ordinary diffusion at equal diffusivities, a much higher Thiele modulus is obtained (compare Eqs. (13.13) and (13.20)), leading to more pronounced diffusion limitations in the case of single-file diffusion.

Although the existence of single-file diffusion has been demonstrated for several ideal cases, structural crystal defects and intracrystalline barriers can diminish the impact of single-file diffusion. For ethane diffusion in $AlPO_4$-5, Gupta *et al.* [50] found single-file diffusion using PFG-NMR, but Jobic *et al.* [51] found normal diffusion by QENS measurements. As reason for this difference, a difference in crystal batches was suggested [51]. Moreover, crystallites used in catalysis are often small and their internal transport may very well be described by the center of mass diffusion, a form of normal diffusion (Eq. (13.18)). Furthermore, the adsorbate concentrations in the zeolite at the relevant catalytic conditions are typically low (high temperature), a regime where single-file diffusion will not be very pronounced, when present.

Evidence for single-file diffusion under catalytic conditions is only indirectly obtained. Zeolite mordenite is a well-known catalyst with a 1D pore structure, for which single-file effects have been claimed in catalysis (e.g., [52, 53]).

As a final remark, besides negative impacts on catalysis, there have also been attempts to benefit from single-file diffusion. There appears to be a theoretical basis to exploit single-file diffusion to enhance the reactivity under certain conditions by this type of molecular traffic control [54]. Selectivity can be enhanced enormously by zeolite coatings on porous materials [55] as well as on zeolite catalysts [56].

### 13.3.3
### Surface Barriers

Surface barriers in zeolite systems have been discussed and used to explain diffusivity results from the late [57]. Recent advances in measuring techniques (Section 13.4), particularly interference microscopy, have provided overwhelming direct evidence for the presence of surface barriers for several zeolite-adsorbate systems (e.g., [26, 58]). The commonly applied assumption that the adsorbed concentration at the boundary of the crystal is in equilibrium with the gas phase concentration is violated in this case. Figure 13.4 shows clearly that in the case of i-butane in silicalite-1 (MFI), no surface barriers are present (Figure 13.4b) and in the case of methanol in ferrierite, (FER) this is clearly the case (Figure 13.4a): the concentration at the boundary of the crystal is not in equilibrium with the gas phase during the uptake.

The exact nature of the surface barrier is not completely clear. Part of the explanation stems from the idea that molecules suffer from restrictions when entering the pores of the zeolite. From this point of view, molecules do not enter the pores of the zeolite directly but a weak surface adsorption step is involved in the adsorption process [60] as already proposed by Barrer [61]. The influence of this type of transport barrier has been reduced by Reitmeier *et al.* [62] by coating an amorphous silica layer (~1–1.5 nm) on top of ZSM-5 crystals. With this coating, the local adsorption at the pore entrance and hence the uptake rate into the zeolite was increased.

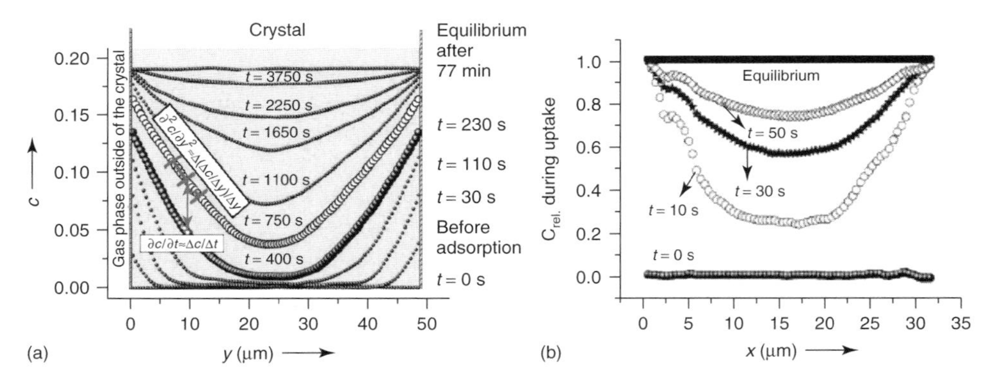


**Figure 13.4** Transient intracrystalline concentration profiles of methanol in FER (a) and i-butane in silicalite-1 (b) measured by interference microscopy. Figures taken from [26] and [59], respectively (Copyright Wiley-VCH Verlag Gmbh & co. KGaA. Reproduced with permission).

This is only one side of the story because also during desorption, surface barriers have been observed [63], indicating that it is not just a matter of entering the crystal. Tzoulaki *et al.* [64] have shown that traces of water form a surface barrier for isobutane entering silicalite-1, indicating that also other adsorbates can introduce additional surface barriers.

Although many of the zeolite–guest systems investigated by interference microscopy have been shown to possess surface barriers, it is not clear when barrier effects are important or not. Crystals that exhibit surface barriers may be treated to remove these [59].

Two concepts have been introduced to describe the barrier effects: surface sticking probability and a surface permeability [65]. The surface permeability describes the transport of a molecule from the surface into the zeolite pore, and the sticking coefficient represents the chance that a molecule that hits the surface also sticks to the surface.

The surface permeability ($\alpha$) relates the flux through the surface to the concentration difference over the barrier, and can be considered as a mass transfer coefficient:

$$N_i^{\text{surf}} = \alpha_i(C_{i,\text{eq}} - C_{i,\text{surf}}) \qquad (13.21)$$

Using this approach, an estimation of the surface permeability can be made using time-resolved concentration profiles generated by Interference Microscopy (IM). It is found that, analogously to the diffusivity, the surface permeability is dependent on the concentration [66], and accurate predictions cannot be made up to now.

The sticking coefficient can vary by orders of magnitude. It can be very close to 1, as observed for *n*-butane in silicalite-[67], or be several orders lower, as found by Jentys *et al.* [68] for benzene, toluene, and xylene on ZSM-5. Clear guidelines for which systems low sticking coefficients can be expected are still lacking. But the sticking probability is clearly a phenomenon that can impose limitations on the mass transport into the zeolite.

Whether surface barriers become apparent is clearly dependent on the relation with the intracrystalline diffusivity. An analogy with the Biot mass number (Bi) for porous catalyst particles can be drawn. It represents the ratio of the resistance against effective diffusion inside a porous catalyst pellet and that of transport through the fictitious fluid film surrounding the particle. Following a similar approach for a zeolite particle, the ratio of the characteristic times of intracrystalline diffusivity and surface permeability can be introduced [69] as

$$\frac{\tau_{\text{intra}}}{\tau_{\text{surf}}} = \frac{L\alpha}{D} = Bi_{\text{zeolite}} \qquad (13.22)$$

So, in relation to reaction, an extra resistance is found besides the internal diffusion resistance: the surface barrier. It is evident that an optimal catalytic process meets the following criterion (provided the parameters are all based on identical concentrations):

$$\begin{gathered} \tau_{\text{intra}}, \tau_{\text{surf}} \ll \tau_{\text{rxn}} \\ \frac{L^2}{D}, \frac{L}{\alpha} \ll \frac{1}{k} \end{gathered} \qquad (13.23)$$

Depending on the nature of the barrier, reduction seems possible in certain cases. As previously noted, Reitmeijer *et al.* [62] succeeded by coating a thin top layer of amorphous silica on top of the crystals. Tzoulaki *et al.* [59] obtained surface barrier free silicalite-1 crystallites by an alkaline treatment, removing a layer of amorphous silica that was the suspected surface barrier. These two examples form a nice contrast, illustrating the need for further elucidation.

The influence of the surface barrier on the catalyst effectiveness will be similar as an external mass transfer limitation: a reduced surface concentration that reduces the catalyst effectiveness. This effect of external barriers can be easily extended to intracrystalline barriers. Catalysis on the subcrystal level will be discussed in more detail in Section 13.3.4 and Chapter 8, where the existence and influence of intracrystalline barriers are nicely demonstrated.

### 13.3.4
### The Thiele Concept: A Useful Approach in Zeolite Catalysis?

The Thiele concept, as described in Section 10.2.6, is very useful and popular because of its conceptual simplicity. In view of the peculiarities of zeolite systems presented in the previous sections, in which they clearly distinguish themselves from meso- and macroporous catalysts, one may wonder if this concept is also useful for zeolite catalysis. The important peculiarities are the loading dependency of diffusivity, single-file diffusion, surface barriers, and mixture diffusion.

The concentration dependence of the diffusivity (Section 13.3.1) can lead to poor effectiveness predictions when a constant diffusivity is applied. Ruthven [70] accounted for the loading dependency in the Thiele modulus and derived some analytical solutions for some types of loading dependency. However, analytical expressions for each type of loading dependency do not seem to be very practical or feasible.

One may wonder, however, if the effect of loading dependency prohibits the use of the Thiele concept. When the diffusivity is measured under practical conditions, an average diffusivity is obtained that may represent the system reasonably well. More importantly, when the loading in the zeolite is low, which is very common for catalytic conversion at elevated temperatures, these loading effects are less significant and a constant diffusivity suffices.

Regarding single-file diffusion, it has been shown that with a modified Thiele modulus, a qualitative description of the catalyst effectiveness could be given (Section 13.3.2).

Surface barriers can be treated as an external mass transport resistance. The surface permeability and sticking coefficient can, however, only be determined by a limited number of techniques. Macroscopic techniques do not seem to be able to discriminate the surface barrier and internal diffusion [26]. A diffusivity determined by macroscopic techniques will then be an apparent quantity, but can be used for engineering purposes to estimate the zeolite's performance.

In Section 13.2.3, it was pointed out that diffusion in mixtures can be significantly different from that in pure components. In the Thiele approach, only the reactant single-component diffusivity is used, neglecting interactions with other adsorbates and counterdiffusion effects in small pores. Competitive adsorption and diffusion effects can negatively impact the catalyst effectiveness [15]. A detailed investigation of these effects was carried out recently by Hansen *et al.* [71] who analyzed the diffusion limitations in the alkylation of benzene over H-ZSM-5, combining different simulation tools. The usual approach to determine effectiveness factors for reactions in porous media, assuming a constant effective diffusivity, may lead to substantial deviations. A possible solution may be to determine the effective diffusivity at conditions relevant for practice.

As a general conclusion, it appears that in many cases, the Thiele concept can be applied well for zeolite catalysts and remains a very valuable tool in the interpretation of reaction data as follows from several studies [72–75] that have successfully applied this approach. This approach should be applied with caution, however, for systems with significant concentration levels in the zeolite where deviations can be expected.

## 13.4
## Pore Structure, Diffusion, and Activity at the Subcrystal Level

One of the main challenges still remaining in the field of molecular transport in zeolites is to match diffusivities measured by different techniques with catalytic performance data. The problem is twofold; first, order-of-magnitude differences between the values of diffusivities determined by different techniques (macro- vs. microscopic) are frequently reported for the same guest–host systems. As a consequence, it is extremely difficult to directly incorporate diffusion data into catalysis studies: no more trends can be extracted from pure diffusion experiments, while the quantitative values relevant for catalysis are still not clear [76].

The second problem arises from the differences between different batches of the same zeolite. In the early 1970s, the sentence "*your ZSM-5 is not mine*" was coined. On the basis of these recent studies, this quote can be extended to the rest of the zeolite topologies: the *time-ago* believed single-crystal zeolites have been shown to be composed of several intergrowth building blocks. The interfaces of these subunits constitute diffusion boundaries due to a potential mismatch in the alignment of the microporous network or a different pore orientation. This mismatch will be different for each crystal and batch of zeolite. This fact has very important consequences for catalysis, since certain regions of the zeolite crystals may be inaccessible for reactant molecules [28].

During the past years, excellent papers have been published in the literature aiming at unraveling the orientation of the complex zeolite channel networks (of mainly, MFI, AFI, and chabasite (CHA)) by identifying the different building blocks [77–79]. Several techniques like AFM, optical and fluorescence microscopy, electron backscatter diffraction/focused ion beam, FTIR, and Raman have been applied (see Chapter 8 for more details). In addition, the catalytic behavior of such crystals have been studied by spatial and time-resolved techniques that complement the structural studies and demonstrate the nonuniform catalytic behavior of such zeolitic crystals [28, 80–85]. These studies indicate a trend that the larger the zeolite crystal, the less perfect its pore structure.

The first indications for building units involved in crystal growth of ZSM-5 were already published in the early 1990s [86]. By combining transmission electron microscopy (TEM) and scanning electron microscopy (SEM). Hay *et al.* reported that most frequently adjoining ZSM-5 crystals are rotated by 90° around a common *c* axis with an intergrowth nucleating from small areas on (010) faces of growing crystals. On (100) faces of large crystals, ramps were also observed in association with impurities. Consistent with these observations, one of the first models was proposed for crystal growth.

Along with the development of optical techniques, their application in the study of zeolite crystals has growth during the last decade. On pioneering work, Koricik *et al.* applied light microscopy to investigate sorption and mass transport phenomena in zeolites together with peculiarities of crystal morphology via coloring of zeolite crystals (MFI) using iodine indicators [87, 88]. Some years later, the group of Kärger started the use of interference microscopy [89] in combination with FTIR [58] to elucidate macroscopic adsorbate distributions and crystal intergrowth. In this approach, the high spatial resolution of interference microscopy is complemented by the ability of FTIR spectroscopy to pinpoint adsorbates by their characteristic IR bands. For the first time, two-dimensional concentration profiles of an unprecedented quality were reported, showing an inhomogeneous distribution of adsorbate [58]. These inhomogeneous profiles were attributed to regular intergrowth effects in CrAPO-5 (AFI structure).

The space and time-resolved study of the detemplation process has been shown as a powerful technique for determining the intergrowth structure on the basis of *in situ* mapping of the template-removal process in individual zeolite crystals. When the formation of light-absorbing and -emitting species during the heating process is monitored by a combination of optical and confocal fluorescence microscopy, as the accessibility of the porous network in the subunits varies, the individual building blocks can be readily visualized by monitoring the template-removal process in time. This concept has been successfully applied to four different zeolite crystals: CrAPO-5 (AFI structure), SAPO-34 (CHA structure), SAPO-5 (AFI structure), and ZSM-5 (MFI structure) [78]; the proposed intergrowth structures are depicted in Figures 13.5 and 13.6.

Because of its high industrial relevance, zeolite ZSM-5 has been widely studied in order to elucidate if its coffin-like crystals are the product of two or three interpenetrating crystals (two- and three-component models, Figure 13.6). The two-component model can be regarded as two interpenetrating crystals rotated by 90° around the common *c* axis [87, 90–92]. This model consists of two central and four side pyramidal subunits. In the so-called three-component model, the crystallographic axes maintain the same orientation across the entire crystal [93]. From the catalysis engineering point of view, there are important consequences arising from the accessibility of the pores, since according to the two-component model and due to the changed orientation within the pyramidal components, the

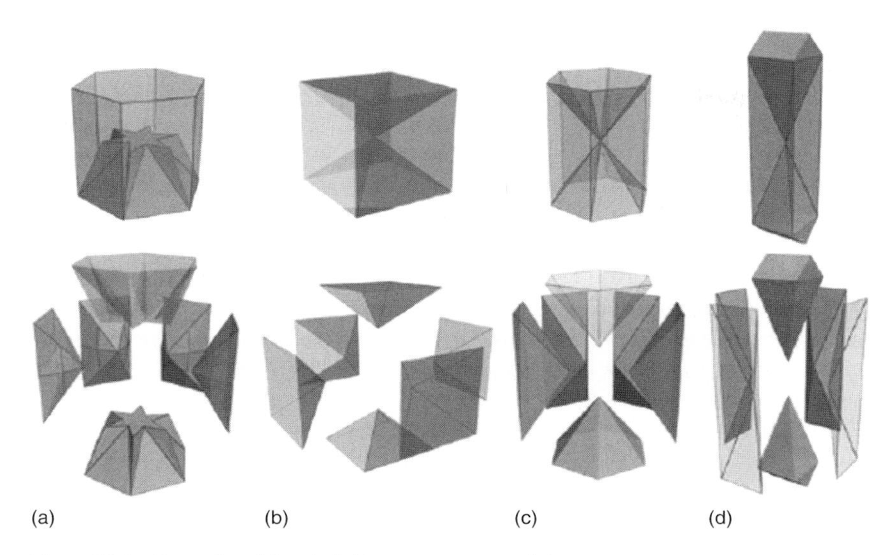


**Figure 13.5** Normal and "exploded" representation of the intergrowth structures of different zeolite crystals as proposed by [78]: (a) CrAPO-5 (front subunits are not shown), (b) SAPO-34, (c) SAPO-5 (front subunits are not shown), and (d) ZSM-5.

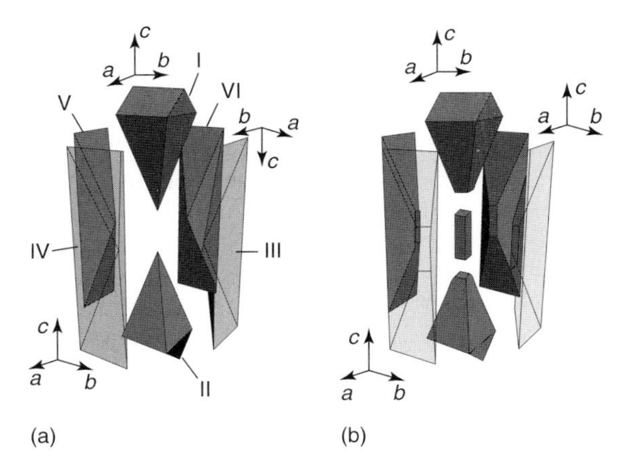


**Figure 13.6** "Exploded" representation of (a) two-component and (b) three-component models of coffin-shaped ZSM-5 crystals. Orientations of crystal axes in the individual subunits are given. Subunits in the two-component models are denoted I–VI (see text). The straight pores align with the *b* crystal axis, whereas the zigzag pores extend along the *a* crystal axis. (Adapted from [77].)

sinusoidal channels appear not only at the (100) faces of the crystallite but also at its hexagonal (010) facets, implying that there is hardly any access to the straight channels from the outer crystal surface.

Experimental results reported by different groups [77, 79] clearly indicate that differences are due to the zeolite batches, that are intrinsically different, confirming the quote "*your zeolite is not mine.*" In view of the unlike pore orientation of the studied samples, probably together with differences in Al distribution along the crystals, differences in the catalytic as well as in the diffusion behavior are expected.

In contrast to many surface science spectroscopic techniques, fluorescence microscopy is capable of studying diffusion and catalysis in zeolite pores of crystals with high 3D spatial and temporal resolution [94]. If fluorogenic probes that are smoothly transformed into fluorescent molecules on chemical transformation are chosen, reaction and diffusion can be followed in a spatial and time-resolved manner [27, 28, 81, 82, 84, 85, 94, 95].

The first examples of application of such techniques dealt with studying the acid-catalyzed oligomerization of furfuryl alcohol on HZSM-5 [84] and H-MOR [94]. Furfuryl alcohol oligomerization starts with alkylation of one furfuryl alcohol molecule by another in an electrophilic aromatic substitution (EAS). After some subsequent acid-catalyzed reaction steps, a family of fluorescent compounds is formed. Especially interesting are the results obtained for H-MOR, fluorescence time lapse measurements (Figure 13.7) clearly show the evolution of catalytic activity from two opposing crystal faces, whereas no light is emitted from the rest of the outer zeolite surface. Transmission images evidenced that the reactive crystal faces are the (001) planes. As the reaction carries on, fluorescence propagates along the (001) direction, corresponding to the 12-ring channels in mordenite. As furfuryl alcohol is too large to access the eight-ring pores, no fluorescence develops from the other faces of the crystal; this result represents an excellent picture of diffusion

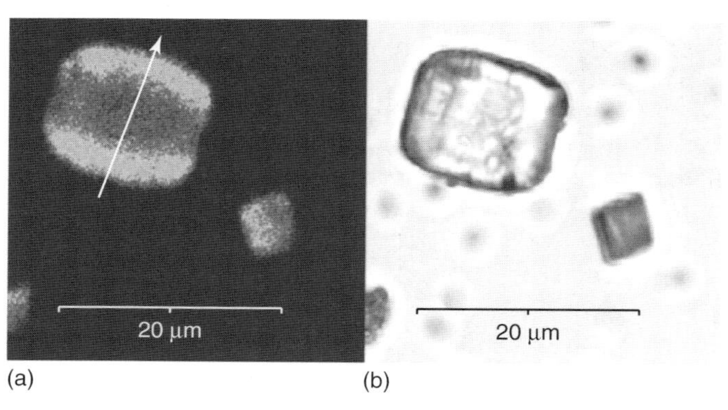


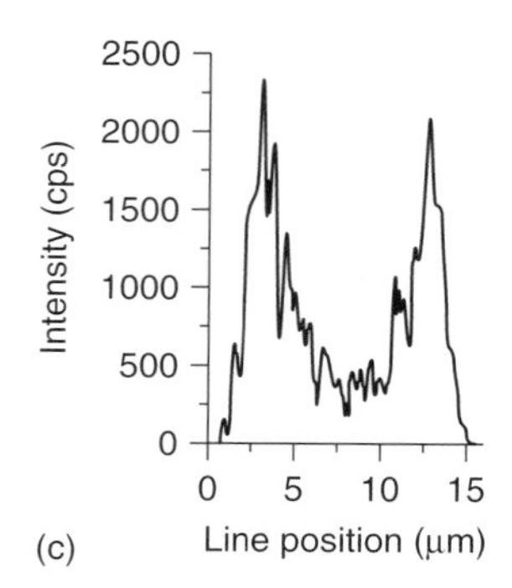


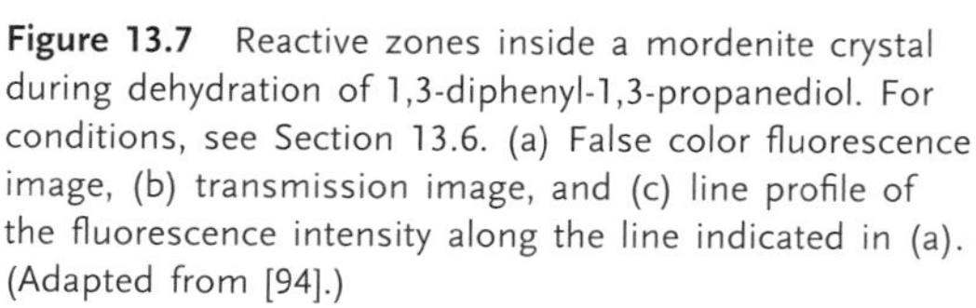

**Figure 13.7** Reactive zones inside a mordenite crystal during dehydration of 1,3-diphenyl-1,3-propanediol. For conditions, see Section 13.6. (a) False color fluorescence image, (b) transmission image, and (c) line profile of the fluorescence intensity along the line indicated in (a). (Adapted from [94].)

limitations associated to the 1D pore system of MOR. When the same reaction was studied in a 3D pore system like ZSM-5, after initiation of the reaction, fluorescence gradually spreads into the crystal starting from the (100) and (101) planes [84].

Other liquid-phase reactions like the acid-catalyzed oligomerization of styrene derivatives inside the pores of ZSM-5 crystals were studied by using optical and fluorescence microspectroscopy and application of polarized light [28, 81, 83], yielding similar insight in diffusion barriers and pore orientations in the crystal.

All these results confirm the spatially nonuniform catalytic behavior of the studied zeolite samples, with specific parts of the crystals being hardly accessible to reactant molecules and pore orientations that deviate from that expected on the basis of the crystal orientation.

## 13.5 Improving Transport through Zeolite Crystals

Limitations due to restricted access, slow transport, and diffusion boundaries provoke a low catalyst utilization. In many cases, zeolites are victims and executioners. When size selectivity is the key of a process, large zeolite crystals are preferred in order to decrease external surface reaction contributions; the methylation of toluene on ZSM-5 [56] is a clear example, where operation under strong diffusion limitation conditions enhances the overall selectivity of the process. This fact brings the accessibility problem to the extreme and limits industrial plants to operate far below their full potential, although it can also be utilized to enhance the selectivity of the process, as shown by Van Vu *et al.* [56], who coated H-ZSM-5 crystals with various Si-to-Al with polycrystalline silicalite-1 layers. When applied to the alkylation of toluene with methanol, the silicalite coating significantly enhanced *para*-selectivity up to 99.9% under all reaction conditions. The enhanced *para*-selectivity may originate from diffusion resistance through the inactive silicalite layer on the H-ZSM-5, resulting in increased diffusion length.

It has become clear from Section 13.3.4 that the classical approach to determine effectiveness factors for reactions in porous media can be inaccurate when applied to zeolites. However, at least from a qualitative point of view, their use may help gaining insight into the performance and the limitations of zeolite-catalyzed reactions, and utilized for the design of *zeotypes* hampered less by diffusion limitations. According to Eq. (13.13), if a small Thiele modulus is needed, two different strategies can be followed: shortening the diffusion length $L$ and/or enhancing the effective diffusivity $D_{\mathrm{eff}}$ in the zeolite pores. The latter strategy has led to the development of ordered mesoporous materials (OMMs) [96], where diffusion is governed by Knudsen or bulk regimes. This approach is valid for processes where bulky molecules are involved that would exceed the size of the pores and cages of the zeolites or when size selectivity is not a priority. However, OMMs still suffer from a poor thermal stability due to their, in general, thin walls. Furthermore, the performance of their active sites is usually far below that of zeolites due to the amorphous character of their walls [96–98].

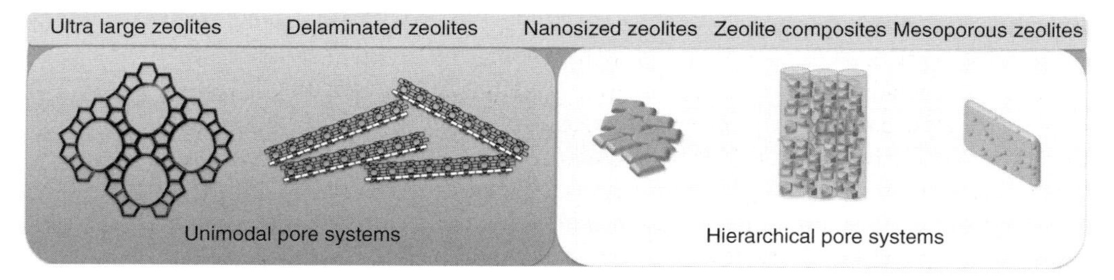


**Figure 13.8** Different zeotypes with enhanced transport characteristics. Ultralarge-pore zeolites (usually 12MR) present an increased effective diffusivity, whereas delaminated, nanosized, composites, and mesoporous zeolites present shorter diffusion path lengths.

Parallel to the development of OMMs, a great effort has been devoted to enhance diffusion in zeolites while maintaining the other intrinsic properties of the material (Figure 13.8). The synthesis of new structures with large and ultralarge pores [99–103], the modification of the textural properties of known frameworks by creating mesopores via synthetic [104, 105] or postsynthetic approaches (mainly acid (dealumination) or basic (desilication) leaching) [106–111], the synthesis of small zeolite crystals with a more convenient external to internal surface ratio [112], the synthesis of micromesoporous composites [113] by using mixed templated systems [114, 115] or by recrystallization [116–119] approaches, and the delamination [120–126] of crystalline-layered structures are the most often followed approaches [127].

Ultralarge-pore zeolites present substantially wider micropores than regular zeolite structures, enhancing the effective diffusivity, whereas in the other approaches, the diffusion path length for reactants and products is shorter: nanosized zeolites have, in addition to nanosized zeolitic pores, intercrystalline pores or voids; zeolite composites are composed of zeolite crystals supported on a material that is typically mesoporous or macroporous. Mesoporous zeolite crystals exhibit intracrystalline mesopores, and delaminated materials are formed by single layers organized in a "*house of cards*"-like structure, where reaction can be considered to take place at the pore mouth, in fact one could no longer distinguish a pore. Thus, ultralarge-pore and delaminated zeolites possess unimodal systems, whereas the other materials are characterized by featuring hierarchical pore systems, since they combine the intracrystalline micropores with larger pores that can be either intercrystalline or intracrystalline. Because of this optimization, the diffusional resistance in these cases is mainly determined by the transport in the larger pores over longer distances.

Ultralarge-pore zeolites like SSZ-53 [102] and delaminated zeolites like ITQ-2 [123] have been successfully applied to the hydrocracking of bulky molecules under mild conditions, showing an outstanding performance due an improved transport of molecules and a higher acidity. The partial conversion of a mesoporous material TUD-1 into BEA or Y-type zeolite and application in alkylation or hydrocracking indicated an effective diffusion improvement by up to 15 times [128].

Some destructive methods like acid leaching or steaming (dealumination) turned out to be less efficient to improve transport than expected. In the case of dealuminated ultrastable Y (USY) pellets (FCC catalyst), the rate of molecular exchange between catalyst particles and their surroundings is primary determined by the intraparticle diffusivity, that is, at the reaction temperature, diffusion is controlled by the macropores and not by the micro- or mesopores [129]. Moreover, when dealuminated crystals are used instead of catalyst pellets, the mesopores do not form an interconnected network, and the diffusion of guest molecules through the crystals via only mesopores is not possible [130].

Another method, desilication [131] seems to be much more effective, yielding to a greatly improved physical transport in the zeolite crystals as revealed by transient uptake experiments of neopentane in ZSM-5 [108] crystals and diffusion studies of *n*-heptane, 1,3-dimethylcyclohexane, *n*-undecane in mesoporous ZSM-12 [132] and diffusion and adsorption studies of cumene in mesopore structured ZSM-5 [133]. Up to 3 orders of magnitude-enhanced rates of diffusion were concluded in the hierarchical systems as compared with their purely microporous precursors due to improved accessibility and a distinct shortening of the micropores. Moreover, Brønsted acidity seems to be preserved, in contrast to dealumination. Catalytic testing of various mesoporous zeolites has shown the effectiveness of the desilication approach in the liquid-phase degradation of HDPE, cumene cracking, and methanol to gasoline on desilicated ZSM-5 [34]. A recent *in situ* microspectroscopic study on the oligomerization of styrene derivatives revealed an enhanced accessibility of the micropores in the hierarchical ZSM-5 zeolites obtained by desilication [81], but still a nonuniform catalytic behavior was discovered, due to a nonuniform distribution of the aluminum over the zeolite crystal, which seems to be an intrinsic phenomenon for this material.

Templated mesoporous zeolites and zeolite nanoparticles deposited on different supports have been widely applied in catalysis. The activation energy of the vapor-phase benzene alkylation with ethylene to ethylbenzene was found to be higher for a carbon-templated ZSM-5 than that of the purely microporous zeolite (77 vs 59 kJ $mol^{-1}$); this fact was attributed to the alleviated diffusion limitation in the case of the mesoporous crystals. Hierarchical mesoporous BEA zeolite templated with a mixture of organic ammonium salts and cationic polymers showed a higher activity in the alkylation of benzene with propan-2-ol than a microporous BEA sample with the same Si/Al ratio [134]. Catalytic test reactions on the oxidation of 1-naphthol over TS-1, Ti-coated MCF, and MCF materials coated with (TS-1) nanoparticles revealed increased 1-naphthol conversion and activity for the TS-1-coated MCF materials compared with the TS-1 zeolite due to the presence of mesopores. Moreover, an increased selectivity, hydrothermal stability, and the absence of titania leaching were observed for the TS-1-coated MCF materials in contrast to the Ti-coated MCF materials because titania was embedded in the zeolitic framework present in the TS-1 nanoparticles [135]. Catalytic tests on MAS-7 and MTS-9 (mesoporous materials made up of zeolite beta and TS-1 precursor particles, respectively) in the cracking and hydroxylation (with $H_2O_2$) of different small and bulky molecules (cumene, phenol, trimethylphosphine (TMP), etc.)

showed high activity. The acylation of different amino derivatives with fatty acids is carried out smoothly and under green conditions when using UL-MFI type (mesoporous ZSM-5) as catalyst [136].

## 13.6 Concluding Remarks and Future Outlook

Zeolites are commonly applied versatile catalysts, but due to their small pores, they are also frequently associated with diffusion limitations. Because of their small pore size, mass transport is characterized by adsorbed phase diffusion (zeolitic diffusion). Since each zeolite has its specific pore connectivity and structure, much effort has been put in measuring and understanding the mass transport phenomena in these systems. Recently, new space and time-resolved measurement techniques, like interference, IR, and fluorescence microscopy, have been introduced and provided a wealth of new insights in zeolite diffusion and catalysis on the (sub)crystal level.

In contrast to gaseous molecular and Knudsen diffusion, zeolitic diffusion involves very narrow confinements and high concentrations, leading to a loading-dependent diffusivity. Additionally, at high concentrations, significant competitive adsorption and hindrance ("friction," "exchange") effects between guest molecules may occur. When the pore structure of a zeolite is 1D and molecules are unable to pass each other, single-file diffusion is the governing transport mechanism, which is much slower than ordinary diffusion. These phenomena, however, are not very prominent at low loadings, which are often in the case for catalysis at elevated temperatures, but dominant in liquid-phase reactions.

Particularly, the microscopic techniques have revealed the existence of surface and internal barriers for several zeolite crystals. This was suggested decades ago but not until recently, it has been demonstrated so clearly for a variety of zeolites. Indications when zeolite–guest system barriers can be expected are not generalized, but their occurrence clearly increases with crystal size, having a twofold detrimental effect on the catalyst effectiveness, due to the combined lower effective diffusivity and the larger particle size.

These new experimental techniques are able to unravel the orientation of the complex zeolite channel networks of "single crystals," deviating from that expected on the basis of their geometry. Often they are composed of subunits with their own pore orientation, explaining important differences between different batches of the same zeolite topology. Since these techniques allow monitoring in time and space, local concentrations inside crystals can be used well for quantification of surface barriers and diffusivities inside zeolite particles.

So, 250 years after the word zeolite was coined and more than 50 years after the first industrial application of such materials was developed, still new insights are obtained at the subcrystal level, impacting the relation "pore structure–diffusion–activity."

A well-established method to account for intraparticle diffusion in porous catalysts and quantify its impact on catalyst performance is the Thiele approach. The application has been very successful in catalysis and seems to be extendable to zeolite catalysts quite well at low concentrations. However, at higher concentrations, the loading dependency of diffusion, competitive adsorption effects, and strong hindrance ("friction") can introduce severe deviations in the description of the catalyst effectiveness.

Shortening the diffusion distance in the zeolite crystal is the best solution to utilize the intrinsic properties of the zeolite to the fullest, and many synthetic and posttreatment approaches are being explored to realize this. The incorporation of the nanosized zeolite structures in meso- and macroscopic bodies may alleviate the diffusional resistance in the zeolite, but those in the catalyst particle may remain, for which diffusion measurements remain essential.

The evolution toward zeolitic materials with improved transport is a well-established field of research, and the proofs of principle have been given. Zeolites are still materials with further design possibilities. In relation to diffusion, a careful analysis of the characteristic times of the various phenomena in a catalyst particle (reaction, diffusion in zeolite-, micro-, meso-, and macropores, barrier transport) may guide the way to compose the optimal hierarchical structure of a catalyst particle for use in practice. This implies careful experimentation and interpretation, including diffusion, on all these aspects, in combination with molecular modeling.

## References


1. Corma, A. (2003) *J. Catal.*, **216**, 298–312.
2. Plank, C.J., Hawthorne, W.P., and Rosinski, E.J. (1964) *Ind. Eng. Chem. Prod. Res. Dev.*, **3**, 165–169.
3. Taramasso, M., Perego, G., and Notari, B. (1983) Preparation of porous crystalline synthetic material comprised of silicon and titanium oxides. US Patent 4,410,501.
4. Song, C. (2002) *Cattech*, **6**, 64–77.
5. Smit, B. and Maesen, T.L.M. (2008) *Chem. Rev.*, **108**, 4125–4184.
6. Ruthven, D.M. (1984) *Principles of Adsorption and Adsorption Processes*, John Wiley & Sons, Inc., New York.
7. Froment, G.F. and Bischoff, K.B. (1990) *Chemical Reactor Analysis and Design*, John Wiley & Sons, Inc., New York.
8. Kapteijn, F., Zhu, W., Moulijn, J.A., and Gardner, T.Q. (2005) Zeolite membranes: modeling and application, in *Structured Catalysts and Reactors*, Chapter 20 (eds A. Cybulski and J.A. Moulijn), Taylor & Francis Group, Boca Raton, pp. 701–747.
9. Krishna, R. and Wesselingh, J.A. (1997) *Chem. Eng. Sci.*, **52**, 861–911.
10. Krishna, R. (1993) *Gas Sep. Purif.*, **7**, 91–104.
11. Kapteijn, F., Moulijn, J.A., and Krishna, R. (2000) *Chem. Eng. Sci.*, **55**, 2923–2930.
12. van de Graaf, J.M., Kapteijn, F., and Moulijn, J.A. (1999) *AIChE J.*, **45**, 497–511.
13. Krishna, R. and van Baten, J.M. (2009) *Chem. Eng. Sci.*, **64**, 870–882. doi: 10.1016/j.ces.2009.03.047.
14. Myers, A.L. and Prausnitz, J.M. (1965) *AIChE J.*, **11**, 121–127.
15. Baur, R. and Krishna, R. (2005) *Catal. Today*, **105**, 173–179.
16. Murthi, M. and Snurr, R.Q. (2004) *Langmuir*, **20**, 2489–2497.

17. Krishna, R. and van Baten, J.M. (2008) *Sep. Purif. Technol.*, **60**, 315–320.
18. Krishna, R. and van Baten, J.M. (2007) *Chem. Phys. Lett.*, **446**, 344–349.
19. Habgood, H.W. (1958) *Can. J. Chem. Rev. Can. Chim.*, **36**, 1384–1397.
20. Fueller, W.N., Schettler, P.D., and Giddings, J.C. (1966) *Ind. Eng. Chem. Res.*, **58**, 19–53.
21. Krishna, R. and van Baten, J.M. (2008) *Microporous Mesoporous Mater.*, **109**, 91–108.
22. Kärger, J. and Ruthven, D.M. (1992) *Diffusion in Zeolites*, John Wiley & Sons, Inc.
23. Coppens, M.O., Keil, F.J., and Krishna, R. (2000) *Rev. Chem. Eng.*, **16**, 71–197.
24. Karger, J. (2003) *Adsorpt. J. Int. Adsorpt. Soc.*, **9**, 29–35.
25. Ruthven, D.M. (2008) in *Introduction to Zeolite Science and Practice* (eds J. Čejka, H. van Bekkum, A. Corma, and F. Schueth), Elsevier, pp. 737–786.
26. Karger, J., Kortunov, P., Vasenkov, S., Heinke, L., Shah, D.R., Rakoczy, R.A., Traa, Y., and Weitkamp, J. (2006) *Angew. Chem. Int. Ed.*, **45**, 7846–7849.
27. Roeffaers, M.B.J., Sels, B.F., Uji-I, H., De Schryver, F.C., Jacobs, P.A., De Vos, D.E., and Hofkens, J. (2006) *Nature*, **439**, 572–575.
28. Kox, M.H.F., Stavitski, E., and Weckhuysen, B.M. (2007) *Angew. Chem. Int. Ed.*, **46**, 3652–3655.
29. Post, M.F.M. *et al.* (1991) in *Introduction to Zeolite Science and Practice* (eds H. van Bekkum, E.M. Flanigen, and J.C. Jansen), Elsevier, Amsterdam, pp. 391–443.
30. Jobic, H., Schmidt, W., Krause, C.B., and Karger, J. (2006) *Microporous Mesoporous Mater.*, **90**, 299–306.
31. Thiele, E.W. (1939) *Ind. Eng. Chem.*, **31**, 916–920.
32. Post, M.F.M., Vanthoog, A.C., Minderhoud, J.K., and Sie, S.T. (1989) *AIChE J.*, **35**, 1107–1114.
33. Post, M.F.M., van Amstel, J., and Kouwenhoven, H.W. (1984) in *Proceedings 6th International Zeolite Conference, Reno, 1983* (eds D. Olson and A. Bisio), Butterworth, Guildford.
34. Perez-Ramirez, J., Christensen, C.H., Egeblad, K., Christensen, C.H., and Groen, J.C. (2008) *Chem. Soc. Rev.*, **37**, 2530–2542.
35. Karger, J. and Pfeifer, H. (1987) *Zeolites*, **7**, 90–107.
36. Skoulidas, A.I. and Sholl, D.S. (2003) *J. Phys. Chem. A*, **107**, 10132–10141.
37. Xiao, J.R. and Wei, J. (1992) *Chem. Eng. Sci.*, **47**, 1143–1159.
38. Pantatosaki, E., Papadopoulos, G.K., Jobic, H., and Theodorou, D.N. (2008) *J. Phys. Chem. B*, **112**, 11708–11715.
39. Krishna, R., van Baten, J.M., Garcia-Perez, E., and Calero, S. (2007) *Ind. Eng. Chem. Res.*, **46**, 2974–2986.
40. Xiao, J.R. and Wei, J. (1992) *Chem. Eng. Sci.*, **47**, 1123–1141.
41. Beerdsen, E. and Smit, B. (2006) *J. Phys. Chem. B*, **110**, 14529–14530.
42. van den Bergh, J., Ban, S., Vlugt, T.J.H., and Kapteijn, F. (2009) *J. Phys. Chem. C*. 17840–17850.
43. Coppens, M.O. and Iyengar, V. (2005) *Nanotechnology*, **16**, S442–S448.
44. Vlugt, T.J.H., Krishna, R., and Smit, B. (1999) *J. Phys. Chem. B*, **103**, 1102–1118.
45. Vlugt, T.J.H., Zhu, W., Kapteijn, F., Moulijn, J.A., Smit, B., and Krishna, R. (1998) *J. Am. Chem. Soc.*, **120**, 5599–5600.
46. Krishna, R. and van Baten, J.M. (2005) *Chem. Phys. Lett.*, **407**, 159–165.
47. Kaerger, J. (2009) *Mol. Sieves*, **7**, 329–366.
48. Karger, J., Petzold, M., Pfeifer, H., Ernst, S., and Weitkamp, J. (1992) *J. Catal.*, **136**, 283–299.
49. Hahn, K. and Karger, J. (1998) *J. Phys. Chem. B*, **102**, 5766–5771.
50. Gupta, V., Nivarthi, S.S., Mccormick, A.V., and Ted Davis, H. (1995) *Chem. Phys. Lett.*, **247**, 596–600.
51. Jobic, H., Hahn, K., Karger, J., Bee, M., Tuel, A., Noack, M., Girnus, I., and Kearley, G.J. (1997) *J. Phys. Chem. B*, **101**, 5834–5841.
52. de Gauw, F.J.M.M., van Grondelle, J., and van Santen, R.A. (2001) *J. Catal.*, **204**, 53–63.
53. Lei, G.D., Carvill, B.T., and Sachtler, W.M.H. (1996) *Appl. Catal. A: Gen.*, **142**, 347–359.
54. Neugebauer, N., Braeuer, P., and Kaerger, J. (2000) *J. Catal.*, **194**, 1–3.

55. Nishiyama, N., Ichioka, K., Park, D.H., Egashira, Y., Ueyama, K., Gora, L., Zhu, W.D., Kapteijn, F., and Moulijn, J.A. (2004) *Ind. Eng. Chem. Res.*, **43**, 1211–1215.
56. van Vu, D., Miyamoto, M., Nishiyama, N., Egashira, Y., and Ueyama, K. (2006) *J. Catal.*, **243**, 389–394.
57. Kocirik, M., Struve, P., Fiedler, K., and Buelow, M. (1988) *J. Chem. Soc., Faraday Trans. 1: Phys. Chem. Condens. Phases*, **84**, 3001–3013.
58. Lehmann, E., Chmelik, C., Scheidt, H., Vasenkov, S., Staudte, B., Karger, J., Kremer, F., Zadrozna, G., and Kornatowski, J. (2002) *J. Am. Chem. Soc.*, **124**, 8690–8692.
59. Tzoulaki, D., Heinke, L., Schmidt, W., Wilczok, U., and Karger, J. (2008) *Angew. Chem. Int. Ed.*, **47**, 3954–3957.
60. Reitmeier, S.J., Mukti, R.R., Jentys, A., and Lercher, J.A. (2008) *J. Phys. Chem. C*, **112**, 2538–2544.
61. Barrer, R.M. (1990) *J. Chem. Soc., Faraday Trans.*, **86**, 1123–1130.
62. Reitmeier, S.L., Gobin, O.C., Jentys, A., and Lercher, J.A. (2009) *Angew. Chem. Int. Ed.*, **48**, 533–538.
63. Kortunov, P., Heinke, L., Vasenkov, S., Chmelik, C., Shah, D.B., Karger, J., Rakoczy, R.A., Traa, Y., and Weitkamp, J. (2006) *J. Phys. Chem. B*, **110**, 23821–23828.
64. Tzoulaki, D., Schmidt, W., Wilczok, U., and Kaerger, J. (2008) *Microporous Mesoporous Mater.*, **110**, 72–76.
65. Karge, H.G. and Karger, J. (2009) *Mol. Sieves*, **7**, 135–206.
66. Heinke, L., Kortunov, P., Tzoulaki, D., and Karger, J. (2007) *Phys. Rev. Lett.*, **99**, 228301–228304.
67. Simon, J.M., Bellat, J.P., Vasenkov, S., and Karger, J. (2005) *J. Phys. Chem. B*, **109**, 13523–13528.
68. Jentys, A., Mukti, R.R., and Lercher, J.A. (2006) *J. Phys. Chem. B*, **110**, 17691–17693.
69. Heinke, L., Kortunov, P., Tzoulaki, D., and Karger, J. (2007) *Adsorpt. J. Int. Adsorpt. Soc.*, **13**, 215–223.
70. Ruthven, D.M. (1972) *J. Catal.*, **25**, 259–264.
71. Hansen, N., Krishna, R., van Baten, J.M., Bell, A.T., and Keil, F.J. (2009) *J. Phys. Chem. C*, **113**, 235–246.
72. Christensen, C.H., Johannsen, K., Toernqvist, E., Schmidt, I., and Topsoe, H. (2007) *Catal. Today*, **128**, 117–122.
73. Haag, W.O., Lago, R.M., and Weisz, P.B. (1981) *Faraday Discuss.*, **72**, 317–330.
74. Al-Sabawi, M., Atias, J.A., and de Lasa, H. (2008) *Ind. Eng. Chem. Res.*, **47**, 7631–7641.
75. Wang, G. and Coppens, M.O. (2008) *Ind. Eng. Chem. Res.*, **47**, 3847–3855.
76. Wloch, J. and Kornatowski, J. (2008) *Microporous Mesoporous Mater.*, **108**, 303–310.
77. Stavitski, E., Drury, M.R., de Winter, D.A.M., Kox, M.H.F., and Weckhuysen, B.M. (2008) *Angew. Chem. Int. Ed.*, **47**, 5637–5640.
78. Karwacki, L., Stavitski, E., Kox, M.H.F., Kornatowski, J., and Weckhuysen, B.M. (2007) *Angew. Chem. Int. Ed.*, **46**, 7228–7231.
79. Roeffaers, M.B.J., Ameloot, R., Baruah, M., Uji-I, H., Bulut, M., De Cremer, G., Muller, U., Jacobs, P.A., Hofkens, J., Sels, B.F., and De Vos, D.E. (2008) *J. Am. Chem. Soc.*, **130**, 5763–5772.
80. Stavitski, E., Kox, M.H.F., Swart, I., de Groot, F.M.F., and Weckhuysen, B.M. (2008) *Angew. Chem. Int. Ed.*, **47**, 3543–3547.
81. Kox, M.H.F., Stavitski, E., Groen, J.C., Perez-Ramirez, J., Kapteijn, F., and Weckhuysen, B.M. (2008) *Chem. Eur. J.*, **14**, 1718–1725.
82. Roeffaers, M.B.J., Ameloot, R., Bons, A.J., Mortier, W., De Cremer, G., de Kloe, R., Hofkens, J., De Vos, D.E., and Sels, B.F. (2008) *J. Am. Chem. Soc.*, **130**, 13516–1351.
83. Stavitski, E., Kox, M.H.F., and Weckhuysen, B.M. (2007) *Chem. Eur. J.*, **13**, 7057–7065.
84. Roeffaers, M.B.J., Sels, B.F., Uji-I, H., Blanpain, B., L'hoest, P., Jacobs, P.A., De Schryver, F.C., Hofkens, J., and De Vos, D.E. (2007) *Angew. Chem. Int. Ed.*, **46**, 1706–1709.
85. Roeffaers, M.B.J., Sels, B.F., Loos, D., Kohl, C., Mullen, K., Jacobs, P.A.,

Hofkens, J., and De Vos, D.E. (2005) *Chemphyschem*, **6**, 2295–2299.
86. Hay, D.G., Jaeger, H., and Wilshier, K.G. (1990) *Zeolites*, **10**, 571–576.
87. Kocirik, M., Kornatowski, J., Masarìk, V., Novak, P., Ziknov·, A., and Maixner, J. (1998) *Microporous Mesoporous Mater.*, **23**, 295–308.
88. Geus, E.R., Jansen, J.C., and van Bekkum, H. (1994) *Zeolites*, **14**, 82–88.
89. Geier, O., Vasenkov, S., Lehmann, E., Karger, J., Schemmert, U., Rakoczy, R.A., and Weitkamp, J. (2001) *J. Phys. Chem. B*, **105**, 10217–10222.
90. Price, G.D., Pluth, J.J., Smith, J.V., Bennett, J.M., and Patton, R.L. (1982) *J. Am. Chem. Soc.*, **104**, 5971–5977.
91. Weidenthaler, C., Fischer, R.X., Shannon, R.D., and Medenbach, O. (1994) *J. Phys. Chem.*, **98**, 12687–12694.
92. Weidenthaler, C., Fischer, R.X., and Shannon, R.D. (1994) *Zeolites and Related Microporous Materials: State of the Art 1994*, Elsevier Amsterdam, New York. vol. 84, pp. 551–558.
93. Agger, J.R., Hanif, N., Cundy, C.S., Wade, A.P., Dennison, S., Rawlinson, P.A., and Anderson, M.W. (2003) *J. Am. Chem. Soc.*, **125**, 830–839.
94. Roeffaers, M.B.J., Hofkens, J., De Cremer, G., De Schryver, F.C., Jacobs, P.A., De Vos, D.E., and Sels, B.F. (2007) *Catal. Today*, **126**, 44–53.
95. Mores, D., Stavitski, E., Kox, M.H.F., Kornatowski, J., Olsbye, U., and Weckhuysen, B.M. (2008) *Chem. Eur. J.*, **14**, 11320–11327.
96. Meynen, V., Cool, P., and Vansant, E.F. (2007) *Microporous Mesoporous Mater.*, **104**, 26–38.
97. Ciesla, U. and Schueth, F. (1999) *Microporous Mesoporous Mater.*, **27**, 131–149.
98. Taguchi, A. and Schueth, F. (2005) *Microporous Mesoporous Mater.*, **77**, 1–45.
99. Lobo, R.F., Tsapatsis, M., Freyhardt, C.C., Khodabandeh, S., Wagner, P., Chen, C.Y., Balkus, K.J., Zones, S.I., and Davis, M.E. (1997) *J. Am. Chem. Soc.*, **119**, 8474–8484.
100. Barrett, P.A., Diaz-Cabanas, M.J., Camblor, M.A., and Jones, R.H. (1998) *J. Chem. Soc., Faraday Trans.*, **94**, 2475–2481.
101. Wessels, T., Baerlocher, C., McCusker, L.B., and Creyghton, E.J. (1999) *J. Am. Chem. Soc.*, **121**, 6242–6247.
102. Tontisirin, S. and Ernst, S. (2007) *Angew. Chem. Int. Ed.*, **46**, 7304–7306.
103. Shvets, O.V., Kasian, N.V., and Ilyin, V.G. (2008) *Adsorpt. Sci. Technol.*, **26**, 29–35.
104. Egeblad, K., Kustova, M., Klitgaard, S.K., Zhu, K., and Christensen, C.H. (2007) *Microporous Mesoporous Mater.*, **101**, 214–223.
105. Zhu, K., Egeblad, K., and Christensen, C.H. (2007) *Eur. J. Inorg. Chem.*, **25**, 3955–3960.
106. Chauvin, B., Boulet, M., Massiani, P., Fajula, F., Figueras, F., and Descourieres, T. (1990) *J. Catal.*, **126**, 532–545.
107. Chauvin, B., Massiani, P., Dutartre, R., Figueras, F., Fajula, F., and Descourieres, T. (1990) *Zeolites*, **10**, 174–182.
108. Groen, J.C., Zhu, W.D., Brouwer, S., Huynink, S.J., Kapteijn, F., Moulijn, J.A., and Perez-Ramirez, J. (2007) *J. Am. Chem. Soc.*, **129**, 355–360.
109. Groen, J.C., Abello, S., Villaescusa, L.A., and Perez-Ramirez, J. (2008) *Microporous Mesoporous Mater.*, **114**, 93–102.
110. Perez-Ramirez, J., Abello, S., Villaescusa, L.A., and Bonilla, A. (2008) *Angew. Chem. Int. Ed.*, **47**, 7913–7917.
111. Perez-Ramirez, J., Abello, S., Bonilla, A., and Groen, J.C. (2009) *Adv. Funct. Mater.*, **19**, 164–172.
112. Wang, X., Qi, G., and Li, G. (2007) Method for Preparing Nano Zeolite Catalyst and its Use in Methylbenzene and Trimethyl Benzene Transalkylation Reaction, CN1850337-A.
113. Čejka, J. and Mintova, S. (2007) *Catal. Rev.*, **49**, 457–509.
114. Wang, J., Groen, J.C., Yue, W., Zhou, W., and Coppens, M.O. (2008) *J. Mater. Chem.*, **18**, 468–474.
115. Wang, J., Groen, J.C., Yue, W., Zhou, W., and Coppens, M.O. (2007) *Chem. Commun.*, 4653–4655.

116. Liu, Y., Zhang, W.Z., and Pinnavaia, T.J. (2000) *J. Am. Chem. Soc.*, **122**, 8791–8792.
117. Zhang, Z.T., Han, Y., Xiao, F.S., Qiu, S.L., Zhu, L., Wang, R.W., Yu, Y., Zhang, Z., Zou, B.S., Wang, Y.Q., Sun, H.P., Zhao, D.Y., and Wei, Y. (2001) *J. Am. Chem. Soc.*, **123**, 5014–5021.
118. Zhang, Z.T., Han, Y., Zhu, L., Wang, R.W., Yu, Y., Qiu, S.L., Zhao, D.Y., and Xiao, F.S. (2001) *Angew. Chem. Int. Ed.*, **40**, 1258–1258.
119. Mazaj, M., Stevens, W.J.J., Logar, N.Z., Ristic, A., Tusar, N.N., Arcon, I., Daneu, N., Meynen, V., Cool, P., Vansant, E.F., and Kaucic, V. (2009) *Microporous Mesoporous Mater.*, **117**, 458–465.
120. Corma, A., Diaz, U., Domine, M.E., and Fornes, V. (2000) *J. Am. Chem. Soc.*, **122**, 2804–2809.
121. Corma, A., Fornes, V., and Diaz, U. (2001) *Chem. Commun.*, 2642–2643.
122. Corma, A., Fornes, V., MartÌnez-Triguero, J., and Pergher, S.B. (1999) *J. Catal.*, **186**, 57–63.
123. Corma, A., Martinez, A., and Martinez-Soria, V. (2001) *J. Catal.*, **200**, 259–269.
124. Galletero, M.S., Corma, A., Ferrer, B., Fornes, V., and Garcia, H. (2003) *J. Phys. Chem. B*, **107**, 1135–1141.
125. Nguyen, C.T., Kim, D.P., and Hong, S.B. (2008) *J. Polym. Sci. A: Polym. Chem.*, **46**, 725–732.
126. Wu, P., Nuntasri, D., Ruan, J.F., Liu, Y.M., He, M.Y., Fan, W.B., Terasaki, O., and Tatsumi, T. (2004) *J. Phys. Chem. B*, **108**, 19126–19131.
127. Shan, Z., Jansen, J.C., Yeh, Y.T., Koegler, J.H., and Maschmeyer, T. (2003) Catalyst containing microporous zeolite in mesoporous support and method for making same. International WO 03/045548 A1.
128. Waller, P., Shan, Z.P., Marchese, L., Tartaglione, G., Zhou, W.Z., Jansen, J.C., and Maschmeyer, T. (2004) *Chem. Eur. J.*, **10**, 4970–4976.
129. Kortunov, P., Vasenkov, S., Karger, J., Elia, M.F., Perez, M., Stocker, M., Papadopoulos, G.K., Theodorou, D., Drescher, B., McElhiney, G., Bernauer, B., Krystl, V., Kocirik, M., Zikanova, A., Jirglova, H., Berger, C., Glaser, R., Weitkamp, J., and Hansen, E.W. (2005) *Chem. Mater.*, **17**, 2466–2474.
130. Kortunov, P., Vasenkov, S., Karger, J., Valiullin, R., Gottschalk, P., Elia, M.F., Perez, M., Stocker, M., Drescher, B., McElhiney, G., Berger, C., Glaser, R., and Weitkamp, J. (2005) *J. Am. Chem. Soc.*, **127**, 13055–13059.
131. Groen, J.C., Caicedo-Realpe, R., Abellò, S., and Pérez-Ramírez, J. (2009) *Mater. Lett.*, **63**, 1037–1040.
132. Wei, X. and Smirniotis, P.G. (2006) *Microporous Mesoporous Mater.*, **97**, 97–106.
133. Zhao, L., Shen, B., Gao, J., and Xu, C. (2008) *J. Catal.*, **258**, 228–234.
134. Xiao, F.S., Wang, L.F., Yin, C.Y., Lin, K.F., Di, Y., Li, J.X., Xu, R.R., Su, D.S., Schlogl, R., Yokoi, T., and Tatsumi, T. (2006) *Angew. Chem. Int. Ed.*, **45**, 3090–3093.
135. Trong-On, D., Ungureanu, A., and Kaliaguine, S. (2003) *Phys. Chem. Chem. Phys.*, **5**, 3534–3538.
136. Musteata, M., Musteata, V., Dinu, A., Florea, M., Hoang, V.T., Trong-On, D., Kaliaguine, S., and Parvulescu, V.I. (2007) *Pure Appl. Chem.*, **79**, 2059–2068.